Handbook of Experimental Pharmacology

Volume 90/II

Catecholamines II

Contributors

M. A. Arnold, J. D. Barchas, G. Bartholini, P. Berger, I. B. Black, F. E. Bloom, M. J. Brownstein, M. E. Conolly, G. M. Jonakait, G. F. Koob, I. J. Kopin, J. B. Martin, J. M. Masserano, U. Otten, M. Palkovits, J. L. Reid, P. C. Rubin, J. M. Saavedra, B. Scatton, J. A. Schulman, A. W. Tank, P. R. Vulliet, N. Weiner, R. M. Weinshilboum, B. Zivkovic

Editors

U. Trendelenburg and N. Weiner

Springer-Verlag
Berlin Heidelberg New York
London Paris Tokyo

U. Trendelenburg, Professor Dr.

Department of Pharmacology and Toxicology
University of Würzburg
Versbacher Str. 9
D – 8700 Würzburg

N. Weiner, Professor Dr.

Department of Pharmacology
Colorado University, Medical Center
4200 East Ninth Street, Denver, CO 80262, USA

With 34 Figures

ISBN-13:978-3-642-73553-0 e-ISBN-13:978-3-642-73551-6
DOI: 10.1007/978-3-642-73551-6

Library of Congress Cataloging-in-Publication Data (Revised for volume 90/II)
Catecholamines. (Handbook of experimental pharmacology; v. 90, I-II) Bibliography: v. 1, p. Includes index. 1. Catecholamines—Metabolism. 2. Catecholamines—Agonists. 3. Catecholamines—Antagonists. 4. Catecholamines—Receptors. I. Bönisch, H. II. Trendelenburg, U. (Ullrich), 1922-. III. Weiner, Norman, 1928-. IV. Series: Handbook of experimental pharmacology; v. 90, I, etc. OP905.H3 vol. 90/1, etc. [QP801.C33] 615'.1 s 88-4652. ISBN-13:978-3-642-73553-0

Softcover reprint of the hardcover 1st edition 1989

2122/3020 - 543210

List of Contributors

M. A. ARNOLD, Implantable Products Division Radionics, Inc., 76 Cambridge Street, Burlington, MA 01803

G. BARTHOLINI, Research and Development Department, Laboratoires D'Etudes et de Recherches Synthelabo (L. E. R. S.), 58, rue de la Glacière, F - 75013, Paris

I. B. BLACK, Department of Neurology, Cornell University Medical College, 525 East 68th Street, New York, NY 07102, USA

F. E. BLOOM, Scripps Clinic and Research Foundation, 10666 North Torrey Pines Road, La Jolla, CA 92037, USA

M. J. BROWNSTEIN, Laboratory of Cell Biology, National Institute, of Mental Health, Building 36, Room 3A-17, Bethesda, MD 20892, USA

M. E. CONOLLY, Division of Clinical Pharmacology, Department of Pharmacology, UCLA School of Medicine, Center for the Health Sciences, Los Angeles, CA 90024, USA

G. M. JONAKAIT, Department of Biological Sciences, Rutgers University, Newark, NJ 07102, USA

G. F. KOOB, Scripps Clinic and Research Foundation, 10666 North Torrey Pines Road, La Jolla, CA 92037, USA

I. J. KOPIN, Division of Intramural Research, National Institute of Neurological and Communicative Disorders and Stroke, Bldg. 10, Room 5N214, Bethesda, MD 20892, USA

J. B. MARTIN, Department of Neurology, Harvard Medical School, Massachusetts General Hospital, Fruit Street, Boston, MA 02114, USA

J. M. MASSERANO, University of Colorado, Health Sciences Center, Department of Pharmacology, C-236, 4200 East 9th Avenue, Denver, CO 80262, USA

M. PALKOVITS, Laboratory of Cell Biology, National Institute of Mental Health, Building 36, Room 3A-17, Bethesda, MD 20892, USA

J. L. REID, Department of Materia Medica, Stobhill General Hospital, Glasgow G21 3UW, Great Britain

P. C. Rubin, Department of Materia Medica, Stobhill General Hospital, Glasgow G21 3UW, Great Britain

J. M. Saavedra, Unit on Preclinical Neuropharmacology, Section on Clinical Pharmacology, Laboratory of Clinical Science, National Institute of Mental Health, Bethesda, MD 20892, USA

B. Scatton, Research and Development Department, Laboratoires D'Etudes et de Recherches Synthelabo (L. E. R. S.), 58, rue de la Glacière, F - 75013, Paris

J. A. Schulman, Scripps Clinic and Research Foundation, 10666 North Torrey Pines Road, La Jolla, CA 92037, USA

A. W. Tank, Department of Pharmacology, University of Rochester Medical Center, 601 Elmwood Ave., Rochester, NY 14643, USA

P. R. Vulliet, Department of Pharmacology and Toxicology, School of Verterinary Medicine, University of California, Davis, CA 95616, USA

N. Weiner, Department of Pharmacology, University of Colorado School of Medicine, 4200 E. Ninth Ave., Denver, CO 80262, USA

R. W. Weinshilboum, Clinical Pharmacology Unit, Mayo Foundation, Mayo Medical School, Rochester, MN 55905, USA

B. Zivkovic, Research and Development Department, Laboratoires D'Etudes et de Recherches Synthelabo (L. E. R. S.), 58, rue de la Glacière, F - 75013 Paris

Preface

The two new volumes *Catecholamines* are actually the third "state-of-the-art" report in the series of *Heffter-Heubner's Handbook of Experimental Pharmacology*. The first (*Adrenalin und adrenalinverwandte Substanzen* by P. Trendelenburg, 1924) covered the subject in 163 pages. *Catecholamines* (edited by H. Blaschko and E. Muscholl, 1972) was published in one volume of slightly more than 1000 pages. The succesor, also called *Catecholamines* now appears in two volumes – and with a further increase in the total number of pages.

It is up to future readers to decide whether present-day authors (able to rely on computers) are more verbose than the earlier ones or whether the undeniable explosion of knowledge accounts for the increase in size. The editors prefer the second of these hypotheses. They should like to draw the sceptical reader's attention to Fig. 1 of the preface of Iversen (1967) which illustrates the explosive increase in yearly publications (quoted by the author) between 1955 and 1965. If that figure suggested that the rate of relevant publications approached a maximum (or V_{max}) at the end of this period, one may legitimately ask whether perhaps this V_{max} characterized certain memory stores rather than the rate of relevant publications. Can any individual store the information contained in 150 publications per year (or 0.41 per day!), expecially when it is highly likely that the publication explosion continued unabatedly throughout the 1970s and 1980s? This high rate of publication is one of the reasons why there is an urgent need for critical reviews that enable the non-expert to enter a new field with a minimum waste of time or the expert to check quickly on progress in neighbouring fields.

The present volumes are a successor to, but *not* a new edition of, *Catecholamines* of 1972. Blaschko (1972) had to acknowledge the sad fact that the editors failed to receive this or that promised chapter. The editors of the present volumes were by no means luckier. Irrespective of these missing chapters, it would have been quite impossible to aim at a presentation of *all* areas in which catecholamines are under active study. Hence, in spite of the increased size of Volume 90, *Catecholamines* (1988) resembles the volume of 1972 in presenting "selected topics". We hope to offer a variety of topics that pleases some and offends very few.

As Szekeres (1980/81) edited the volumes *Adrenergic Activators and Inhibitors* (also in this series), the present editors felt entitled to concentrate on "basic mechanisms", with relatively little emphasis on "organ- or system-specific actions".

The editorial board of *Heffter-Heubner's Handbook* selected two persons inexperienced in the formidable task of editing such a wealth of information. Hence (and for reasons of health), fruition of our editorial efforts came later than originally anticipated. This fact is mentioned in order to convince the contributors to these volumes that our thanks for their patience and unflagging enthusiasm are very sincere.

Any selection of topics *must* (to some degree) reflect the editors' bias. Moreover, if such a volume has a long gestation period, topics of very recent interest cannot be included. Within these limitations, we tried to follow the splendid example set by Blaschko and Muscholl with *Catecholamines* (1972): experts were asked to review areas of active research, irrespective of whether they deal with basic mechanisms or clinical problems. It is, of course, in the hospitals that our increasing knowledge of the catecholamines should eventually lead to improved treatment of patients.

Thus, whether researchers or clinicians are in love with central or peripheral catecholamines, we hope to have provided them with interesting and helpful reading material.

As successors to the 1972 volume, the present volumes should be regarded as a unit. Although some attempt was made to arrange the table of contents in a certain order, the publication in *two* volumes arose much more from the publisher's concern about the formidable weight of one single volume than from any wish of the editors to subdivide this area into two separate parts. Moreover, the late arrival of this or that chapter necessitated deviations from an optimal order in the table of contents. In recognition of the fact that the two volumes constitute a unity, the subject index at the end of the second volume covers *both* volumes. The subject index at the end of the first volume is meant to be helpful for those who buy the first volume only.

U. TRENDELENBURG and N. WEINER

References

Blaschko H (1972) Introduction – Catecholamines 1922–1971. In: Blaschko H, Muscholl E (eds) Catecholamines. Springer, Berlin Heidelberg New York, 33:1–15 (Handbook of Experimental Pharmacology Vol 33)

Blaschko H, Muscholl E (1972) Catecholamines. Handbook of Experimental Pharmacology, Vol 33. Springer, Berlin Heidelberg New York

Iversen LL (1967) The uptake and storage of noradrenaline in sympathetic nerves. Cambridge University Press, Cambridge

Szekeres L (1980/81) Adrenergic Activators and Inhibitors. Handbook of Experimental Pharmacology Vol 54/I and 54/II. Springer, Berlin Heidelberg New York

Trendelenburg P (1924) Adrenalin und adrenalinverwandte Substanzen. In: Handbuch der Exp. Pharmakologie Heffter A, (ed.) 2:1130-1293, Springer Berlin Heidelberg

Contents

CHAPTER 14

β-Phenylethylamine, Phenylethanolamine, Tyramine and Octopamine

CHAPTER 17

Catecholamines and Blood Pressure

Contents of Companion Volume 90, Part I

List of Abbreviations (both volumes)

AADC	L-aromatic-amino acid decarboxylase
ACh	acetylcholine
ACTH	corticotropin
ADP	adenosine diphosphate
AMP	adenosine monophosphate
ATP	adenosine triphosphate
cAMP	cyclic adenosine monophosphate
CAT	choline acetyl transferase
cGMP	cyclic guanosine monophosphate
CNS	central nervous system
COMT	catechol-O-metyhl transferase
CRF	corticotropin-releasing factor
CSF	cerebrospinal fluid
CTP	cytidine triphosphate
DBH	dopamine-β-hydroxylase
DOMA	dihydroxymandelic acid
DOPAC	dihydroxyphenylacetic acid
DOPEG	dihydroxyphenylethylene glycol
DOPET	dihydroxyphenylethanol
EC_{50}	half-maximally effective concentration
EDTA	ethylenediaminetetraacetate
EGTA	ethyleneglycol-bis-(β-aminoethyl ether)N,N,N′,N′-tetraacetic acid
FAD	flavin-adenine dinucleotide
FRL	fractional rate of loss
FSH	follicle-stimulating hormone
GABA	gamma-aminobutyric acid
GDP	guanosine diphosphate
GTP	guanosine triphosphate
5-HT	5-hydroxytryptamine, serotonin
HVA	methoxyhydroxyphenylacetic acid (homovanillic acid)
IC_{50}	half-maximally inhibitory concentration
ITP	inosine triphosphate
LH	luteinizing hormone
LHRH	LH-releasing hormone
MAO	monoamine oxidase
MN	metanephrine

MOPEG	methoxyhydroxyphenylethylene glycol
NADH	nicotinamide-adenine dinucleotide
NA/DOPEG	noradrenaline/dihydroxyphenylethylene glycol
NGF	nerve growth factor
NMN	normetanephrine
OMI	3-O-methyl-isoprenaline
PC 12 cells	phaeochromocytoma cells
PG	prostaglandin
PNMT	phenylethanolamine-N-methyl transferase
SCG	superior cervical ganglion
SIF cells	small, intensely fluorescent cells
TH	tyrosine hydroxylase
TRH	thyrotropin-releasing hormone
TSH	thyrotropin
UDP	uridine diphosphate
$uptake_1$	neuronal noradrenaline uptake
$uptake_2$	extraneuronal noradrenaline uptake
UTP	uridine triphosphate
VMA	methoxyhydroxymandelic acid (vanillylmandelic acid)

CHAPTER 10

Catecholamines in the Central Nervous System

M. PALKOVITS and M. J. BROWNSTEIN

A. Central Catecholamine-Containing Neuronal Systems

I. Introduction

Histochemical and biochemical studies have shown that catecholamines are present in almost all brain areas. Similarly, immunocytochemical studies have shown tyrosine hydroxylase (TH), dopamine-β-hydroxylase (DBH) and phenylethanolamine N-methyl transferase (PNMT) immunoreactivity everywhere in the CNS in fiber-networks and nerve terminals. The catecholaminergic innervation of the brain may be characterized as a neuronal network with "open" (i.e. nonspecific) synapses as opposed to a chain of "point to point", specific neuronal connections. Catecholamine-containing neurons in the brain have widespread efferent trajectories, rich aborization, abundant axon collaterals, and particularly large numbers of nerve terminals. A single aminergic cell may have anywhere from ten to one hundred thousand nerve terminals and innervate several cells in the CNS. Furthermore, a single neuron in the brain may be innervated by several catecholaminergic cells. Thus, chemical or mechanical lesions of individual catecholaminergic cell groups never result in the complete disappearance of catecholamines from other brain areas. Because of the overlapping pattern of catecholaminergic innervation in many brain regions, it is not easy to determine the origin and final destination of central catecholamine-containing fibers. Of course, there are pathways or projections along which catecholamine-containing axons travel. These are called dopaminergic, noradrenergic or adrenergic systems. To summarize the topography of brain catecholaminergic systems, we have adopted this method of classification (Table 1), but overlaps in both the origins and terminations of these "systems" should be noted. Detailed neuroanatomy of brain catecholamines has been summarized by JACOBOWITZ and PALKOVITS (1974), PALKOVITS and JACOBOWITZ (1974), and recently by BJÖRKLUND and LINDVALL (1984), MOORE and CARD (1984) and HÖKFELT et al. (1984a, b).

1. **Dopaminergic Systems** (Table 1)

a. Tuberoinfundibular Dopamine System

These cell bodies are principally located in the arcuate nucleus. A few cells are found along the third ventricle in the medial basal hypothalamus outside the confines of the arcuate nucleus (A12-cell group). Axons of these cells

Table 1. Major catecholaminergic systems in the rat brain

Pathways	Cell Groups	References
Dopamine (DA) (Fig. 1)		
Tuberoinfundibular DA-system	A12	1-4
Mesocortical DA-system	A9, A10	5-9
Mesostriatal DA-system	A9, A8, A10	5-9
Mesolimbic DA-system	A10	5-8, 10, 11
Incerto-hypothalamic DA-system	A13, A11	3, 7, 8, 12
Periventricular DA-system	A14, A11	1, 3, 7, 8
Descending DA-fibers	A8, A9, A10	7
Olfactory DA-system	A15	11, 13
Noradrenaline (NA) (Fig. 2)		
Ventral NA-bundle	A1, A2, A5, A6, A7	5, 7, 14, 15
Dorsal NA-bundle	A6	5, 7, 10, 11, 14, 16
Dorsal periventricular system	A6, A7, A2	7, 8
Ventral periventricular system	A1, A5, A7	7, 8
Cerebellar NA-pathway	A4, A5, A6	14, 17, 18
Bulbospinal NA-pathways	A2, A6	19
Adrenaline (A) (Fig. 3)		
Ascending A-fibers	C1, C2	20, 21

References: 1 — DAHLSTRÖM and FUXE (1964); 2 — FUXE and HÖKFELT (1966); 3 — BJÖRKLUND and NOBIN (1973); 4 — BJÖRKLUND et al. (1973); 5 — UNGERSTEDT (1971); 6 — FALLON and MOORE (1978 b); 7 — LINDVALL and BJÖRKLUND (1978); 8 — LINDVALL (1979); 9 — ANDÉN et al. (1966); 10 — FALLON et al. (1978); 11 — FALLON and MOORE (1978 a); 12 — BJÖRKLUND et al. (1975); 13 — HALÁSZ et al. (1977); 14 — MOORE and BLOOM (1979); 15 — JACOBOWITZ (1978); 16 — PALKOVITS and JACOBOWITZ (1974); 17 — OLSON and FUXE (1971); 18 — PASQUIER et al. (1980); 19 — NYGREN and OLSON (1977); 20 — HÖKFELT et al. (1974); 21 — PALKOVITS et al. (1980 a).

form the tuberoinfundibular dopamine tract that innervates the median eminence, pituitary stalk, and neural and intermediate lobes of the pituitary gland. These very short fibers (0.5-1.5 mm in the rat) comprise a network rather than a well defined pathway.

b. Mesocortical Dopamine System

The perikarya are found in the ventral tegmental area (A10-cell group) and in the medial portion of the compact zone of the substantia nigra (A9-cell group). Most of these fibers course rostrally through the lateral part of the medial forebrain bundle (MFB) to arborize at the level of the septum (Fig. 1).

c. Mesostriatal Dopamine System

This is the largest dopamine-system; it is made up of two-thirds of the dopaminergic neurons in the CNS. These cell bodies are found in the A10- and

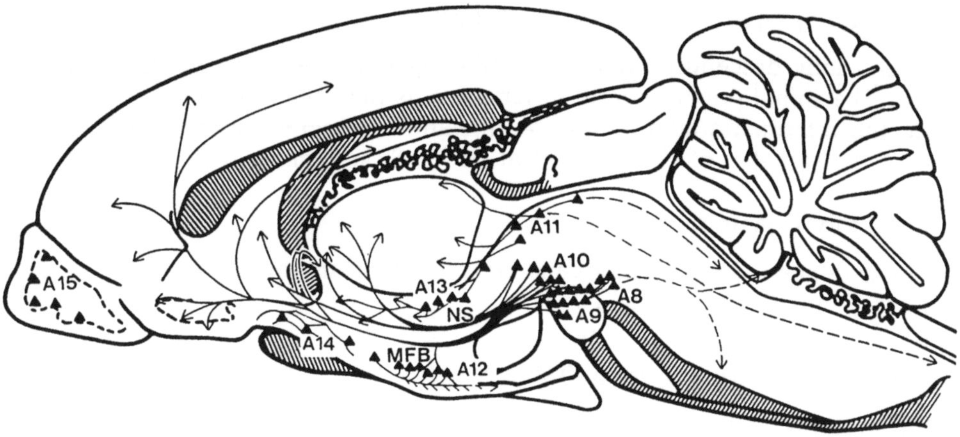

Fig. 1. Dopaminergic cell groups (A8–A15 according to BJÖRKLUND and LINDVALL 1984; HÖKFELT et al. 1984b) and pathways. Midsagittal section of the rat brain. (See further anatomical details on Fig. 3.) NS — nigro-striatal dopaminergic pathway, MFB — medial forebrain bundle

A9-cell groups and also in the lateral part of the substantia nigra in the A8-cell groups. There is a specific topographical relation between the terminal pattern in the striatum and the localization of the perikarya. Fibers of the nigrostriatal pathway ascend in the ansa lenticularis, enter the dorsolateral MFB, and make their way as far as the globus pallidus, where they then pass through the internal capsule to radiate into the striatum (Fig. 1).

d. Mesolimbic Dopamine System

The axons of this system arise mainly from cells in the ventral tegmental area and run through the lateral hypothalamus along with the other ascending dopaminergic fibers. Fibers leave the lateral hypothalamus in small bundles, go to the habenula, amygdala, bed nucleus of the stria terminalis, septum, and olfactory tubercle (Fig. 1).

e. Incerto-Hypothalamic Dopamine System

The cells are located in the zona incerta and Forel Hl-field (A13-cell group) and in the caudal portion of the thalamus and hypothalamus (A11-cell group). The fibers form a short pathway that mainly runs to the dorsal portion of the medial hypothalamus (Fig. 1).

f. Periventricular Dopamine System

The cells and fibers of this system are close to the midline along the third ventricle and the cerebral aqueduct. The most rostral neurons are found in the preoptic area (A14-cell group), the most caudal ones in the midbrain (A11-cell group).

g. Descending Spinal Dopaminergic Fibers

This is a long descending uncrossed fiber system arising from the A11 cell group. The spinal dopaminergic innervation is mainly confined to the dorsal horn, the intermedio-lateral cell column and associated parts of the central gray of the spinal cord (HÖKFELT et al. 1979; SKAGERBERG et al. 1982).

h. Olfactory Dopamine System

Intrinsic dopamine-containing cells are present in the olfactory bulb, among the periglomerular cells (Fig. 1).

2. The Noradrenergic System (Table 1)

NA-containing perikarya are found exclusively in the pons and medulla oblongata. The major cell group is the locus coeruleus (A6-cell group) in which about 45% of all the brain noradrenergic cells are present. Many noradrenergic cells are located ventral to the locus coeruleus, mostly in the parabrachial nuclei. The most rostral noradrenaline-containing cells are a group in the dorsolateral portion of the midbrain reticular formation (A7-cell group). Several noradrenergic cells are found in the reticular formation, along the ventrolateral corner of the pons (A5-cell group) and the medulla (A1-cell group). A small portion of the noradrenergic cells are found within the vagal nuclei (nucleus of the solitary tract and dorsal vagal nucleus) and adjacent to them (A2-cell group).

The brainstem noradrenergic neurons give rise to two major systems: the locus coeruleus noradrenaline-containing cells and their projections make up one of these, and the remaining noradrenergic neurons and their projections constitute the others, the lateral tegmental noradrenergic-system. In both systems ascending and descending noradrenergic bundles can be distinguished (Fig. 2).

a. Ventral Noradrenergic Bundle
(also called the ventral tegmental tract)

This bundle originates in the medulla oblongata, and all of the noradrenaline-cell groups of the brainstem participate in its formation (Fig. 2). The most caudal fibers enter the central tegmental tract and run rostrally in the dorsal part of the reticular formation. They enter the pons by passing under the internal genu of the facial nerve and then they move ventrolaterally. The fibers next pass the ventrolateral portion of the pontine and midbrain reticular formation, run rostrally into the MFB and enter the lateral hypothalamus (Fig. 2).

b. Dorsal Noradrenergic Bundle

Originating mainly from the locus coeruleus and from neighboring cells, this bundle ascends in the lateral side of the central gray matter. At the diencephalon-midbrain junction most of the fibers form ventrally to join the ventral noradrenergic-bundle, others run into the thalamus, habenula or decussate in

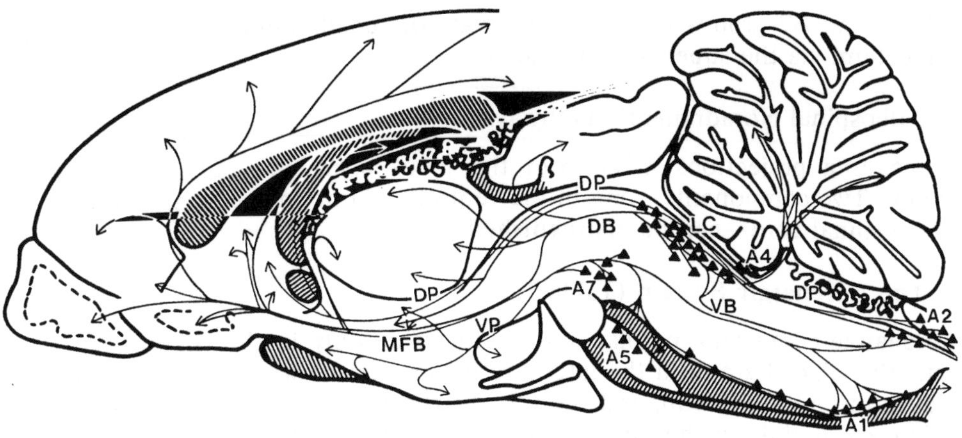

Fig. 2. Noradrenaline-containing cell groups (A1, A2, A4, A5, A7 and LC according to Moore and Card, 1984; Hökfelt et al. 1984b) and pathways. Midsagittal section of the rat brain. (See further details on Fig. 3.) *Abbr.*: DB — dorsal noradrenergic bundle, DP — dorsal periventricular noradrenergic system, MFB — medial forebrain bundle, VB — ventral noradrenergic bundle, VP — ventral periventricular noradrenergic system

the posterior commissure and terminate in the tectum. The ascending fibers in the MFB (together with fibers from the ventral noradrenergic bundle) run rostrally to terminate in the forebrain (Fig. 2).

c. Dorsal Periventricular Noradrenergic System

Most of the caudal fibers that comprise this system are contributed by the A2-cell group and run rostrally underneath the fourth ventricle. The system may also contain descending fibers from the locus coeruleus which go to the nucleus of the solitary tract. The locus coeruleus contributes ascending fibers to this system as well. Fibers traverse the midbrain in the lateral part of the central gray matter, independent of the dorsal noradrenergic bundle, and enter the dorsal region of the hypothalamus near the third ventricle. They then join those axons of the dorsal noradrenergic bundle that are destined to innervate the hypothalamus and axons of the ventral periventricular pathway to form a dense fibers network beside the third ventricle (Fig. 2).

d. Ventral Periventricular Noradrenergic System

This pathway represents the periventricular projection of the ventral noradrenergic bundle. It forms part of the noradrenergic hypothalamic-preoptic periventricular neuronal network (Fig. 2).

e. Cerebellar Noradrenergic Pathway

This pathway originates mainly from cells in the locus coeruleus, but the A4-cell group (cells in the area of the superior medullary velum and the

A5-cell and A7 groups may also innervate the cerebellum. Fibers reach the cerebellum via the superior peduncle.

f. Bulbospinal Noradrenergic Pathways

Descending projections arise from the locus coeruleus, A1- and A2-cell groups. Two components—one in the ventral and one in the lateral funiculus of the spinal cord—can be distinguished.

3. The Adrenergic System (Table 1)

Adrenaline-containing perikarya seem to reside only in the medulla oblongata: the C1-cell group in the lateral reticular nucleus with cells situated medial to the A1-cell group, and the C2-cell group in the dorsomedial medulla oblongata, mainly in the nucleus of the solitary tract (Fig. 3).

It was at first assumed that ascending adrenergic fibers traveled in the ventral noradrenergic bundle. Recent studies, however, have indicated that the adrenaline-containing fibers might run more ventrally, probably together with other fibers along the ventromedial edge of the pons and the medulla oblongata (Fig. 3).

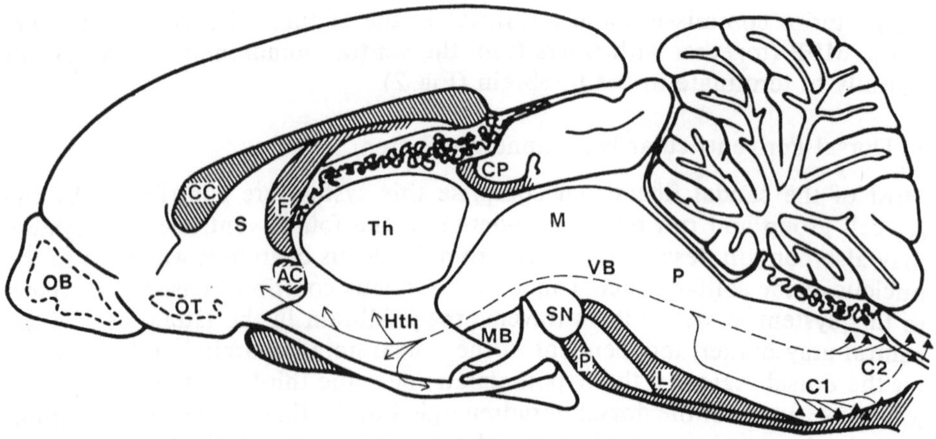

Fig. 3. Adrenaline-containing cell groups (C1, C2 according to Hökfelt et al. 1974, 1984a) and pathways. Midsagittal section of the rat brain. *Abbr.*: AC — anterior commissure, CC — corpus callosum, CP — posterior commissure, F — fornix, Hth — hypothalamus, L — medial lemniscus, M — midbrain, MB — mamillary body, OB — olfactory bulb, OT — olfactory tubercle, p — pons, P — pyramidal tract, S — septum, SN — substantia nigra, Th — thalamus, VB — ventral catecholaminergic bundle

B. Regional Concentrations of Catecholamines in the Brain

To investigate the distribution of catecholamines in the CNS histological (histofluorescence, immunohistochemistry) and chemical (gas chromatography, mass spectrometry, radiometric enzymatic assays) techniques can be employed. The histological and analytical studies complement one another. Histological techniques are suitable for topographical mapping of catecholamine-containing cells, pathways and fibers; the neurochemical assays in microdissected brain nuclei provide quantitative information about the concentrations of amines in individual cell groups in the CNS. In the past decade, a number of workers have determined the regional distributions of catecholamines and their biosynthetic enzymes. Data obtained by these workers in rat are summarized in Tables 2-4. Far fewer studies of other species, including the human, have been published to date. These are listed in Table 5.

I. Dopamine

Biochemical data show that dopamine is present in all brain nuclei. The dopamine detected may either be acting as a neurotransmitter in its own right or as a precursor of noradrenaline. In structures such as the pineal gland that receive a pure noradrenergic innervation, the ratio of noradrenaline to dopamine is about 6 to 1. When a ratio much lower than this is observed in a particular brain area, it is likely that this area is innervated by both noradrenergic and dopaminergic neurons. The distribution of dopamine is rather uneven; the ratio between the richest and poorest regions is over 500:1 (Table 2). Fairly large differences in tyrosine hydroxylase activity of various brain areas are also seen (Table 4).

The caudate putamen, accumbens nucleus, olfactory tubercle and median eminence are especially rich in dopamine. They have dopamine concentrations in excess of 60 ng/mg protein in rat. The TH-activity of these regions, especially in the caudate putamen is also very high (Table 4). High dopamine concentrations (between 20 and 60 ng/mg) have been measured in the nucleus tractus diagonalis, globus pallidus, lateral septal nucleus, central amygdaloid nucleus, substantia nigra, ventral tegmental area, locus coeruleus and in the organum vasculosum laminae terminalis (Table 2). Relatively high dopamine levels are present in the lateral amygdaloid, the periventricular thalamic nuclei and in the nucleus of the solitary tract.

The dopamine concentrations in the septum, preoptic region, and hypothalamus, in general, are higher than in other major brain regions.

II. Noradrenaline

Noradrenaline (Tables 2 and 3) and DBH activity (Table 4) have been detected in all CNS regions studied.

The hypothalamus, preoptic region, locus coeruleus, midbrain reticular formation and vagal nuclei are particularly rich in noradrenaline. Certain cell groups within the hypothalamus are especially well endowed with noradrena-

Table 2. Distribution of catecholamines in the rat brain

	Dopamine ng/mg protein	Noradrenaline ng/mg protein	Adrenaline ng/mg protein
Rhinencephalon			
olfactory bulb	0.8- 2.8	3.2- 4.7	-
ant. olfact. nucl.	0.5	6.1	
olfactory tubercle	60.5-114.4	4.5- 7.2	
nucl. tractus diagonalis	21.0- 42.8	22.8-28.8	-
hippocampus	0.3- 0.8	3.2- 3.8	-
dentate gyrus	0.1	1.9	
Cerebral Cortex			
frontal	0.2- 1.0	2.2- 4.1	-
frontopolar	0.8	2.4	
parietal	0.2- 1.3	3.0- 3.5	-
insular	0.4- 1.3	4.9	
temporal	0.3	2.2	
occipital	0- 0.3	1.8- 2.2	
cingulate	0.9- 4.5	2.1- 4.5	-
piriform	0.8- 2.9	4.5- 5.8	
entorhinal	0- 2.6	1.5- 5.8	-
subiculum	0.2	1.9	
pre- and parasubiculum	0.2	1.6	
Basal Ganglia			
caudate nucleus	70.5-133.9	1.1- 3.1	
caudate-putamen	88.4	1.3	
nucleus accumbens	87.2-100.5	3.6- 9.3	
globus pallidus	39.6	5.3	
Septum			
dorsal nucl.	8.1- 13.2	4.5- 6.6	0.9
intermediate nucl.	19.3- 19.5	6.0- 8.3	-
medial nucl.	8.6- 14.0	9.9-10.0	-
lateral nucl.	21.2- 21.4	11.0-12.3	-
fimbrial nucl.	2.6- 4.3	8.7-10.9	0.8
triangular nucl.	4.1	17.4	
Amygdala			
anterior area		9.5	
nucl. lat. olfact. tract	6.6- 8.2	10.0-12.8	-
lateral nucl.	11.5- 17.5	10.8-12.0	-
medial nucl.	3.6- 7.5	11.4-12.6	-
medial-post. nucl.	1.4- 1.6	4.3-10.0	-
central nucl.	15.0- 24.1	13.4-17.8	-
cortical nucl.	3.1- 5.1	5.5- 8.5	-
basal nucl.	7.6- 8.6	9.1- 9.7	0.7
posterior nucl.	1.6- 1.9	2.5- 8.4	-
Thalamus			
antero-ventral nucl.	1.2- 7.3	12.7-25.9	-
reticular nucl.	-	11.0	-
ventral nucl.	0.2- 2.1	2.9- 3.7	

Table 2. (continued)

	Dopamine ng/mg protein	Noradrenaline ng/mg protein	Adrenaline ng/mg protein
medial nuclei	1.0	9.9	
midline nuclei	4.0	11.7	1.3
periventricular nucl.	18.6	15.7	
posterior nucl.	2.3	13.7	0.3
Habenula	1.9- 3.9	3.2- 9.7	0.2
Geniculate Bodies			
medial	2.2- 3.3	3.0- 7.9	-
lateral	1.0	10.3	0.4
Subthalamic Nucleus	5.3	10.6	-
Zona Incerta	5.5	13.8	-
Preoptic Region			
medial nucl.			1.1-2.3
suprachiasmatic nucl.	11.7	39.5	
lateral area (MFB)	4.3- 7.1		0.6-2.0
bed nucl. stria term. (NIST)	16.3	20.1	
ventral part	7.7- 13.9	27.2-80.2	2.0
dorsal part	15.4- 20.1	16.0-47.6	
bed nucl. stria med.	10.4- 12.9	28.6-37.7	0.6
Hypothalamus			
periventricular nucl.	7.1- 10.9	32.7-75.1	2.5-4.9
suprachiasmatic nucl.	7.6- 9.0	20.5-29.2	-
supraoptic nucl.	3.7- 8.1	22.4-34.4	2.1
paraventricular nucl.	5.8- 10.0	49.7-66.6	2.2
anterior nucl.	4.5- 6.1	16.2-31.0	0-2.1
retrochiasmatic area	3.7- 15.1	25.5-48.0	
median eminence	44.0- 84.0	15.9-41.4	1.8
arcuate nucl.	7.0- 28.2	12.0-35.9	1.4
ventromedial nucl.	5.1- 10.0	16.1-38.2	-
dorsomedial nucl.	8.5- 11.9	32.3-88.9	2.3
perifornical nucl.	6.0- 6.2	17.9-23.4	-
dorsal premamillary nucl.	3.9- 4.0	14.2-22.2	-
ventral premamillary nucl.	3.4- 3.8	16.3-30.0	-
posterior nucl.	4.3- 7.2	13.9-32.8	-
medial forebrain bundle	4.0- 11.0	20.2-38.1	0-1.1
Mamillary Body			
medial nucl.	2.6- 4.3	13.0-22.5	0.9
lateral nucl.	2.5- 2.7	17.0-20.5	1.5
Mesencephalon			
central gray matter	3.9- 4.7	10.4-24.5	-
superior collicle	1.1	4.7- 5.9	-
inferior collicle	0.4- 2.9	1.8- 2.5	-
substantia nigra			
zona compacta	19.5- 22.2	5.5-11.7	-

Table 2. (continued)

	Dopamine ng/mg protein	Noradrenaline ng/mg protein	Adrenaline ng/mg protein
zona reticularis	4.9- 7.8	2.5- 4.1	-
pars lateralis	7.8	8.2	1.0
ventral tegmental area (A10)	21.9- 22.4	10.7-15.2	1.0
interpeduncular nucl.	8.9	13.0	-
cuneiform nucl.	2.9	22.9	0.9
A8-cell group	17.5	33.1	
red nucleus	5.8	12.7	-
dorsal raphe nucl.	4.0- 5.7	10.0-20.4	-
nucl. cent. sup. (midbrain raphe)	7.5	3.2	
nuclei lineares	2.0	3.1	
Pons			
pontine nuclei	1.4	11.4	-
raphe pontis	1.0	2.1	
reticular pontine nuclei	1.4	6.8	-
parabrachial nuclei	1.6	26.6	-
locus coeruleus	16.7- 23.9	53.1-76.5	0.3-0.6
dorsal tegm. nucl.	1.4	13.6	-
superior olive	1.1	10.0	-
A5-cell group	1.8	15.1	0.5
dorsal NA-bundle	1.5- 2.4	6.6- 6.7	
Cerebellum			
cortex	0.2- 1.3	1.6- 2.9	0-0.2
nuclei	0.8	3.5	0.3
Medulla			
gigantocell. ret. nucl.	1.1	11.3	-
parvocell. ret. nucl.	1.6	13.0	-
ret. nucl. med. obl.	1.9	14.7	0.5
lateral ret. nucl. (A1-cell group)	1.1- 3.3	12.7-15.2	0.2
nucl. tract. sp. Vth.	2.6	8.4	-
facial motor nucl.	1.6	15.0	0.5
inferior olive	0.8- 3.4	11.5-13.3	0.2
raphe magnus	1.2- 1.5	6.2-13.3	-
raphe obscurus	0.9	2.7	
raphe pallidus	1.4	2.8	
cochlear nuclei	0.6	7.4	-
vestibular nuclei	0.9	5.7	-
nucl. solit. tract			
rostral part	3.2	14.1	0.6
A2-cell group	7.3	21.8-36.5	0.4-1.4
commissural part	10.1- 10.8	38.3-83.0	1.5-2.2
dorsal vagal nucl.	0.7	21.9	0.7
hypoglossal motor nucl.	0.5- 2.5	14.6-15.7	0-0.2

Table 2. (continued)

	Dopamine ng/mg protein	Noradrenaline ng/mg protein	Adrenaline ng/mg protein
Circumventricular Organs			
organum vasc. lam. term.	21.0	11.0	
subfornical organ	7.0	10.0	
subcommissural organ	9.3	6.5	
area postrema	8.9	8.6	

References: VERSTEEG et al. (1976); BROWNSTEIN et al. (1974); KIZER et al. (1976a); van der GUGTEN et al. (1976); O'DONOHUE et al. (1979); ROIZEN et al. (1976); PALKOVITS et al. (1974, 1979, 1980a, 1980b); KOBAYASHI et al. (1974); MOYER et al. (1978); ST. LAURENT et al. (1975); KOSLOW and SCHLUMPF (1974); KVETŇANSKÝ et al. (1977); BEART et al. (1979); SAAVEDRA et al. (1976a, 1976b)

Table 3. Distribution of catecholamines in the spinal cord of rats and rabbits (from ZIVIN et al. 1975)

	Rat	Rabbit		
	Noradrenaline	Noradrenaline	Dopamine	Adrenaline
Cervical				
Ventral horn	770 ± 126	462 ± 35.9	40.3 ± 8.9	44.2 ± 17.0
Central	516 ± 55.8	325 ± 38.4	91.5 ± 15.2	32.5 ± 10.7
Dorsal horn	542 ± 100	240 ± 17.9	68.1 ± 16.0	37.5 ± 8.3
White matter	77.6 ± 62.8	92.6 ± 9.4	8.5 ± 8.9	2.5 ± 4.3
Thoracic				
Ventral horn	476 ± 88.4	359 ± 28.2	113 ± 29.2	26.2 ± 14.0
Lateral horn	654 ± 64.3	465 ± 39.8	141 ± 24.8	50.9 ± 16.6
Dorsal horn	328 ± 71.7	175 ± 20.8	120 ± 27.3	40.5 ± 15.7
White matter	53.5 ± 54.0	60.4 ± 99.1	17.3 ± 7.8	10.1 ± 4.7
Lumbar				
Ventral horn	784 ± 135	387 ± 21.6	54.9 ± 9.9	72.7 ± 15.6
Central	659 ± 78.3	311 ± 25.6	76.4 ± 23.2	31.9 ± 17.6
Dorsal horn	387 ± 128	225 ± 30.2	52.2 ± 7.6	39.4 ± 17.8
White matter	173 ± 49.5	82.8 ± 10.0	20.5 ± 10.9	9.4 ± 7.1

shown are means ± S.E.M.

line: the paraventricular, periventricular, dorsomedial nuclei. High noradrenaline levels have also been found in a few telencephalic areas, such as the bed nucleus of the stria terminalis (NIST), bed nucleus of the stria medullaris, nucleus tractus diagonalis, and central amygdaloid nucleus. In the diencephalon except the hypothalamus, only the antero-ventral thalamic nucleus has a higher noradrenaline concentration than the brain's average (Table 2).

Table 4. Distribution of catecholamine synthesizing enzymes in the rat brain

	TH	DBH	PNMT
	nmol/mg protein/h		pmol/mg protein/h
Rhinencephalon			
olfactory tubercle	7.6	0.2	-
nucl. tract. diagonalis	2.5	0.8	
hippocampus	0.7	0.2	-
Cerebral Cortex			
frontal	0.2	0.5	-
parietal	0.6	0.2	-
cingulate	0.6	0.4	-
piriform	1.9- 3.6	0.3	-
entorhinal	1.2	0.2	
Basal Ganglia			
caudate nucleus	19.6	0.1	
caudate-putamen	53.4-54.5		
nucl. accumbens	7.1	0.2	
Septum			6.3
dorsal nucl.	2.2	0.4	
intermediate nucl.	3.9	0.3	
medial nucl	0.5	0.4	
lateral nucl.	3.4	0.4	
fimbrial nucl.	1.7	0.3	
triangular nucl.	0.7	0.4	
Amygdala			-
anterior area	1.8	0.4	
nucl. lat. olfact. tract	3.0	0.3	
lateral nucl.	3.4	0.1	
medial nucl.	0.7- 1.4	0.2	
medial-post. nucl.	0.3	0.3	
central nucl.	1.3-13.1	0.2	
cortical nucl.	1.1	0.2	
basal nucl.		0.2	
medial part	1.1		
lateral part	4.0- 5.6		
posterior part	1.6	0.2	
posterior nucl.	0.4	0.2	
Habenula			
medial nucl.	2.2	0.2	
lateral nucl:	2.5	0.2	
Preoptic Region			
medial nucl.			4.0- 7.7
lateral area (MFB)	3.3- 5.7	0.8-0.9	1.9- 6.2
bed nucl. stria term. (NIST)			2.1- 6.7
dorsal part	1.8	0.4	
ventral part	1.5	0.9	

Table 4 (continued)

	TH	DBH	PNMT
	nmol/mg protein/h		pmol/mg protein/h
bed nucl. stria med. (NISM)	3.7	0.6	
bed nucl. com. hipp. (NICH)	0.7	0.2	
Hypothalamus			
periventricular nucl.	4.6	4.2	8.0-11.2
suprachiasmatic nucl.	1.8	0.8	5.6
supraoptic nucl.	0.9	1.8	5.2
paraventricular nucl.	3.1	3.6	15.0
anterior nucl.	1.5	0.7	4.1- 5.9
retrochiasmatic area			9.7
median eminence	18.1	1.3	15.3
arcuate nucl.	4.3	0.8	10.3
ventromedial nucl.	1.8	1.0	8.3
dorsomedial nucl.	2.1	3.1	10.9
perifornical nucl.	4.5	0.5	8.1
dorsal premamillary nucl.	0.4	0.5	4.1
ventral premamillary nucl.	0.9	0.6	4.9
posterior nucl.	1.5	0.6	5.1
medial forebrain bundle	7.2-14.0	0.8-1.0	1.5- 5.3
Mamillary Body			
medial nucl.	1.8	0.2	
lateral nucl.	0.7	0.1	
Mesencephalon			
dorsal raphe	1.7	0.4	
nucl. cent. sup. (midbrain raphe)	1.4	0.2	
nuclei lineares	2.5	0.2	
Pons			
raphe pontis	1.3	0.6	
locus coeruleus			12.0
A7-cell groups			5.6
A5-cell groups			6.1
Cerebellum			
Medulla			
lat. ret. nucl. (A1 cell group)			41.5
raphe magnus	0.9	0.3	
raphe obscurus	0.4	0.3	
raphe pallidus	0.7	0.2	
nucl. solit. tract			
rostral part			3.6- 6.9
A2-cell group			40.3
commissural part			12.0

Table 4 (continued)

	TH	DBH	PNMT
	nmol/mg protein/h		pmol/mg protein/h
dorsal vagal nucl.			11.8
hypoglossal motor nucl.			4.8
Circumventricular Organs			
organum vas. lam. term.	1.3	0.9	10.4
subfornical organ	1.5	0.6	5.1
subcommissural organ	1.6	0.8	5.1
area postrema	3.2	0.8	15.9

References: SAAVEDRA et al. (1974a, b; 1976a, b); SAAVEDRA and ZIVIN (1976); BEN-ARI et al. (1978); BEART et al. (1979)
MFB — medial forebrain bundle
NIST — nucleus interstitialis striae terminalis
NISM — nucleus interstitialis striae medullaris
NICH — nucleus interstitialis commissurae hippocampi

In the other brain areas the noradrenaline levels are moderate, except in the cerebral cortex, basal ganglia spinal cord and the cerebellum, and there are a few areas, where the noradrenaline concentrations are below 10 ng/mg protein (Tables 2 and 3).

III. Adrenaline

Adrenaline is present in low concentrations in the CNS. These amount to only one tenth of the noradrenaline levels (Tables 2 and 3). Relatively high concentrations of adrenaline have been measured in the hypothalamus, especially in the periventricular, dorsomedial, paraventricular and supraoptic nuclei (over 2 ng/mg protein). No adrenaline could be detected—probably due to the limited sensitivity of the assays—in the cerebral cortex, rhinencephalon, basal ganglia, and most of the amygdaloid nuclei. All together 92 brain regions have been studied (VAN DER GUGTEN et al. 1976) and adrenaline has only been detected in 29 of them (Table 2).

The regional distribution of PNMT has been determined by means of a specific and sensitive radiometric technique. The results parallel those for adrenaline, except in the case of the lower brain stem (C1- and C2-cell groups). The greatest enzyme activity in the brain was measured in these two areas. Thus adrenaline-containing cell bodies appear to have high PNMT activity. The adrenaline concentrations of these cell bodies is moderate. This same pattern—i.e. high levels of biosynthetic enzymes and moderate levels of amines—is seen in other aminergic cell groups. The PNMT activity is unevenly distributed in the rat brain (Table 4).

Table 5. Biochemical studies on brain catecholamines and related enzymes in various mammals and the human

Mouse	TIZABI et al. (1979)	NA, DA
Sheep	BERTLER and ROSENGREN (1959)	NA, DA
	LAVERTY and SHARMAN (1965)	NA, DA
Goat	LAVERTY and SHARMAN (1965)	NA, DA
Rabbit	LAVERTY and SHARMAN (1965)	NA, DA
	METCALF (1974)	NA, DA
	ZIVIN et al. (1976)	TH, DBH, PNMT
Pig	BERTLER and ROSENGREN (1959)	NA, DA
Cow	KIZER et al. (1976a)	NA, DA, TH, DBH, PNMT
Cat	BERTLER and ROSENGREN (1959)	NA, DA
	MCGEER et al. (1963)	NA, DA
	LAVERTY and SHARMAN (1965)	NA, DA
Dog	BERTLER and ROSENGREN (1959)	NA, DA
	LAVERTY and SHARMAN (1965)	NA, DA
Monkey	WADA and MCGEER (1966)	NA, DA
	LEW et al. (1977)	PNMT
	BJÖRKLUND et al. (1978)	NA, DA
	BROWN et al. (1979)	NA, DA
Human	VOGEL et al. (1976)	PNMT
	FARLEY and HORNYKIEWICZ (1977)	NA, DA
	LEW et al. (1977)	PNMT
	MEFFORD et al. (1978)	A
	KOPP et al. (1979)	PNMT
	FARLEY et al. (1977)	DA

NA — noradrenaline
DA — dopamine
A — adrenaline

C. Biochemical Studies on the Neuronal Projections of the Catecholaminergic Cells

Specific chemical or electrolytic lesions of catecholaminergic cell groups, or transections of identified pathways have been used to unravel projections of catecholamine-synthesizing neurons. Both histo- and biochemical techniques are suitable for detecting the consequences of the operations. Histochemical observations demonstrate the disappearance of catecholamines from axons distal to the lesions. These observations, however, are only qualitative and in regions where catecholaminergic fibers or terminals are scattered, they are difficult and subjective. On the other hand, post-operative biochemical measurements provide quantitative data about catecholamine levels in discrete brain regions. Biochemical studies alone have real limitations though. One cannot tell whether changes in catecholamine levels result from transection and de-

Table 6. Biochemical studies on the neuronal projections of midbrain dopamine-containing cell groups (A8, A9, A10)

Pathways	Fibers via	to the	Lesions				References
			A9	A10	A9+10	A8+9+10	
Mesocortical DA-system	MFB	cerebral cortex	++	++		++	1–3
	external capsule	thalamus (perivent.)				++	1
		hypothalamus				+	1
Mesostriatal DA-system	ansa lenticularis	caudate-putamen	++	+			4
	internal capsule	nucl. accumbens		++			4
Mesolimbic DA-system	MFB	amygdala		++		++	1, 3, 5
	ansa lenticularis	habenula				++	1
	diagonal band	septum				+++	1
	septohypoth. tract	olfactory tub.		++	+++	++	1, 2
	medial olfactory tract	entorhinal cortex	++				3
Descending DA-fibers		cerebellum				++	1

\+ = 25–50 % decrease from control DA or TH levels
++ = 50–75% decrease
+++ = > 75 % decrease
DA — dopamine
References: 1 — Kizer et al. (1976b); 2 — Emson and Koob (1978); 3 — Fallon et al. (1978); 4 — Koob et al. (1975); 5 — Emson et al. (1979).

generation of catecholaminergic fibers or from degeneration of other neuronal inputs that are important for catecholamine synthesis, reuptake, or metabolism. Furthermore, the interpretation of the effects (or lack thereof) of lesions on catecholamine levels is made more difficult by the fact that the various catecholaminergic neurons have many targets in common. Thus, one cell group seems to be able to compensate for destruction of another by sprouting or increasing its synthesis of amine. Consequently, biochemical data may help one to understand catecholaminergic systems (origin and distribution of fibers), but these data should be complemented by histochemical observations.

I. Dopamine

Histochemical studies have been carried out to characterize the neuronal projections from dopaminergic cell groups. In addition, biochemical measurements of dopamine in various brain areas following lesions of the A8-, A9- and A10-cell groups have been made, and are summarized in Table 6.

Individual lesions of the A9-cell group (the compact zone of the substantia nigra) resulted in a decrease of more than 50% in dopamine concentrations of the cerebral cortex and caudate-putamen. This cell group contributes to both the mesocortical and mesostriatal dopaminergic systems.

Substantial falls in dopamine levels in the cortex as well as in the nucleus accumbens, amygdala and olfactory tubercle have been observed following lesions of the ventral tegmental area (A10-cell group). These cells appear to project preferentially to the mesolimbic and mesocortical systems since damaging them produces only a 25–50% decrease in the dopamine level of the caudate putamen.

Large mesencephalic lesions destroying all of the A8-, A9- and A10-cells caused dopamine depletion from brain areas that were not affected by smaller lesions: the periventricular thalamic nucleus, median eminence, habenula, septum and cerebellum (Table 6). Most of the dopamine of the medial hypothalamus is intra-hypothalamic. Deafferentiation of the area fails to induce substantial alterations in the dopamine level. The bulk of dopaminergic terminals in the median eminence originates in arcuate dopamine-containing cells (tuberoinfundibular dopaminergic system).

II. Noradrenaline

Histofluorescence studies first demonstrated the extensive innervation of telencephalic structures by locus coeruleus noradrenergic cells. These studies have subsequently been confirmed by biochemical analyses of forebrain areas after lesions of the locus coeruleus or transections of the dorsal noradrenergic bundle (Table 7). Noradrenaline levels and DBH activity fell by more than 75% in various cerebral cortical areas, olfactory lobe and hippocampus after destruction of the locus coeruleus. Less marked, but still over 50% of decrease occurred in the thalamus und amygdala, less than 50% in the habenula. Mid-

Table 7. Biochemical studies on the neuronal projections of the locus coeruleus and neighboring noradrenergic cell groups

Pathways	Fibers via	to the	Lesions			References
			LC		Dorsal Bundle	
Dorsal NA-bundle		pons	+-++	(+++)		1, 2
		midbrain	+-+++	(++)		1–3
	fasciculus retroflexus	habenula	+			3
	ex. medull. lam. thal.	thalamus	+	(++)		1, 3
	ansa lent.-vent. amygd. bund.	amygdala	+-+++	(++)		1, 4, 5
	diagonal band	septum		(++)	+	1
		hippocampus	++	(+++)		1, 3, 6, 7
	rostral MFB	olfactory lobe	+-+++	(+++)	+-++	1, 4, 5
	cingulum	cerebral cortex	+++	(+++)	+-++	1, 2, 5, 7, 8
Dorsal periventricular tract	dorsal long. fascicle	central gray mat.	+	(++)		1, 2
		hypothalamus (NArc, NDM, NPV, NSO, NPE)	+	(+)		1, 3, 4, 7, 9
	dorsal tegm. tract	medulla	+-+++			2
		spinal cord	+	(++)		1, 6, 10
Cerebellar NA-pathway	superior cereb. pedunc.	cerebellum	++	(+)		1, 3, 10

\+ = 25–50 % decrease from control NA or DBH (circled levels)
++ = 50–75 % decrease
+++ = > 75 % decrease

NA — noradrenaline; NPV — paraventricular nucleus;
MFB — medial forebrain bundle; NSO — supraoptic nucleus;
NArc — arcuate nucleus; NPE — periventricular nucleus
NDW — dorsomedial nucleus;

References: 1 — Ross and Reis (1974); 2 — Levitt and Moore (1979); 3 — Kobayashi et al. (1974); 4 — Záborszky et al. (1977); 5 — Fallon et al. (1978); 6 — Ader et al. (1979); 7 — Roizen et al. (1976); 8 — van der Gugten et al. (1976); 9 — Palkovits et al. (1980b); 10 — Commissiong et al. (1978).

brain and pontine projections of the locus coeruleus have been demonstrated by both histofluorescence and biochemical microassays. All the above regions are innervated by noradrenaline fibers of the dorsal noradrenergic bundle (Table 7).

There are other ascending and descending axons from the locus coeruleus. Noradrenaline-containing fibers from the locus coeruleus and its neighboring cells may reach the hypothalamus through the dorsal periventricular tract, and slight, but significant decreases in noradrenaline levels or DBH activity have been found in the dorsomedial, periventricular, paraventricular, supraoptic and arcuate nuclei. These decreases, however, never exceeded 50% of the total noradrenaline or DBH in these nuclei (Table 7). The existence of descending locus coeruleus noradrenergic fibers has been verified by depletion of noradrenaline and DBH in the medulla, spinal cord and cerebellum after lesions of the locus coeruleus. Fibers to the medulla and spinal cord run in or along the dorsal tegmental or dorsal periventricular tract, fibers to the cerebellum run within the superior cerebellar peduncle.

Table 8. Biochemical studies on the neuronal projections of the tegmental noradrenergic cell groups

Pathways	Fibers via	to the	Lesions					References
			A1	A5	A7	A2	VB	
A. Ventrolateral NA-Cell Groups (A1, A5, A7)								
Ventral NA-bundle	central tegmental tract							
	MFB	NIST (ventral)		+			+	1, 2
		septum					++	2
		amygdala (cent.)					+	3
	periventricular sys.	hypothalamus	+	+	+		+-++ (++) (circled)	1, 2, 4, 5
Bulbospinal NA-system	ventral funiculus	spinal cord						
	lateral funiculus							
B. Dorsal tegmental NA-cell group (A2)								
Dorsal periventricular tract	dors. long. fasc.	hypothalamus				+		4

\+ = 25-50%
++ = 50-75% decrease of NA levels or DBH activity (circled)
MFB — medial forebrain bundle; NIST — bed nucleus of stria terminalis;
VB — ventral noradrenergic bundle.

References: 1 — Speciale et al. (1978); 2 — O'Donohue et al. (1979); 3 — Fallon et al. (1978); 4 — Palkovits et al. (1980b); 5 – Kizer et al. (1976c).

Biochemical assays performed following individual lesions of the brain stem noradrenergic cell groups as well as histofluorescent observations indicate that all noradrenergic cell groups participate in the innervation of the hypothalamus (Tables 7 and 8). All fibers other than those originating from the locus coeruleus ascend rostrally in the ventral noradrenergic bundle which also takes up the fibers of the dorsal noradrenergic bundle before entering the lateral hypothalamus. Transections of the ventral noradrenergic bundle result in decreases in hypothalamic noradrenaline content and DBH activity (Table 8). Transections close to the hypothalamus were more effecitve in this respect than ones more caudal. However, no lesion of an individual noradrenergic cell group nor any single brainstem transection induced as great a reduction of noradrenaline in the hypothalamus as that observed following complete deafferentiation of the hypothalamus (PALKOVITS et al. 1980b).

Ventral noradrenergic bundle fibers also innervate the limbic system. After surgical interruption of these fibers 25–75 % decreases in noradrenaline levels were detected in the septum, central amygdaloid nucleus and ventral portion of the bed nucleus of the stria terminalis (Table 8). These fibers may run from the lower brainstem rostrally in the ventral noradrenergic bundle and arrive at their final destinations via the medial forebrain bundle.

Based on histofluorescent observations it seems that the ventrolateral noradrenergic groups also participate in providing the spinal cord its noradrenergic innervation (bulbospinal noradrenergic system). Biochemical studies designed to confirm and extend these observations have not been performed to the best of our knowledge.

III. Adrenaline

The projections of the dorsal tegmental (C2-cell group) and lateral tegmental (C1-cell group) adrenaline-containing neurons have yet to be characterized. Most of the adrenaline nerve terminals in the hypothalamus originate from the lower brainstem. Following total deafferentiation of the hypothalamus, PNMT activity is reduced by 60 %. About 60–80 % decrease of adrenaline content was found in the para- and periventricular nuclei as well as in the median eminence after unilateral transection of the brain-stem. The fibers originate mainly from the C1-cell group. Lesions of these nuclei decreased hypothalamic adrenaline levels by 60–75 % (paraventricular nucleus: 64–71 %; periventricular nucleus: 68 %; arcuate nucleus: 65–74 %; median eminence: 62 %). Lesions of the C2-cell group resulted in a 35 % decrease of adrenaline in the median eminence (PALKOVITS et al. 1980a).

The origins of adrenaline nerve terminals in the septum, amygdala, midbrain, pons and spinal cord have not been investigated yet (Table 9). Until now, adrenaline cell bodies in the CNS have been recognized only in the lower brainstem, so these cells are the best candidate to be the sources of the brain adrenaline innervation. This hypothesis, however, still needs to be proven.

Table 9. Biochemical studies on the projections of the brainstem adrenaline-containing cell groups (C1, C2)

Pathway	Fibers via	to the	Lesions		Reference
			C1	C2	
Ventral NA-bundle	Central tegm. tract Lat. tegm. tract	hypothalamus septum amygdala midbrain pons	++	+	1
Bulbospinal A-system		spinal cord			

\+ = 25–50 %
++ = 50–75 % decrease in A levels
NA — noradrenaline
A — adrenaline
Reference: PALKOVITS et al. (1980a)

D. Regional Turnover of Catecholamines

A number of methods have been used to study turnover of dopamine, noradrenaline, and adrenaline in the central nervous system. These have involved **1)** blocking the biosynthesis of the amines and measuring their rates of disappearance from brain tissue or the build-up of their precursors (KIZER et al. 1976c; DEMAREST et al. 1979), **2)** measuring changes in amine metabolites, **3)** determining the rates of conversion of radioactive or heavy isotope-labeled precursors into the amines, **4)** labeling the amine pools with radioactive catecholamines and determining their rates of disappearance, and **5)** detecting changes in amines in the extracellular space by means of selective electrodes. All of these methods have shortcomings and few of them have been applied to microdissected brain samples or small, discrete brain regions. The method that has been used most extensively for this purpose employs α-methyl-para tyrosine or its methyl ester to inhibit TH. At various times after the drug is administered, the animals are killed and their brains are processed for catecholamine determinations. Assuming that the rate of change of the amine is proportional to the amount of amine remaining in the brain at any time, the log of the amine concentration should be inversely proportional to the duration of the drug treatment. In applying this technique, one assumes that the drug itself will have no effect on the turnover rates measured, an assumption that may not be warranted. The studies that have been published to date indicate that noradrenaline (or dopamine) fractional rate constants are fairly constant from one region to the next. Since the concentration of amine varies from one area to the next, however, the total amount of noradrenaline or dopamine synthesized per mg tissue per hour is variable (VERSTEEG et al. 1975).

Under certain circumstances, the rates of synthesis and utilization of the amines seem to change. Thus, stress (MOYER et al. 1978) or isolation (THOA et al. 1977) have been reported to alter noradrenaline and dopamine turnover rates in specific nuclei. Studies of this sort, when done well, give useful clues about the role of catecholamines in mediating neuroendocrine or behavioral phenomena.

E. References

Ader J-P, Postema F, Korf J (1979) Contribution of the locus coeruleus to the adrenergic innervation of the rat spinal cord: a biochemical study. J Neural Transm 44:159-173

Andén N-E, Carlsson A, Dahlström A, Fuxe K, Hillarp N-A, Larsson K (1966) Ascending monoamine neurons to the telencephalon and diencephalon. Acta Physiol Scand 67:313-326

Beart PM, Prosser D, Louis WJ (1979) Adrenaline and phenylethanolamine-N-methyltransferase in rat medullary and anterior hypothalamic-preoptic nuclei. J Neurochem 33:947-950

Ben-Ari Y, Zigmond RE, Moore KE (1978) Regional distribution of tyrosine hydroxylase, norepinephrine and dopamine within the amygdaloid complex of the rat. Brain Res 87:96-101

Bertler A, Rosengren E (1959) Occurrence and distribution of catecholamines in brain. Acta Physiol Scand 47:350-361

Björklund A, Lindvall O (1984) Dopamine-containing systems in the CNS. In: Björklund A, Hökfelt T (Eds) Classical Transmitters in the CNS, part I. Handbook of Chemical Neuroanatomy, vol 2. Elsevier, Amsterdam New York Oxford, pp 55-122

Björklund A, Nobin A (1973) Fluorescence histochemical method and microfluorimetric mapping of dopamine and noradrenaline cell groups in the rat diencephalon. Brain Res 51:193-205

Björklund A, Moore RY, Nobin A, Stenevi U (1973) The organization of tuberohypophyseal and reticuloinfundibular catecholamine neuron systems in the rat brain. Brain Res 51:121-191

Björklund A, Lindvall O, Nobin A (1975) Evidence of an incertohypothalamic dopamine neurone system in the rat. Brain Res 89:29-42

Björklund A, Divac I, Lindvall O (1978) Regional distribution of catecholamines in monkey cerebral cortex, evidence for a dopaminergic innervation of the primate prefrontal cortex. Neurosci Lett 7:115-119

Brown RM, Crane AM, Goldman PS (1979) Regional distribution of monoamines in the cerebral cortex and subcortical structures of the rhesus monkey: concentrations and in vivo synthesis rates. Brain Res 168:133-150

Brownstein M, Saavedra JM, Palkovits M (1974) Norepinephrine and dopamine in the limbic system of the rat. Brain Res 79:431-436

Commissiong JW, Hellstrom SO, Neff NH (1978) A new projection from locus coeruleus to the spinal ventral columns: histochemical and biochemical evidence. Brain Res 148:207-213

Dahlström A, Fuxe K (1964) Evidence for the existence of monoamine-containing neurons in the central nervous system. I. Demonstration of monoamines in the cell bodies of brain stem neurons. Acta Physiol Scand Suppl 232:1-55

Demarest KT, Alper RH, Moore KE (1979) Dopa accumulation is a measurement of dopamine synthesis in the median eminence and posterior pituitary. J Neural Transm 46:183-193

Emson PC, Koob GF (1978) The origin and distribution of dopamine-containing afferents to the rat frontal cortex. Brain Res 142:249-267

Emson PC, Björklund A, Lindvall O, Paxinos G (1979) Contributions of different afferent pathways to the catecholamine and 5-hydroxytryptamine innervation of the amygdala: a neurochemical and histochemical study. Neuroscience 4:1347-1357

Fallon JH, Moore RY (1978a) Catecholamine innervation of the basal forebrain. III. Olfactory bulb, anterior olfactory nuclei, olfactory tubercle and piriform cortex. J Comp Neurol 180:533-544

Fallon JH, Moore RY (1978b) Catecholamine innervation of the basal forebrain. IV. Topography of the dopamine projection to the basal forebrain and neostriatum. J Comp Neurol 180:545-580

Fallon JH, Koziell DA, Moore RY (1978) Catecholamine innervation of the basal forebrain. II. Amygdala, suprarhinal cortex and entorhinal cortex. J Comp Neurol 180:509-532

Farley IJ, Hornykiewicz O (1977) Noradrenaline distribution in subcortical areas of the human brain. Brain Res 126:53-62

Farley IJ, Price KS, Hornykiewicz O (1977) Dopamine in the limbic regions of the human brain: normal and abnormal. Adv Biochem Psychopharmacol 16:57-64

Fuxe K, Hökfelt T (1966) Further evidence for the existence of tuberoinfundibular dopamine neurons. Acta Physiol Scand 66:245-246

van der Gugten J, Palkovits M, Wijnen HLJM, Versteeg DHG (1976) Regional distribution of adrenaline in rat brain. Brain Res 107:171-175

Halász N, Ljungdahl Å, Hökfelt T, Johansson O, Goldstein M, Park D, Biberfeld P (1977) Transmitter histochemistry of the rat olfactory bulb I. Immunohistochemical localization of monoamine synthesizing enzymes, supports for intrabulbar, periglomerular dopamine neurons. Brain Res 126:455-476

Hökfelt T, Fuxe K, Goldstein M, Johansson O (1974) Immunohistochemical evidence for the existence of adrenaline neurons in the rat brain. Brain Res 66:235-251

Hökfelt T, Philipson O, Goldstein M (1979) Evidence for a dopaminergic pathway in the rat descending from A11 cell group to the spinal cord. Acta Physiol Scand 107:393-395

Hökfelt T, Johansson O, Goldstein M (1984a) Central catecholamine neurons as revealed by immunohistochemistry with special reference to adrenaline neurons. In: Björklund A, Hökfelt T (Eds) Classical Transmitters in the CNS, part I, Elsevier, Amsterdam New York Oxford, pp 157-276 Handbook of Chemical Neuroanatomy, vol 2

Hökfelt T, Mårtensson R, Björklund A, Kleinau S, Goldstein M (1984b) Distributional maps of tyrosine-hydroxylase-immunoreactive neurons in the rat brain. In: Björklund A, Hökfelt T (Eds) Classical Transmitters in the CNS, Handbook of Chemical Neuroanatomy, vol 2 part I, Elsevier, Amsterdam New York Oxford, pp 277-379

Jacobowitz DM (1978) Monoaminergic pathways in the central nervous system. In: Lipton MA, DiMascio A, Killam KF (Eds) Psychopharmacology: A Generation of Progress, Raven Press, New York, pp 119-129

Jacobowitz DM, Palkovits M (1974) Topographic atlas of catecholamine and acetylcholinesterase-containing neurons in the rat brain. I. Forebrain (Telencephalon, Diencephalon). J Comp Neurol 157:13-28

Kizer JS, Palkovits M, Tappaz M, Kebabian J, Brownstein MJ (1976a) Distribution of releasing factors, biogenic amines, and related enzymes in the bovine median eminence. Endocrinology 98:685-695

Kizer JS, Palkovits M, Brownstein MJ (1976b) The projections of the A8, A9 and A10 dopaminergic cell bodies: evidence for a nigral-hypothalamic-median eminence dopaminergic pathway. Brain Res 108:363–370
Kizer JS, Muth E, Jacobowitz DM (1976c) The effect of bilateral lesions of the ventral noradrenergic bundle on endocrine-induced changes of tyrosine hydroxylase in the rat median eminence. Endocrinology 98:886–893
Kobayashi RM, Palkovits M, Kopin IJ, Jacobowitz DM (1974) Biochemical mapping of noradrenergic nerves arising from the rat locus coeruleus. Brain Res 77:269–279
Koob GF, Balcom JG, Meyerhoff JL (1975) Dopamine and norepinephrine levels in the nucleus accumbens, olfactory tubercle and corpus striatum following lesions in the ventral tegmental area. Brain Res 94:44–45
Kopp N, Denoroy L, Renaud B, Pujol JF, Tabib A, Tommasi M (1979) Distribution of adrenaline-synthesizing enzyme activity in the human brain. J Neurol Sci 41:397–409
Koslow SH, Schlumpf M (1974) Quantitation of adrenaline in rat brain nuclei and areas by mass fragmentography. Nature 251:530–531
Kvetňanský R, Palkovits M, Mitro A, Torda T, Mikulaj L (1977) Catecholamines in individual hypothalamic nuclei of acutely and repeatedly stressed rats. Neuroendocrinology 23:257–267
Laverty R, Sharman DF (1965) The estimation of small quantities of 3,4-dihydroxyphenylethylamine in tissues. Br J Pharmacol Chemother 24:538–548
Levitt P, Moore RY (1979) Origin and organization of brainstem catecholamine innervation in the rat. J Comp Neurol 186:505–528
Lew JY, Matsumoto Y, Pearson J, Goldstein M, Hökfelt T, Fuxe K (1977) Localization and characterization of phenylethanolamine N-methyltransferase in the brain of various mammalian species. Brain Res 119:199–210
Lindvall O (1979) Dopamine pathways in the rat brain. In: Horn AS, Korf J, Westerink BHC (Eds) The Neurobiology of Dopamine. Academic Press, London New York San Francisco, pp 319–342
Lindvall O, Björklund A (1978) The organization of the ascending catecholamine neuron systems in the rat brain. Acta Physiol Scand 412 (Suppl):1–48
McGeer PL, McGeer EG, Wada JA (1963) Central aromatic amine levels and behavior. II. Serotonin and catecholamine levels in various cat brain areas following administration of psychoactive drugs or amine precursors. Arch Neurol (Chicago) 9:81–89
Mefford I, Oke A, Keller R, Adams RN, Jonsson G (1978) Epinephrine distribution in human brain. Neurosci Lett 9:227–231
Metcalf G (1974) The use of stereotaxic dissection followed by fluorimetric assay, to determine the distribution of noradrenaline, dopamine and 5-HT in the preoptic hypothalamic area of rabbit brain: an alternative approach to histochemistry. Br J Pharmacol 50:189–195
Moore RY, Bloom FE (1979) Central catecholamine neuron systems: anatomy and physiology of the norepinephrine and epinephrine systems. Annu Rev Neurosci 2:113–168
Moore RY, Card JP (1984) Noradrenaline-containing neuron system. In: Björklund A, Hökfelt T (Eds). Classical Transmitters in the CNS, Handbook of Chemical Neuroanatomy, vol 2 part I, Elsevier, Amsterdam New York Oxford, pp 123–156
Moyer JA, Herrenkohl LR, Jacobowitz DM (1978) Stress during pregnancy: effect on catecholamines in discrete brain regions of offspring as adults. Brain Res 144:173–178

Nygren L-G, Olson L (1977) A new major projection from locus coeruleus: the main source of noradrenergic nerve terminals in the ventral and dorsal columns of the spinal cord. Brain Res 132:85-93
O'Donohue TL, Crowley WR, Jacobowitz DM (1979) Biochemical mapping of the noradrenergic ventral bundle projection sites: evidence for a noradrenergic-dopaminergic interaction. Brain Res 172:87-100
Olson L, Fuxe K (1971) On the projections from the locus coeruleus noradrenaline neurons: the cerebellar innervation. Brain Res 28:165-171
Palkovits M, Jacobowitz DM (1974) Topographic atlas of catecholamine and acetylcholinesterase-containing neurons in the rat brain. II. Hindbrain (Mesencephalon, Rhombencephalon). J Comp Neurol 157:29-42
Palkovits M, Brownstein M, Saavedra JM, Axelrod J (1974) Norepinephrine and dopamine content of hypothalamic nuclei of the rat. Brain Res 77:137-149
Palkovits M, Záborszky L, Brownstein MJ, Fekete MIK, Herman JP, Kanyicska B (1979) Distribution of norepinephrine and dopamine in cerebral cortical areas of the rat. Brain Res Bull 4:593-601
Palkovits M, Mezey É, Záborszky L, Feminger A, Versteeg DHG Wijnen HJLM, de Jong W, Fekete MIK, Herman JP, Kanyicska B (1980a) Adrenergic innervation of the rat hypothalamus. Neurosci Lett 18:237-243
Palkovits M, Záborszky L, Feminger A, Mezey É, Fekete MIK, Herman JP, Kanyicska B, Szabo D (1980b) Noradrenergic innervation of the rat hypothalamus: experimental biochemical and electron microscopic studies. Brain Res 191:161-171
Pasquier DA, Gold MA, Jacobowitz DM (1980) Noradrenergic perikarya (A5-A7, sub coeruleus) projections to the rat cerebellum. Brain Res 196:270-275
Roizen MF, Kobayashi RM, Muth EA, Jacobowitz DM (1976) Biochemical mapping of noradrenergic projections of axons in the dorsal noradrenergic bundle. Brain Res 104:384-389
Ross RA, Reis DJ (1974) Effects of lesions of locus coeruleus on regional distribution of dopamine-β-hydroxylase activity in rat brain. Brain Res 73:161-166
Saavedra JM, Zivin J (1976) Tyrosine hydroxylase and dopamine-β-hydroxylase: distribution in discrete areas of the rat limbic system. Brain Res 105:517-524
Saavedra JM, Brownstein M, Palkovits M, Kizer S, Axelrod J (1974a) Tyrosine hydroxylase and dopamine-β-hydroxylase: distribution in the individual rat hypothalamic nuclei. J Neurochem 23:869-871
Saavedra JM, Palkovits M, Brownstein MJ, Axelrod J (1974b) Localization of phenylthanolamine N-methyl transferase in the rat brain nuclei. Nature 248:695-696
Saavedra JM, Grobecker H, Zivin J (1976a) Catecholamines in the raphe nuclei of the rat. Brain Res 114:339-345
Saavedra JM, Brownstein MJ, Kizer JS, Palkovits M (1976b) Biogenic amines and related enzymes in the circumventricular organs of the rat. Brain Res 107:412-417
Skagerberg G, Björklund A, Lindvall O, Schmidt RH (1982) Origin and termination of the diencephalo-spinal dopamine system in the rat. Brain Res Bull 9:237-244
Speciale SG, Crowley WR, O'Donohue TL, Jacobowitz DM (1978) Forebrain catecholamine projections of the A5 cell groups. Brain Res 154:128-133
St. Laurent J, Roizen MF, Miliaressis E, Jacobowitz DM (1975) The effects of self-stimulation on the catecholamine concentration of discrete areas of the rat brain. Brain Res 99:194-200
Thoa NB, Tizabi Y, Jacobowitz DM (1977) Effect of isolation on catecholamine concentration and turnover in discrete areas of rat brain. Brain Res 131:259-269

Tizabi Y, Thoa NB, Maengwyn-Davies GD, Kopin IJ, Jacobowitz DM (1979) Behavioral correlation of catecholamine concentration and turnover in discrete brain areas of three strains of mice. Brain Res 166:199–205
Ungerstedt U (1971) Stereotaxic mapping of the monoamine pathways in the rat brain. Acta Physiol Scand 367 (Suppl):1–48
Versteeg D, van der Gugten J, van Ree JH (1975) Regional turnover and synthesis of catecholamine in rat hypothalamus. Nature 256:502–503
Versteeg DHG, van der Gugten J, de Jong W, Palkovits M (1976) Regional concentrations of noradrenaline and dopamine in rat brain. Brain Res 113:563–574
Vogel WH, Lewis LE, Boehme DH (1976) Phenylethanolamine-N-methyltransferase activity in various areas of human brain, tissues and fluids. Brain Res 115:357–359
Wada JA, McGeer EG (1966) Central aromatic amines and behavior. III. Correlative analysis of conditioned approach behavior and brain levels of serotonin and catecholamines in monkeys. Arch Neurol (Chicago) 14:129–142
Záborszky L, Brownstein MJ, Palkovits M (1977) Ascending projections to the hypothalamus and limbic nuclei from the dorsolateral pontine tegmentum: a biochemical and electron microscopic study. Acta Morphol Acad Sci Hung 25:175–188
Zivin JA, Reid JL, Saavedra JM, Kopin KJ (1975) Quantitative localization of biogenic amines in the spinal cord. Brain Res 99:293–301
Zivin JA, Reid JL, Tappaz ML, Kopin KJ (1976) Quantitative localization of tyrosine hydroxylase, dopamine-β-hydroxylase, phenylethanolamine-N-methyl transferase, and glutamic acid decarboxylase in spinal cord. Brain Res 105:151–156

CHAPTER 11

Catecholamines and Behavior

F. E. BLOOM, J. A. SCHULMAN, and G. F. KOOB

A. Introduction

Research on the neural function of catecholamines spans all levels of active investigation from the molecular to the behavioral. In this chapter we consider the behavioral actions of catecholamines and their related drugs in respect to the molecular actions, cellular structure, and the behavioral repertoire of central catecholamine neurons. Each of these aspects of central catecholamine neurons is an extensive independent body of research. However, we emphasize evidence for the relationships between the molecular level and the behavioral level. We first review those aspects of cellular structure and receptor mechanisms which are essential to an integrative assessment of function for the dopamine and noradrenaline systems. We next analyse the central dopaminergic neurons and the results of experimental perturbations on their actions, indicating the general gaps and overlaps in these studies. We conclude by considering the behavioral implications of the electrophysiological properties of noradrenergic neurons in awake, behaving animals in order to develop a basis on which to evaluate the behavioral effects of drugs and lesions which alter the function of the noradrenergic system.

I. Catecholamine Hypotheses of Behaviors

Before delving into the molecular and cellular pharmacological evidence relating catecholamine systems with behavioral phenomena, some important, but experimentally practical, considerations are necessary. For the past two decades brain catecholamines have been implicated as "the" critical chemical mediators of a variety of physiological-behavioral outputs of the brain, ranging from feeding, drinking, thermoregulation, and sexual behavior to such abstract actions as pleasure, reinforcement, attention, motivation, memory, learning-cognition, the major psychoses and their chemotherapy. While many such hypotheses have been proposed, there has still been no conclusive proof that a monoamine "mediates" any behavior. Furthermore, no unifying hypothesis as to how any catecholamine cellular system could possibly be legitimately involved in so many global actions has yet been put forth.

II. Theoretical Constructions Linking Catecholamines and Behavior

Two major experimental paradigms have been employed to attempt to link catecholamine systems with behavioral phenomena: 1) the "endocrinological" paradigm and 2) the "correlational" paradigm. In the "endocrinological" paradigm the functional involvement of a brain component (or endocrine organ) is assessed by focussing on what happens (or does not) when that circuit is activated or removed. In the classical endocrine experiments, the importance of a gland's secretions could be verified by subsequent reversal of deficits through administration of glandular extracts; however, brain lesions cannot be so easily overcome by administration of transmitter agonists, and their immediate effects can be devastating. The adaptation of the endocrine paradigm to the functional analysis of brain systems, especially in the context of behavioral research, requires the further assumption that a given behavior is a unitary phenomenon, and that only one system, cleanly amenable to a lesion, is necessary and sufficient for its execution.

In the correlational paradigm, alterations in the basal activity of a system (i.e. metabolic turnover, firing rate, release or receptor regulation) during elicitation of a behavior are used to infer the participation of the system in that behavior. Although the correlational paradigm cannot *per se* establish the necessary and sufficient roles of a given chemical system in a given behavioral phenomenon, the resultant correlations of changes in activity with specific behaviors can be used to set constraints on the functional relationships in which that system is involved (i.e. changes in firing or metabolism with "sleep" or "reward" or "fear" suggest possible involvements). These constraints can then be used in turn to devise an appropriate set of contingencies for an "endocrinological" experiment such that elimination or simulation of the action of a specific chemical circuit will be expected to alter the behavior in a predictable fashion.

In attempts to test at the behavioral level the specific role of a given catecholamine system, these two contrasting experimental paradigms have each been applied with varying success. The failure to incriminate the noradrenaline system in learning behaviors generally, by lesions to the locus coeruleus or its major pathways (see below) has forced a retrenchment to determine the behavioral repertoire of these neurons by unit recordings in behaviorally intact animals; new predictions of its role in behavior based on correlated changes in cell firing now await "endocrinological" type testing. Conversely, significant disruptions of motor behaviors have been found to result from manipulations of the dopamine systems, but in this case, the lack of direct correlations between unit discharge patterns and the evocation and execution of the motor responses beclouds precise description of the way in which these dopamine systems operate to modify normal motor behavior.

It would, therefore, seem useful in assessing the behavioral effects of central catecholamine neurons to distinguish between 4 categories of behavioral relationship: **1)** direct mediation (e.g. the specific catecholamine system is the exclusive mediator of a given behavior); **2)** involvement (one or more catecholamine, and other transmitter-specific, systems are able to modify the elicitation or execution of a behavioral phenomenon); **3)** parallel mediation (e.g. the catecholamine circuit can mediate the behavior, but so can other systems, such that elimination of the catecholamine system does not eliminate the behavioral act); **4)** remote mediation (e.g. the catecholamine circuit's involvement depends upon the recognition of a more abstract, and non-specific role than the specific behavior being tested).

For purposes of the present review, we have therefore surveyed the central catecholamine systems, especially the ventrotegmental dopaminergic systems and the locus coeruleus noradrenergic system in order to highlight their structural, physiological and biochemical properties which could be regarded as pertinent to their participation in behavioral phenomena.

B. Dopamine Systems

The dopamine neuron systems are more complex in their anatomy, more diverse in localization and apparent function, and more numerous, both in terms of definable systems and in numbers of neurons, than the adrenergic or noradrenergic systems (see MOORE and BLOOM 1978; BJÖRKLUND and LINDVALL 1986). Dopamine neuron systems are principally located in the upper midbrain and hypothalamus. Their shape varies from neurons without axons (retina, olfactory bulb) and with very restricted projections, to neurons with extensive axonal arborizations. With the introduction of the glyoxylic acid fluorescence histochemical method (LINDVALL et al. 1974a, b), the increasing application of ultrastructural analysis (e.g. AJIKA and HÖKFELT 1973; ARLUISON et al. 1978a, b; BLOOM 1973), and the advent of new, powerful neuroanatomical methods (see BJÖRKLUND and LINDVALL 1986), there has been a very rapid, marked increase in our understanding of the extent and organization of dopamine neuron systems. One other chapter in this volume (see PALKOVITS and BROWNSTEIN, Chap. 10) discusses the organization of two major dopamine systems investigated for behavioral relevance, namely the nigrostriatal projection and the ventrotegmental projections to the cerebral cortex and to the nucleus accumbens septi.

Dopamine neurons in the upper midbrain form a continuous group extending rostrally from the mammillary complex to the level of the decussation of the pyramidal tracts. The cell groups are actually contiguous in the horizontal plane and, while not readily falling into distinct anatomically separated groupings, the original separate nuclear designations A8, A9 and A10 continue to be used in many studies (DAHLSTRÖM and FUXE 1964). It is possible, however, to discriminate two distinct components to the system. The first, arising largely in the substantia nigra and terminating in the neostriatum, is referred to as the nigrostriatal system. The second, arising in both the substantia nigra and the ventral tegmental area and projecting widely to nucleus accumbens, frontal and rhinencephalic cortical regions of the telencephalon, is termed the mesocortical or mesolimbic dopamine system. Ontogenetic studies (OLSON and SEIGER 1972; SEIGER and OLSON 1973; also see SCHLUMPF et al. 1980) indicate that these dopamine neuronal groups emerge from a single embryonic anlage.

I. Nigrostriatal System

1. Anatomical Foundations

Axons arising from the pars compacta of the substantia nigra and the medial ventral tegmental area run dorsally and ascend through the medial tegmentum to the dorsal part of the lateral hypothalamic area (UNGERSTEDT 1971a; MOORE et al. 1971; see BJÖRKLUND and LINDVALL 1986). As the nigrostriatal bundle courses rostrally in the diencephalon, it occupies a position partly in the lateral hypothalamic area and partly on the medial portion of the internal capsule. Further rostrally, the fibers run dorsally into the internal capsule and distribute throughout the caudate and putamen. The pathway is characterized by very thin (< 0.1 μm), nonvaricose fibers. The plexus of dopamine fibers terminating in caudate and putamen is so dense that in fluorescence histochemical material the neostriatum appears to be virtually covered with fine dopamine varicosities (MOORE and BLOOM 1978; LINDVALL et al. 1974b). As the dopamine axons of the nigrostriatal projection enter the neostriatum they undergo a massive collateralization, each branch containing numerous small varicosities. In ultrastructural studies, dopamine axon terminals have been identified predominantly on the small dendritic branches of small neurons of the neostriatum (HÖKFELT and UNGERSTEDT 1969; IBATA et al. 1973; HATTORI et al. 1973; ARLUISON et al. 1978a, b). The terminals are small, about 0.5-0.7 μm in diameter, containing 400-500 Å pleomorphic synaptic vesicles that do not contain a dense core with the usual fixation methods. Occasional large granular vesicles are evident. Electron-microscopic autoradiographic studies (HATTORI et al. 1973; MCGEER et al. 1975; ARLUISON et al. 1978a, b) demonstrate that dopamine-labeled terminals make synaptic contacts. These contacts are virtually all of the asymmetric (Gray type I) variety on dendritic spines.

There is a striking internal topographic organization between distinct dopamine cell bodies and their terminal fields (FALLON and MOORE 1976b). The most medial and rostral portion of the neostriatum is innervated by dopamine neurons of the lateral ventral tegmental area. This overlaps the area innervating the nucleus accumbens septi. More lateral areas of neostriatum are innervated by progressively more lateral nigral dopamine neurons with the most caudal and ventral portion of the neostriatum innervated by the most caudal meso-telencephalic dopamine neurons. There is also a rostro-caudal topography and a dorso-ventral topography of the projection. Dorsally placed cells in the mesencephalic dopamine cell group project ventrally on the neostriatum, whereas ventrally placed cells project dorsally (FALLON and MOORE 1976b). This morphological feature of the nigrostriatal system was not recognized in the early mapping studies but is quite pertinent to a critical analysis of the system's physiological and behavioral actions.

2. Cellular Actions

The cellular function of dopamine nigral and ventro-tegmental neurons on their target cells in the striatum and forebrain have been studied by three general approaches: 1) Release of dopamine from axons is simulated by iontophoretic applications and the effects evaluated by extracellular or intracellular electrophysiologic indices (CONNER 1968, 1970; YORK 1975; SIGGINS 1977); **2)** the effect of the dopamine-containing tracts is evaluated by stimulation of dopamine neurons and observation of the electrophysiologic consequences on target cells; **3)** the function of dopamine fibers is inferred from the dysfunction observed after electrolytic or chemical lesioning of the dopamine neurons. All three methods have been employed extensively and, with certain important exceptions, the evidence strongly supports the original iontophoretic experiments which indicated that dopamine inhibits the spontaneous or induced activity of neostriatal neurons (see SIGGINS 1977; MOORE and BLOOM 1978). Nevertheless, the continuing controversy over the cellular action of dopamine in the neostriatum and other terminal areas deserves further attention in this review because the primary cellular action and its underlying molecular mechanisms provide additional insight into the possible behavioral properties of this system.

a. Iontophoretic Actions

The proportion of cells inhibited by dopamine, regardless of species, anesthetic, or iontophoretic technique, vary from a low of 78% (BRADSHAW et al. 1973) to a high of 100% (FELTZ and DE CHAMPLAIN 1972). The inhibition typically appears some 2 to 15 sec after the onset of iontophoresis and may persist for minutes after the termination of the ejection current. Dopamine applied by iontophoresis inhibits spontaneous activity (BLOOM et al. 1965; MCLENNAN and YORK 1967; HERZ and ZIEGLGÄNSBERGER 1968; CONNER 1970; BUNNEY and AGHAJANIAN 1973; SIGGINS et al. 1974; SPEHLMANN 1975; STONE and BAILEY 1975; STONE 1976) and inhibits activity produced by the iontophoresis of excitatory amino acids (HERZ and ZIEGLGÄNSBERGER 1968; FELTZ and DE CHAMPLAIN 1972; GONZALEZ-VEGAS 1974; SIGGINS et al. 1974; SPENCER and HAVLICEK 1974; SPEHLMANN 1975; STONE and BAILEY 1975; STONE 1976). Dopamine iontophoresis also inhibits activity induced by stimulation of orthodromic or antidromic axon projections (HERZ and ZIEGLGÄNSBERGER 1968; MCLENNAN and YORK 1967; FELTZ and DE CHAMPLAIN 1972; GONZALEZ-VEGAS 1974).

While these data are consistent with an overall inhibitory effect, it has also been reported that caudate neurons are *excited* by stimulating substantia nigra of the cat (MCLENNAN and YORK 1967; FELTZ and DE CHAMPLAIN 1972) and rat (GONZALEZ-VEGAS 1974), but in these cases the result may be non-dopaminergic since it is unaffected by chemical destruction of the nigra neurons (see the following).

b. Intracellular Recordings of Dopamine Action

Further controversy on the qualitative analysis of dopamine action arose when intracellular recordings of cat neostriatal neurons were conducted in combination with nigra stimulation or dopamine iontophoresis and the results reported as very-rapid-onset depolarizing responses (KITAI et al. 1976). These results, which are incompatible with virtually all other analyses of dopamine action in striatum, have nevertheless received considerable attention, since the frequently reported dopamine-induced behavioral activation (i.e. motor activity) is, in some minds, more easily explained if dopamine excites neurons in striatum rather than inhibits.

c. Actions of the Pathways

The most direct approach to elucidating the function of dopaminergic circuits would be to determine the cellular effects produced on the proper target cells by stimulating the dopamine pathway. However, tests of this experimental idea have yielded a variety of results with different methodologies. With intracellular recording methods, single stimuli in the substantia nigra have been found to have depolarizing actions (EPSPs) or EPSPs followed by a hyperpolarization (IPSPs), but rarely produce IPSPs alone (FRIGYESI and PURPURA 1967; HULL et al. 1974; KITAI et al. 1975). Conversely, trains of stimuli delivered to the nigra generally lead to inhibitory responses (ISPSs or long hyperpolarizations) (see SIGGINS 1977); excitatory responses, however, have also been seen with multiple nigra stimuli (KITAI et al. 1975, 1976). These inhibitory and excitatory striatal effects of nigral stimulation generally show latencies of 5 to 25 msec.

The important question is: which—if any—of these effects reveals the function of the dopaminergic projection? If the effects of stimulating the nigra were dopamine-mediated, pharmacological verification requires that known dopamine antagonists block the effect. Although very high concentrations of haloperidol or phenothiazines have been reported to block some excitations (see FELTZ 1971) even higher doses failed to block other nigral-related excitations (BEN-ARI and KELLY 1976). The inhibitions by dopamine and nigral stimulation are, however, antagonized by α-methyl dopamine (CONNER 1970) but this is not a widely tested dopamine antagonist. Furthermore, neither the nigra-induced excitations nor the inhibitions could be eliminated when the dopamine pathway is destroyed by 6-OHDA (FELTZ and DE CHAMPLAIN 1972; SIGGINS 1978; SIGGINS and GRUOL 1986), although the latter study indicated the inhibitions to be sensitive to GABA antagonists. The likely explanation for these ranges of responses is that none have as yet tested the effects of dopamine cell activation selectively due to the failure to eliminate activation of lower-threshold, fast-conducting, pathways which lie in the vicinity of the nigra (such as the medial lemniscus, the cerebral peduncles, and other components of the ascending central tegmental system). Furthermore, no stimulation study has been reconciled with the nigro-striatal topography using stimulating parameters which are most likely to activate the fine unmyelinated dopamine fibers. Even if the adjacent non-dopaminergic myelinated systems were neglected, selective stimulation of the nigra would seem an ambitious goal, which is further complicated by probable parallel, fine fiber non-dopaminergic pathways coursing between nigra and striatum such as GABA- (DRAY et al. 1976) or substance P-containing fibers (HÖKFELT et al. 1978).

The great majority of the striatal dopamine fibers are very fine unmyelinated C-fibers with diameters of 0.1-0.2 μm or less (FUXE et al. 1964; HÖKFELT

1968). Their conduction velocities and response latency times can only be further lengthened by the extensive branching that occurs after the dopamine fibers penetrate the neostriatum. Such fine branching dopamine axons could not conduct faster than 0.1 m/sec, according to the properties of such fine fibers in the peripheral nervous system (NISHI et al. 1965; see also SIGGINS et al. 1971b). Thus, response latencies of less than 30 msec in rodent brain must be suspected of being non-dopaminergic.

The electrophysiological mechanism of central dopamine receptor actions is not yet easily explicable. Although dopamine stimulates cyclic AMP formation in the caudate both in vivo and in vitro (see reviews and this volume) there are at least 2 major classes of dopamine receptors (see KEBABIAN and CALNE 1979); therefore it is unlikely that all dopamine effects are cyclic AMP-associated actions. Despite these concerns, more than 90% of unidentified caudate neurons in rat show reduced spontaneous discharge with iontophoretic administration of cyclic AMP (SIGGINS et al. 1974, 1976b; also see SIGGINS 1978; SIGGINS and GRUOL 1986) and the effects of both dopamine and of cyclic AMP are enhanced by phosphodiesterase inhibitors such as papaverine and isobutylmethylxanthine (SIGGINS et al. 1974). Adenosine may also participate in the cyclic AMP-linked effects of dopamine in caudate, since adenosine and 2-chloro-adenosine (a derivative which cannot be converted to ATP or cyclic AMP; STURGILL et al. 1975) also inhibit the spontaneous firing of caudate neurons (SIGGINS et al. 1976a). However, unlike the effects of cyclic AMP and of dopamine, the depressant actions of adenosine are antagonized by methylxanthine phosphodiesterase inhibitors, suggesting that the molecular site of the adenosine interaction is distinct from that for dopamine and cyclic AMP. Other dopamine-rich CNS areas have been studied in less detail for cyclic nucleotide related actions of dopamine; inhibitory responses to both dopamine and cyclic AMP have been reported in olfactory tubercle and nucleus accumbens (BUNNEY and AGHAJANIAN 1973), interpeduncular nucleus (OBATA and YOSHIDA 1973) and in the deeper cortical layers of the rat frontal cortex (BUNNEY and AGHAJANIAN 1977). Nevertheless, one awaits with interest the results of target cell-specific tests, especially in caudate nucleus, of the newer synthetic dopamine agonists lacking ability to activate adenylate cyclase. Given the strong relationship of β-adrenoceptors (see below) and of some dopamine receptors to adenylate cyclase activation, it would be reasonable to consider at least for the present, that molecular mechanisms of action of specific catecholamine receptors within their own terminal fields are similar.

d. Functional Inferences from Lesion Studies

Since neither iontophoresis nor electrical stimulation of the nigra may yet have provided definitive information on the function of the dopamine pathway, we can consider the cellular deficits emerging after transection of the nigrostriatal dopamine circuit. HULL et al. (1974) reported that electrolytic lesions of the nigra, or of the dopamine projection, produce no statistically significant increases in the firing rates of neurons ipsilateral to the lesion, but did produce significant decreases in firing in the caudate on the side contralateral to the lesion. This effect was not correlated with the degree of depletion of striatal dopamine. This result is difficult to interpret; it may reflect the complex nigro-nigral interactions revealed from in vivo release studies (NIEOULLON et al. 1977; GLOWINSKI 1979). Neuronal discharge rate, per se, may not reflect the degree to which underlying dopamine function has been impaired by total destruction of nigra, because the changes that follow the loss of dopamine fibers in the straitum may be masked by the loss of non-dopamine fibers either to the striatum or to other structures projecting to the striatum.

A far more selective tool for producing lesions in dopamine systems is the use of the chemical neurotoxin, 6-OHDA (see BLOOM 1975c). Firing rates of striatal neurons increase ipsilateral to lesions produced by the injection of 6-OHDA into the "dopamine bundle" or the substantia nigra in both rat (ARBUTHNOTT 1974; SIGGINS et al. 1976b) and cat (FELTZ and DE CHAMPLAIN 1972). However, this result also does not directly document the inhibitory nature of the nigrostriatal dopamine system, because the effect could theoretically arise through the loss of a tonic dopamine inhibitory projection to more remote brain regions that in turn project to the striatum.

Such objections cannot be leveled at iontophoretic studies of dopamine sensitivity following 6-OHDA lesions. Here, receptor sensitivity provides a cellular index to the degree of denervation supersensitivity that results from the lesion (see UNGERSTEDT 1971b). Under these conditions, striatal neurons show enhanced sensitivity to the inhibitory actions of dopamine and the dopamine-agonist apomorphine (FELTZ and DE CHAMPLAIN 1972; SIGGINS et al. 1974). In neither study was there any increased incidence of dopamine excitations. Long-term treatment with the presumptive dopamine antagonist haloperidol (YARBROUGH 1975) also produced an apparent "supersensitivity" that was again selective for depressant response to dopamine or apomorphine (but not GABA). Here again the inference that dopamine is inhibitory for caudate neurons is sustained. We conclude on the basis of all the lines of converging information that qualitatively dopamine should be regarded as inhibiting the activity of its cellular target.

e. Functional Implications from Behavioral Correlations of Dopamine Neuron Firing Patterns

The behavioral implications of the functional changes that follow either stimulation or lesioning of the dopaminergic cell groups are presumed to represent a marked accentuation of the functional sequelae of their natural firing patterns. More direct insight into the possible behavioral "repertoire" of the dopamine neurons emerges from single unit recording studies of the degree to which identifiable dopamine neurons, recorded in freely moving and behaviorally responsive cats and monkeys, can be correlated with motoric or other responses to environmental stimuli. In freely moving cats (STEINFELS et al. 1983a, b; STRECKER et al. 1983) dopamine neurons showed their highest overall activity during alert waking stages, and showed significantly slower firing rates in quiet waking, slow wave sleep and rapid eye movement sleep stages of behavior; however, there were no obvious correlates between non-alerted stages and those of diminished responsiveness. The constancy of the feline dopamine neuron firing rates was further manifest by their general invariance during periods of fasting, feeding, or augmented blood glucose levels (STRECKER et al. 1983). Although these feline dopamine neurons did respond to auditory stimuli during alert waking behavior with a latency of approximately 60 msec, this response diminished during slow wave sleep, and the neurons proved unresponsive during rapid eye movement sleep stages (STEINFELS et al. 1983b). From these data in the cat, one might conclude that the dopaminergic neurons are relatively monotonic in their responsiveness to spontaneous behavior.

In contrast, studies of the dopaminergic neurons discharge properties in the awake, behaviorally responsive monkey (SCHULTZ et al. 1983; AEBISCHER and SCHULTZ 1984) show considerably greater correlations. In these studies, monkeys were trained to make specific hand and arm movements under control of external signals that indicated both anticipatory preparations and rapid executions of these movements. Under these conditions, dopaminergic neurons were found to increase their firing during and just prior to the execution of large reaching movements of the contralateral arm. The duration of their changed firing continued beyond that of the arm movements. However, these cells did not change in response to the sensory signals, with spontaneous movements, or with movements that required only the distal arm to respond. The results of the primate study suggest that dopaminergic neurons may subserve the general function associated with the behavioral activation that occur prior to and during certain intentional motor responses. These latter views show a better correlation with other forms of behavioral functional inference, described below.

3. Neuropsychopharmacology

a. Motor Behavior

Progress in the study of the nigrostriatal dopamine system rests largely on three factors. First, the anatomical studies described above separated the major forebrain dopamine circuits into circumscribed projections; this anatomical separation of dopamine fibers has allowed discrete lesions of both the terminal regions and cell bodies of origin of the dopamine system using 6-OHDA. Second, many drugs permit relatively powerful perturbation of the catecholamine systems. For example, psychomotor stimulants, such as amphetamine and cocaine, modify the release, reuptake, or metabolism of dopamine (GLOWINSKI et al. 1966) whereas almost all neuroleptics block the catecholamine receptors (JANSSEN et al. 1968; MØLLER-NIELSEN et al. 1973). Third, in marked contrast with the noradrenaline system, the behavioral changes following activation or inhibition of the dopamine system can be readily measured as a direct, easily observable change in unconditioned motor behavior. This general motor activation has to be reflected in any behavior the animal performs. As we shall see, this general effect has also led to a number of probably false functional conceptualizations.

b. Sensory-Motor Integration

Early work using 6-OHDA demonstrated that almost total destruction of the nigrostriatal dopamine system produced a virtually "Parkinsonian" rat, i.e. one which is ataxic, aphagic, adipsic and generally un-responsive to the environment (Ungerstedt 1971c). In addition, Ungerstedt demonstrated that rats sustaining unilateral destruction of the nigrostriatal dopamine system would show strong drug-induced rotational behavior whose direction was determined by the nature of the presumed neuropharmalogical action of the drug on the central dopamine systems (Ungerstedt 1971a, c). For example, rats turn contralateral to the denervated striatum, following apomorphine administration, but ipsilateral to the denervated striatum following amphetamine administration. The amphetamine effect was interpreted as presynaptic release from the intact dopamine system; the apomorphine effect was interpreted as a direct action on the "supersensitive" denervated postsynaptic receptors. Later studies revealed that bilateral 6-OHDA lesions of the substantia nigra blocked the motor behavior induced by psychomotor stimulants (Creese and Iversen 1973; Fibiger et al. 1973a). These results were consistent with earlier work indicating that amphetamine-induced stereotyped behavior was mediated by central nervous system dopamine (Randrup and Munkvad 1960; Weisman et al. 1966; Scheel-Krüger and Randrup 1967). In contrast, extensive depletions of noradrenaline in the cortex, hippocampus and hypothalamus failed to influence either spontaneous or stimulant-induced motor behavior (Roberts et al. 1975).

However, the identification of separate functional properties for each of the forebrain dopamine terminal fields remained to be established. Dopamine (200 μg/rat) injected directly into the corpus striatum of rats, pretreated with a monoamine oxidase inhibitor to prevent destruction of the dopamine, produced clear stereotyped behavior characterized by continuous sniffing and biting without grooming or other normal activities (Fog et al. 1967). Similar results were seen with p-hydroxyamphetamine (200 μg/rat) and dopamine (400 μg/rat) injected into the striatum of reserpine-pretreated rats (Fog and Pakkenberg 1971). In addition, injection of quaternary antipsychotics (chlorprothixen, flupenthixol) into the corpus striatum blocked the stereotyped behavior produced by amphetamine (Fog et al. 1968). In lesion studies, ablation of the corpus striatum blocked the stereotyped behavior induced by amphetamine and apomorphine (Amster 1923; Fog et al. 1970; Naylor and Olley 1972). When 6-OHDA lesions were restricted to the corpus striatum, there was a block of the stereotyped behavior induced by amphetamine, but not the locomotor response (Asher and Aghajanian 1974; Creese and Iversen 1974; Kelly et al. 1975). No blockade of stereotyped behavior was observed with similar lesions to the olfactory tubercles (Asher and Aghajanian 1974; Creese and Iversen 1974; Kelly et al. 1975). Interestingly, as would be predicted from Ungerstedt's turning model using nigrostriatal bundle lesions (Ungerstedt 1971b) with striatal 6-OHDA lesions, apomorphine produced an exaggerated response consisting of sniffing, licking and gnawing. In this regard it is interesting to note that, although turning can be obtained by a 6-OHDA lesion to the corpus striatum alone, the addition of a second lesion to the nucleus accumbens appears to potentiate this turning (KELLY and MOORE 1976), a finding not particularly surprising in view of the role of the nucleus accumbens in locomotion itself (see below, Sec. III.A.2.a).

Bilateral 6-OHDA lesions of the substantia nigra also produce a pronounced aphagia and adipsia not unlike that observed after lateral hypothalamic lesions (UNGERSTEDT 1971c; FIBIGER et al. 1973a, b, 1974; MARSHALL et al. 1974). As with the lateral hypothalamic syndrome, many of the deficits in ingestive behavior caused by these lesions seemed to reflect more fundamental deficits in responding to sensory stimulation (MARSHALL et al. 1974; STRICKER and ZIGMOND 1974). For example, when rats with bilateral 6-OHDA injections into the nigrostriatal bundle became aphagic and adipsic, they progress through the same stages of recovery as in the lateral hypothalamic syndrome and show similar deficits in response to regulatory challenges (such as failing to eat after deoxy-D-glucose or failing to drink after cellular dehydration; MARSHALL and TEITELBAUM 1973; MARSHALL et al. 1974). Sensorimotor impairments such as akinesia, limb dysfunction and impairments in orientation to sensory stimuli may also contribute to many of these feeding deficits.

While these deficits have long been associated with nigrostriatal damage, more recent work shows that this syndrome of aphagia and adipsia may require destruction of both the mesolimbic and nigrostriatal dopamine system (KOOB et al. 1984). Rats with separate 6-OHDA lesions of the frontal cortex, nucleus accumbens and corpus striatum failed to show aphagia and adipsia or any deficit in learning of active avoidance. However, rats with a combined 6-OHDA lesion of the nigrostriatal and mesolimbic systems showed the classical aphagia and inability to "learn" an active avoidance response (KOOB et al. 1984).

Although such global deficits may require an overall massive depletion of forebrain dopamine, it is clear that limited removal of dopamine (< 60%) can result in deficits in certain motor tasks. Rats trained in a reaction time task where a lever had to be pressed and released as fast as possible following a light cue showed a neuroleptic-like decrease in performance following 6-OHDA lesions restricted to the corpus striatum (AMALRIC and KOOB 1987). Similar deficits were not observed in rats with nucleus accumbens 6-OHDA lesions (see also CARLI et al. 1985).

In summary, for motor behavior, the cumulative evidence suggests that the nigrostriatal system is not involved in the actual induction of locomotion itself, but is more important for the stereotypies (e.g. compulsive sniffing, head bobbing, licking and biting) observed with a high degree of activation such as with high doses of psychomotor stimulants. In non-drugged animals destruction restricted to the nigrostriatal system is associated with deficits in reaction time performance.

II. Mesocortical and Mesolimbic Systems

1. Anatomical Foundations

The term mesocortical (or less accurately, mesolimbic) system covers the dopamine projection from the substantia nigra and ventral tegmental area to telencephalic areas other than the basal ganglia. This dopamine system has four major branches (LINDVALL et al. 1974b; LINDVALL and BJÖRKLUND 1978 for reviews): **1)** to the amygdala, the ventral entorhinal area and the perirhinal and piriform cortex; **2)** to the septal nuclei, the nucleus accumbers septi, and the interstitial nucleus of the stria terminalis; **3)** to the mesial frontal and anterior cingulate cortex; **4)** to the olfactory tubercle, the anterior olfactory nucleus and olfactory bulb. The density and distribution of the mesocortical dopamine innervation varies markedly from area to area. The densest innervation is uniformly present in the nucleus accumbens, in the olfactory tubercle in the medial part of the lateral septal nucleus, and in the central amygdaloid nucleus (UNGERSTEDT 1971; LINDVALL et al. 1974b; LINDVALL 1975; FALLON and MOORE 1976a). In each case the density of innervation is much like that to neostriatum.

Like the nigrostriatal projection, the mesocortical projection appears to have a distinct topography. The projection appears to arise from both the ventral tegmental area and the substantia nigra; in the latter the most dorsally situated cells of the pars compacta appear to give rise to the mesocortical projection (FALLON and MOORE 1976b). As with the nigrostriatal projection, there is precision of topography in all 3 anatomical planes: dorsoventrally, mediolaterally, and rostro-caudally.

2. Cellular Actions

At the present time, there have been no reported instances of attempts to evaluate electrophysiologically the effects of stimulating the dopamine projection to any cortical region. However, the reports published on iontophoretic effects of dopamine in cortical regions are in general agreement. In neurons of the cat amygdaloid complex, BEN-ARI and KELLY (1976) have reported potent inhibitory actions of iontophoretically applied dopamine; these actions were partially antagonized by either intravenous or iontophoretically administered α-flupenthixol, a potent antipsychotic drug (BEN-ARI and KELLY 1976). However, large intravenous doses of either pimozide or haloperidol, two other potent antipsychotics for which neurochemical evidence of dopamine antagonism had previously been reported (IVERSEN 1975), failed to block the inhibitory effects of iontophoretic dopamine.

In the frontal and cingulate cortex, in which a substantial dopamine innervation has been noted, the actions of iontophoretic dopamine are also reported to be inhibitory (BUNNEY and AGHAJANIAN 1977). Pharmacologically the inhibitory effects of dopamine here can be distinguished from the similarly inhibitory β-noradrenergic responses found in the more superficial layers of the cingulate cortex.

3. Neuropsychopharmacology

a) Motor Behavior

The first evidence suggesting a functional distinction between the different anatomic projections of the dopamine system came from the studies of the effects of direct intracerebral injections of ergometrine and dopamine (5 μg/rat) into the region of the nucleus accumbens in rats (PIJNENBURG et al. 1973; PIJNENBURG and VAN ROSSUM 1973). Here, in contrast to the stereotyped behavior resulting from dopamine injections into the corpus striatum, dopamine produced a dose-related hyperactivity. Similar effects were observed with direct injections into the nucleus accumbens of apomorphine and amphetamine; however, dopamine metabolites such as methoxytyramine, homovanillic acid or 3,4 dihydroxyphenylacetic acid, were ineffective, as were noradrenaline and 5-HT in rats not pretreated with monoamine oxidase inhibitors (PIJNENBURG et al. 1976).

Consistent with these findings, extensive destruction of the terminal areas of the mesolimbic dopamine projection resulted in a blockade of the locomotor stimulating effects of amphetamine and other psychomotor stimulants (KELLY et al. 1975; KELLY and IVERSEN 1976). The stereotyped behavior produced by high doses of (d)-amphetamine, however, was unaffected, although the locomotor activity associated with these higher doses was blocked (SEGAL et al. 1979). In addition, there is some pharmacological specificity in that the locomotor activating effects of caffeine and scopolamine are not altered by this mesolimbic dopamine lesion (JOYCE and KOOB 1981). In non-drug treated rats, this same lesion produces hypoactivity during habituation (IVERSEN and KOOB 1977), hypoactivity during a restricted feeding situation (KOOB et al. 1978a), and hypoactivity in the open field (JOYCE and IVERSEN 1978). A similar but more extensive lesion produced decreases in exploratory behavior (FINK and SMITH 1979).

Although lesion studies have clearly implicated the ventral tegmental area in motor activation, arousal and even the general organization of behavior (LE MOAL et al. 1969, 1975, 1977; STINUS et al. 1977, 1978; SESSIONS et al. 1980), the specific role of dopamine remained clouded until recently. As described above, 6-OHDA lesions to the terminal regions of the mesolimbic system produced hypoactivity and hyporesponsiveness to stimulant drugs. However, in earlier work, radiofrequency lesions to the region of the dopamine neurons in the ventral tegmental area were reported to result in a permanent behavioral syndrome characterized by hyperactivity and a general disorganization of behavior. These animals were chronically hyperactive whether tested in a circular corridor or in the open field (LE MOAL et al. 1969), or in photocell cages (SESSIONS et al. 1980). These rats also demonstrated deficits in behavioral suppression (LE MOAL et al. 1969) and impairments in tests that require temporal ordering or sequencing (GALEY et al. 1976). In addition, these animals increased their baseline hyperactivity after amphetamine (SESSIONS et al. 1980), while they showed a decrease in this hyperactivity following low doses of apomorphine (STINUS et al. 1977). 6-Hydroxydopamine lesions of the ventral tegmental area also produced hyperactivity, although not as pronounced as that observed after radiofrequency lesions (le Moal et al. 1975).

Thus, lesions to the ventral tegmental area in the region of the dopamine cell bodies produced a set of behavioral consequences that appeared to be markedly different from those produced by 6-OHDA lesions of the terminal regions of these same cells. This paradox was resolved, at least for locomotor activity, in a study designed to establish the critical substrate for the hypo- and hyper-activity associated with destruction of different parts of the mesolimbic dopamine system (Koob et al. 1981). Using a combined lesion approach, the addition of a 6-OHDA lesion to the dopamine terminals in the nucleus accumbens reversed the pronounced hyperactivity established by radiofrequency lesions of the ventral tegmental area. In addition, the hyperactivity induced by low doses of 6-OHDA to the ventral tegmental area could be reversed by simply increasing the dose of 6-OHDA. In both cases, the more severely depleted groups showed a blockade of the hyperactivity produced by low doses of amphetamine. These results suggested an essential role for dopamine in the expression of spontaneous and stimulant-induced activity. In addition, the much greater increase in spontaneous activity produced by the radiofrequency lesion as compared to the hyperactivity produced by a small dose of 6-OHDA in the ventral tegmental area suggested the presence of an as yet unidentified powerful inhibitory influence on the mesolimbic dopamine system within the midbrain tegmentum.

b. Motivational Effects

Although no behavioral regulatory deficits (i.e. aphagia or adipsia or akinesia) have been observed with mesolimbic dopamine lesions, 6-OHDA lesions to the nucleus accumbens did increase food intake in a restricted feeding test, and decreased food-associated drinking (Koob et al. 1978a). Such 6-OHDA lesions to N. accumbens also blocked the development of adjunctive drinking while not altering normal deprivation-induced drinking (Robbins and Koob 1980). These results were interpreted as reflecting an inability to switch from one behavioral activity to another (Koob et al. 1978a) and would be in direct contrast with the disruption of stereotyped behavior induced by high doses of stimulant drugs as observed with 6-OHDA lesions of the corpus striatum.

III. General Behavioral Correlations: Dopamine and Reward

The support for an important role of dopamine circuits in reward is based upon pharmacological and anatomical lesion studies, and rests largely on studies using direct electrical stimulation of the brain as the reward or reinforcer. Stimulation through a chronic electrode implanted into the brain of rat is operationally defined as rewarding or reinforcing because an animal will perform an operant response to obtain that stimulation.

The basic assumption was that the anatomical region underlying the electrode was part of a general brain "reward" system. Intracranial self stimulation

(ICSS) is typically obtained from most regions of the limbic system. Although electrodes placed in a midbrain-forebrain system which courses through the lateral hypothalamus produce the highest rates of responding, ICSS has also been produced in regions as far removed from the classical limbic system as the cerebellum and nucleus solitarius.

a. Anatomical Studies

Early work, which clearly demonstrated a correspondence between the cell bodies of origin and projections of the dopamine system and ICSS (German and Bowden 1974), formed the basis for the hypothesis that neurons in the mesolimbic and nigrostriatal dopamine system served as substrates for ICSS. This "local dopamine reward circuit" hypothesis was using a movable electrode preparation. Here, high rates of ICSS were found to correspond well with the distribution of dopamine in the ventral tegmental area (Corbett and Wise, 1980). However, several observations make a hypothesis involving the dopamine neurons directly under the electrode tip less tenable. First, other areas in the brain support ICSS at high rates, including areas devoid of dopamine, such as the dorsal raphe (Simon et al., 1976). Second, most electrophysiological assessments of the firing properties of the dopamine neurons suggested that they normally fire at lower rates (<10 Hz) than the optimal frequency used in most ICSS experiments (60–100 Hz). Third, evidence using pulse pairing techniques suggested that a major proportion of the "reward" signal is processed by fast conducting, descending neurons in the medial forebrain bundle (Shizgal 1980).

Further erosion of support for the "local dopamine reward circuit" can be found in the lesion studies using 6-OHDA. With different pretreatments, intraventricular 6-OHDA can be used to deplete preferentially either dopamine or noradrenaline. Selective depletion of noradrenaline failed to alter ICSS in the substantia nigra, lateral hypothalamus or locus coeruleus (Cooper et al. 1974, 1978), whereas dopamine depletions produced significant disruption of ICSS at all these sites. Further experiments designed to remove that specific portion of the dopamine system directly underlying the actual electrode tip have shown mainly negative results. ICSS continues in the medial frontal cortex and substantia nigra after complete removal of the dopamine neurons in these regions (Clavier et al., 1980).

Attempts to dissociate reward deficits from motor deficits in ICSS with dopamine lesions, by using an ipsilateral-versus-contralateral approach, have succeeded only when the dopamine lesion itself approached a unilateral destruction of dopamine in all its midbrain projections on one side (Koob et al. 1978b; Carey 1980). Unilateral destruction of the midbrain dopamine systems produced transient deficits in ICSS of rats with electrodes ipsilateral but not contralateral to the lesion (Koob et al. 1978b; Phillips et al. 1976; Phillips and Fibiger 1978). Other studies have failed to obtain even a transient difference using the ipsilateral/contralateral design (Ornstein and Huston 1975; Clavier and Fibiger 1977; Carey 1980; Mitchell et al. 1981). What importance laterality in the dopamine system may have for these effects remains to be seen (see UGURU-OKORIE and ARBUTHNOTT 1981; MITTLEMAN and VALENSTEIN 1982).

However, in one study rats with electrodes in the nucleus accumbens, frontal cortex and ventral tegmental area were subjected to 6-OHDA lesions into the ascending pathways between the rewarding electrode placements at the level of the lateral hypothalamus. With massive depletions of dopamine (>95 %) self-stimulation of the ventral tegmental area was greatly reduced and remained depressed for a 21 day post-lesion test period (PHILLIPS and FIBIGER 1978). Self-stimulation in the nucleus accumbens and frontal cortex recovered either partially or completely. These results suggest that self-stimulation in the ventral tegmental area, itself, may depend largely on stimulation of dopaminergic ascending fibers. For further developments in this hypothesis of separate brain reward circuits see PHILLIPS (1984).

A revised hypothetical explanation for the above results would be that the midbrain dopamine system acts as a global unit to facilitate the initiation of motor responses to reinforcing stimuli. Thus, an intact dopamine system is required for the expression of self-stimulation behavior regardless of where or on what side of the brain the ICSS electrode is placed. This hypothesis would predict that a reinforcement paradigm involving less of a requirement for the initiation of behaviour would be less affected by disruption of the dopamine system. Preliminary results examining the effects of the dopamine antagonist, α-flupenthixol, appear to support this hypothesis. Sniffing for electrical brain stimulation was much less affected by α-flupenthixol than was lever pressing for electrical brain stimulation (ETTENBERG et al. 1981). In addition, as noted above, low rates of responding, e.g. ICSS from electrodes in the area of the locus coeruleus, were much less affected by "dopaminectomy" than ICSS from electrodes in the lateral hypothalamic area (KOOB et al. 1978b).

b. Pharmacological Studies

Extensive pharmacological evidence also suggested that the combined total "catecholamine systems" played an important role in the maintenance of ICSS (for reviews see GERMANN and BOWDEN 1974; FIBIGER 1978; WISE 1978a). In subsequent work functional dissection of these systems has been attempted. As far as such pharmacological studies are useful, a general role for dopamine can be substantiated. Neuroleptics such as the phenothiazines and butyrophenones blocked ICSS at very low doses, in correspondence with their efficacy in blocking dopamine receptors. For example, the ED_{50} for blockade of ICSS in the posterior hypothalamus was 2, 0.05, and 0.01 mg/kg for chlorpromazine, haloperidol and spiroperidol, respectively (DRESSE 1966). Haloperidol also blocked lateral hypothalamic ICSS with an ED_{50} of 0.055 mg/kg and pimozide with an ED_{50} of 0.220 mg/kg (WAUQUIER and NIEMEGEERS 1972). Other studies have shown that these effects are not restricted to hypothalamic ICSS. Pimozide blocked ICSS from electrodes in the substantia nigra (LIEBMAN and BUTCHER 1974) as well as from the dorsal noradrenergic bundle (PHILLIPS et al. 1975). In addition, neuroleptics appeared to raise the threshold for ICSS in rate-independent measures, in which the subject has access to two levers, one of which delivered ICSS only in a continuously decreasing sequence of current intensities, while the other lever simply reset the sequence to its initial intensity. Chlorpromazine (STEIN and RAY 1960), haloperidol (SCHAEFFER and HOLTZMAN 1979), and pimozide (ZAREWICS and SETLER 1979) increased the average current at which the reset occurred. Thus, it was clear that based on pharmacological studies the early anatomic observations suggesting a role for noradrenaline in reward could be largely attributed to actions on dopamine systems (see also below).

Later work using the dopamine antagonist, pimozide, on behavior motivated by amphetamine or food reward, as well as direct ICSS, has given rise to the hypothesis that the midbrain dopamine neurons are the actual "reward neurons" in the brain (YOKEL and WISE 1975; WISE 1978a). This hypothesis is based largely on the observation that the change in responding produced by a dopamine receptor antagonist resembles the extinction of ICSS seen with the omission of ICSS or drug reward (YOKEL and WISE 1975; WISE et al. 1978; WISE 1978b).

WISE and colleagues have also observed that low doses of pimozide increased rates of self-administration of amphetamine similarly to the effects of to reward reduction, i.e. reducing the amount of amphetamine delivered per lever press, whereas a higher dose (0.5 mg/kg) produced an extinction-like effect i.e. a burst of responding followed by response termination (YOKEL and WISE 1975). Similar time-dependent decrements in responding following pimozide were seen with both food reward and ICSS (FOURIEZOS and WISE 1976).

Although there has been some support for this dopamine reward hypothesis (FRANKLIN and McCOY 1979), other investigators question the interpretation of a "reward" effect simply based on behavioral evidence. For example, the effects of pimozide injected concurrently with regular non-reward extinction were found to be additive, i.e. animals treated with pimozide and placed into extinction ceased responding more quickly than animals subjected to either pimozide or extinction alone (ETTENBERG et al. 1979; MASON et al. 1980). In addition, WISE has published evidence that pimozide does *not* produce the prolonged extinction seen after removal of reward (partial reinforcement effect) in rats maintained on a partial reinforcement schedule (GRAY and WISE 1980). As one would predict simply from fatigue or motor decrement, rats show a dose-dependent decrease in responding. Finally, the extinction-like burst of responding before cessation of self-administration of amphetamine observed with high doses of pimozide (YOKEL and WISE 1975) may simply be an artefact of the time course of the onset of the pharmacological action of pimozide. For example, the latency of onset of the cataleptic action of pimozide at a dose of 0.63 mg/kg is 2.6 hours (JANSSEN et al. 1968). Thus, in the YOKEL and WISE (1975) experiment the early burst of responding may simply reflect the well established phenomenon of increased self-administration of drugs following low doses of receptor blockers.

It is, however, a well documented observation that presynaptic activation of the dopamine system with a psychomotor stimulant such as cocaine or amphetamine is indeed "rewarding", by definition, since animals will readily self administer these drugs intravenously (Thompson and Pickens 1970). As noted above, dopamine receptor blockade using low doses of neuroleptic will increase stimulant self-administration (Wilson and Schuster 1972; Risner and Jones 1976). Recent lesion studies using 6-OHDA have provided evidence that the reinforcing properties of these drug stimulations are mediated by the mesolimbic projection of the dopamine system (Roberts et al. 1977, 1980; Lyness et al. 1979). A specific role for the nigro-striatal system in stimulant drug self administration has not yet been determined. This hypothesis of a role for dopamine in reward is more limited in scope than the view that the dopamine system is the "reward" neuron of the brain. If, as hypothesized above, the dopamine system acts as a global unit to facilitate the initiation of motor responses, then this limited view would suggest that under certain circumstances the energizing of this system would also be rewarding, such as following administration of a psychomotor stimulant. This limited view would also predict, however, that reward and learning could occur with a total functional elimination of the dopamine system, despite an inability to move.

IV. Dopamine and Learning

A natural extension of any hypotheses regarding a role for dopamine in reward or reinforcement would be a role for a dopamine circuit or system in learning. However, as we shall see, similar questions to those regarding motor deficits and response initiation can also be extended to any learning hypothesis. It is well established that neuroleptics produce a dose-dependent impairment in the acquisition of a conditioned avoidance response, but not escape responses (Miller et al. 1957; Herz 1960; Cook and Kelleher 1963). In fact, suppression of the conditioned avoidance response is regularly used as a screening procedure for new neuroleptic drugs (Janssen et al. 1968; Møller-Nielsen et al. 1973). Bilateral 6-OHDA lesions of the nigrostriatal projection produce virtually identical results (Echavarria-Mage et al. 1972; Fibiger et al. 1974; Neill et al. 1974; Price and Fibiger 1975; Delacour et al. 1977) including 6-OHDA lesions of the corpus striatum (Neill et al. 1974; Cooper et al. 1974; Delacour et al. 1977). These deficits appear to be independent of any involvement of noradrenaline, in that rats receiving 6-OHDA with desipramine to protect noradrenaline fibers were still impaired, while rats with noradrenaline lesions alone actually showed a facilitation of acquisition (Cooper et al. 1973). These deficits also appear to be independent of nutritional debilitation (Delacour et al. 1977) and can be reversed by treatment with L-dopa (Zis et al. 1974).

Several hypotheses have been suggested to explain these deficits in avoidance learning. One possibility is that neuroleptic treatment or 6-OHDA lesions produce a generalized learning deficit. This, however, does not appear to be the case. Rats that showed a deficit in acquisition of an active avoidance response were unimpaired in acquisition of a passive avoidance response (Cooper et al. 1973) and rats that showed a deficit in acquisition of an active avoidance response were also unimpaired in the acquisition of a light discrimination shock escape test (Price and Fibiger 1975). In an analogous situation, RANJE and UNGERSTEDT (1977) treated rats with either spiroperidol or 6-OHDA and then tested them in an underwater swim maze brightness or spatial discrimination test. The rats treated with spiroperidol learned spatial discrimination, but were impaired in the acquisition of brightness discrimination (RANJE and UNGERSTEDT 1977). However, these rats clearly showed a time-related improvement; their error rate dropped from 80% to 30 to 40% over 4 sessions. As discussed by others (BENINGER et al. 1980) the deficit in learning rates in these experiments might be due to motor impairments while performing the task.

An alternative hypothesis proposed originally by POSLUNS (1962) and later by FIBIGER and associates (PRICE and FIBIGER 1975; FIBIGER et al. 1975; BENINGER et al. 1980) was that dopamine dysfunction disrupts the conditioned avoidance response not by blocking associative learning, but by blocking or delaying the initiation of voluntary motor responses similar to the known motor initiation deficits observed in patients with Parkinson's disease. This hypothesis was based on experiments showing chlorpromazine, by delaying the initiation of motor acts, reinstated a behavioral sequence during avoidance training that resembled initial learning (see POSLUNS 1962 for details). Indeed, rats treated with high doses of pimozide were unable to acquire a conditioned avoidance response, but could learn a conditioned suppression task in which non-responding was the learning index (BENINGER et al. 1980).

As noted above, the deficit in avoidance learning following neuroleptic or 6-OHDA treatment seemed likely to be the result of dysfunction in both nigrostriatal and mesolimbic dopamine systems (KOOB et al. 1984).

Other studies have shown that the mesolimbic and mesocortical dopamine projections may be involved in another specific learning task, one which is more cognitive in nature, the delayed alternation test (BROZOSKI et al. 1979; SIMON et al. 1980; SIMON 1980). In this situation, two goal boxes (with rats in a maze) or two goal objects (in monkey tests) are presented and the animal must alternate his choice from trial to trial. 6-OHDA lesions in a circumscribed area of association cortex (pre-frontal area) in rhesus monkeys produced an impairment in the performance of a spatial delayed alternation task (BROZOSKI et al. 1979). This deficit was reversible with L-dopa and apomorphine. Virtually identical results have been observed with 6-OHDA lesions of the frontal cortex or with 6-OHDA lesions of the cell bodies of the mesolimbic and mesocortical dopamine systems in rats (SIMON 1980; SIMON et al. 1980). Since the delayed alternation task is particularly sensitive in demonstrating cognitive deficits after ablation-type lesions of the prefrontal cortex in mammals, these similar deficits with 6-OHDA suggest a possible role for cortical dopamine in these deficits. It will be intriguing to determine if similar dysfunctions shown in the frontal cortical ablation-type lesions exist following the use of neuroleptics in the treatment of schizophrenia.

V. Summary and Conclusions

The midbrain dopamine neurons clearly have an important role in behavior. Activation of the dopamine system produces an activation of the organism as is reflected by increased spontaneous locomotion and, at least in rats, by increases in species-specific actions, such as sniffing, rearing, and gnawing. Conversely, functional inactivation of the dopamine system by receptor blockade or lesions produces a general behavioral inactivation that is reflected by decreases in spontaneous locomotion and in severe cases by aphagia, adipsia, and even in reduced responsiveness to sensory stimuli. These deficits after destruction of the dopamine system can be magnified by administration of stimulant drugs known to release dopamine presynaptically. These drug/lesion interactions provide strong evidence for a specificity of function of the terminal projections of the dopamine system. The mesolimbic dopamine system appears responsible for the enhanced locomotion and reinforcement produced by psychomotor stimulants, whereas the nigrostriatal dopamine system appears responsible for the stereotyped behavior patterns resulting from high doses of stimulant drugs. Deficits in motor activation to various stimuli are observed in animals with mesolimbic dopamine lesions, whereas reaction time performance is decreased significantly following corpus striatum dopamine lesions. Thus, the midbrain dopamine neurons appear to be involved in reward, not as reward neurons per se, but rather as a general response initiation system. Similar conclusions can be drawn from studies aimed at examining the role of dopamine in learning. Learning situations requiring the animal to initiate an integrated and coordinated skeletal response to a specific contingency are much more likely to be disrupted by inactivation of the dopamine system than situations where learning does not require a motor response. Learning situations requiring switching between two or more cognitive response strategies may be attributed to the mesolimbic and mesocortical dopamine projections. However the deficits observed in acquisition of the conditioned avoidance response may involve dopamine terminals in both the corpus striatum and the nucleus accumbens.

In conclusion, the dopamine system may very well be the system in the brain that enables sensory input to interact with the motor subroutines required for action. An analogous cellular "enabling" function has been proposed as an abstract explanation of the behavioral actions of noradrenaline (see the following). The terminal projections of the ventrotegmental dopamine system may be viewed as a "go button" to allow sensory input to elicit appropriate organized responses. This hypothesis fits nicely with the deficits in response initiation observed in Parkinson's disease. Less clear is how an overactivity of the dopamine system could produce the symptoms associated with schizophrenia such as loss of affect. One possibility is that such a "go button" could also be involved in the initiation of cognitive processing as well as in the initiation of motor acts, and this functional deficit may be the conceptual abstraction linking Parkinson's disease and schizophrenia.

C. Noradrenergic Systems

I. Anatomical Foundations

The major group of central noradrenergic neurons resides in the locus coeruleus, a prominent nucleus located in the brainstem reticular formation at the level of the isthmus and, in rats and primates, composed entirely of noradrenaline-producing neurons. Their detailed morphological features are also reviewed by PALKOVITS and BROWNSTEIN (this volume, Chapt. 10). The neurons of the locus coeruleus form a distinct, compact cell group largely contained within the central gray of the pons. In the cat, the nucleus is much more widely dispersed than in other mammalian species that have been described (CHU and BLOOM 1974; JONES and MOORE 1974). In the rat, the locus coeruleus is composed of approximately 1,500 neurons on each side of the brainstem (SWANSON 1976). In Golgi preparations (SWANSON 1976) locus coeruleus neurons appear to contain predominantly multipolar dendrites that extend well beyond the limits of the nucleus. As in the case of the dopamine neurons of the nigra, there is both morphological and electrophysiological evidence (GROVES and WILSON 1980; see FOOTE et al. 1983) of dendro-dendritic interactions for the locus coeruleus neurons, including ultrastructural evidence that the locus coeruleus dendrites contain clusters of the small granular vesicles regarded as the organelles of transmitter storage and release.

Five major pathways originate from the locus coeruleus. The projection into the mesencephalic tegmentum, termed by UNGERSTEDT (1971a) the dorsal noradrenaline bundle, is by far the largest ascending projection, serving nearly all telencephalic areas. A second projection enters the central gray and ascends as a component of the dorsal longitudinal fasciculus. A third component ascends ventrally from the locus coeruleus to cross through the ventral tegmental area into the medial forebrain bundle. Two other major groups of fibers leave the locus coeruleus. One ascends in the superior cerebellar peduncle to innervate the cerebellum, and the last descends in the central tegmental bundle through the brainstem to enter the ventral portion of the lateral column of the spinal cord.

Although casual inspection of the neurons of the locus coeruleus could suggest that this small cluster of neurons projects its axons in a non-specific manner to its general and diffuse series of targets within the mammalian brain, detailed analyses of this issue suggest otherwise. Three-dimensional reconstructions of this nucleus in the rat, and comparison of these integrated structures across multiple individual animals (LOUGHLIN et al. 1986a) shows that cells with distinct morphological and packing density features consistently populate distinct sub-regions of the nuclear complex. When combined with distinguishable retrograde marker tracers, these subsets show distinct, and generally non-overlapping efferent target systems (LOUGHLIN et al. 1986a) as had been suggested by earlier, less quantitative studies (MASON and FIBIGER 1979; WATERHOUSE et al. 1983). Furthermore, the suggestions of selective target systems for the locus coeruleus in the rodent brain are far more strikingly demonstrated in terms of projections within the primate brain, and is exceedingly well defined in the primate visual system (see MORRISON and FOOTE 1986) where the noradrenergic terminal fields and those of other monoamine systems are well separated. In these higher primate brains, the noradrenergic projections are largely devoted to the extra-striate, or diffuse, movement, detection system, rather than to the detailed, visual projections of the geniculo-striate pathway. Such a distribution would perhaps be compatible with the view that the locus coeruleus is an early detector of novel stimuli, regardless of the nature of the stimuli (see FOOTE et al. 1983; ASTON-JONES and BLOOM 1981b).

Locus coeruleus axons have a distinctive morphology in fluorescence histochemical material prepared by any of the glyoxylic acid methods (LINDVALL et al. 1974b; BLOOM and BATTENBERG 1976; see BJÖRKLUND and LINDVALL 1986). In a terminal field, the preterminal fibers branch into a highly collateralized network with common features in all areas innervated (BJÖRKLUND and LINDVALL 1986); the primary common feature is the axonal morphology, with typically fine axons, exhibiting regularly spaced, round, intensely fluorescent varicosities approximately 1-2 μm in diameter (see PALKOVITS and BROWNSTEIN this volume, Chap. 10; FOOTE et al. 1983).

One of the most controversial aspects of locus coeruleus fibers is the nature of their connections with potential target cells, and the interpretation of whether these fibers form "true" synapses or end in a diffuse endocrine type of relationship with many unspecified target cells. The controversy stems from differences in the methods used to visualize the terminals and the region of the brain in which they are studied (see FOOTE et al. 1983 for more extensive discussion).

The resolution of the synaptic controversy is functionally important. If (as DESCARRIES et al. 1977, suggest) locus coeruleus axons have no specific cells for their postsynaptic elements, stimulation of the locus coeruleus would be presumed to result in the release of noradrenaline from all terminals and the effects of this stimulation would depend solely upon the presence of appropriate receptors in nearby neural elements. This organization would not be unlike that of the peripheral sympathetic axon terminals and the structures they innervate (HÖKFELT 1968).

Other alternatives must be considered (see MOORE and BLOOM 1979) including that which arises from the morphological analysis of neocortical noradrenaline fibers revealed by immunocytochemistry with anti-dopamine-β-hydroxylase staining (GRZANNA et al. 1977; MORRISON et al. 1978; OLSCHOWSKA et al. 1980). In three areas known to be innervated by locus coeruleus neuron axons, and in which there is little or no other catecholamine neuron innervation, such as the cerebellum (BLOOM et al. 1971; LANDIS and BLOOM 1975), hippocampal formation (KODA and BLOOM 1977), and parietal cortex (DESCARRIES and LAPIERRE 1973; LAPIERRE et al. 1973; DESCARRIES et al. 1977; OLSCHOWSKA et al. 1980, 1981) noradrenaline terminals have been examined by one or more selective histochemical methods (see BLOOM 1973; 1988; MOORE and BLOOM 1979). With detailed immunohistochemical localization of neocortical noradrenaline acons (MORRISON et al. 1978), montage reconstructions in all planes reveal a fairly precise matrix-like interlacing of cortical neurons and noradrenaline fibers in layers I and II. Fibers in layers V, VI were thought to appear more "terminal-like" because of their clustering around the cell bodies and proximal dendrites of pyramidal neurons. A similar precision has been observed in more recent studies of the primate neocortex (MORRISON et al. 1982 a, b).

These data indicate that the fibers in the outer neocortical molecular layer, which were labeled most heavily by the topically applied exogenous noradrenaline in the studies by DESCARRIES et al. (1977), may not be typical terminal noradrenaline fibers. By the immunocytochemical staining approach, which does not require an external labelling procedure, noradrenaline terminals exhibiting positive anti-dopamine-β-hydroxylase immunoreactivity exhibit typical synaptic specializations in more than 50% of the cases, but such terminals are much more frequent in layers V and VI of the neocortex (OLSCHOWSKA et al. 1980, 1981).

Other studies also support a synaptic termination of locus coeruleus axons on specific target neurons. In the cerebral cortex of the immature rat 5-OHDA administered parenterally crosses the blood-brain barrier and allows identification of presumptive noradrenaline terminals on the basis of small granular vesicles; such labelled terminals also show a very high incidence of specialized synapse-like contacts (COYLE and MOLLIVER 1977). These synaptic-like contacts (COYLE and MOLLIVER 1977) were seen mainly in the deeper cortical layers, rather than in the superficial layers, which had been most densely labeled in the autoradiographic studies by DESCARRIES et al. (1977).

In the hippocampus of the rat, KODA and BLOOM (1977) and their colleagues (KODA et al. 1978 a, b) made a detailed examination of the endogenous noradrenaline-containing boutons revealed by permanganate fixation; they concluded that the small vesicle boutons correlated closely with the topographic distribution of glyoxylic acid-induced fluorescent boutons, and that these presumptive noradrenaline boutons showed specialized junctional contacts (in about 18 to 20% of the boutons) as frequently as did all other boutons (KODA et al. 1978 b).

As in all other cases of attempted ultrastructural-physiological correlation, the assumption that the specialized contact zone is the site of synaptic transmission (i.e. the actual release and response site) remains to be documented with certainty. It is clear that sympathetic fibers of the peripheral autonomic nervous system can transmit to smooth muscle without such specializations. However sympathetic neurons do exhibit specialized sections at intraganglionic (neuron-to-neuron) contacts (GRILLO 1966). BEVAN (1977) has suggested that receptor distribution and post-junctional sensitivity are closely correlated with the distance between release and response sites. The greater the separation, the greater is the junctional sensitivity. Thus, an accurate determination of the precise cellular location of central adrenoreceptors with reference to the specialized junctions of identified noradrenergic boutons (see PALACIOS and KUHAR 1980; STRADER et al. 1983) may be ultimately required to settle the issue of post-synaptic specificity for these noradrenergic fibers.

II. Cellular Interactions Between Central Catecholamine Systems and Other Synaptic Systems

Multiple previous reviews have considered the continually accumulating evidence that noradrenaline despresses the spontaneous activity of test cells in virtually every region of the brain (see Moore and Bloom 1979; Bloom 1988; Foote et al. 1983) and that these effects are antagonized by β-adrenoceptor blockers, prostaglandins, certain antipsychotic agents and acute application of lithium. Recent evidence suggests that the actions of iontophoretic noradrenaline on principal cells of the rat lateral geniculate are antagonized by α-adrenoceptor blockers (Rogawski and Aghajanian 1980), as were responses of mitral cells in the olfactory bulb (Bloom et al. 1964).

In general, central noradrenaline actions in vivo after iontophoretic application are homogeneously inhibitory. In contrast to the multiple sources of dopamine fibers, an experimental advantage of studies on the noradrenaline system is the ability to study the effects of locus coeruleus stimulation on the several classes of target cells. Since the locus coeruleus is the sole source of forebrain and cerebellar noradrenaline fibers, the target cells of this projection can be studied directly. Interactions between the locus coeruleus and these targets, and the more complex effects of this noradrenergic regulation on the overall responsiveness of the target cells to their other afferent synaptic connections can also be tested.

Purkinje cells show remarkably uniform inhibitory responses to stimulation of the locus coeruleus with a latency of approximately 125 msec (Hoffer et al. 1973). Antipsychotic phenothiazines (Freedman and Hoffer 1975), cobalt, lanthanum, or lead (Freedman et al. 1975), and lithium (Henriksen, Siggins, and Bloom, unpublished), as well as prostaglandins (see Hoffer et al. 1973 a, b) are all able to antagonize these effects of locus coeruleus stimulation. Intracellular recording from some Purkinje cells during stimulation of locus coeruleus with single shocks reveals long latency (around 150 msec), and relatively small hyperpolarizations. With trains of pulses to locus coeruleus, large hyperpolarizations were observed that outlasted the stimulation period. Input resistance either increased or did not change during the locus coeruleus-evoked hyperpolarizations. Identical effects, hyperpolarization of the membrane potential, with either minimal change or increased input impedance of the membrane also characterize the actions of noradrenaline during intracellular recordings from cerebellar Purkinje cells (Siggins et al. 1971 c, d; Hoffer et al. 1973 a), spinal motorneurons (Marshall and Engberg 1979) and hippocampal pyramidal neurons (Oliver and Segal 1974; Segal 1980). In the brainstem, iontophoretic application of noradrenaline produces only slight hyperpolarizing effects with minimal changes in membrane impedance (Vandermaelen and Aghajanian 1980). An extensive review of this property is found in Siggins and Gruol (1986).

In studies on hippocampus (Segal and Bloom 1976 a, b), septum (Segal 1976) and spinal trigeminal nucleus (Sasa et al. 1974, 1977), like the effects described in the cerebellum, the effects of locus coeruleus activation are generally inhibitory.

III. Molecular Mechanisms of Central Noradrenergic Synapses: Role of Cyclic AMP

The mechanism of this apparent inhibitory response to noradrenaline and the locus coeruleus has been analyzed extensively. In both cerebellum and hippocampus, cyclic AMP mimicked the ability of noradrenaline to depress spontaneous activity (Hoffer et al. 1969, 1973a, b; Siggins et al. 1971c, d; Segal and Bloom 1974a, b; Oliver and Segal 1974), fulfilling some of the criteria for cyclic AMP serving as a central "second messenger" (Robison et al. 1971) for central noradrenaline synapses (see Hoffer et al. 1969, 1971a, b,1973a, b; Bloom 1975a, b; Siggins and Gruol 1986).

Biochemical evidence had already suggested that noradrenaline could elevate cerebellar cyclic AMP levels through β-adrenoceptors (see Rall 1972; also see references in Bloom 1975b). Recently, using methods for direct microscopic localization of β-receptors, Palacios and Kuhar (1980) showed that Purkinje neurons in the cerebellum, and pyramidal neurons in the hippocampus and cerebral cortex, were the major cell classes possessing such receptors. Furthermore, parenteral administration of phosphodiesterase inhibitors such as aminophylline or theophylline potentiated noradrenaline depressions of Purkinje cells, while iontophoretic administration of these methylxanthines, and of papaverine, converted weak excitant actions of iontophoretic cyclic AMP into pronounced depressions (Hoffer et al. 1969, 1971a, b, 1973a, b; Siggins et al. 1971a). These observations led to the proposal that the actions of noradrenaline (Hoffer et al. 1969), and those of the noradrenaline-mediated locus coeruleus synaptic projections to Purkinje cells (Siggins et al. 1971c, d) could be mediated by cyclic AMP.

Subsequently, the proposal has been strengthened by observations that noradrenaline (Siggins et al. 1971c), the noradrenaline pathway (Hoffer et al. 1973; Siggins et al. 1971d), and cyclic AMP (Siggins et al. 1971c) all hyperpolarize Purkinje cells through similar membrane actions in which conductance to passive ion flow is decreased or unchanged. The cyclic AMP mediation of the noradrenaline actions also finds support from the observations that prostaglandins and nicotinate (Siggins et al. 1971a, b) selectively block noradrenaline effects on Purkinje cells and hippocampal pyramidal cells (Segal and Bloom 1974a,b) as they do in the cyclic AMP-mediated adrenergic responses of adipocytes (Robinson et al. 1971). In fact, all the substances shown to block the response of Purkinje neurons to iontophoretic noradrenaline, also show very good antagonism toward the activation of cerebellar adenylate cyclase by noradrenaline (see Nathanson 1977). Even more direct confirmation of the second messenger hypothesis stems from the observation that both noradrenaline iontophoresis and stimulation of the noradrenaline pathway will increase the number of Purkinje cells reacting positively to an immunocytochemical method detecting bound intracellular cyclic AMP (see Bloom 1975b).

The alternative suggestion that the actions of cyclic AMP are mediated by conversion to adenosine has been disproven by the observations in cerebellum and the cerebral cortex that methylxanthine phosphodiesterase inhibitors potentiate noradrenaline and cyclic AMP yet block the effect of adenosine or 5'AMP (see SIGGINS et al. 1971a; HOFFER et al. 1973b; STONE and TAYLOR 1977). With cultured Purkinje neurons, GAHWILER (1976) has observed that phosphodiesterase inhibitors potentiate depression of cell firing with noradrenaline and cyclic AMP. He also observed that the thresholds of responses to cyclic AMP were 100-1000 times higher than for responses to noradrenaline (applied by superfusion); such a ratio is in keeping with predictions from other instances of second messenger mediation (see ROBISON et al. 1971; BLOOM 1975b; NATHANSON 1977), and indicates that direct effects of cyclic AMP may require high concentrations of the cyclic nucleotides.

Finally, cyclic AMP is known to activate a class of enzymes, termed protein kinases (see GREENGARD 1978), which phosphorylate specific enzymatic and structural protein substrates. This effect of cyclic AMP may be the mode by which changes in intracellular cyclic AMP levels mediate functional changes in cellular activity (see GREENGARD 1978). In cerebellum, chemical analogs of cyclic AMP mimic the inhibitory effects of noradrenaline in direct correlation to their ability to activate brain cyclic AMP-dependent protein kinase (SIGGINS and HENRIKSEN 1975).

The hyperpolarizing effect of iontophoretic noradrenaline and of locus coeruleus stimulation, accompanied by increased membrane resistance, is in direct contrast to changes seen with classical inhibitory postsynaptic potentials (IPSPs) or with iontophoresis of GABA (see SIGGINS et al. 1971c). The classical inhibitory pathways and inhibitory amino acid transmitters are thought to operate exclusively through mechanisms that increase conductance to ionic species whose equilibrium potentials are more negative than the resting membrane potential. In such cases, the hyperpolarization is associated with a decrease in membrane resistance. Intracellular recordings indicate that the hyperpolization produced by noradrenaline, on the other hand, is accompanied by, and perhaps generated by, a decrease in passive ionic conductance across the membrane (that is, an increase in membrane resistance) (SIGGINS et al. 1971d). This noradrenergic effect may be due to a decrease in conductance to some ion such as sodium or calcium, or to activation of an electrogenic pump (PHILLIS et al. 1973). However, the depressant actions of noradrenaline do not require extracellular calcium (GELLER and HOFFER 1977). Furthermore, substances, which also "antagonize" calcium actions, fail to interfere selectively with noradrenaline inhibitory effects (FREEDMAN et al. 1975).

The effect of noradrenaline might therefore be mediated in one of three ways: 1) a decrease in resting sodium conductance which hyperpolarizes the cell membrane, 2) activation of an electrogenic ion pump to produce the hyperpolarization (SIGGINS et al. 1971a, b, c,), 3) alteration of voltage-sensitive catecholamine conductances involved in pacemaker generation in analogy with the effects of noradrenaline on cyclic AMP-mediated pacemaker mechanisms in the heart (also see SIGGINS and GRUOL 1986).

The evidence for cyclic nucleotide-mediated effects on passive ionic conductances in the vertebrate sympathetic ganglion has become difficult to interpret (MCAFEE et al. 1971; GALLAGHER and SHINNICK-GALLAGHER 1977; DUN et al. 1977; BUSIS et al. 1978). Cyclic nucleotide effects on these membrane conductances appear to be subtler and more complex than originally proposed (e.g. KOBAYASHI et al. 1978), and no case for mediation of passive conductance changes in nerve cells by cyclic AMP has been convincingly established. Although some evidence suggesting that synaptic transmission can regulate electrogenic ion pumps has appeared (SASTRY and PHILLIS 1977; AKASU and KOKETSU 1977; YARBROUGH 1976), the role of electrogenic ion pumps in the generation of synaptic potentials remains controversial (see SIGGINS and GRUOL 1986). Moreover, we are unaware of a convincing demonstration of cyclic AMP involvement in synaptic regulation of electrogenic ion pumps in mammalian neurons.

Cyclic nucleotide effects on voltage-sensitive conductances, however, have been established in at least two invertebrate neuronal system. KLEIN and KANDEL (1978) have shown that 5-HT-acts to broaden the spike in a sensory neuron in Aplysia, and that this broadening is due to an effect on voltage-sensitive calcium conductance. They have presented evidence suggesting that this effect may be mediated by cyclic AMP (also see TSIEN 1973). Moreover, LEVITAN and NORMAN (1980) have shown that cyclic nucleotides modify bursting activity in invertebrate neurons. These findings indicate that, at least in some systems, catecholamines may act through direct effects on pacemakers and suggest that the effects may be cyclic nucleotide-mediated. Of these three mechanisms, current evidence favors the possible effect on endogenous pacemaker activity.

IV. Cerebellar Synaptic Systems as a Model for Noradrenergic Integrative Actions

The coeruleo-cerebellar noradrenergic circuit to Purkinje neurons offers the most extensive data on which to construct an analysis of the interactions between the noradrenergic and other afferent systems. According to the classical view, simple spike activity in Purkinje cells is only the result of mossy fiber input, relayed by the granule cells (ECCLES et al. 1967). However, several lines of evidence suggest that a pacemaker accounts for at least part of the simple spike activity of Purkinje cells. ECCLES et al. (1967), for example, showed that ongoing simple spike activity could be recorded in the cerebellum after complete deafferentation. WOODWARD et al. (1974) showed that simple spike activity was normal in animals which had no granule cells at all. The activity of Purkinje cells in grafts grown in the anterior chamber of the eye of host rats, in which Purkinje cells do not receive their normal extracerebellar innervation, also approaches normal patterned simple spike activity (HOFFER et al. 1975).

The fact that locus coeruleus reduces endogenous or "background" simple spike activity, while facilitating evoked activity, has led to the proposal that locus coeruleus may act to increase the "signal-to-noise" ratio of transmission via the Purkinje cells (FREEDMAN et al. 1977). The same proposal was made earlier for effects of noradrenaline on auditory cortical neurons in the monkey (FOOTE et al. 1975). Such an interpretation seems consistent with the apparent role of the locus coeruleus in signalling or regulating levels of arousal (FOOTE et al. 1980, 1983; ASTON-JONES and BLOOM 1981 a, b).

An alternative hypothesis concerning the function of the coeruleo-cerebellar projection can be adduced from examination of the intracerebellar connections. Mossy fiber input to cerebellar cortex is relayed by the parallel fibers and excites those Purkinje cells which lie on the discrete narrow fiber connections, or "beam" of fibers by which parallel fibers innervate Purkinje neurons. Parallel fibers also excite basket and stellate cells, which exert an inhibitory effect on Purkinje cells which is offset from the directly exciting parallel fiber "beam." This off-beam system has been suggested to provide an "inhibitory surround" for incoming excitatory signals (ECCLES et al. 1967). This effect is considered (ECCLES et al. 1967) to be analogous to the lateral inhibition found in the eye, for example, where it serves a contrast-enhancing function (HARTLINE and RATLIFF 1957). Since noradrenaline appears to potentiate both the on-beam excitatory effects, and off-beam GABA-mediated inhibition (see MOISES and WOODWARK 1980; MOISES et al. 1981), locus coeruleus may also serve to regulate the strength of lateral inhibition in the cerebellar cortex.

Lateral inhibitory systems appear to provide for sharp separation of inputs to neighboring portions of the system (HARTLINE and RATLIFF 1957). Activation of locus coeruleus might therefore restrict the area of cerebellar cortex activated by a given mossy fiber input, and enhance the spatial precision with which mossy fiber systems could act on the cerebellar cortex. Likewise, a decrease in locus coeruleus activity would be predicted to decrease the precision of, and increase the interaction between, neighboring mossy fiber projections. Under this hypothesis, locus coeruleus could act to change the character of the signal and the way it is processed, rather than simply favoring the transmission of evoked activity over putative "noise." If a similar generalization could be ascribed to catecholamine-mediated actions in other CNS regions, then the predicted global effects would be the transient emphasis of brief periods of elicited activity.

V. Hypotheses of Noradrenergic Cellular Integration: Enabling

The material reviewed earlier clearly indicates that locus coeruleus activity produces a reduction of target cell "spontaneous" activity in several brain regions. The fact that the mechanism of synaptic action of locus coeruleus on its target cells has some rather unconventional aspects, such as coupling to cyclic AMP formation and possible selective effects on ionic channels and pacemaker activity, may be the reason that locus coeruleus has more complex consequences for the postsynaptic integration of synaptic signals than "conventional" synaptic inhibition. Various hypotheses have been advanced which describe the net impact of the cellular effects produced by the locus coeruleus system. Some of these emphasize the enhancement of evoked activity relative to spontaneous activity for afferent systems simultaneously active with locus coeruleus (FOOTE et al. 1975; SEGAL and BLOOM 1976b; WOODWARD et al. 1979). Others describe this process as "enabling" (BLOOM 1988; MOORE and BLOOM 1979), by which co-activity in locus coeruleus terminals enables other systems converging on the same target neurons to transmit more effectively during the period of simultaneous activity. As noted above, YEH et al. (1981) and MOISES et al. (1981) present evidence that such enhancement by noradrenaline may occur only with those transmitters which actually project to targets of locus coeruleus. Thus, locus coeruleus will potentiate GABA, a transmitter in cerebellum, but will not potentiate glycine which is not a transmitter in cerebellum. The implication here is one of selective interactions between endogenous receptors.

SCHULMAN (1981; also see DISMUKES 1979) has noted that one possible ramification of this signal enhancement is sharpened spatial localization of activity, within any area of diffuse afferent activation, to targets convergent with those of locus coeruleus. In cerebellum, for example, since noradrenaline appears to potentiate both on-beam excitatory effects and off-beam GABA-mediated inhibition (see above), locus coeruleus may serve to regulate the strength of lateral inhibition in the cerebellar cortex. Activation of locus coeruleus might therefore restrict the area of cerebellar cortex activated by a given mossy fiber input and enhance the spatial precision with which mossy fiber systems could act on the cerebellar cortex. FLICKER et al. (1981) have related aminergic activation to response plasticity of target neurons, suggesting the involvement of cyclic AMP as a second messenger.

The molecular and ionic mechanisms responsible for these complex effects of noradrenaline on various aspects of target neuron activity are not yet understood, although the locus coeruleus effect on Purkinje cells involves an apparent decrease in membrane ionic conductance. It has been shown in Aplysia sensory neurons (KANDEL et al. 1981) and in neurons of the bullfrog sympathetic ganglion (SCHULMAN and WEIGHT 1976) that generation of conductance decreases by one synaptic pathway results in increased transmission *via* other synaptic pathways, if the latter function by classical conductance-increase mechanisms. It has also been shown that facilitatory effects of noradrenaline and of locus coeruleus stimulation outlast effects on firing rates, and can be demonstrated even when locus coeruleus stimulation does not detectably alter the firing rate of the cerebellar Purkinje cell. The demonstration that noradrenaline potently induces glycogenolysis in cortical slices (MAGISTRETTI et al. 1981), may also be relevant to the induced changes in neuronal responsiveness.

VI. Behavioral Correlates of Central Catecholamine Neurons

We must now approach the question of when (i.e. during what sorts of behavioral conditions) central catecholamine neurons fire to execute these complex regulatory functions on their targets. Two types of information are needed: 1) knowledge of the functional and neurochemical properties of the systems which can regulate locus coeruleus activity and 2) knowledge of the behavioral repertoire of these neurons (i.e. the conditions of the internal or external world which reproducibly elicit enhanced firing).

1. Anatomy of Afferents to Locus Coeruleus

Neuroanatomical data employing early orthograde and retrograde labelling methods (see PALKOVITS and BROWNSTEIN, this volume, Chap. 10; FOOTE et al. 1983) indicated that locus coeruleus neurons receive an extremely rich array of inputs. After large horseradish peroxidase (HRP) injections into locus coeruleus the locations of retrogradely labelled cells indicate that several forebrain and brainstem sites project into, or adjacent to, locus coeruleus (SAKAI et al. 1977; CEDARBAUM and AGHAJANIAN 1978). Areas containing substantial numbers of labelled cells in the rat include the bed nucleus of the stria terminalis, central nucleus of the amygdala, several hypothalamic areas, central grey, and the lateral reticular nucleus; a few labelled neurons were also observed in the fastigial nuclei and the marginal zones of the spinal dorsal horns. Furthermore, after injections of radioactively labelled amino acids into various telencephalic and brainstem nuclei, "terminal" labelling within or adjacent to locus coeruleus has been reported to arise from the ventromedial nucleus of the hypothalamus (SAPER et al. 1976), nucleus cuneiformis (EDWARDS 1975), dorsal raphe nucleus (PIERCE et al. 1976), preoptic area (CONRAD and PFAFF 1976a; SWANSON 1976), and anterior hypothalamus (CONRAD and PFAFF 1976b). In addition, immunocytochemical techniques revealed reactive "terminals" within or adjacent to locus coeruleus for β-endorphin (BLOOM et al. 1978), Substance P (HÖKFELT et al. 1978), tryptophan hydroxylase (PICKEL et al. 1977), and PNMT (HÖKFELT et al, 1974). Except for projections from contralateral locus coeruleus (CEDARBAUM and AGHAJANIAN 1978), caudal noradrenaline (CEDARBAUM and AGHAJANIAN 1978; SILVER and JACOBOWITZ 1979) and adrenaline cell groups (HÖKFELT et al. 1974), and midbrain raphe (CEDARBAUM and AGHAJANIAN 1978; PICKEL et al. 1977), that data do not yet overlap sufficiently to permit combined anatomical and histochemical characterization of locus coeruleus afferents.

While these initial studies suggested that the rodent locus coeruleus, and by implication the locus coeruleus of all species, was likely to be the convergent target of a large number of afferent sources of information, recent careful re-examinations of this same question have led to a different conclusion (ASTON-JONES et al. 1986). Using one of the more sensitive methods of retrograde circuit detection, namely the focal iontophoretic injection of the bean lectin, wheat germ agglutinin, conjugated to the marker enzyme horseradish peroxidase, ASTON-JONES and colleagues found that there were a much smaller number of sources for afferent projections to the rat locus coeruleus: the paragigantocellularis and prepositus hypoglossi nuclei of the brain stem, with much more limited inputs from the paraventricular hypothalamus and spinal intermediate grey column. Many of the other presumptive sources of afferents in earlier studies may therefore be interpreted now as projecting to regions adjacent to, but not actually within the boundaries of the locus coeruleus. A similar construction of chemical inputs may also be expected with better definition of the actual boundaries of this nuclear group.

2. Firing Patterns in Anesthetized Paralyzed Preparations

In anesthetized rats, noradrenaline-locus coeruleus cells have been reported to discharge spontaneously in a slow (1 to 6 Hz) tonic fashion (GRAHAM and AGHAJANIAN 1971; NYBÄCK et al. 1975; SVENSSON et al. 1975; NAKAMURA and IWAMA 1975; CEDARBAUM and AGHAJANIAN 1976; FAIERS and MOGENSON 1976; AGHAJANIAN et al. 1977; GUYENET and AGHAJANIAN 1977; BIRD and KUHAR 1977; TAKIGAWA and MOGENSON 1977; AGHAJANIAN 1978). The only peripheral sensory stimuli initially reported to affect this pattern were strongly noxious, painful stimuli. Pressure applied to the tail or paws evokes biphasic (excitatory-inhibitory) discharge responses (CEDARBAUM and AGHAJANIAN 1976). Mild body strokes, bright flashes, and loud auditory stimuli in these preparations failed to reliably elicit responses (CEDARBAUM and AGHAJANIAN 1976). However, discharge responses were consistently evoked by noxious peripheral stimuli but not by other visual, auditory or non-noxious somatosensory stimuli regardless of the anesthetic agent used; increased sensitivity to somatosensory stimuli was reported for unanesthetized, paralyzed preparations (KORF et al. 1974). Electrical stimulation in peripheral (vagal, splanchnic and sciatic nerves) and central (bed nucleus of the striata terminalis, olfactory bulb, preoptic region, ventromedial nucleus of the hypothalamus, ventral tegmentum, and central grey) sites has been reported to produce converging, orthodromic excitatory and inhibitory influences onto noradrenaline-locus coeruleus neurons (AGHAJANIAN et al. 1977; TAKIGAWA and MOGENSON 1977).

Noradrenaline-locus coeruleus neurons have also been studied after antidromic activation from target areas (NAKAMURA and IWAMA 1975; FAIERS and MOGENSON 1976; GERMAN and FETZ 1976; AGHAJANIAN et al. 1977; TAKIGAWA and MOGENSON 1977). This technique not only helps identify these cells, but also permits analysis of various physiological properties. Locus coeruleus neurons in rat conduct slowly, in the range of 0.4–1.3 m/sec (NAKAMURA and IWAMA 1975; FAIERS and MOGENSON 1976; GERMAN and FETZ 1976; AGHAJANIAN et al. 1977; TAKIGAWA and MOGENSON 1977). Antidromic stimulation reliably elicits inhibitory responses at slightly longer latencies than driven spikes, even in cases where the recorded cell is not one that is antidromically activated. This post-excitatory pause in discharge activity has been attributed to a collateral inhibition among noradrenergic neurons either by recurrent axons (AGHAJANIAN et al. 1977), or perhaps via dendro-dendritic interactions (KODA et al. 1980; GROVES 1983).

Numerous studies have described the effects of various pharmacological agents on noradrenaline-locus coeruleus discharge activity in anesthetized rats. Systemically administered amphetamine (GRAHAM and AGHAJANIAN 1971), morphine (AGHAJANIAN 1978), clonidine (AGHAJANIAN 1978) or tricyclic antidepressants (NYBÄCK et al. 1975) slowed discharge rates, while systemic (SVENSSON et al. 1975) piperoxane, an adrenoceptor antagonist, increased impulse activity (CEDARBAUM and AGHAJANIAN 1976).

3. Firing Patterns in Unanesthetized Behaving Animals

There have been few published studies of locus coeruleus discharge activity in unanesthetized animals. Locus coeruleus discharge rates have been analyzed as a function of the sleep-waking cycle in cat, a species whose noradrenaline-containing neurons are loosely interdigitated with non-noradrenaline neurons (Chu and Bloom 1973, 1974; McCarley and Hobson, 1975; Steriade and Hobson 1976; Sakai 1980). In general, these studies were not wholly consistent; they suggested that many, but not all, locus coeruleus neurons discharge fastest in waking and slowest during REM sleep periods (REM-off cells). However, the neurochemical heterogeneity among neurons in cat locus coeruleus prevents specifying REM-off properties to noradrenaline (or to non-noradrenaline) neurons. In addition, none of these studies systematically measured discharge activity during waking as a function of arousal changes, sensory stimulation, or behavior.

Far more insight into the behavioral correlates of noradrenaline cells has recently been obtained from single unit recording studies on the awake behaving squirrel monkey (Foote et al. 1980) and rat (Aston-Jones and Bloom 1981a,b); this insight is possible because the locus coeruleus in these species contains virtually 100% noradrenaline-containing neurons, and therefore recording sites localized to locus coeruleus are, by this definition, recordings from noradrenaline neurons. The general results on these two species may be summarized as follows: Spontaneous discharge activity co-varied with stages of the sleep-waking cycle yielding rates directly related to vigilance levels; that is, the more alert the animal, the more active its locus coeruleus. Spontaneous discharge anticipated phasic cortical events (e.g. cortical spindle activity during slow wave sleep) as well as tonic cortical epochs (slow wave to wake transition stages). The activity was not linked to general behavioral movement, although arousal-behaviors were typically accompanied by phasic, robust discharge. Responses in rat locus coeruleus activity were predictably evoked by mild, non-noxious environmental stimuli, and the magnitudes of sensory-evoked responses also varied directly with arousal level. Spontaneous and sensory-evoked discharge decreased during grooming and consummatory behaviors, similar to results obtained during sleep.

4. A Behavioral Hypothesis of Locus Coeruleus Firing Correlates

These behavioral correlates of locus coeruleus firing, suggesting a strong association with level of arousal, form a provocative functional complement with the cellular activity changes produced by locus coeruleus neurons on their targets. Taken together these data indicate that salient environmental stimuli or spontaneous abrupt arousals can activate the locus coeruleus to release noradrenaline. Under the enabling hypothesis put forward above, this would then enhance CNS activity in terminal fields concurrently receiving transient bursts of input stimuli and would further simultaneously dampen activity in regions receiving only inhibitory inputs. The recent rat and squirrel monkey data indicate that these locus coeruleus cells are most heavily influenced by three distinct input systems: 1) excitatory afferents, reflecting salient environmental events; 2) inhibitory afferents, reflecting internally generated stimuli linked to tonic biological homeostasis; 3) inhibitory, recurrent intracoerulean collaterals, modulating overall epochs of discharge activity. Noradrenaline-locus coeruleus activity may at all times be a function of the relative intensities of these general classes of input systems. Locus coeruleus neuron activity could also reflect varying degrees of tonic inhibition and phasic event-related excitations. Based on reports of REM-off cells in cats and rats, the most intense tonic inhibition may occur during paradoxical sleep. However, because locus coeruleus discharge typically anticipates interruptions of such states and since intense afferent excitation can overcome this strong tonic inhibition, this suggests that robust noradrenaline-locus coeruleus activity has a role in generating the associated behavioral reorientation during early awakening.

Thus, noradrenaline-locus coeruleus activity may be controlled by phasic, environmentally determined input as well as tonic, endogenously generated input. Such powerful gating suggests specific behavioral consequences of noradrenaline-locus coeruleus discharge. The working hypothesis proposed by Aston-Jones (see Aston-Jones and Bloom 1981b) states that the system is arranged to bias global behavioral attention toward novel, rapidly changing external signals and away from tonic internal signals. Considering an animal within a period of changing external environmental signals, the activity within locus coeruleus and within target cells receiving convergent sensory information would, under the enabling hypothesis described above, enhance the responsiveness of those targets while simultaneously depressing targets which were only weakly affected by the novel stimuli. This selected enhancement of the responsiveness of targets receiving sensory inputs during this transient novelty period would have the effect of favoring orientation toward effective responses and depressing competitive acts from weakly actived output systems. Conversely, during periods of stable external environmental signals, diminished noradrenaline-locus coeruleus activity would permit the performance of endogenously triggered homeostatic behaviors (e.g. grooming, consuming, etc.).

VII. Neuropsychopharmacology

1. The Locus Coeruleus and Dorsal Noradrenergic Bundle

The locus coeruleus has intrigued many neuroscientists, and inspired a plethora of hypotheses regarding its functional significance. The locus coeruleus and its ascending forebrain projection have been implicated in stress, reward, anxiety, attention and even learning and memory. Unfortunately, the behavioral data to support these views, in animals manipulated to remove the noradrenergic neurons, are limited by many interpretative problems.

a. The "Stress" Connection

One of the first roles assigned to brain noradrenaline was one that paralleled its putative function in the peripheral autonomics, i.e. a role in "stress". In the earliest studies, exposure to cold and electric shocks to the feet caused decreases in brain and adrenal noradrenaline in rats (BARCHAS and FREEDMAN 1963; MAYNERT and LEVI 1964). Similar results have been seen with aggregation stress (BLISS and AILION 1969), hypoxic stress (KOOB and ANNAU 1974), muscular exertion (GORDON et al. 1966), and exhaustion stress (STONE 1973). These decreases in brain noradrenaline during stress were accompanied by increased synthesis of noradrenaline (GORDON et al. 1966; BLISS et al. 1968), increased turnover of noradrenaline (THIERRY et al. 1968a, b; BLISS et al. 1968; BLISS and AILION 1969; STONE 1971; 1973; STOLK et al. 1974) and increases in the levels of two major metabolites of noradrenaline, MOPEG and DOPEG (CESAR et al. 1974; STONE 1975). These results are all consistent with the hypothesis that "stress" can activate central noradrenergic neurons. Of some interest is the observation that a more severe stress is needed to activate peripheral sympathetic neurons to the adrenal gland and spleen (BLISS and AILION 1969). KORF and colleagues have linked the increased turnover of noradrenaline in the rat forebrain to an ipsilateral activation of the locus coeruleus itself (KORF et al. 1973). Cold stress also increased tyrosine hydroxylase activity in the locus coeruleus (ZIGMOND et al. 1974).

Other monoamine systems must also be involved in stress-induced changes. The turnover of dopamine and 5-HT is also facilitated by "stress" (BLISS et al. 1968; THIERRY et al. 1968a, 1976; FADDA et al. 1978; LAVIELLE et al. 1979). However, it is not particularly surprising that such a general phenomenon as "stress" would mobilize many different neuronal systems.

b. Reward and Learning

Many different experimental observations have focussed on the role of one or another central catecholamine system as playing an "important role" in certain behaviors, a conceptual level one step removed from a catecholaminergic "mediation" of a behavior. By virtue of the successful efforts to describe the fundamental neurochemistry of catecholamines and by employing pharmacological tools shown to have considerable specificity, it has been relatively simple to investigate the reversible deficits or actions which arise from interference with or simulation of chemical or receptive processes attributable to these neurons. However, despite the apparent ease of testing, interpretation has remained difficult.

The early evidence regarding a role for the catecholamines in reward has been discussed elsewhere in detail (GERMAN and BOWDEN 1974; WISE 1978a; FIBIGER 1978) and is discussed above in regard to the dopamine systems. It was mainly the similarity of the anatomical distribution of the catecholamine systems and the sites in the brain supporting ICSS that led to the first hypotheses regarding a role for the locus coeruleus in reward. As originally proposed by CROW and ARBUTHNOTT (1972) "Activation of (catecholamine) neurons may facilitate motor responses and thus mobilize 'appetitive' or approach behaviours." Noradrenaline neurones from the locus coeruleus were hypothesized to be activated when a behavioral sequence has been 'consummated'; thus these neurons were viewed as part of the mechanism for learning adaptive motor responses, or "reinforcement" (CROW and ARBUTHNOTT 1972). Unfortunately, this elegant hypothesis has been largely unsupported by experimental studies. The strongest pharmacological evidence in support of the original noradrenaline hypothesis was the observation that the DBH inhibitors, disulfiram and diethyldithiocarbamate (DDC) decreased ICSS and this effect could be reversed by intraventricular injection of noradrenaline (WISE and STEIN 1969). However, ROLLS et al. (1974) later showed that these particular DBH inhibitors also had general sedative effects and that the exogenous noradrenaline simply reversed these sedative effects. This latter interpretation was supported when different DBH inhibitors such as FLA-63 and U-14,624, drugs which lacked the sedative side effects of disulfiram and DDC, failed to block ICSS (LIPPA et al. 1973; STINUS et al. 1976).

However, lesion studies using the synthetic neurotoxin, 6-OHDA, have provided the most devastating evidence against a role for the locus coeruleus in ICSS. Some of the earliest studies utilizing 6-OHDA showed that repeated intraventricular injections of 6-OHDA produced marked deficits in ICSS that were accompanied by a depletion of noradrenaline (BREESE et al. 1971; STEIN and WISE 1971). Subsequently, rats receiving such treatments recover their ICSS within 5 to 7 days, although the noradrenaline depletion remained at 10% of controls (LIPPA et al. 1973). To provide further information about the relative importance of dopamine versus noradrenaline, BREESE and coworkers utilized different 6-OHDA treatments to preferentially deplete one or the other amine (COOPER et al. 1974; 1978). Preferential depletion of noradrenaline did not result in a significant decrease in ICSS in the substantia nigra, lateral hypothalamus or locus coeruleus. However, acute depressions in ICSS were observed in rats preferentially depleted of dopamine.

Other experiments aimed at observing intracranial self stimulation after destruction of specific noradrenaline pathways have yielded similar negative results. Locus coeruleus lesions facilitate, instead of inhibit, ICSS in the hypothalamus (KOOB et al. 1976) and ICSS in the locus coeruleus is unaffected by lesions of the dorsal noradrenergic bundle (CLAVIER et al. 1976; CORBETT et al. 1977). Further, ICSS in the area of the dorsal noradrenergic bundle was unaffected by locus coeruleus lesions (CLAVIER and ROUTTENBERG 1976), although it was severely disrupted by lesions of the medial forebrain bundle. Finally, near complete depletion of hippocampal and olfactory bulb noradrenaline by chemical lesions had no significant effect on ICSS in these sites (PHILLIPS et al. 1977).

The most detailed statement of a catecholamine hypothesis in memory was presented by KETY (1970). Building upon emerging evidence that noradrenaline or other central monoamines might mediate certain emotional and behavioral states, KETY proposed the following hypothetical events might account for an adaptive mechanism which could consolidate experiences significant for survival: **1)** an aroused state induced by a novel stimulus genetically recognized as significant affects synapses throughout the central nervous system, suppressing most, but permitting or even accentuating activity in those which were transmitting novel or significant stimuli; **2)** this state favors the development of persistent facilitatory changes in all synapses which are or have recently been active; **3)** a widely distributed network of synapses with sufficient complexity exists between the neurons which would mediate the primitive and genetically endowed adaptive responses of the many other neurons necessary for appropriate action. Kety noted that the morphology of noradrenergic neurons could represent the type of widespread arousal system which might influence synapses throughout the brain, and that coupling of adrenoceptors to cyclic AMP formation may favor consolidation of learning by stimulating protein synthesis.

Subsequently, CROW (1968) proposed an even more detailed version of a similar hypothesis, suggesting that noradrenergic neurons of the locus coeruleus deliver a special type of signal to target neurons of the cerebral cortex and thus symbolize at the cellular level the outcome of a particular behavioral sequence. The consequences of this signal were proposed to initiate the long-term synaptic changes presumed to underlie learning. CROW (1968), ANLEZARK et al. (1973), CROW and WENDLANDT (1976) and STEIN et al. (1975) have continued to present evidence that pharmacological or electrolytic inactivation of locus coeruleus circuits can impair acquisition and retention of passive avoidance responses, and adversely affect apparent acquisition within an L-shaped maze for food reward.

Although this specific runway deficit has been replicated (SESSIONS et al. 1976; KOOB et al. 1978c), results using other tests have revealed that rats without an intact locus coeruleus forebrain projection can still learn a wide variety of tasks (AMARAL and FOSS 1975; MASON and IVERSEN 1975; SESSIONS et al. 1976; ROBERTS et al. 1976; KOOB et al. 1978).

However, an extensive series of experiments have suggested that the forebrain projections from the locus coeruleus have a role in non-reward (for review and references see Mason and IVERSEN 1979). 6-OHDA lesions to the dorsal noradrenergic bundle produce a resistance to extinction when reward is withheld in a number of learned tasks. This phenomenon, termed the "dorsal bundle extinction effect", has been attributed to an attentional deficit where rats with this lesion have a deficit in filtering out irrelevant stimuli. Consistent with this hypothesis were the findings that rats with these dorsal bundle lesions: 1) are impaired in acquisition and reversal of stimulus discrimination tasks; 2) show deficits in some complex tests of attention and 3) show improvement in the acquisition of a learning task where learning about irrelevant stimuli would ultimately benefit an animal (non-reversal shift task) (see MASON 1980 for review and references). Significant controversy exists as to the reproducibility of the 6-OHDA-induced attentional deficits (PISA and FIBIGER 1983a, b) and even the "dorsal bundle extinction effect," itself (TOMBAUGH et al. 1983). Indeed, more recent studies have failed to show impairment in delayed alternation (PISA and FIBIGER 1983b), attenuation of latent inhibition (TSALTAS et al. 1984b), or an enhancement of non-reversal shift (PISA and FIBIGER 1983a). However, several investigators have observed that rats with depletion of forebrain noradrenaline either by 6-OHDA-induced dorsal bundle lesions or locus coeruleus electrolytic lesions are more distractable than controls (ROBERTS et al. 1976; KOOB et al. 1978c; MASON and FIBIGER 1978; OKE and ADAMS 1978). Indeed more recent work suggests that learning deficits can be observed in rats with locus coeruleus or dorsal noradrenergic bundle lesions, but only under conditions with significant stress or aversive contingencies. When discrimination performance is made particularly difficult rats with dorsal noradrenergic bundle lesions show deficits (EVERITT et al. 1983); rats receiving a 6-OHDA dorsal noradrenergic bundle lesion were impaired on acquisition of both on- and off- the baseline conditioned suppression (COLE and ROBBINS 1987). These results suggest that the earlier work relating central noradrenergic function to stress may be relevant for understanding the role of this system in behavioral change, a role not incompatible with work involving electrophysiological recording in freely moving animals (see below).

c. Anxiety

Recently, a series of studies have appeared which suggest that the locus coeruleus has a role in anxiety (REDMOND 1977; REDMOND and HUANG 1979). This hypothesis is based on observations using physiological and pharmacological studies of locus coeruleus function in primates, where low intensity electrical stimulation of the locus coeruleus in the awake restrained stump-tailed monkey (Macaca arctoides) elicited behavioral responses which appeared similar to the behavior observed in the monkeys following direct threatening confrontation by humans (REDMOND et al. 1976). Lesions of the locus coeruleus decreased the natural occurrence of these behaviors in a social situation and in response to similar direct threatening confrontation by humans (REDMOND and HUANG 1979). Marshalled also as support for this hypothesis was pharmacological evidence that drugs which increased or decreased firing of the locus coeruleus produced behavioral effects consistent with the electrical stimulation or lesion effects. For example, piperoxane, an α_2-adrenoceptor antagonist, which activates the locus coeruleus (CEDARBAUM and AGHAJANIAN 1976), produced behavioral effects in monkeys similar to stimulation of the locus coeruleus (REDMOND and HUANG 1979). Diazepam, a benzodiazepine used clinically for treatment of anxiety, reduced the effects of locus coeruleus stimulation on the same behaviors (REDMOND 1977).

All these observations led REDMOND (1977) to propose that activation of the locus coeruleus produced various manifestations of anxiety in the projection areas at which the anti-anxiety drugs such as benzodiazepines acted. The benzodiazepines, in turn, act via inhibition of the locus coeruleus to produce their anti-anxiety effects. In a similar vein, others have hypothesized that the anti-opiate withdrawal effects of clonidine, an α-adrenoceptor agonist, were also mediated by the locus coeruleus (GOLD et al. 1982). If these hypotheses are valid, and the inhibiton of locus coeruleus mediates the anti-anxiety effects of benzodiazepines, as well as the anti-opiate withdrawal effects of opiates or clonidine, it should be possible to test this view by asking how these latter effects are influenced by removal of locus coeruleus.

In fact, however, when this theory is tested, 6-OHDA lesions of the ascending locus coeruleus projections do not at all influence the dose-response effects of benzodiazepines in the sensitive Geller-Seifter test for anxiolytic behavior (GELLER and SEIFTER 1960; KOOB et al. 1984). Also in earlier work, noradrenaline injected intraventricularly actually facilitated the anti-anxiety action of the benzodiazepine, oxazepam (Stein et al. 1973). Nor do identical lesions of the ascending locus coeruleus projection alter the withdrawal responses of opiate-dependent rats nor their response to the anti-withdrawal actions of clonidine (Thatcher et al. 1984). There are reports that clonidine can induce a "hypokinetic" syndrome which may blunt responsiveness in some tests of sensory responsiveness.

The conflict model has proved useful in predicting clinical anxiolytic efficacy, since an excellent correlation exists between the minimum effective dose in the rat anti-conflict procedure and the average clinical daily dose for treating psychoneuroses (Cook and Sepinwall 1975). Furthermore, animal studies using other conflict models of anxiety have provided no evidence at all to support the hypothetical role for the locus coeruleus in anxiety. If anything, the results of lesion studies in the rat have provided evidence opposite to that presented by Redmond and colleagues. For example, 6-OHDA-induced destruction of the dorsal noradrenergic bundle in rats failed to alter responding in an operant conflict test (Tye et al. 1977; Koob et al. 1984a). Rats with 6-OHDA-induced destruction of cortical noradrenaline neurons did not alter their response to social contacts in an unfamiliar situation, considered an anxiety test (Crow et al. 1978); lesioned rats were also unimpaired in a drinking anxiety test (File et al. 1979). Further, 6-OHDA-induced destruction of the dorsal noradrenergic bundle actually produced what may be termed an "anxiogenic-like effect", that is, increased neophobia to a number of novel tastes in rats and increased measures of neophobia in novel environments (Mason et al. 1978; Britton et al. 1984).

In summary, the putative role for the locus coeruleus in anxiety is based largely on ethological observations in monkeys and pharmacological correlations between the cellular and behavioral action of drugs. Evidence in rats using 6-OHDA-induced destruction of the dorsal noradrenergic projections fails to support this hypothesis and suggests that the anti-anxiety properties of decreased locus coeruleus function may be more related to an inability to respond appropriately to certain types of stress or novelty.

2. Ventral Noradrenergic Bundle

a. Feeding Behavior

A large number of studies have explored the possibility that noradrenaline particularly in the hypothalamus, has a role in the regulation of feeding. In early studies, noradrenaline injected into the hypothalamus produced feeding in satiated rats (Grossman 1960, 1962a, b; Miller et al. 1964; Bovt 1968; Leibowitz 1970), an effect apparently evoked from placements between the medial and lateral hypothalamus at the level of the paraventricular and dorsomedial nuclei. Subsequent work showed that eating could also be reliably elicited by very large noradrenaline injections (1–3 μg doses) into extrahypothalamic structures (Coury 1967; Booth 1967). Even larger amounts of noradrenaline (14–28 μg) injected into perifornical hypothalamus suppressed feeding, and phentolamine induced feeding (Margules 1969). A *β*-adrenoceptor blocker, propranolol, had no effect (MARGULES 1969; 1970). These doses of catecholamine are several orders of magnitude above the entire hypothalamic content and make these effects difficult to interpret functionally.

The intervening decade and a half has witnessed continued efforts to localize responsive sites, minimally effective doses, and the circuitry of the systems involved. This era of work has been summarized comprehensively (see HOEBEL and LEIBOWITZ 1981).

More recent research has concentrated on identifying the specific neuroanatomical basis for this phenomenon of noradrenaline-induced feeding. The critical variable appears to be the brain site of the injection, since injections into the region of the paraventricular nucleus produce the most reliable feeding (LEIBOWITZ 1978a; MATTHEWS et al. 1978). Using a dose of 6.8 μg of noradrenaline dissolved in saline, LEIBOWITZ found that essentially all sites outside the hypothalamus were ineffective and sites in the dorsomedial part of the hypothalamic paraventricular nucleus gave the largest increase in feeding: >4 g in 15 min, with the shortest latency (2.5 min). LEIBOWITZ (1978b) has reported reliable feeding from injections of noradrenaline into the paraventricular nucleus at doses as low as 0.004 μg (0.8 g food eaten per 30 min) to 1 μg (3.5 g food eaten per 30 min) using a descending order of presentation of doses in rats that had already demonstrated a reliable feeding response to 3 μg.

At least one study has suggested that noradrenaline induces feeding by altering the metabolic rate of cells regulating food intake, via indirect actions such as glycogenolysis, increased release of insulin or the release of free fatty acids from adipose tissue (all established metabolic effects of noradrenaline) rather than directly activating a neural net involved in feeding behavior (DAVIS and KEESEY 1971). This hypothesis was derived from the observation that insulin injected subcutaneously prior to intracranial noradrenaline markedly potentiated the noradrenaline-induced feeding (DAVIS and KEESEY 1971) and from the observation of a long latency for the feeding response to noradrenaline. Consistent with this type of hormonal interaction are the observations that vagotomy, hypophysectomy and adrenalectomy have been reported to attenuate noradrenaline-induced feeding (LEIBOWITZ 1978a, b). Since the hypothalamus, particularly the paraventricular nucleus, is thought to be involved in the release of adrenocorticotrophic hormone (see VALE et al. 1983) it is possible that this hormonal connection somehow mediates the noradrenaline response. This becomes particularly interesting since feeding can be elicited in animals by nonspecific arousal, such as mild pinch to the tail in rats (ANTELMAN and SZECHTMAN 1975).

Further extension of the role of noradrenaline in some aspect of feeding has been provided by a series of lesion studies. As mentioned above, ventral medial hypothalamic and midbrain electrolytic lesions have long been known to produce hyperphagia (HETHERINGTON and RANSON 1940; SKULTETY and GARY 1962). Later work attempting to localize the anatomical substrate for these effects in the hypothalamus has only succeeded in showing that the effective sites for this overeating converge on the medial hypothalamus at the level of the paraventricular nucleus. In fact, parasagittal knife cuts aimed just lateral to the paraventricular nucleus produce hyperphagia (GOLD et al. 1977).

The dense noradrenergic innervation of the paraventricular nucleus originates in fibers from the ventral aspect of the central tegmental tract, and specific destruction of this projection with 6-OHDA injected into the ventral midbrain produces chronic increases in food intake (Ahlskog and Hoebel 1973; Ahlskog et al. 1975). Subsequent work attributed this to nonspecific effects of 6-OHDA (Oltmans et al. 1977). However, in an extensive study, Hernandez and Hoebel (1982) showed that the increase in food intake produced by 6-OHDA lesions in the ventral midbrain correlated highly with the dose of 6-OHDA and the percent depletion of hypothalamic noradrenaline. Even more impressive was the observation that pre-injection of desipramine into the midbrain (specific noradrenaline reuptake blocker) prevented the hyperphagia and partially blocked the depletion of noradrenaline (Hernandez and Hoebel 1982). These results suggest that previous observations of hyperphagia due to a ventral midbrain lesion may be attributed at least in part to the catecholamine pathway located therein.

Interestingly these same 6-OHDA lesions attenuated the anorectic potency of systemically administered amphetamine, suggesting that 6-OHDA destroyed at least one of the substrates by which amphetamine produces anorexia, presumably by some presynaptic action (Ahlskog 1974; Leibowitz and Brown 1980). Consistent with this hypothesis postsynaptically acting agonists such as adrenaline increased in potency post-lesion (Leibowitz and Brown 1980). Unknown at this time is whether this ventral midbrain system is noradrenergic or adrenergic; also unknown are the precise locations of the cell bodies for this "satiety" system or what role it plays in normal feeding.

In summary, injections of noradrenaline into the region of the paraventricular nucleus of the hypothalamus produce feeding in sated rats. The feeding response can be obtained with doses usually above 1 μg and appears to be α-adrenergic in that α-adrenoceptor blockers inhibit the response. Removal of a major noradrenergic input to the paraventricular nucleus, the ventral part of the central tegmental tract, causes hyperphagia and partially attenuates amphetamine anorexia. The link to brainstem visceral afferent systems via the noradrenergic central tegmental tract suggests that the noradrenaline feeding response other than those concerned with the primary direct behavioral control of feeding, may be secondary to changes in the pituitary adrenal system.

b. Sexual Behavior

A substantial amount of evidence supports the hypothesis that noradrenaline, particularly in the hypothalamus, also has a role in the neuroendocrine regulation of sexual behavior (Everitt et al. 1975; Crowley et al. 1976). For example, drugs activating noradrenergic transmission potentiate lordotic behavior while phenoxybenzamine, an α-receptor antagonist, abolished sexual receptivity (Everitt et al. 1975; Crowley et al. 1976). More recent work involving relatively specific destruction of the ventral portion of the ascending noradrenergic pathways produced a striking dissociation of two major elements of sexual behavior in female rats (Hansen et al. 1980, 1981). Lesions that produced approximately 70% reduction in ^{3}H-noradrenaline uptake in the hypothalamus, but little reduction (8 %) in noradrenaline uptake in cortex virtually eliminated the lordotic posture of female rats (receptive posture) when mounted by a male (Hansen et al. 1980, 1981). These rats, however, retained undisturbed the characteristic proceptive elements of female sexual behavior, i.e. the typical hopping, darting and ear-vibratory responses characteristic of pre-mounting behavior which serves presumably to attract the male.

One hypothesis to explain this result suggested by Everitt and colleagues (Hansen et al. 1980; Robbins and Everitt 1982) is that the ventral noradrenergic bundle processes somatosensory information essential for this lordotic response. Ventral bundle lesions also abolish the analgesic effects of vaginal stimulation (Crowley et al. 1977) further suggesting a role for the ventral noradrenergic bundle in the processing of somatosensory information. This work with sexual behavior has led Robbins and Everitt (1982) to propose that the noradrenergic subcortical projections have a role in an organism's responsiveness to external stimuli, particularly when they are tactile in nature. How this hypothesis would extend to other sensory modalities and behavioral responses remains to be determined.

3. Summary and Conclusions

Research using the classical neuropsychopharmacological techniques of electrical and chemical stimulation, lesions and drug administration has provided little evidence towards a unified hypothesis regarding the function of the central noradrenergic neurons and particularly those arising from the locus coeruleus. The central noradrenergic systems are clearly activated during "stress", but so are other neurochemical systems. In addition the lesion studies have not provided much support for putative roles of the locus coeruleus in reward, learning and anxiety.

Feeding induced by noradrenaline can be reliably produced by injections into the paraventricular nucleus of the hypothalamus and lesions to the ventral noradrenegic bundle cause a mild hyperphagia and weight gain in rats. However, the mechanism by which this occurs is unclear, as is its relationship to normal homeostatic feeding mechanisms. This becomes particularly important since this feeding appears to be modulated by changes in the pituitary adrenal axis. Noradrenaline injections also alter sexual behavior and ventral noradrenergic bundle lesions virtually eliminate the female lordotic response and the analgesia associated with vaginal stimulation. How this apparent lack of central processing of somatosensory stimuli for sexual behavior is related to the alterations in feeding behavior is unclear; however, it is interesting to note that the noradrenergic input to the parventricular nucleus of the hypothalamus overlaps significantly with the vasopressinergic cells therein (Swanson and Sawchenko 1980). This input to the paraventricular nucleus appears to arise from cell bodies in the dorsal vagal complex (Swanson and Hartman 1980).

Indeed, one of the more intriguing clues from the neuropsychopharmacological evidence regarding the functional significance of the central noradrenergic systems may center around the inability of animals with noradrenergic lesions to respond appropriately to stress or aversive situations. This is particularly interesting in regards to the hypotheses derived from chronic unit recordings in awake animals on the cellular properties of the dorsal noradrenergic system during physiological states. How these hypotheses are related, if at all, to the increased liberation of noradrenaline during stress or to noradrenaline-induced feeding remains to be determined.

D. Overall Conclusions

In order to draw together the descriptive structural and functional properties of central catecholamine neurons and their possible function at the behavioral level we must return to the theoretical construct points raised in the Introduction. From a purely hypothetical view, one could argue that behavioral tests of a central catecholamine circuit do not require a data base of molecular and cellular attributes. In contrast, a cellularly-based neurobiologist would argue that until hypotheses of behavioral level functions integrate themselves with the known anatomy and cellular physiology of these transmitter systems, the behavioral interpretations cannot be validated at the cellular level. Thus, a behavioral event lacking intrinsically specified operations for particular cellular sequences is a relatively untestable hypothesis in terms of whether the cells and transmitter system are, in fact, essential for the behavior.

For example, with the locus coeruleus system, the recent direct electrophysiological observations in the behaviorally responsive animal provide a far more detailed base of data on which to formulate specific testable hypotheses regarding a noradrenaline cellular role in specific types of behavior. These cellular data indicate that the locus coeruleus system fires with the occurrence of novel external sensory events, rather than with learning or extinction per se, and that the neurons are under strong inhibitory influences during vegetative acts such as eating, grooming or sleeping. If we were to take a "bottom up" approach to analysing noradrenergic relevant behavior based on these cellular analyses, the suggestion might be that tasks involving contingencies of sensory discrimination of novel objects or during significant environmental demand (stress) might be the most relevant conditions in which to demonstrate behavioral perturbations.

In regards to the dopaminergic systems, the cellular correlative approach to behavioral function has so far been rather unrewarding. Better clues to be followed in characterizing the dopaminergic neurons may, in contrast, be a "top down" approach, in which the optimal conditions for seeking cellular correlates would be based upon the results of the behavioral observations. The behavioral observations suggest that a response initiation task would be most critically dependent upon dopaminergic function. If this view were valid, cellular correlates of neuronal firing should then become apparent in such tasks.

Among the many profound problems awaiting future research is the riddle of how to evolve these tentative behavioral consequences in the normal brain to clarify the possible relevance of the central catecholamine neurons to the behavioral symptoms of the major psychopathologies. Although that extension currently seems unlikely to be realized in the near future, continued attention to a multidisciplinary, multi-level approach to central catecholamine neuron function may eventually provide such answers.

E. References

Aebischer P, Schultz W (1984) The activity of pars compacta neurons of the monkey substantia nigra is depressed by apomorphine. Neurosci Lett 50:25-29
Aghajanian GK (1978) Tolerance of locus coeruleus neurones to morphine and suppression of withdrawal response by clonidine. Nature 276:186-188
Aghajanian GK, Cedarbaum JM, Wang RY (1977) Evidence for norepinephrine-mediated collateral inhibition of locus coeruleus neurons. Brain Res 136:570-577
Ahlskog JE (1974) Food intake and amphetamine anorexia after selective forebrain norepinephrine loss. Brain Res 82:211-240
Ahlskog JE, Hoebel BG (1973) Overeating and obesity from damage to a noradrenergic system in the brain. Science 182:166-169
Ahlskog JE, Randall DI, Hoebel BG (1975) Hypothalamic hyperphagia: dissociation from hyperphagia following destruction of noradrenergic neurons. Science 190:399-401
Ajika K, Hökfelt T (1973) Ultrastructural identification of catecholamine neurones in the hypothalamus periventricular-arcuate nucleus-median eminence complex with special reference to quantitative aspects. Brain Res 57:97-117
Akasu T, Koketsu K (1977) Effects of dibutyryl cyclic adenosine 3′,5′-monophosphate and theophylline on the bullfrog sympathetic ganglion. Br J Pharmacol 60:331-336
Amalric M, Koob GF (1987) Depletion of dopamine in the caudate nucleus but not nucleus accumbens impairs reaction time performance in rats. J Neuroscience, in press
Amaral DG, Foss JA (1975) Locus coeruleus and learning. Science 188:377-378
Amsler C (1923) Über einige Wirkungen des Apomorphins. Naunyn-Schmiedeberg's Arch Exp Pathol Pharmacol 97:14
Anlezark GM, Grow TJ, Greenway AP (1973) Impaired learning and decreased cortical norepinephrine after bilateral locus coeruleus lesions. Science 181:582-684
Antelman SM, Szechtman H (1975) Tail-pinch induces eating in sated rats which appears to depend on nigrostriatal dopamine. Science 189:731-733
Arbuthnott GW (1974) Spontaneous activity of single units in the striatum after unilateral destruction of the dopamine input. J Physiol (Lond) 239:121-122
Arluison M, Agid Y, Javoy F (1978a) Dopaminergic nerve endings in the neostriatum of the rat. I. Identification by intracerebral injections of 5-hydroxydopamine. Neurosci 3:657-673
Arluison M, Agid Y, Javoy F (1978b) Dopaminergic nerve endings in the neostriatum of the rat. 2. Radioautographic study following local microinjections of tritiated dopamine. Neurosci 3:675-684
Asher IM, Aghajanian GK (1974) 6-Hydroxydopamine lesions of olfactory tubercles and caudate nuclei: effect on amphetamine-induced stereotyped behavior in rats. Brain Res 82:1-12
Aston-Jones G, Bloom FE (1981a) Activity of norepinephrine-containing locus coeruleus neurons in behaving rats anticipates fluctuations in the sleep-waking cycle. J Neurosci 1:876-886
Aston-Jones G, Bloom FE (1981b) Norepinephrine-containing locus coeruleus neu-

rons in behaving rats exhibit pronounced responses to non-noxious environmental stimuli. J Neurosci 1:887-900
Aston-Jones G, Ennis M, Pieribone VA, Nickell W, Thompson, Shipley MT (1986) The brain nucleus locus coeruleus: restricted afferent control of a broad efferent network. Science 234:734-737
Barchas JD, Freedman DX (1963) Brain amines: response to physiological stress. Biochem Pharmacol 12:1232-1235
Ben-Ari Y, Kelly JS (1976) Dopamine-evoked inhibition of single cells of the feline putamen and basolateral amygdala. J Physiol (Lond) 256:1-21
Beninger RJ, Mason ST, Phillips AG, Fibiger HC (1980) The use of conditioned suppression to evaluate the nature of neuroleptic induced avoidance deficits. J Pharmacol Exp Ther 213:623-627
Bevan JA (1977) Some functional consequences of variation in adrenergic synaptic cleft width and in nerve density and distribution. Fed Proc 36:2439-2443
Bird SJ, Kuhar MJ (1977) Iontophoretic application of opiates to the locus coeruleus. Brain Res 122:523-533
Björklund A, Lindvall O (1986) Catecholaminergic brain stem regulatory systems. In: Bloom FE (ed) Handbook of Physiology, Amer. Physiol. Soc. vol. IV. Bethesda, Maryland, pp 155-236
Bliss EL, Ailion J (1969) Response of neurogenic amines to aggregation and strangers. J Pharmacol Exp Ther 168:258-263
Bliss EL, Ailion J, Zwanziger J (1968) Metabolism of norepinephrine, serotonin and dopamine in rat brain with stress. J Pharmacol Exp Ther 164:122-134
Bloom FE (1973) Ultrastructural identification of catecholamine-containing central synaptic terminals. J Histochem Cytochem 21:333-348
Bloom FE (1975a) Amine receptors in CNS: I. Norepinephrine. In: Iversen LL, Iversen SD, Snyder SH (Eds) Handbook of Psychopharmacology. Raven Press, New York, pp 1-22
Bloom FE (1975b) The role of cyclic nucleotides in central synaptic function. Rev Phys B 74:1-103
Bloom FE (1975c) Monoaminergic neurotoxins: are they selective? J Neural Trans 37:183-187
Bloom FE (1988) Neurotransmitters: past, present and future directions. FASEB J. 2:32-41
Bloom FE, Battenberg ELF (1976) A rapid, simple and more sensitive method for the demonstration of central catecholamine-containing neurons and axons by glyoxylic acid induced fluorescence: II. A detailed description of methodology. J Histochem Cytochem 24:651-671
Bloom FE, von Baumgarten R, Oliver AP, Costa E, and Salmoiraghi GC (1964) Microelectrophoretic studies on adrenergic mechanisms of rabbit olfactory bulb neurons. Life Sci 3:131-136
Bloom FE, Costa E, Salmoiraghi GC (1965) Anesthesia and the responsiveness of individual neurons of the cat's caudate nucleus to acetylcholine, norepinephrine, and dopamine administered by microelectrophoresis. J Pharmacol Exp Ther 150:244-252
Bloom FE, Hoffer BJ, Siggins GR (1971) Studies on norepinephrine-containing afferents to Purkinje cells of rat cerebellum. I. Localization of the fibers and their synapses. Brain Res 25:501-521
Bloom F, Battenberg E, Rossier J, Ling N, Guillemin R (1978) Neurons containing β-endorphin in rat brain exist separatly from those containing enkephalin: immunocytochemical studies. Proc Natl Acad Sci USA 75:1591-1595

Booth DA (1967) Localization of the adrenergic feeding system in the rat diencephalon. Science 158:515-517

Booth DA (1968) Mechanism of action of norepinephrine in eliciting an eating response on injection into the rat hypothalamus. J Pharmacol Exp Ther 160:336-348

Bradshaw CM, Szabadi E, Roberts MHT (1973) Kinetics of the release of noradrenaline from micropipettes: interaction between ejecting and retaining currents. Br J Pharmacol 49:667-677

Breese GR, Howard J, Leaky P (1971) Effect of 6-hydroxydopamine on electrical self-stimulation of brain. Brit J Pharmacol 43: 255-257

Britton DR, Ksir C, Thatcher Britton K, Young D, Koob GF (1984) Brain norepinephrine depleting lesions selectively enhance behavioral responsiveness to novelty. Physiol Behav 33:473-478

Brozoski TJ, Brown RM, Rosvold HE, Goldman PS (1979) Cognitive deficits caused by regional depletion of dopamine in prefrontal cortex of rhesus monkey. Science 205:929-932

Bunney BS, Aghajanian GK (1973) Electrophysiological effects of amphetamine in dopaminergic neurons. In: Usdin E, Snyder S (Eds) Frontiers in Catecholamine Research. Pergamon Press, Oxford, pp 957-962

Bunney BS, Aghajanian GK (1977) Studies on cerebral cortex neurons. In Costa E, Trabucchi M, Gessa GL (Eds) Pharmacology of Non-striatal Dopaminergic Neurons. Raven Press New York, pp 65-70

Busis NA, Weight FF, Smith PA (1978) Synaptic potentials in sympathetic ganglia: are they mediated by cyclic nucleotides? Science 200:1079-1081

Carey RJ (1980) Anticholinergic drugs promote recovery from self-stimulation deficits produced by bilateral but not unilateral dopamine lesions. Neurosci Abstr 6:368

Carli M, Everden JL, Robbins TW (1985) Depletion of unilateral striatal dopamine impairs initiation of contralateral actions and not sensory attention. Nature 313:679-682

Cedarbaum JM, Aghajanian GK (1976) Noradrenergic neurons of the locus coeruleus: inhibition by epinephrine and activation by the alpha-antagonist piperoxane. Brain Res 112:413-419Ot

Cedarbaum JM, Aghajanian GK (1978) Activation of locus coeruleus neurons by peripheral stimuli: modulation by a collateral inhibitory mechanism. Life Sci 23:1383-1392

Cesar PM, Hague P, Sharman DF, Werdinius (1974) Studies on the metabolism of catecholamines in the central nervous system of the mouse. Br J Pharmacol 51:187-195

Chu N-S, Bloom FE (1973) Norepinephrine-containing neurons: Changes in spontaneous discharge patterns during sleeping and waking. Science 179:908-910

Chu N-S, Bloom FE (1974) Activity patterns of catecholamine-containing pontine neurons in the dorso-lateral tegmentum of unrestrained cats. J Neurobiol 5:527-544

Clavier RM, Fibiger HC (1977) On the role of ascending catecholaminergic projections in intracranial self-stimulation of the substantia nigra. Brain Res 131:271-286

Clavier RM, Routtenberg A (1976) Brain stem self-stimulation attenuated by lesions of the medial forebrain bundle but not by lesions of the brain stem norepinephrine systems. Brain Res 101:251-272

Clavier RM, Fibiger HC, Phillips AG (1976) Evidence that self-stimulation of the region of the locus coeruleus in rats does not depend upon noradrenergic projections to telencephalon. Brain Res 113:71-81

Clavier RM, Gerfen CR, Henkelman DH (1980) The contribution of nigral efferents to substantia nigra self-stimulation. Neurosci Abstr 6:422

Cole BJ, Robbins TW (1987) Dissociable effects of cortical and hypothalamic noradrenaline depletion on the acquisition, performance, and extinction of aversive conditioning. Behavioral Neuroscience, 101:476-488

Conner JD (1968) Caudate unit responses to nigral stimuli: evidence for a possible nigro-neostriatal pathway. Science 160:899-900

Conner JD (1970) Caudate nucleus neurones: correlation of the effects of substantia nigra stimulation with iontophoretic dopamine. J Physiol (Lond) 208:691-703

Conrad LCA, Pfaff DW (1976a) Efferents from medial basal forebrain and hypothalamus in the rat. I. An autoradiographic study of the medial preoptic area. J Comp Neurol 167:185-220

Conrad LCA, Pfaff DW (1976b) Efferents from medial basal forebrain and hypothalamus in the rat. II. An autoradiographic study of the anterior hypothalamus. J Comp Neurol 167:221-262

Cook L, Kelleher RT (1963) Effects of drugs on behavior. A Rev. Pharmacol 3:205-222

Cook L, Sepinwall J (1975) Behavioral analysis of the effects and mechanisms of action of benzodiazepines. In: Costa E, Greengard P (Eds) Mechanism of Action of Benzodiazepines. Raven New York, pp 1-28

Cooper BR, Breese GR, Grant LD, Howard JL (1973) Effects of 6-hydroxydopamine treatments on active avoidance responding: evidence for involvement of brain dopamine. J Pharmacol Exp Ther 185:358-370

Cooper BR, Cott JM, Breese GR (1974) Effects of catecholamine depleting drugs and amphetamine on self-stimulation of the brain following various 6-hydroxydopamine treatments. Psychopharmacologia 37:235-248

Cooper BR, Konkol RJ, Breese GR (1978) Effects of catecholamine depleting drugs and d-amphetamine on self-stimulation of the substantia nigra and locus coeruleus. J Pharmacol Exp Ther 204:592-605

Corbett D, Wise RA (1980) Intracranial self-stimulation in relation to the ascending dopaminergic systems of the midbrain: a moveable electrode mapping study. Brain Res 185: 1-15

Corbett D, Skelton RW, Wise, RA (1977) Dorsal bundle lesions fail to disrupt self-stimulation from the region of locus coeruleus. Brain Res 133:37-44

Coury JN (1967) Neural correlates of food and water intake in the rat. Science 156: 1763-1764

Coyle JT, Molliver ME (1977) Major innervation of newborn rat cortex by monoaminergic neurons. Science 196:444-447

Creese I, Iversen SD (1973) Blockade of amphetamine induced motor stimulation and stereotypy in the adult rat following neonatal treatment with 6-hydroxydopamine. Brain Res 55:369-382

Creese I, Iversen SD (1974) The role of forebrain dopamine systems in amphetamine induced stereotyped behavior in the rat. Psychopharmacologia 39:345-357

Crow TJ (1968) Cortical synapses and reinforcement: A hypothesis. Nature 219:736-737

Crow TJ, Arbuthnott GW (1972) Function of catecholamine-containing neurones in mammalian central nervous system. Nature 238:245-246

Crow TJ, Wendlandt S (1976) Impaired acquisition of a passive avoidance response after lesions induced in the locus coeruleus by 6-OH dopamine. Nature 259:42-44

Crow TJ, Deakin JFW, File SE, Longdon A, Wendlandt S (1978) The locus coeruleus noradrenergic system—evidence against a role in attention, habituation, anxiety and motor activity. Brain Res 155:249-261

Crowley WR, Feder HH, Morin LP (1976) Role of monoamines in sexual behavior of the female guinea pig. Pharmacol Biochem Behav 4:67-71

Crowley WR, Rodriguez-Sierra JF, Komisaruk BR (1977) Monoamine mediation of the antinociceptive effect of vaginal stimulation in rats. Brain Res 137:67-84

Dahlström A, Fuxe K (1964) Evidence for the existence of monoamine-containing neurons in the central nervous system. I. Demonstration of monoamines in the cell bodies of brain stem neurons. Acta Physiol Scand Suppl 62 (232):1-55

Davis JR, Keesey RE (1971) Norepinephrine-induced eating—its hypothalamic locus and an alternative interpretation of action. J Comp Physiol 77:394-402

Delacour J, Echavarria MJ, Senault B, Houcine O (1977) Specificity of avoidance deficits produced by 6-hydroxydopamine lesions of the nigrostriatal system of the rat. J Comp Physiol Psychol 91:875-885

Descarries L, Lapierre Y (1973) Noradrenergic axon terminals in the cerebral cortex of rat. I. Radioautographic visualization after topical application of DL-(^{3}H)-norepinephrine. Brain Res 51:141-160

Descarries L, Watkins KC, Lapierre Y (1977) Noradrenergic axon terminals in the cerebral cortex of rat. III. Topometric ultrastructural analysis. Brain Res 133:197-222

Dismukes RK (1979) New concepts of molecular communication among neurons. The Behav and Brain Sci 2:409-448

Dray A, Gonye TJ, Oakley NR (1976) Caudate stimulation and substantia nigra activity in the rat. J Physiol (London) 259:825-849

Dresse A (1966) Influence de 15 neuroleptiques (butyrophenones et phenothiazines) sur les variations de la teneur du cerveau en noradrenaline et activite du rat dans le test d'autostimulation. Arch int Pharmacodyn 159:353-365

Dun JN, Kaibara K, Karczmar AG (1977) Dopamine and adenosine 3',5'-monophosphate responses of single mammalian sympathetic neurons. Science 197:778-788

Eccles JC, Ito M, Szentagothai J (1967) The Cerebellum as a Neuronal Machine. Springer, New York

Echavarria-Mage MT, Senault B, Delacour J (1972) Effets de microinjections de 6-hydroxydopamine dans le système nigro-strié sur un aprentissage chez la rat blanc. Comptes-Rendus de l'Academie des Sciences 275:1155-1158

Edwards SB (1975) Autoradiographic studies of the projections of the midbrain reticular formation: descending projections of nucleus cuneiformis. J Comp Neurol 162:341-358

Ettenberg A, Cinsavich SA, White N (1979) Performance effects with repeated-response measures during pimozide-produced dopamine receptor blockade. Pharmacol Biochem Behav 11:557-561

Ettenberg A, Koob GF, Bloom FE (1981) Response artifact in the measurement of neuroleptic-induced anhedonia. Science 213:357-359

Everitt BJ, Fuxe K, Hökfelt T, Jonsson G (1975) Role of monoamines in the control by hormones of sexual receptivity in the female rat. J Comp Physiol Psychol 89:556-572

Everitt BJ, Robbins TW, Gaskin M, Fray PJ (1983) The effects of lesions to ascending noradrenergic neurons on discrimination learning and performance. Neuroscience 10:397-410

Fadda F, Argiolas A, Melis ME, Tissary AM, Onali PL, Gessa GL (1978) Stress-induced increase in 3,4-dihydroxyphenylacetic acid (DOPAC) levels in the cerebral cortex and in N. accumbens: reversal by diazepam. Life Sci 23:2219-2224

Faiers AA, Mogenson GJ (1976) Electrophysiological identification of neurons in locus coeruleus. Exptl Neurol 53:254-266

Fallon JH, Moore RY (1976a) Catecholamine neurons innervation of the rat amygdala. Anat Rec 184:399

Fallon JH, Moore RY (1976b) Dopamine innervation of some basal forebrain areas in the rat. Neurosci Abstr 2:486

Feltz P (1971) Sensitivity to haloperidol of caudate neurones excited by nigral stimulation. Eur J Pharmacol 14:360-364
Feltz P, de Champlain J (1972) Enhanced sensitivity of caudate neurons to microiontophoretic injections of dopamine in 6-hydroxydopamine treated cats. Brain Res 43:601-605
Fibiger HC (1978) Drugs and reinforcement mechanisms: a critical review of the catecholamine theory. Ann Rev Pharmacol 18:37-56
Fibiger HC, Fibiger HP, Zis AP (1973a) Attenuation of amphetamine induced motor stimulation and stereotypy by 6-hydroxydopamine in the rat. Brit J Pharmacol 4:683-692
Fibiger HC, Zis AP, Mcgeer EG (1973b) Feeding and drinking deficits after 6-hydroxydopamine administration in the rat: similarities to the lateral hypothalamic syndrome. Brain Res 55:135-148
Fibiger HC, Phillips AG, Zis AP (1974) Deficits in instrumental responding after 6-hydroxydopamine lesions of the nigro-neostriatal dopaminergic projection. Pharmacol Biochem Behav 2:87-96
Fibiger HC, Zis AP, Phillips AG (1975) Haloperidol-induced disruption of conditioned avoidance responding: attenuation by prior training or by anticholinergic drugs. Eur J Pharmacol 30:309-314
File SE, Deakin JFW, Longden A, Crow TJ (1979) An investigation of the role of the locus coeruleus in anxiety and agonistic behavior. Brain Res 169:411-420
Fink JS, Smith GP (1979) Decreased locomotor and investigatory exploration after denervation of catecholamine terminal fields in the forebrain of rats. J Comp Physiol Psychol 43:34-65
Flicker C, McCarley RW, Hobson JA (1981) Aminergic neurons: state control and plasticity in three model systems. J Cell Molec Neurobiol 1:123-166
Fog R, Pakkenberg H (1971) Behavioral effects of dopamine and p-hydroxyamphetamine injected into corpus striatum of rats. Exper Neurol 31: 75-86
Fog RL, Randrup A, Pakkenberg H (1967) Aminergic mechanisms in corpus striatum and amphetamine-induced stereotyped behavior. Psychopharm 11:179-183
Fog RL, Randrup A, Pakkenberg H (1968) Neuroleptic action of quaternary chlorpromazine and related drugs injected into various brain areas in rats. Psychopharm 12:428-432
Fog RL, Randrup A, Pakkenberg H (1970) Lesions in corpus striatum and cortex of rat brain and the effect on pharmacologically induced stereotyped, aggressive and cataleptic behavior. Psychopharm 18:346-350
Foote SL, Freedman R, Oliver AP (1975) Effects of putative neurotransmitters on neuronal activity in monkey auditory cortex. Brain Res 86:229-242
Foote SL, Aston-Jones G, Bloom FE (1980) Impulse activity of locus coeruleus neurons in awake rats and monkeys is a function of sensory stimulation and arousal. Proc Natl Acad Sci USA 77:3033-3037
Foote SL, Bloom FE, Aston-Jones G (1983) The nucleus locus coeruleus: new evidence of anatomical and physiological specificity. Physiol Reviews, 63:844-914
Fouriezos G, Wise RA (1976) Pimozide-induced extinction of intracranial self-stimulation: response patterns rule out motor or performance deficits. Brain Res 103:377-380
Franklin KBJ, McCoy SN (1979) Pimozide-induced extinction in rats; stimulus control of responding rules out motor deficit. Pharmacol Biochem Behav 11:71-75
Freedman R, Hoffer BJ (1975) Phenothiazine antagonism of the noradrenergic inhibition of cerebellar Purkinje neurons. J Neurobiol 6:277-288
Freedman RJ, Hoffer BJ, Woodward DJ (1975) A quantitative microiontophoretic an-

alysis of the responses of central neurones to noradrenaline interactions with cobalt, manganese, verapamil and dichloroisoprenaline. J Pharmacol 54:529–539

Freedman R, Hoffer BJ, Woodward DJ, Puro D (1977) A functional role for the adrenergic input to the cerebellar cortex: interaction of norepinephrine with activity evoked by mossy and climbing fibers. Exp Neurol 55:269–288

Frigyesi TL, Purpura DP (1967) Electrophysiological analysis of reciprocal caudate-nigral relations. Brain Res 6:440–456

Fuxe K, Hökfelt T, Nilsson O (1964) Observations on the localization of dopamine in the caudate nucleus of the rat. Z Zellforsch Mikrosk Anat 63:701–706

Gahwiler BH (1976) Inhibitory action of noradrenaline and cyclic AMP in explants of rat cerebellum. Nature 259:483–484

Galey D, Jaffard R, Le Moal M (1976) Spontaneous alternation disturbance after lesions of the ventral mesencephalic tegmentum in the rat. Neurosci Lett 3:65–69

Gallagher JP, Shinnick-Gallagher P (1977) Cyclic nucleotides injected intracellularly into rat superior cervical ganglion cells. Science 198:851–852

Geller HM, Hoffer BJ (1977) Effect of calcium removal on monoamine-elicited depressions of cultured tuberal neurons. J Neurobiol 8:43–55

Geller I, Seifter J (1960) The effects of meprobamate, barbiturates, d-amphetamine and promazine on experimentally induced conflict in the rat. Psychopharmacologia 1:482–492

Gerfen CR, Clavier RM, Henkelman DH (1981) Intracrania self-stimulation from the sulcal prefrontal cortex in the rat: the effect of 6-hydroxydopamine or kainic acid lesion at the site of stimulation. Brain Research 774:791–304

German DC, Bowden DM (1974) Catecholamine systems as the neural substrate for intracranial self-stimulation: a hypthesis. Brain Res 73:381–419

German DC, Fetz EE (1976) Responses of primate locus coeruleus and subcoeruleus neurons to stimulation at reinforcing brain sites and to natural reinforcers. Brain Res 109:497–514

Glowinski J (1979) Some properties of the ascending dopaminergic pathways: interactions of the nigrostriatal dopaminergic system with other neuronal pathways. In Schmitt FO, Worden FG (Eds) The Neurosciences Fourth Study Program. Cambridge MIT Press, pp 1069–1083

Glowinski J, Axelrod J, Iversen LL (1966) Regional studies of catecholamines in the rat brain. Effects of drugs on the disposition and metabolism of ^{3}H-norepinephrine and ^{3}H-dopamine. J Pharmacol Exp Ther 153:30–41

Gold RM, Jones AP, Sawchenko PE, Kapatos G (1977) Paraventricular area: critical focus of a longitudinal neurocircuitry mediating food intake. Physiol Behav 18:1111–1119

Gold MS, Pottash ALC, Extein I (1982) Clonidine: inpatient studies from 1978–1981. J Clinical Psychiat 43:35–38

Gonzalez-Vegas JA (1974) Antagonism of dopamine-mediated inhibition in the nigro striatal pathway: modes of action of some catatonic-inducing drugs. Brain Res 80:219–228

Gordon R, Spector S, Sjoerdsma A, Udenfriend S (1966) Increased synthesis of norepinephrine and epinephrine in the intact rat during exercise and exposure to cold J Pharmacol Exp Ther 153:440–447

Graham AW, Aghajanian GK (1971) Effects of amphetamine on single cell activity in a catecholamine nucleus, the locus coeruleus. Nature 234:100–102

Gray T, Wise RA (1980) Effect of pimozide on lever pressing behavior maintained on an intermittent reinforcement schedule. Pharmacol Biochem Behav 12:931–935

Greengard P (1978) Cyclic Nucleotides, Phosphorylated Proteins, and Neuronal Function. Raven: New York

Grillo MA (1966) Electron microscopy of sympathetic tissues. Pharmacol Rev 18:387-400
Grossman SP (1960) Eating or drinking elicited by direct adrenergic or cholinergic stimulation of hypothalamus. Science 132:301-302
Grossman SP (1962 a) Direct adrenergic and cholinergic stimulation of hypothalamic mechanisms. Am J Physiol 202:872-882
Grossman SP (1962 b) Effects of adrenergic and cholinergic blocking agents on hypothalamic mechanisms. Am J Physiol 202:1230-1236
Groves PM (1983) A theory of the functional organization of the neostriatum and the neostriatal control of voluntary movement. Brain Res Reviews 5:109-132
Groves PM, Wilson CJ (1980) Monoaminergic presynaptic axons and dendrites in rat locus coeruleus seen in reconstructions of serial sections. J Comp Neurol 193:853-862
Grzanna R, Morrison JH, Coyle JT, Molliver ME (1977) The immunohistochemical demonstration of noradrenergic neurons in the rat brain: the use of homologous antiserum to dopamine-β-hydroxylase. Neurosci Letters 4:127-134
Guyenet PG, Aghajanian GK (1977) Excitation of neurons in the nucleus locus coeruleus by substance P and related peptides. Brain Res 136:178-184
Hansen S, Stanfield EJ, Everitt BJ (1980) The role of ventral bundle noradrenergic neurones in sensory components of sexual behavior and coitus-induced pseudopregnancy. Nature 286:152-154
Hansen S, Stanfield EJ, Everitt BJ (1981) The effects of lesions of lateral tegmental noradrenergic neurons on components of sexual behavior and pseudopregnancy in female rats. Neurosci 6:1105-1117
Hartline HK, Ratliff F (1957) Inhibitory interaction of receptor units in the eye of Limulus. J Gen Physiol 40:357-376
Hattori T, Fibiger HC, McGeer PL, Maler L (1973) Analysis of the fine structure of the dopaminergic nigrostriatal projection by electron microscopic autoradiography. Exp Neurol 41:599-611
Hernandez L, Hoebel BG (1982) Overeating after midbrain 6-hydroxydopamine: prevention by central injection of catecholamine reuptake blockers. Brain Res 245:333-343
Herz A (1960) Drugs and the conditioned avoidance response. Int Rev Neurobiol 2:229-277
Herz A, Zieglgänsberger W (1968) The influence of microelectrophoretically applied biogenic amines, cholinomimetics and procaine on synaptic excitation in the corpus striatum. Int J Neuropharmacol 7:221-230
Hetherington AW, Ranson SW (1940) Hypothalamic lesion and adiposity in the rat. Anat Rec 78: 149-172
Hoebel BG, Leibowitz SF (1981) Brain monoamines in the modulation of self-stimulation feeding and body weight. In Weiner H, Hofer MA, Stunkard AJ (Eds) Brain Behavior and Bodily Disease. Raven Press, New York, pp 103-142
Hoffer BJ, Siggins GR, Bloom FE (1969) Prostaglandins E1 E2 antagonize norepinephrine effects on cerebellar Purkinje cells: microelectrophoretic study. Science 166:1418-1420
Hoffer BJ, Siggins GR, Bloom FE (1971 a) Studies on norepinephrine-containing afferents to Purkinje cells of rat cerebellum. II. Sensitivity of Purkinje cells to norepinephrine and related substances administered by microiontophoresis. Brain Res 25:523-534
Hoffer BJ, Siggins GR, Woodward DJ, Bloom FE (1971 b) Spontaneous discharge of Purkinje neurons after destruction of catecholamine-containing afferents by 6-hydroxydopamine. Brain Res 30: 425-430

Hoffer BJ, Siggins GR, Oliver AP, Bloom FE (1973) Activation of the pathway from locus coeruleus to rat cerebellar Purkinje neurons: pharmacological evidence of noradrenergic central inhibition. J Pharmacol Exp Ther 184:553-569

Hoffer B, Olson L, Seiger A, Bloom F (1975) Formation of a functional adrenergic input to intra-ocular cerebellar grafts: ingrowth of inhibitory sympathetic fibers. J Neurobiol 6:565-585

Hökfelt T, (1968) In vitro studies on central and peripheral monoamine neurons at the ultrastructural level. Z Zellforsch 91:1-74

Hökfelt T, Ungerstedt U (1969) Electron and fluorescence microscopical studies on the nucleus caudatus putamen of the rat after unilateral lesions of nigro-neostriatal dopamine neurons. Acta Physiol Scand 76:415-426

Hökfelt T, Fuxe K, Goldstein M, Johansson O (1974) Immunohistochemical evidence for the existence of adrenaline neurons in the rat brain. Brain Res 66:235-251

Hökfelt T, Elde R, Johansson O, Ljungdahl A, Schultzberg N, Fuxe K, Goldstein M, Nilsson G, Pernow B, Terenius L, Ganten D, Jeffcote FL, Rehfeld J, Faid S (1978) The distribution of peptide containing neurons in the CNS. In: Lipton MA, Killam KF, Dimasio A (Eds) Psychopharmacology—A Generation of Progress, Raven Press, New York, pp 39-66

Hull CD, Levine MS, Buchwald NA, Heller A, Browning RA (1974) The spontaneous firing pattern of forebrain neurons. I. The effects of dopamine and non-dopamine depleting lesions on caudate unit firing patterns. Brain Res 73:241-262

Ibata Y, Nojyo Y, Matsuura T, Sano Y (1973) Nigro-neostriatal projection. Z Zellforsch Mikrosk Anat 138:333-344

Iversen LL (1975) Dopamine receptors in the brain. Science 188:1084-1089

Iversen SD, Koob GF (1977) Behavioral implications of dopaminergic neurons in the mesolimbic system. In E Costa, GL Gessa (Eds) Nonstriatal Dopamine Mechanisms, Adv Biochem Psychopharmacol, Raven, New York

Janssen PAJ, Niemegeers CJE, Schellekens KHL, Dresse A, Lenaerts FM, Pinchard A, Schaper WKA, van Nueten JM, Verbruggen FJ (1968) Pimozide, a chemically novel, highly potent and orally long-acting neuroleptic drug. Arzneimittel-Forschung 18:261-279

Jones BE, Moore RY (1974) Catecholamine-containing neurons of the nucleus locus coeruleus in the cat. J Comp Neur 157:43-52

Joyce EM, Koob GF (1981) Amphetamine-, scopolamine- and caffeine-induced locomotor activity following 6-hydroxydopamine lesions of the mesolimbic dopamine system. Psychopharmacol 73:311-313

Joyce EM, Iversen SD (1978) The effect of 6-hydroxydopamine lesions to mesolimbic dopamine terminals on spontaneous behaviour in the rat. Neuroscience Letters Suppl 2:289

Kandel ER, Klein M, Bailey CH, Hawkins RD, Castellucci VF, Lubit BW, Schwartz JH (1981) Serotonin, cyclic AMP, and the modulation of the calcium current during behavioral arousal. In: Jacobs BL, Gelperin A (Eds) Serotonin Neurotransmission and Behavior. Cambridge, MIT Press pp 211-254

Kebabian JW, Calne CB (1979) Multiple receptors for dopamine. Nature 277:93-96

Kelly PH, Iversen SD (1976) Selective 6-OHDA induced destruction of mesolimbic dopamine neurons: abolition of psychostimulant induced locomotor activity in rats. Eur J Pharmacol 40:45-56

Kelly PH, Moore KE (1976) Mesolimbic dopaminergic neurons in the rotational model of nigrostriatal function. Nature 263:695-696

Kelly PH, Seviour P, Iversen SD (1975) Amphetamine and apomorphine responses in the rat following 6-OHDA lesions of the nucleus accumbens septi and corpus striatum. Brain Res 94:507-522

Kety SS (1970) The biogenic amines in the central nervous system: their possible roles in arousal, emotion and learning. In: Schmitt FO (Ed) The Neurosciences: Second Study Program. Rockefeller University Press, New York pp 324-336

Kitai ST, Wagner A, Precht W, Ohno T (1975) Nigro-caudate and caudato-nigral relationship: an electrophysiological study. Brain Res 85:44-48

Kitai ST, Sugimori M, Kocsis JC (1976) Excitatory nature of dopamine in the nigrocaudate pathway. Exp Brain Res 24:351-363

Klein M, Kandel ER (1978) Presynaptic modulation of voltage-dependent Ca^{++} current: mechanism for behavioral sensitization in *Aplysia californica*. Proc Natl Acad Sci USA 75:3512-3516

Kobayashi H, Hashiguchi T, Ushiyama NS (1978) Postsynaptic modulation of excitatory process in sympathetic ganglia by cyclic AMP. Nature 271:268-270

Koda LY, Bloom FE (1977) A light and electron microscopic study of noradrenergic terminals in the rat dentate gyrus. Brain Res 120:327-335

Koda LY, Wise RA, Bloom FE (1978a) Light and electron microscopic changes in the rat dentate gyrus after lesions or stimulation of the ascending locus coeruleus pathway. Brain Res 144:363-368

Koda LY, Schulman JA, Bloom FE (1978b) Ultrastructural identification of noradrenergic terminals in the rat hippocampus: unilateral destruction of the locus coeruleus with 6-hydroxydopamine. Brain Res 145:190-195

Koda LY, Aston-Jones G, Bloom FE (1980) Small granular vesicles in the locus coeruleus may indicate dendritic release of norepinephrine. Soc Neurosci Abstr 6:446

Koob GF, Annau Z (1974) Behavioral and neurochemical alterations induced by hypoxia in rats. Am J Physiol 227:73-78

Koob GF, Balcom GJ, Meyerhoff JL (1976) Increases in intracranial self-stimulation in the posterior hypothalamus following unilateral lesions in the locus coeruleus. Brain Res 101:554-560

Koob GF, Riley SJ, Smith SC, Robbins TW (1978a) Effects of 6-hydroxydopamine of the nucleus accumbens septi and olfactory tubercle on feeding, locomotor activity and amphetamine anorexia in the rat. J Comp Physiol Psychol 92:917-927

Koob GF, Fray PJ, Iversen SD (1978b) Self-stimulation at the lateral hypothalamus and locus coeruleus after specific unilateral lesions of the dopamine system. Brain Res 146:123-140

Koob GF, Kelley AE, Mason ST (1978c) Locus coeruleus lesions: learning and extinction. Physiol Behav 20:709-716

Koob GF, Stinus L, Le Moal M (1981) Hyperactivity and hypoactivity produced by lesions to the mesolimbic dopamine system. Behavioural Brain Res 3:341-359

Koob GF, Thatcher-Britton K, Britton D, Roberts DCS, Bloom FE (1984a) Destruction of the locus coeruleus or the dorsal noradrenergic bundle does not alter the release of punished responding by ethanol and chlordiazepoxide. Physiol Behav 33:479-485

Koob GF, Simon H, Herman JP, Le Moal M (1984b) Neuroleptic-like disruption of the conditioned avoidance response requires destruction of both the mesolimbic and nigro-striatal dopamine systems. Brain Res 303:319-329

Korf J, Aghajanian GK, Roth RH (1973) Increased turnover of norepinephrine in the rat cerebral cortex during stress: role of the locus coeruleus. Neuropharmacol 12:933-938

Korf J, Bunney BS, Aghajanian GK (1974) Noradrenergic neurons: morphine inhibition of spontaneous activity. Eur J Pharmacol 25:165-169

Landis SC, Bloom FE (1975) Ultrastructural identification of noradrenergic boutons in mutant and normal mouse cerebellar cortex. Brain Res 96:299-305

Lapierre Y, Beaudet A, Demianczuk N, Descarries L (1973) Noradrenergic axon terminals in the cerebral cortex of rat. II. Quantitative data revealed by light and electron microscope autoradiography of the frontal cortex. Brain Res 63:175-182

Lavielle S, Tassin JP, Theirry AM, Blanc G, Herve D, Barthelemy C, Glowinski J (1979) Blockade by benzodiazepines of the selective high increase in DA turnover induced by stress in mesocortical dopaminergic neurons of the rat. Brain Res 168:585-594

Leibowitz SF (1970) Reciprocal hunger-regulating circuits involving alpha- and beta-adrenergic receptors located, respectively, in the ventromedial and lateral hypothalamus. Proc Natl Acad Sci USA 67:1063-1070

Leibowitz SF (1978a) Paraventricular nucleus: A primary site mediating adrenergic stimulation of feeding and drinking. Pharmacol Biochem Behav 8:163-175

Leibowitz SF (1978b) Adrenergic stimulation of the paraventricular nucleus and its effects on ingestive behavior as a function of drug rise and time injection in the light-dark cycle. Brain Res Bull 3:357-363

Leibowitz SF, Brown LL (1980) Histochemical and pharmacological analysis of catecholaminergic projections to the perifornical hypothalamus in relation to feeding inhibition. Brain Res 201:289-314

Le Moal M, Cardo B, Stinus L (1969) Influence of ventral mesencephalic lesions on various spontaneous and conditioned behaviors in the rat. Physiol Behav 4:567-572

Le Moal M, Galey D, Cardo B (1975) Behavioral effects of local injections of 6-hydroxydopamine in the medial tegmentum in the rat. Possible role of the mesolimbic dopaminergic system. Brain Res 88:190-194

Le Moal M, Stinus L, Simon H, Tassin JP, Thierry AM, Blanc G, Glowinski J, Cardo B (1977) Behavioral effects of a lesion in the ventral mesencephalic tegmentum: evidence for involvement of A10 dopaminergic neurons. In: Costa E, Gessa GL (Eds) Nonstriatal Dopaminergic Neurons. Adv Biochem Psychopharmacol 16:237-245

Levitan IB, Norman J (1980) Different effects of cAMP and cGMP derivatives on the activity of an identified neuron: biochemical and electrophysiological analysis. Brain Res 187:415-429

Liebman JM, Butcher LL (1974) Comparative involvement of dopamine and noradrenaline in rate-free self-stimulation in substantia nigra, lateral hypothalamus and mesencephalic central gray. Naunyn-Schmiedeberg's Arch Pharmacol 284:167-194

Lindvall O (1975) Mesencephalic dopaminergic afferents to the lateral septal nucleus of the rat. Brain Res 87:89-95

Lindvall O, Björklund A (1978) Organization of catecholamine neurons in the rat central nervous system. In Iversen L, Iversen S, Snyder SH (Eds) Handbook of Psychopharmacology ol 9. Plenum Press, New York

Lindvall O, Björklund A, Moore RY, Steveni U (1974a) Mesencephalic dopamine neurons projecting to neocortex. Brain Res 81:325-331

Lindvall O, Björklund A, Nobin A, Stenevi U (1974b) The adrenergic innervation of the rat thalamus as revealed by the glyoxylic acid fluorescence method. J Comp Neurol 154:317-348

Lippa AS, Antelman SM, Fisher AE, Canfield DR (1973) Neurochemical mediation of reward: a significant role for dopamine? Pharmacol Biochem Behav 1:23-28

Loughlin SE, Foote SL, Bloom FE (1986a) Efferent projections of nucleus locus coeruleus: topographic organization of cells of origin demonstrated by three-dimensional reconstruction. Neuroscience 18:291-306

Loughlin SE, Foote SL, Grzanna R (1986b) Efferent projections of nucleus locus coer-

uleus: Morphologic subpopulations have different efferent targets. Neuroscience 18:307-319
Lyness WH, Friedle NM, Moore KE (1979) Destruction of dopaminergic nerve terminals in nucleus accumbens: effect on d-amphetamine, self-administration. Pharmacol Biochem Behav 11:553-556
Magistretti PJ, Morrison JH, Shoemaker WJ, Sapin V, Bloom FE (1981) Vasoactive intestinal polypeptide induces glycogenolysis in mouse cortical slices: a possible regulatory mechanism for the local control of energy metabolism. Proc Natl Acad Sci USA 78:6535-6539
Margules DL (1969) Noradrenergic synapses for the suppression of feeding behavior. Life Sci 8:693-704
Margules DL (1970) Alpha-adrenergic receptors in hypothalamus for the suppression of feeding behavior by satiety. J Comp Physiol Psychol 73:1-12
Marshall KC, Engberg I (1979) Reversal potential for noradrenaline-induced hyperpolarization of spinal motoneurons. Science 205:422-424
Marshall, JF, Teitelbaum PA (1973) Comparison of the eating in response to hypothermic and glucoprivic challenges after nigral 6-hydroxydopamine and lateral hypothalamic electrolytic lesions in rats. Brain Res 55:229-233
Marshall JF, Richardson JS, Teitelbaum P (1974) Nigrostriatal bundle damage and the lateral hypothalamic syndrome. J Comp Physiol Psychol 87:808-830
Mason ST (1980) Noradrenaline and selective attention: a review of the model and the evidence. Life Sci 27:627-631
Mason ST, Fibiger HC (1978) Evidence for a role of brain noradrenaline in attention and stimulus sampling. Brain Res 159:421-426
Mason ST, Fibiger HC (1979) Regional topography within noradrenergic locus coeruleus as revealed by retrograde transport of horseradish peroxidase. J Comp Neurol 187:703-724
Mason ST, Iversen SD (1975) Learning in the absence of forebrain noradrenaline. Nature 258:422-424
Mason ST, Iversen SD (1979) Theories of the dorsal bundle extinction effect. Brain Res Dev 1:107-137
Mason ST, Roberts DCS, Fibiger HC (1978) Noradrenaline and neophobia. Physiol Behav 21:353-361
Mason ST, Beninger RJ, Fibiger HC, Phillips AG (1980) Pimozide-induced suppression of responding: evidence against a block of food reward. Pharmac Biochem Behav 12:917-923
Matthews JW, Booth DA, Stolerman IP (1978) Factors influencing feeding elicited by intracranial noradrenaline in rats. Brain Res 141:119-128
Maynert EW, Levi R (1964) Stress-induced release of brain norepinephrine and its inhibition by drugs. J Pharmacol Exp Ther 143:90-95
McAfee DA, Schorderet M, Greengard P (1971) Adenosine 3′,5′-monophosphate in nervous tissue: increase associated with synaptic transmission. Science 171:1156-1158
McCarley RW, Hobson JA (1975) Neuronal excitability modulation over the sleep cycle: a structural and mathematical model. Science 189:58-60
McGeer EG, Hattori T, McGeer PL (1975) Electron microscopic localization of labeled norepinephrine transported in nigrostriatal neurons. Brain Res 86:478-482
McLennan H, York DH (1967) The action of dopamine on neurons of the caudate nucleus. J Physiol (London) 189:393-402
Miller RE, Murphy JV, Minsky IA (1957) The effect of chlorpromazine on fear-motivated behavior in rats. J Pharmacol Exp Ther 120:379-387

Miller NE, Gottesman KS, Emery N (1964) Dose response to carbachol and norepinephrine in rat hypothalamus. Am J Physiol 206:1384-1388

Mitchell MJ, Wright AK, Arbuthnott GW (1981) The role of dopamine in pontine intracranial self-stimulation: a re-examination of the problem. Neurosci Letters 26:169-175

Mittleman G, Valenstein ES (1982) Mesostriatal dopamine systems and eating and drinking evoked by hypothalamic stimulation: differences between the dominant and non-dominant hemispheres. Neurosci Abstr 8:894

Moises HC, Woodward DJ (1980) Potention of GABA inhibitory action in cerebellum by locus coeruleus stimulation. Brain Res 182:327-344

Moises HC, Waterhouse BD, Woodward DJ (1981) Locus coeruleus stimulation potentiates Purkinje cell responses to afferent input: the climbing fiber system. Brain Res 222:43-64

Møller-Nielsen I, Pedersen V, Nymark M, Franck KF, Boeck V, Fjalland B, Christsen AV (1973) The comparative pharmacology of flupenthixol and some reference neuroleptics. Acta Pharmacol et Toxicol 33:353-362

Moore RY, Bloom FE (1978) Central catecholamine neuron systems: anatomy and physiology of the dopamine systems. Ann Rev Neurosci 1:129-169

Moore RY, Bloom FE (1979) Central catecholamine neuron systems: anatomy and physiology of the norepinephrine and epinephrine systems. Ann Rev Neurosci 2:113-168

Moore RY, Bhatnagar RK, Heller A (1971) Anatomical and chemical studies of a nigro-neostriatal projection in the cat. Brain Res 30:119-135

Morrison JH, Foote SL (1986) Noradrenergic and serotoninergic innervation of cortical, thalamic, and tectal visual structures in old and new world monkeys. J Comp Neurol 243:117-138

Morrison JH, Grzanna R, Molliver ME, Coyle JT (1978) The distribution and orientation of noradrenergic fibers in neocortex of the rat: an immunofluorescence study. J Comp Neur 181:17-40

Morrison JH, Foote SL, Molliver ME, Bloom FE, Lidov HGW (1982a) Noradrenergic and serotonergic fibers innervate complementary layers in monkey primary visual cortex: an immunohistochemical study. Proc Natl Acad Sci USA 79:2401-2405

Morrison JH, Foote SL, O'Connor D, Bloom FE (1982b) Laminar, tangential and regional organization of the noradrenergic innervation of monkey cortex: dopamine-β-hydroxylase immunohistochemistry. Brain Res Bull 9:309-319

Nakamura S, Iwama K (1975) Antidromic activation of the rat locus coeruleus neurons from hippocampus, cerebral and cerebellar cortices. Brain Res 99:372-376

Nathanson JA (1977) Cyclic nucleotides and nervous system function. Physiol Rev 57:157-256

Naylor RJ, Olley JE (1972) Modification of the behavioral changes induced by amphetamine in the rat by lesions in the caudate nucleus, the caudate-putamen and globus pallidus. Neuropharmacol 11:91-99

Neill DB, Bogan WO, Grossman SP (1974) Impairment of avoidance performance by intrastriatal administration of 6-OHDA. Pharmacol Biochem Behav 2:97-103

Nieoullon A, Cheramy A, Glowinski J (1977) Release of dopamine from terminals and dendrites of the two nigrostriatal dopaminergic pathways in response to unilateral sensory stimuli in the cat. Nature 269:340-341

Nishi S, Soeda H, Koketsu K (1965) Studies in sympathetic B and C neurons and patterns of preganglionic innervation. J Cell Comp Physiol 66:19-32

Nybäck HV, Walters JR, Aghajanian GK, Roth RH (1975) Tricyclic antidepressants: effects on the firing rate of brain noradrenergic neurons. Eur J Pharmacol 32:302-312

Obata K, Yoshida M (1973) Caudate-evoked inhibition and actions of GABA and other substances on cat pallidal neurons. Brain Res 64:455-459

Oke AF, Adams RN (1978) Selective attention dysfunctions in adult rats neonatally treated with 6-hydroxydopamine. Pharmacol Biochem Behav 9:429-432

Oliver AP, Segal M (1974) Transmembrane changes in hippocampal neurons: hyperpolarizing actions of norepinephrine, cyclic AMP, and locus coeruleus. Soc Neurosci Abstr 4:361

Olschowska JA, Grzanna R, Molliver ME (1980) The distribution and incidence of synaptic contacts of noradrenergic varicosities in the rat neocortex: an immunocytochemical study. Soc Neurosci Abstr 6:352

Olschowska JA, Molliver ME, Grzanna R, Rice FL, Coyle JT (1981) Ultrastructural demonstration of noradrenergic synapses in the rat central nervous system by dopamine-β-hydroxylase immunocytochemistry. J Histochem Cytochem 29:271-280

Olson L, Seiger A (1972) Early prenatal ontogeny of central monoamine neurons in the rat: fluorescence histochemical observations. Z Anat EntwGesch 137:301-316

Oltmas GA, Lorden JF, Margules DL (1977) Food intake and body weight: effects of specific and non-specific lesions in the midbrain path of the ascending noradrenergic neurons of the rat. Brain Res 128:293-308

Ornstein D, Huston JP (1975) Influence of 6-hydroxydopamine injections in the substantia nigra on lateral hypothalamic reinforcement. Neuroscience Lett 116:339-342

Palacios JM, Kuhar MJ (1980) Beta-adrenergic-receptor localization by light microscopic autoradiography. Science 208:1378-1380

Phillips AG, (1984) Brain reward circuitry: the case for separate systems. Brain Res Bull 12:195-201

Phillips AG, Fibiger HC (1978) The role of dopamine in maintaining intracranial self-stimulation in the ventral tegmentum, nucleus accumbens and medial prefrontal cortex. Canad J Psychol 32:58-66

Phillips AG, Brooke SM, Fibiger HC (1975) Effects of amphetamine isomers and neuroleptics on self-stimulation from the nucleus accumbens and dorsal noradrenergic bundle. Brain Res 85:13-22

Phillips AG, Carter DA, Fibiger HC (1976) Dopaminergic substrates of intracranial self-stimulation in the caudate-putamen. Brain Res 104:221-222

Phillips AG, van der Kooy D, Fibiger HC (1977) Maintenance of intracranial self-stimulation in hippocampus and olfactory bulb following depletion of noradrenaline. Neurosci Lett 4:77-84

Phillis JW, Lake N, Yarbrough G (1973) Calcium mediation of the inhibitory effects of biogenic amines on cerebral cortical neurones. Brain Res 53:465-469

Pickel VM, Joh TH, Reis DJ (1977) A serotonergic innervation of noradrenergic neurons in nucleus locus coeruleus: demonstration by immunocytochemical localization of the transmitter specific enzymes tyrosine and tryptophan hydroxylase. Brain Res 131:197-214

Pierce ET, Foote WE, Hobson JA (1976) The efferent connection of the nucleus raphe dorsalis. Brain Res 107:137-144

Pijnenburg AJJ, van Rossum JM (1973) Stimulation of locomotor activity following injection of dopamine into the nucleus accumbens. J Pharm Pharmacol 25:1003-1005

Pijnenburg AJJ, Woodruff GN, van Rossum JM (1973) Ergometrine induced locomotor activity following intracerebral injection into the nucleus accumbens. Brain Res 59:289-302

Pijnenburg AJJ, Honig WMM, van der Heyden JAM, van Rossum JM (1976) Effects of chemical stimulation of the mesolimbic dopamine system upon locomotor activity. Eur J Pharmacol 35:45-58

Pisa M, Fibiger HC (1983 a) Evidence against a role of the rat's dorsal noradrenergic bundle in selective attention and place memory. Brain Res 272:319-329

Pisa M, Fibiger HC (1983 b) Intact selective attention in rats with lesions of the dorsal noradrenergic bundle. Behavioral Neuroscience 97:519-529

Posluns D (1962) An analysis of chlorpromazine-induced suppression of the avoidance response. Psychopharmacologia 3:361-373

Price MTC, Fibiger HC (1975) Discriminated escape learning and response to electric shock after 6-hydroxydopamine lesions of the nigro-neostriatal dopaminergic projection. Pharmacol Biochem Behav 3:285-290

Rall TW (1972) Role of adenosine 3′,5′-monophosphate (cyclic AMP) in actions of catecholamines. Pharmacol Rev 24:399-409

Randrup A, Munkvad I (1960) Role of catecholamines in the amphetamine excitatory response. Nature 211:540

Ranje C, Ungerstedt U (1977) Lack of acquistion in dopamine denervated animals tested in our underwater Y-maze. Brain Res 134:95-111

Redmond DE (1977) Alterations in the function of the nucleus locus coeruleus: a possible model for studies of anxiety in aninal models. In: Usdin E, Hanin I (Eds) Psychiatry and Neurology. Pergamon, New York pp 293-305

Redmond DE, JR, Huang YH, Snyder DR, Maas JW (1976) Behavioral effects of stimulation of the locus coeruleus in the stump tail monkey (Macaca arctoides). Brain Res 116:502-510

Redmond DE, Huang YH (1979) New evidence for a locus coeruleus-norepinephrine connection with anxiety. Life Sci 25:2149-2162

Risner ME, Jones BE (1976) Role of noradrenergic and dopaminergic processes in amphetamine self-administration. Pharmacol Biochem Behav 5:477-482

Robbins TW, Everitt BJ (1982) Functional studies of the central catecholamines. In: International Review of Neurobiology, Vol 23, Academic Press, Inc, pp 303-365

Robbins TW, Koob GF (1980) Selective disruption of displacement behavioral by lesions of the mesolimbic dopamine system. Nature 285:409-412

Roberts DCS, Zis AP, Fibiger HC (1975) Ascending catecholamine pathways and amphetamine induced locomotor activity: importance of dopamine and apparent non-involvement of norepinephrine. Brain Res 93:441-454

Roberts DCS, Price MTC, Fibiger HC (1976) The dorsal tegmental noradrenergic projection: analysis of its role in maze learning. J Comp Physiol Psychol 90:363-372

Roberts DCS, Corcoran ME, Fibiger HC (1977) On the role of ascending catecholamine systems in intravenous self-administration of cocaine. Pharmacol Biochem Behav 6:615-620

Roberts DCS, Koob GF, Klonoff P, Fibiger HC (1980) Extinction and recovery of cocaine self-administration following 6-hydroxydopamine lesions of the nucleus accumbens. Pharmacol Biochem Behav 12:781-787

Robison GA, Butcher RW, Sutherland EW (1971) Cyclic AMP. Academic Press, New York London

Rogawski MA, Aghajanian GK (1980) Activation of lateral geniculate neurons by norepinephrine: mediation by an α-adrenergic receptor. Brain Res 182:345-359

Rolls ET, Kelly PH, Shaw SG (1974) Noradrenaline dopamine and brain stimulation reward. Pharmacol Biochem Behav 2:735-740

Sakai K (1980) Some anatomical and physiological properties of pontomesencephalic tegmental neurons with special reference to the PGO waves and postural atonia during paradoxical sleep in the cat. In: Hobson J, Brazier M (Eds) The Reticular System Revisited. Raven Press, New York, pp 427-448

Sakai K, Touret M, Salvert D, Leger L, Jouvet M (1977) Afferent projection to the cat locus coeruleus as visualized by the horseradish peroxidase technique. Brain Res 119:21-41

Saper CB, Swanson LW, Cowan WM (1976) The efferent connections of the ventromedial nucleus of the hypothalamus of the rat. J Comp Neur 169:409-442

Sasa M, Munekiyo K, Takaoir S (1974) Impairment by 6-hydroxydopamine of locus coeruleus-induced monosynaptic potential in the spinal trigeminal nucleus. Jap J Pharmacol 24:863-868

Sasa M, Igarashi S, Takaoir S (1977) Influence of the locus coeruleus on interneurons in the spinal trigeminal nucleus. Brain Res 125:369-375

Sastry BSR, Phillis JW (1977) Antagonism of biogenic amine-induced depression of cerebral cortical neurones by Na^+, K^+,-ATPase inhibitors. Can J Physiol Pharmacol 55:170-179

Schaefer GJ, Holtzman SG (1977) Dose and time-dependent effects of narcotic analgetics on intracranial self-stimulation in the rat. Psychopharmacol 53:227-234

Schaefer GT, Holtzman SG (1979) Free-operation and autotitration brain stimulation procedures in the rat. A comparison of drug effects. Pharmacol Biochem and Behavior 10:127-135

Scheel-Krüger J, Randrup A (1967) Stereotyped hyperactive behavior produced by dopamine in the absence of noradrenaline. Life Sci 6:1389-1398

Schlumpf M, Shoemaker WJ, Bloom FE (1980) Innervation of embryonic rat cerebral cortex by catecholamine-containing fibers. J Comp Neur 192:361-376

Schulman JA (1981) Organization and regulation of neuronal activity in the inferior olive and cerebellum. Ph D Thesis Univ Calif San Diego

Schulman JA, Weight FF (1976) Synaptic transmission: long-lasting potentiation by a postsynaptic mechanism. Science 194:1437-1439

Schultz W, Ruffieux A, Aebischer P (1983) The activity of pars compacta neurons of the monkey substantia nigra in relation to motor activation. Experimental Brain Res 51:377-387

Segal M (1976) Brain stem afferents to the rat medial septum. J Physiol (Lond) 261:617-631

Segal M (1980) The noradrenergic innervation of the hippocampus. In: Hobson JA, Brazier MAB (Eds) The Reticular Formation Revisited. pp 415-425

Segal M, Bloom FE (1974a) The action of norepinephrine in the rat hippocampus. I. Iontophoretic studies. Brain Res 72:79-97

Segal M, Bloom FE (1974b) The action of norepinephrine in the rat hippocampus. II. Activation of the input pathway. Brain Res 72:99-114

Segal M, Bloom FE (1976a) The action of norepinephrine on the rat hippocampus. III Hippocampal cellular responses to locus coeruleus stimulation in the awake rat. Brain Res 107:499-511

Segal M, Bloom FE (1976b) The action of norepinephrine on the rat hippocampus. IV. The effects of locus coeruleus stimulation on evoked hippocampal activity. Brain Res 107:513-525

Segal DS, Kelly PH, Koob G, Roberts DCS (1979) Nonstriatal dopamine mechanisms in the response to repeated d-amphetamine administration. Usdin E, Kopin IJ, Barchas J (Eds) In: Catecholamines: Basic and Clinical Frontiers. Pergamon Press, New York

Seiger A, Olson L (1973) Late prenatal ontogeny of central monoamine neurons in the rat: fluorescence histochemical observations. Z Anat Entwickl-Gesch 140:281-318

Sessions GR, Kant GJ, Koob GF (1976) Locus coeruleus lesions and learning in the rat. Physiol Behav 17:853-859

Sessions GR, Meyerhoff JL, Kant GJ, Koob GF (1980) Effects of lesions of the ventral medial tegmentum on locomotor activity, biogenic amines and response to amphetamine in rats. Pharmacol Biochem Behav 12:603-608

Shizgal P, Bielajew C, Kiss I (1980) Anodal hyperpolarization block technique provides evidence for rostro-caudal conduction of reward related signals in the medial forebrain bundle. Neurosci Abstr 6:422

Siggins GR (1977) The electrophysiological role of dopamine in striatum: excitatory or inhibitory? In: Lipton MA, Dimascio A, Killam KF (Eds) Psychopharmacology—a generation of progress. Raven Press, New York, pp 143-157

Siggins GR (1978) Electrophysiological role of dopamine in striatum: excitatory or inhibitory? In Lipton MA, DiMascio A, Killam KF (Eds) Psychopharmacology: A Generation of Progress. Raven Press, New York, pp 143-157

Siggins GR, Gruol DL (1986) Mechanisms of transmitter action in the vertebrate central nervous system In: Bloom FE (Ed) Handbook of Physiology. American Physiological Society, Bethesda, MD. USA, pp 1-114

Siggins GR, Henriksen SJ (1975) Inhibition of rat Purkinje neurons by analogues of cyclic adenosine monophosphate: correlation with protein kinase activation. Science-189:557-560

Siggins GR, Hoffer BJ, Bloom FE (1971a) Studies on norepinephrine-containing afferents to Purkinje cells of rat cerebellum. III. Evidence for mediation of norepinephrine effects by cyclic 3′,5′-monophosphate. Brain Res 25:535-553

Siggins GR, Hoffer BJ, Oliver AP, Bloom FE (1971b) Activation of a central noradrenergic projection to cerebellum. Nature 233:481-483

Siggins GR, Hoffer BJ, Bloom FE (1971c) Cyclic adenosine monophosphate and norepinephrine: effect on purkinje cells in rat cerebellar cortex. Science 174:1258-1259

Siggins GR, Oliver AP, Hoffer BJ, Bloom FE (1971d) Cyclic adenosine monophosphate and norepinephrine: effects on transmembrane properties of cerebellar Purkinje cells. Science 171:192-194

Siggins GR, Hoffer BJ, Ungerstedt U (1974) Electrophysiological evidence for involvement of cyclic adenosine monophosphate in dopamine responses of caudate neurons. Life Sci 16:779-792

Siggins GR, Hoffer BJ, Bloom FE, Ungerstedt U (1976a) In: Yahr MD (Ed) The Basal Ganglia. New York: Raven Press pp 227-248

Siggins GR, Henriksen SJ, Landis SC (1976b) Electrophysiology of Purkinje neurons in the Weaver mouse: iontophoresis of neurotransmitters and cyclic nucleotides, and stimulation of the nucleus locus coeruleus. Brain Res 114:53-70

Silver MA, Jacobowitz DM (1979) Specific uptake and retrograde flow of antibody to dopamine-β-hydroxylase by central nervous system noradrenergic neurons in vivo. Brain Res 167:65-75

Simon H (1980) Neurones dopaminergiques A10 et système frontal. J Physiol (Paris) 77:81-95

Simon H, Le Moal M, Cardo B (1976) Intracranial self-stimulation from the dorsal raphe nucleus of the rat: effects of the injection of para-chlorophenylalanine and of alpha-methylparatyrosine. Behav Biol 16:353

Simon H, Scatton B, Le Moal M (1980) Dopaminergic A10 neurones are involved in cognitive functions. Nature 286:150-151

Skultety FM, Gary TM (1962) Experimental hyperphagia in cats following destructive midbrain lesions. Neurology 12:394-401

Spehlmann R (1975) The effects of acetylcholine and dopamine on the caudate nucleus depleted of biogenic amines. Brain 98:219-230

Spencer HJ, Havlicek V (1974) Alterations by anesthetic agents of the responses of rat striatal neurons to iontophoretically applied amphetamine, acetylcholine, noradrenaline and dopamine. Can J Physiol Pharmacol 52:808-813

Stein L, Ray OS (1960) Brain stimulation reward "thresholds" self-determined in rat. Psychopharmacologia 1:251-256

Stein L, Wise CD (1971) Possible etiology of schizophrenia: progressive damage to the noradrenergic reward system by 6-hydroxydopamine. Science 171:1032-1036

Stein L, Wise, CD, Berger BD (1973) Antianxiety action of benzodiazepines: decrease in activity of serotonin neurons in the punishment system In: Garattini S, Mussini E, Randall LO (Eds) The Benzodiazepines. Raven Press, New York pp 29-44

Stein L, Belluzzi JD, Wise CD (1975) Memory enhancement by central administration of norepinephrine. Brain Res 84:329-335

Steinfels GF, Heym J, Strecker RE, Jacobs BL (1983a) Behavioral correlates of dopaminergic unit activity in freely moving cats. Brain Res 258:217-228

Steinfels GF, Heym J, Strecker RE, Jacobs BL (1983b) Response of dopaminergic neurons in cat to auditory stimuli presented across the sleep-waking cycle. Brain Res 277:150-154

Steriade M, Hobson JA (1976) Neuronal activity during the sleep-waking cycle. Prog Neurobiol 6:155-376

Stinus L, Thierry AM, Cardo B (1976) Effects of various inhibitors of tyrosine hydroxylase and dopamine beta-hydroxylase on rat self-stimulation after reserpine treatment. Psychopharmacologia 45:287-294

Stinus L, Gaffori O, Simon H, Le Moal M (1977) Small doses of apomorphine and chronic administration of d-amphetamine reduce locomotor hyperactivity produced by radiofrequency lesions of dopaminergic A10 neurons area. Biol Psych 12:719-732

Stinus L, Simon H, Le Moal M (1978) Disappearance of hoarding and disorganization of eating behavior after ventral mesencephalic tegmentum lesions in rats. J Comp Physiol Psychol 92:289-296

Stolk JM, Conner RL, Levine S, Barchas JD (1974) Brain norepinephrine metabolism and shock induced fighting behavior in rats: diffferential effects of shock and fighting on the neurochemical response to a common footshock stimulus J Pharmacol Exp Ther 190:193-209

Stone EA (1971) Hypothalamic norepinephrine after acute stress. Brain Res 35:260-263

Stone EA (1973) Accumulation and metabolism of norepinephrine in rat hypothalamus after exhaustive stress. J Neurochem 21:589-601

Stone EA (1975) Effect of stress on sulfated gylcol metabolites of brain norepinephrine. Life Sci 16:1725-1730

Stone TW (1976) Responses of neurones in the cerebral cortex and caudate nucleus to amantadine, amphetamine and dopamine. Br J Pharmcol 56:101-110

Stone TW, Bailey EV (1975) Responses of central neurones to amantadine: comparison with dopamine and amphetamine. Brain Res 85:126-129

Stone TW, Taylor DA (1977) Microiontophoretic studies of the effects of cyclic nucleotides on excitability of neurones in the rat cerebral cortex. J Physiol (Lond) 266:523-543

Strader CD, Pickel VM, Joh TJ, Strohsacker MW, Shorr RG, Lefkowitz RJ, Caron MG (1983) Antibodies to the beta-adrenergic receptor: attenuation of catecholamine-sensitive adenylate cyclase and demonstration of postsynaptic receptor localization in brain. Proc Natl Acad Sci USA 80:1840-1844

Strecker RE, Steinfels GF, Jacobs BL (1983) Dopaminergic unit activity in freely moving cats: lack of relationship to feeding satiety, and glucose injections. Brain Res 260:317-321

Stricker EM, Zigmond MJ (1974) Effects on homeostasis of intraventricular injections of 6-hydroxydopamine in rats. J Comp Physiol Psychol 86:973-994

Sturgill TW, Schrier BK, Gilman AG (1975) Stimulation of cyclic AMP accumulation by 2-chloroadenosine: lack of incorporation of nucleoside into cyclic nucleotides. J Cyclic Nucleotide Res 1:21-30

Svensson TH, Bunney BS, Aghajanian GK (1975) Inhibition of both noradrenergic and serotonergic neurons in brain by the α-adrenergic agonist clonidine. Brain Res 92:291-306

Swanson LW (1976) The locus coeruleus: a cytoarchitectonic, golgi and immuohistochemical study in the albino rat. Brain Res 110:39-56

Swanson LW, Hartman BK (1980) Biochemical specificity in central pathways related to peripheral and intracerebral homeostatic functions. Neurosci Lett 16:55-60

Swanson LW, Sawchenko PE (1980) Paraventricular nucleus: a site for the integration of neuroendocrine and autonomic mechanisms. Neuroendocrinology 37:410-417

Takigawa M, Mogenson GJ (1977) A study of inputs to antidromically identified neurons of the locus coeruleus. Brain Res 135:217-230

Thatcher-Britton K, Svensson T, Schwartz J, Bloom FE, Koob GF (1984) Dorsal noradrenergic bundle lesions fail to alter opiate withdrawal or the suppression of opiate withdrawal by clonidine. Life Sci 34:133-139

Thierry AM, Tassin JP, Blanc G, Glowinski J (1976) Selective activation of the mesocortical dopaminergic system by stress. Nature 263:242-244

Thierry AM, Javoy F, Glowinski J, Kety SS (1968a) Effects of stress on the metabolism of norepinephrine, dopamine and serotonin in the central nervous system of the rat. I. Modifications of norepinephrine turnover. J Pharmacol Exp Ther 163:163-171

Thierry AM, Fekete M, Glowinski J (1968b) Effects of stress on the metabolism of noradrenaline, dopamine and serotonin (5HT) in the central nervous system of the rat (II). Modifications of serotonin metabolism. Eur J Pharmacol 4:384-389

Thompson T, Pickens R (1970) Stimulant self-administration by animals: some comparisons with opiate self-administration. Fed Proc 29:6-12

Tombaugh TN, Pappas BA, Roberts DCS, Vickers GJ, Szostak C (1983) Failure to replicate the dorsal bundle extinction effect: telencephalic norepinephrine depletion does not reliably increase resistance to extinction but does augment gustatory neophobia. Brain Res 261:231-242

Tsaltas E, Gray JA, Fillenz M (1984a) Alleviation of response suppression to conditioned aversive stimuli by lesions of the dorsal noradrenergic bundle. Behavioral Brain Res 13:115-127

Tsaltas E, Preston GC, Rawlins JN, Winocur G, Gray JA (1984b) Dorsal bundle lesions do not affect latent inhibition of conditioned suppression. Psychopharmacol 84:549-555

Tsien RW (1973) Adrenaline-like effects of intracellular iontophoresis of cyclic AMP in cardiac Purkinje fibres. Nature (Lond) 245:120-122

Tye NC, Everitt BJ, Iversen SD (1977) 5-Hydroxytryptamine and punishment. Nature 268:741-743

Uguru-Okorie DC, Arbuthnott GW (1981) Altered paw preference after unilateral 6-hydroxydopamine injections into lateral hypothalamus. Neuropsychologia 19:463-467

Ungerstedt U (1971a) Stereotaxic mapping of the monoamine pathways in the rat brain. Acta Physiol Scand 82, Suppl 367:1-48

Ungerstedt U (1971b) Postsynaptic supersensitivity after 6-hydroxydopamine induced degeneration of the nigro-striatal dopamine system. Acta Physiol Scand 82, Suppl 367:69-93

Ungerstedt U (1971c) Adipsia and aphagia after 6-hydroxydopamine induced degeneration of nigro-striatal dopamine system. Acta Physiol Scand 82, Suppl 367:95-122

Vale WW, Rivier C, Spiess J, Brown MR, Rivier J (1983) Corticotropin releasing factor. In Krieger D, Brownstein M, Martin J (Eds) Brain Peptides. John Wiley and Sons, New York, pp 961-974

Vandermaelen CP, Aghajanian GK (1980) Intracellular studies showing modulation of facial motoneurons excitability by serotonin. Nature 287:346-347

Waterhouse BC, Lin C-S, Burne RA, Woodward DJ (1983) The distribution of neocortical projection neurons in the locus coeruleus. J Comp Neurol 217:418-431.

Wauquier A, Niemegeers CJE (1972) Intracranial self-stimulation in rats as a function of various stimulus parameters. II. Influence of haloperidol, pimozide and pipamperone on medial forebrain bundle stimulation with monopolar electrodes. Psychopharmacologia 27:191-202

Weissman A, Koe B, Tenen S (1966) Antiamphetamine effects following inhibition of tyrosine hydroxylase. J Pharmacol Exp Ther 151:339-352

Wilson MC, Schuster CR (1972) The effects of chlorpromazine on psychomotor stimulant self-administration in the rhesus monkey. Psychopharmacology 26:115-126

Wise CD, Stein L (1969) Facilitation of brain self-stimulation by central administration of NE. Science 163:299-301

Wise RA (1978a) Catecholamine theories of reward: a critical review. Brain Res 152:215-247

Wise RA (1978b) Neuroleptic attenuation of intracranial self-stimulation: reward or performance deficits. Life Sci 22:535-543

Wise RA, Spindler J, Dewit H, Gerber GJ (1978) Neuroleptic-induced "anhedonia" in rats: pimozide blocks the reward quality of food. Science 201:252-264

Woodward DJ, Hoffer BJ, Altman J (1974) Physiological and pharmacological properties of Purkinje cells in rat cerebellum degranulated by postnatal X-0 irradiation. J Neurobiol 5:283-304

Woodward DJ, Moises HC, Waterhouse BD, Hoffer BJ, Freedman R (1979) Modulatory actions of norepinephrine in the central nervous system. Fed Proc 38:2109-2116

Yarbrough GG (1975) Supersensitivity of caudate neurons after repeated administration of haloperidol. Eur J Pharmacol 31:367-369

Yarbrough GG (1976) Ouabain antagonism of noradrenaline inhibitions of cerebellar Purkinje cells and dopamine inhibitions of caudate neurones. Neuropharmacol 15:335-338

Yeh HH, Moises HC, Waterhouse BD, Woodward DJ (1981) Modulatory interactions between norepinephrine and taurine, beta-alanine, gamma-aminobutyric acid and muscimol, applied iontophoretically to cerebellar Purkinje cells. Neuropharmacol 20:549-560

Yokel RA, Wise RA (1975) Increased lever pressing for amphetamine after pimozide in rats: implications for a dopamine theory of reward. Science 187:547-549

Zarevics P, Setler PE (1979) Simultaneous rate-independent and rate-dependent assessment of intracranial self-stimulation: evidence for the direct involvement of dopamine in brain reinforcement mechanisms. Brain Res 169:499-512

Zigmond RE, Schon F, Iversen LL (1974) Increased tyrosine hydroxylase activity in the locus coeruleus of rat brainstem after reserpine treatment and cold stress. Brain Res 70:547-552

Zis AP, Fibiger HC, Phillips AG (1974) Reversal by L-dopa of impaired learning due to destruction of the dopaminergic nigro-neostriatal projection. Science 185:960-962

CHAPTER 12

Central Control of Anterior Pituitary Function

M. A. ARNOLD and J. B. MARTIN

A. Introduction

Although the anterior pituitary lacks neuronal innervation, its hormonal secretions are controlled precisely by the overlying brain. The anatomic observation by WISLOCKI and KING (1936) of a blood circulation from the mediobasal hypothalamus to the anterior pituitary prompted HARRIS (1948), on the basis of direct observations of the direction of blood flow, to formulate the portal vessel-chemotransmitter hypothesis. This proposal statet that the hypothalamus secretes regulatory factors into the portal capillaries of the median eminence which are transported to the anterior pituitary to regulate its hormonal output. The hypothalamic regulatory factors are released from tuberoinfundibular neurons which are distributed throughout the basal hypothalamus—the term *hypophysiotropic area* defines this area. In research spanning four decades, five of the hypothalamic hormones have been identified as small peptides. These include thyrotropin-releasing hormone (TRH), luteinizing hormone releasing hormone (LHRH), somatotropin-release inhibiting factor (somatostatin or SRIF), growth hormone releasing factor (GRF), and corticotropin-releasing factor (CRF). Dopamine, a catecholamine, is now accepted as a prolactin release-inhibiting factor (PIF), to comprise the sixth factor identified. In addition, extracts from hypothalamus have been reported to have activity which indicates the presence of a prolactin releasing factor (PRF).

Regulation of anterior pituitary hormone secretion is achieved with specific hypothalamic releasing hormones and, for some pituitary hormones, by feedback inhibition. In the case of thyrotropin, corticotropin, and the gonadotropins, hormonal output of the target organs inhibits the release of the respective trophic hormone from the pituitary (e.g. glucocorticoids inhibit corticotropin release). In the case of prolactin and growth hormone (GH), both an excitatory and inhibitory drive is exerted by the hypothalamus. There is good evidence that prolactin can inhibit its own secretion by stimulating the release of dopamine into hypophyseal portal blood. A similar "short-loop" feedback control has been proposed for GH, although a mechanism for GH autoregulation has not been conclusively demonstrated.

The control of pituitary function by the central nervous system is provided by neuronal afferents that synapse on hypothalamic peptidergic releasing factor cells. These releasing factor cells have been termed "neuroendocrine transducers" by WURTMAN (1973), since they translate a neuronal (neurotransmitter) input into a hormonal output (hypothalamic hormone). Neurons may either stimulate or inhibit neuroendocrine transducer cells through axo-axonic interactions in the median eminence or by axodendritic or axosomatic influences at other brain sites. A host of putative neurotransmitter substances are under investigation for their role in anterior pituitary regulation. This review will focus primarily on three of these neurotransmitter candidates, noradrenaline, dopamine and 5-hydroxytryptamine (5-HT). Attention will be drawn to identifying possible interactions of these neurotransmitters with particular hypothalamic peptidergic cells.

B. Growth Hormone

I. Normal Patterns of Growth Hormone (GH) Secretion

Growth hormone, or somatotropin, is localized to acidophilic cells in the anterior pituitary, where it is synthesized as a precursor molecule (pregrowth hormone) and rapidly converted *in situ* to its final form (SUSSMAN et al. 1976). The pulsatile release of GH from the pituitary causes dramatic fluctuations in plasma levels of the hormone during day and night in man and in experimental animals. In man, plasma GH levels may surge to 40–60 ng/ml, the largest bursts occurring during the first part of sleep (QUABBE et al. 1966). A variety of other species including the rhesus monkey (JACOBY et al. 1974), sheep (CHAMLEY et al. 1974) and rabbit (MACINTYRE and ODELL 1974) have spontaneous GH secretory bursts. In rats (TANNENBAUM and MARTIN 1976) and in baboons

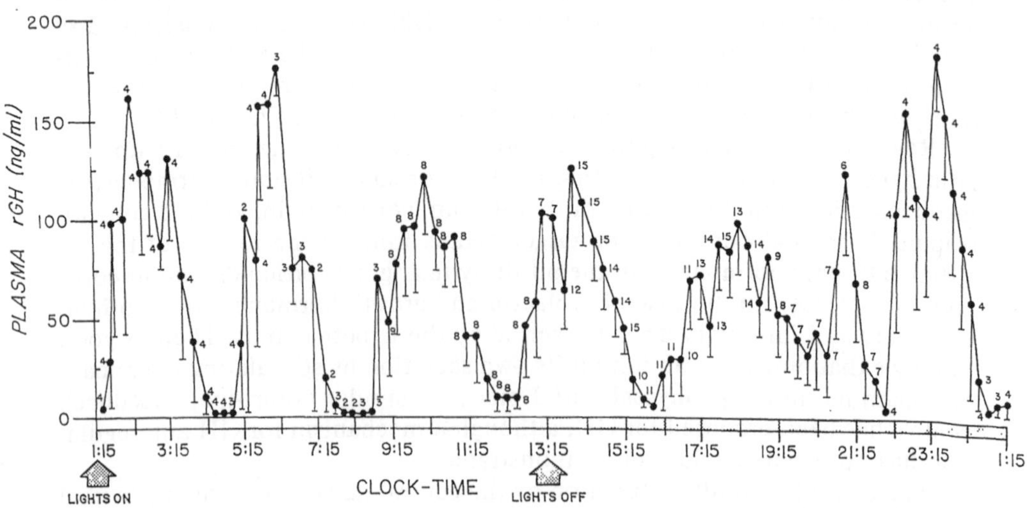

Fig. 1. Mean plasma GH levels of 24 rats sampled for 4–24 h periods during different phases of a 12 h light-dark cycle. Number of animals sampled at each time point are indicated. Vertical lines represent standard error of the mean (SEM). (From TANNENBAUM and MARTIN, 1976, with permission).

(STEINER et al. 1978) episodic GH release has been demonstrated to occur on a regular 3-4 hour cycle (Fig. 1). As in man, the baboon shows increased GH release during sleep (PARKER et al. 1972).

It has been shown, in man, that the physiological basis for these variations in GH secretion is complex, for it may be related to sleep (TAKAHISHI et al. 1968), exercise (HUNTER et al. 1965), stress (SCHALCH and REICHLIN 1968) or post-prandial glucose decline (PARKER and ROSSMAN 1973). A number of agents when administered in pharmacological doses may also stimulate GH release in man. These include insulin, ACTH, vasopressin, estrogens and the amino acids arginine and leucine. Attempts, however, to correlate plasma GH levels to physiological changes in blood glucose, amino acid, or non-esterified fatty acid levels have been largely unsuccessful in humans (REICHLIN 1974) and in rats (MULLER et al. 1972; TANNENBAUM et al. 1976). Since no good correlation could be made with the 24 hour pattern of GH secretion in man and potential metabolic regulators, it was suggested that GH secretion is controlled primarily by neural mechanisms (FINKELSTEIN et al. 1972).

II. Hypothalamic Regulation of GH Secretion

The best evidence that the hypothalamus exerts primarily a stimulatory influence on GH release is that pituitary stalk section or hypothalamic destruction impairs secretion. In man, hypothalamic lesions block the release of GH that accompanies hypoglycemia and sleep (KRIEGER and GLICK 1974). These stimuli are therefore thought to mediate their effects on the pituitary via the central nervous system. In other species, such as the squirrel monkey, lesions of the median eminence or midline basal hypothalamus block insulin-induced (ABRAMS et al. 1971) or stress-induced (BROWN et al. 1971) GH secretion. In young female rats ventromedial nucleus lesions cause growth retardation and a decrease in pituitary and plasma GH levels. Lesions of the ventromedial nucleus are also effective at reducing pulsatile GH release in adult male rats (MARTIN et al. 1974a).

Electrical stimulation studies also show the importance of the ventromedial nucleus-arcuate region with regard to GH secretion. Both unilateral (FROHMAN et al. 1968) and bilateral (MARTIN 1972) stimulation of the mediobasal hypothalamus in pentobarbital-anesthetized male rats elicits GH secretion. Stimulation of the lateral or anterior hypothalamus (FROHMAN et al. 1972; MARTIN 1972), or of the supraoptic or paraventricular nuclei (MARTIN 1972; CHENG et al. 1972) have no effect on GH release. On the other hand, stimulation of the preoptic area inhibits GH secretion (MARTIN et al. 1978a).

III. Hypothalamic Factors Involved in GH Regulation

Over 20 years ago KRULICH et al. (1965) isolated an acid-extractable fraction from ovine hypothalamus which depletes bioassayable pituitary GH content *in vivo*. In the following decade and a half a number of other investigators reported similar results, although the structure of a GRF eluded identification. Recently, GUILLEMIN et al. (1982) isolated and identified a 44 residue peptide from a human pancreatic tumor with potent GH-releasing activity. Soon afterward Spiess et al. (1983) reported the characterization of rat hypothalamic GRF with potent GH-releasing activity *in vitro.* The rat GRF was identified as a 43 residue peptide (Spiess et al. 1983), and has much structural similarity to the N-terminus of the pancreatic tumor peptide with GRF activity described above. Further work will be necessary to define the physiological role for GRF in the regulation of GH secretion.

In their attempt to isolate GRF, Brazeau et al. (1973) succeeded in isolating a compound from ovine hypothalamic fragments with potent GH-inhibiting properties both *in vivo* and *in vitro.* This substance—named somatostatin—was characterized by amino acid analyses as a cyclic tetradecapeptide. Subsequent studies have indicated a wide distribution of this peptide throughout the brain and spinal cord as well as in the gastrointenstinal tract. In the brain, somatostain may function both as a hypothalamic hormone and as a neurotransmitter. Renaud et al. (1975) first demonstrated that microiontophoretically applied somatostatin depresses the firing rate of central nervous system neurons. In addition to effects on GH secretion, somatostatin inhibits the relase of another pituitary hormone, TSH (Vale et al. 1974), and pancreatic insulin and glucagon (Devane et al. 1973; Koerker et al. 1974). The administration of somatostatin to rats blocks pulsatile GH secretion (Martin et al. 1974a) and inhibits GH release induced by pentobarbital (Brazeau et al. 1973), or morphine (Martin et al. 1975). Somatostatin also reduces sleep-related GH secretion in humans (Yen et al. 1974) and decreases the elevated plasma GH levels in acromegaly (Hall et al. 1973). These pharmacological studies, however compelling, still fail to demonstrate a clear role for somatostatin in mediating the physiological secretion of GH.

Experiments using animals passively immunized with somatostatin antiserum have demonstrated that endogenous somatostatin is important in physiological GH regulation. The administration of somatostatin antiserum to rats reverses inhibition of GH release induced by stress (Arimura et al. 1976; Terry et al. 1976) or by prolonged (72 hour) food deprivation (Tannenbaum et al. 1978). Ferland et al. (1976) reported that basal GH release is enhanced by the administration of somatostatin antiserum. This finding differs somewhat from that of Terry et al. (1977), who found that somatostatin antiserum elevates trough GH levels without a change in mean 6 hour GH levels. A role for endogenous hypothalamic somatostatin is suggested by the experiments of Chihara et al. (1978a), who found an inverse relationship between somatostatin levels in hypophyseal portal blood and GH levels in systemic blood. In addition Chihara et al. (1978a) reported that the somatostatin level in hypophyseal portal blood was significantly higher than in systemic blood after various anesthetic treatments. These findings support the notion that somatostatin of hypothalamic origin can inhibit the physiologic secretion of GH.

IV. Monoaminergic Control of GH Secretion

The release of GH from the pituitary is regulated by a dual system comprised of an inhibitory (somatostatin) and an excitatory (GRF) hypothalamic hormone. The secretion of these hypothalamic hormones is regulated by converging connections from other brain neurons—including inputs from cells that use the monoamines dopamine, noradrenaline or 5-HT as neurotransmitters. These conclusions are supported by a multitude of studies of the effects of drugs thought to either enhance or diminish transmission across particular monoaminergic synapses in the brain.

In man, as well as in other primates, brain neurons that use dopamine, noradrenaline, or 5-HT are generally thought to have a stimulatory influence on GH secretion. Thus, the administration of L-dopa, the precursor of both dopamine and noradrenaline, stimulates GH release in man (Boyd et al. 1970), rhesus monkeys (Jacoby et al. 1974), and in dogs (Lovinger et al. 1974). The stimulatory effect of L-dopa in man is blocked by phentolamine (Kansel et al. 1972), an α-adrenoceptor antagonist, suggesting mediation by noradrenaline receptors. The systemic administration of α-adrenoceptor antagonists to man (Blackard and Heidingsfelder 1968; Imura et al 1971), or to unanesthetized rats (Martin et al. 1978b; Arnold and Fernstrom 1980) inhibits GH secretion. Similarly, the administration of the tyrosine hydroxylase inhibitor α-methyl-p-tyrosine to rats markedly inhibits pulsatile GH release (Durand et al. 1977)—an effect reversed by clonidine but not by apomorphine. The administration of the α-adrenoceptor agonist clonidine also stimulates GH release in man (Lal et al. 1975), dogs (Lovinger et al. 1976), and in rhesus monkeys (Chambers and Brown 1976). On the other hand, the administration of the β-adrenoceptor antagonist propranolol has been shown to potentiate exercise-induced GH release in man (Sutton and Lazarus 1974), but not to affect pulsatile GH secretion in unanesthetized rats (Martin et al. 1978b).

Dopamine neurons in the brain have also been implicated in GH regulation in some species. For instance, apomorphine, a centrally-active dopamine receptor agonist, stimulates GH release in man (Lal et al. 1973) and in rats (Mueller et al. 1976a), but has no effect in the rhesus monkey (Chambers and Brown 1976) or the dog (Lovinger et al. 1976). The administration of haloperidol decreases plasma GH levels in rats (Mueller et al. 1976a), whereas treatment with another dopamine receptor antagonist, (+)-butaclamol, results in only minor suppression of pulsatile GH secretion in this species (Willoughby et al. 1977). In man, glucose administration blocks the rise in plasma GH following L-dopa (Mims et al. 1973) or apomorphine (Ettigi et al. 1975), suggesting that glucoreceptor stimulation can override the catecholaminergic stimuli for GH secretion.

Considerable evidence has accumulated in support of the notion that the release of 5-HT by brain neurons has a stimulatory influence on GH secretion. For instance, the administration of the 5-HT precursor tryptophan to humans has been reported to slightly increase (Muller et al. 1974) or to substantially increase (Woolf and Lee 1977; Fraser et al. 1979; Glass et al.

1979; Koulu and Lammintausta 1979) plasma GH levels. In one study (Fraser et al. 1979) pretreatment with the 5-HT antagonist cyproheptadine blocked the rise in plasma GH following oral tryptophan to women. The administration of the immediate precursor of 5-HT, 5-hydroxytryptophan also stimulates GH secretion in man (Imura et al. 1973; Nakai et al. 1974; Lancranjan et al. 1977). As 5-hydroxytryptophan (but not tryptophan) increases 5-HT synthesis in serotoninergic and adrenergic neurons, caution is advised in interpreting these data.

The putative 5-HT receptor antagonists cyproheptadine or methysergide are effective in blocking GH release following oral 5-hydroxytryptophan (Nakai et al. 1974), insulin-induced hypoglycemia (Bivens et al. 1973; Smythe and Lazarus 1974) and during sleep (Chihara et al. 1976). Caution must be exercised in interpreting studies using 5-HT receptor antagonists, however, as two distinct populations of central 5-HT receptors have been identified (Peroutka et al. 1981).

In rats, the intraventricular injection of 5-HT (Collu et al. 1972), or the intraperitoneal injection of 5-hydroxytryptophan (Smythe and Lazarus 1973) stimulates GH release. The pulsatile release of GH by the pituitary was shown to be diminished by treatment with 5-HT receptor antagonists (Arnold and Fernstrom 1978; Martin et al. 1978 b), while enhanced by treatment with tryptophan (Arnold and Fernstrom 1979).

Systemic administration of neuropharmacologic agents in most of the previously described studies does not provide information regarding the site of action of these drugs. A direct effect on the pituitary is unlikely, however, as application of monoamines on the pituitary *in vitro* does not modify GH release (MacLeod et al. 1969; Birge et al. 1970). Most likely, neuropharmacologic agents act on central neurons, which in turn modulate GH secretion via their effects on hypothalamic neuroendocrine transducer cells (somatostatin or GRF-containing cells).

There is evidence that the release of catecholamines from nerve terminals in the hypothalamus modifies GH release. Toivala and Gale (1972) showed that the microinjection of noradrenaline into the ventromedial nucleus of conscious baboons stimulates GH release. Further they showed that other hypothalamic areas were unresponsive to noradrenaline infusion—including the preoptic area/anterior hypothalamus, the mammillary bodies, and the area dorsal to the ventromedial nucleus (Toivola and Gale 1971). In their studies dopamine infusion into the ventromedial nucleus inhibited GH secretion (Toivola and Gale 1972). It is likely that hypothalamic α-adrenoceptors mediate the effect of noradrenaline since phentolamine infusion into the third ventricle or anterior hypothalamus reduces GH secretion in this species (Toivola et al. 1972).

We have shown that in the rat GH release after ventromedial nucleus stimulation is not prevented by agents that block catecholamine synthesis (Martin et al. 1973). On the other hand, the secretion of GH following electrical stimulation of several extrahypothalamic areas is dependent upon the release of catecholamines by brain neurons. For instance, GH release after stimulation of the hippocampus or the basolateral amygdala is blocked by pretreatment with α-methyl-p-tyrosine. Further, amygdaloid-induced GH release is blocked by phenoxybenzamine but not by propranolol (MARTIN et al. 1978a). These studies indicate that α-adrenoceptors mediate some extrahypothalamic influences on GH regulation.

Final control of GH secretion is most likely modulated by the interaction of a GRF and somatostatin with the pituitary somatotroph. The precise role of each of these factors in generating the endogenous GH secretory rhythm in the rat is not clear. However, there is evidence that somatostatin functions to regulate the trough periods of low basal GH secretion. Anterior hypothalamic lesions that destroy somatostatin-containing fibers and deplete median eminence somatostatin cause more frequent bursts of GH release and elevated trough values (WILLOUGHBY and MARTIN 1978). In addition, TERRY et al. (1977) demonstrated that passive immunization of rats with somatostatin antiserum elevates trough GH levels without affecting peak hormone levels. A recent study has shed some light on the mechanism by which monoamine neurons in the brain modify GH secretion in unanesthetized animals. ARNOLD and FERNSTROM (1980) showed that the injection of antisomatostain serum to rats does not affect the inhibition of episodic GH release caused by an a-adrenoceptor antagonist (yohimbine). Further, they showed that inhibition of growth hormone secretion by a 5-HT receptor antagonist (metergoline) is abolished by treatment with antisomatostatin serum. A likely interpretation of these findings is that (i) yohimbine blocks GH release by suppressing the secretion of hypothalamic GRF but that (ii) metergoline blocks GH release by stimulating somatostatin secretion without impairing GRF release. Thus, episodic GH secretion in the rat may occur as the result of an α-adrenoceptor-modulated GRF release and 5-HT receptor-modulated somatostatin secretion. Similar evidence for adrenoceptor modulation of GRF release in rats has been reported (WAKABAYASHI et al. 1980).

Other approaches have been used to determine the effect of putative neurotransmitter substances on hypothalamic somatostatin release. The intraventricular administration of dopamine, noradrenaline, or acetylcholine was reported to increase the concentration of somatostatin in the hypophyseal portal blood of urethane-anesthetized rats (CHIHARA et al. 1979). In this study plasma GH levels were not reported, so the effect of these agents on GRF release cannot be estimated. It has also been observed that dopamine releases somatostatin from rat median eminence fragments (NEGRO-VILAR et al. 1978), hypothalamic fragments (MAEDA and FROHMAN 1980), and synaptosomal preparations (WAKABAYASHI et al. 1977). On the other hand, EPELBAUM et al. (1979) did not find a significant effect of dopamine on somatostatin release from rat mediobasal hypothalamus *in vitro*. Comparing these conflicting *in vitro* reports with *in vivo* GH data is difficult because of differences in the experimental preparations.

Other neurotransmitter mechanisms undoubtedly are important for GH regulation in this species. However, their influence on hypothalamic releasing factors *in vivo* is unknown.

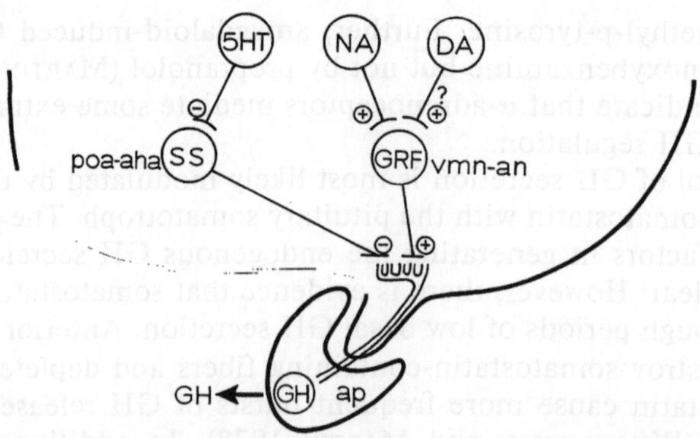

Fig. 2. Diagram of GH regulatory system including possible sites at which monoaminergic neurons might interact with hypothalamic releasing factor neurons. Abbreviations: 5-HT, 5-hydroxytryptamine; NA, noradrenaline; DA, dopamine; SS, somatostatin; GRF, growth hormone releasing factor; GH, growth hormone; poa-aha, preoptic area-anterior hypothalamic area; vmn-an, ventromedial nucleus-arcuate nucleus; ap, anterior pituitary.

V. Conclucions

Pituitary GH secretion (Fig. 2) is controlled by a stimulatory (GRF) and an inhibitory hypothalamic hormone (somatostatin). In man and in animals GH is released episodically by the pituitary, resulting in dramatic surges of the hormone in plasma. The release of noradrenaline, dopamine, and 5-HT by brain neurons is generally thought to stimulate GH secretion by interactions with hypothalamic releasing hormones. There is evidence that episodic GH secretion may occur as the result of the modulation of GRF release via α-adrenoceptors and the modulation of somatostatin release via 5-HT receptors.

C. Prolactin

I. Normal Patterns of Prolactin Secretion

Among its important actions prolactin is a potent stimulus for mammary development and lactogenesis in man and in other mammals. In the rodent the role of prolactin in reproductive function may be somewhat broader as it has luteotrophic activity (Gospodarowicz and Legault-Demare 1963) in addition to effects on testicular (Bartke et al. 1975) and prostatic function (Negro-Vilar et al. 1977).

In man, prolactin is released episodically from the pituitary with plasma levels ranging generally from 0.6–40 ng/ml (Frantz et al. 1972). Women tend to have slightly higher plasma prolactin levels, probably due to mildly stimulated release of the hormone during the late follicular and luteal phases of the menstrual cycle (Vekemans et al. 1977). Humans normally exhibit a nocturnal rise in plasma prolactin although this increase is not correlated with any stage of sleep (SASSIN et al. 1972). Other stimuli for prolactin release in man include pregnancy (HWANG et al. 1971), lactation (HWANG et al. 1971; FRANTZ et al. 1972), and major stresses like surgery (FRANTZ et al. 1972).

In the rhesus monkey (QUADRI and SPIES 1976) and rat (KIZER et al. 1975), there is also evidence for nocturnal surges of prolactin secretion. In the female rat plasma prolactin levels show minimal fuctuations during the estrous cycle except for a dramatic rise during the afternoon of proestrus (NEILL 1972a; SAUNDERS et al. 1976). SMITH et al. (1975) showed that increases in LH, FSH, progesterone, and estradiol accompany the proestrus surge in prolactin. Further, NEILL et al. (1971) demonstrated that passive immunization of rats with antisera to estradiol prevents the increase in prolactin on the afternoon of proestrus. Thus, proestrus prolactin release is dependent on increased circulating levels of estrogens in this species. The administration of estrogens to male rats has been shown to stimulate prolactin release (NEILL 1972b). Prolactin secretion increases 4–6 h. prior to parturition and declines during delivery (GROSVENOR and TURNER 1960; AMENOMORI et al. 1970; SAUNDERS et al. 1976). As in man, the suckling stimulus rapidly elevates plasma prolactin in the rat (SAR and MEITES 1969; SAUNDERS et al. 1976). Stress, whether from restraint (EUKER et al. 1973), handling (BROWN and MARTIN 1974), or ether (NEILL 1970), also elicits the release of prolactin in the rat.

II. Hypothalamic Influence on Prolactin secretion

The inhibitory influence of the mammalian hypothalamus on prolactin secretion was first suggested by EVERETT (1954) who found that relocation of the pituitary to the kidney capsule in the rat resulted in prolonged maintenance of the corpus luteum and pseudopregnancy. Direct evidence for hypothalamic inhibition, however, awaited TALWALKER et al. (1963) who showed that the high rate of prolactin release *in vitro* is suppressed by application of crude hypothalamic extracts. In studies using a radioimmunoassay for prolactin determinations, the tonic prolactin-inhibiting influence of the hypothalamus can also be demonstrated. CHEN et al. (1970) showed that implantation of a pituitary under the kidney capsule of a hypophysectomized rat results in elevated serum prolactin levels. Further, it was shown that the administration of hypothalamic extracts either systemically (AMENOMORI and MEITES 1970) or into hypophyseal portal vessels (KAMBERI et al. 1971a) depresses serum prolactin levels.

Although the dominant influence of the mammalian hypothalamus on prolactin release is inhibitory, there is also some evidence for a stimulatory component in this system. MEITES et al. (1960) showed that the injection of acid extracts of rat hypothalamus into estrogen-primed rats induces milk secretion. Later, NICOLL et al. (1970a) provided *in vitro* evidence for prolactin-releasing activity. In these studies incubation of anterior pituitaries with hypothalamic extracts first inhibits, but later stimulates prolactin secretion. Similar findings have been reported by VALVERDE et al. (1972) who found that methanol extracts of rat or porcine stalk-median eminence tissue stimulate prolactin release *in vivo*.

Other indirect evidence for hypothalamic prolactin-releasing activity is provided by an analysis of the nature of physiological prolactin secretion. The rapid and dramatic increases in prolactin release that follow lactation, stressful stimuli or ether anesthesia are consistent with such an hypothesis. Additional indirect evidence for prolactin-releasing activity has been provided recently by SHIN (1979), who reports that ether stress stimulates prolactin release during a constant infusion of dopamine, a putative prolactin release inhibitory factor. Thus, prolactin release *in vivo* cannot always be attributed to the lack of an inhibiting factor (PIF).

III. Hypothalamic Factors Involved in Prolactin Regulation

1. Prolactin Release-Inhibiting Factor (PIF)

It is now generally believed that dopamine of hypothalamic origin serves as a physiological PIF in experimental animals. Although speculation about this notion (VAN MAANEN and SMELIK 1968) has persisted for over a decade, only recently has convincing physiological evidence been presented to indicate a direct inhibitory action of hypothalamic dopamine on the pituitary lactotroph.

Early evidence for catecholaminergic inhibition of prolactin secretion was provided by KANEMATSU et al. (1963), who showed that reserpine, an agent that depletes vesicular storage of catecholamines, produces lactation and reduces pituitary prolactin concentrations in rabbits. Later it was shown (LU et al. 1970) that reserpine administration to rats elevates serum prolactin concentrations. The inhibition of tyrosine hydroxylase with α-methyl-p-tyrosine was also found to increase serum prolactin concentrations in rats (LU and MEITES 1971). In each of these studies the depletion of central catecholamine stores appears to stimulate the release of prolactin from the pituitary.

The antipsychototic drugs which appear to exert their behavioral effect by blocking dopamine receptors are potent stimuli for prolactin release. Perphenazine, a phenothiazine, elicits large increases in serum prolactin levels in rats (LU et al. 1970). The chronic administration of chlorpromazine, perphenazine, or haloperidol to male or female psychiatric patients is also associated with high levels of plasma prolactin (FRANTZ and KLEINBERG 1970). Pimozide, a specific dopamine receptor antagonist, implanted into the median eminence-arcuate region elicits prolactin release; a small rise also occurs with implants in the anterior pituitary (OJEDA et al. 1974). Chlorpromazine treatment in rats with extensive hypothalamic deafferentation increases prolactin secretion (OHGO et al. 1976) but has no effect on augmented prolactin levels in the stalk-sectioned rhesus monkey (DIEFENBACH et al. 1976). Similarly, in median eminence-lesioned rats α-methyl-p-tyrosine has no effect on augmented plasma prolactin concentrations (DONOSO et al. 1973). In each of the studies these agents could be modifying prolactin secretion via effects on the hypothalamus, although a direct action on the pituitary cannot be eliminated. In support of the latter argument are data demonstrating that haloperidol or pimozide block the inhibitory action of dopamine on prolactin release *in vitro* (MACLEOD and LEHMEYER 1974).

The systemic administration of L-dopa decreases prolactin secretion in the rat (DONOSO et al. 1971) and inhibits the increase induced by lactation (CHEN et al. 1974), stress (MELTZER and FANG 1976), TRH (CHEN and MEITES 1975), median eminence lesions (DONOSO et al. 1971), or during the afternoon of proestrus (WIGGINS and FERNSTROM 1977). In man, L-dopa treatment is also associated with a fall in plasma prolactin levels (MALARKEY et al. 1971). The effect of L-dopa is not blocked by diethyldithiocarbamate, a DBH inhibitor (DONOSO et al. 1971, 1973), which indicates that the suppressive effect is not mediated by noradrenaline. The locus at which L-dopa inhibits prolactin secretion has been contested. LU and MEITES (1972) argue that L-dopa stimulates the synthesis and release of a hypothalamic PIF, which, in turn, acts on the pituitary to inhibit prolactin secretion. On the other hand, L-dopa, after conversion to dopamine, could have a direct action on the pituitary. In rats with a pituitary transplant under the kidney capsule, L-dopa decreases prolactin secretion but has no effect when given with a peripheral decarboxylase inhibitor, even though the dopamine content of the hypothalamus is greatly increased (DONOSO et al. 1974). DIEFENBACH et al. (1976) showed that L-dopa decreases prolactin secretion in normal and stalk-sectioned rhesus monkeys and blocks prolactin release induced by TRH. These studies suggest a pituitary site of action for L-dopa.

A host of dopamine receptor agonists have also been tested for their effects on prolactin secretion *in vivo* and *in vitro*. Apomorphine has been shown to inhibit prolactin secretion in rats (SMALSTIG et al. 1974) and in humans (MARTIN et al. 1974b). Piribedil has also been reported to reduce prolactin secretion in rats (MUELLER et al., 1976a) and in humans (BESSER et al. 1972). The ergot alkaloid derivatives ergotamine, ergocornine, and bromocriptine are potent inhibitors of prolactin release *in vivo*; this effect can be attributed to a direct action of these agents on the pituitary (NICOLL et al. 1970b). These dopamine receptor agonists interfere with the ability of haloperidol, pimozide, or perphenazine to stimulate prolactin release *in vivo* (MACLEOD and LEHMEYER 1974), strengthening the argument that the effect of these agents on prolactin secretion is mediated by dopamine receptors in the pituitary.

The observation by HÖKFELT (1967) of tuberoinfundibular dopaminergic nerve endings in the median eminence adjacent to hypophyseal portal vessels provides anatomical evidence for direct control of prolactin release by dopamine. The notion that dopamine is a PIF is supported further by the work of TAKAHARA et al. (1974) who showed that dopamine infused into the hypophyseal protal vessels *in vivo* inhibits the release of prolactin from the pituitary. SHAAR and CLEMENS (1974) reported that removal of catecholamines from hypothalamic extracts by incubation with monoamine oxidase or by binding catecholamines to aluminum oxide eliminates PIF activity. In this study PIF activity was resistant to enzymatic degradation by pepsin.

Crucial physiological evidence to indicate that dopamine is a PIF has been presented recently. Dopamine is present in portal blood at higher concentrations than found in arterial blood from the same rat (BEN-JONATHAN et al. 1977), indicating secretion of dopamine by the median eminence. The level of dopamine in portal blood is sufficient to inhibit prolactin release *in vivo* (GIBBS and NEILL 1978) and *in vitro* (SHAAR and CLEMENS 1974). BEN-JONATHAN et al. (1977) have demonstrated that dopamine levels in portal blood from female rats are highest during estrus, diestrus, and during pregnancy, and lowest during proestrus, showing an inverse relationship with expected systemic prolactin levels in female rats. DE GREEF and NEILL (1979) report the dopamine levels in portal blood from female rats are reduced 36% during cervical stimulation (a manipulation which induces prolactin secretion). However, this reduction in dopamine is considered to be insufficient to fully account for the surges of prolactin observed during cervical stimulation (DE GREEF and NEILL 1979). The postulation of additional prolactin-releasing factors is generally considered necessary to describe adequately prolactin regulation by the hypothalamus.

Hypothalamic compounds other than dopamine are also under investigation as possible PIFs. ENJALBERT et al. (1977) have evidence for PIF activity in hypothalamic extracts that have catecholamines removed by treatment with aluminum oxide. This PIF activity *in vitro* was not diminished by co-incubation with dopamine receptor antagonists. Other groups have also demonstrated PIF activity distinct from the catecholamines in porcine hypothalamic extracts (SCHALLY et al. 1976a, 1977). One such compound was identified as GABA and was shown to inhibit prolactin release *in vivo* and *in vitro* (SCHALLY et al. 1977). A recent study (RACAGNI et al. 1979) has indicated that the *in vivo* prolactin response to domperidone (a dopamine receptor antagonist) is diminished by increasing brain GABA levels with the intraventricular injection of ethanolamine-O-sulfate (a GABA-transaminase inhibitor). In this study prolactin inhibition is attributed to increased delivery of GABA to the pituitary by portal blood, although hypothalamic GABA could also be inhibiting prolactin secretion by blocking the release of a hypothalamic PRF. Clearly, additional study will be necessary to implicate hypothalamic GABA as a physiological PIF.

2. Prolactin Releasing Factor (PRF)

The possibility that a hypothalamic compound other than TRH functions to stimulate prolactin secretion is supported by some chemical evidence. NICOLL et al. (1970a) showed that incubation of anterior pituitaries with hypothalamic extracts first inhibits but later stimulates prolactin secretion. Methanol extracts of rat and porcine stalk-median eminence tissue stimulate prolactin release *in vivo* (VALVERDE et al. 1972). This material was devoid of TRH, oxytocin, and vasopressin activity, and could be chromatographically separated into two distinct fractions. Similar findings were also reported by DULAR et al. (1974), who showed that a chromatographic fraction from acetic acid extracts of bovine stalk-median eminence stimulates prolactin release from bovine pituitaries *in vitro*. This fraction was also reported to be devoid of TRH activity.

Prolactin secretion in animals has been elicited by the administration of a number of peptides including β-endorphin (RIVIER et al. 1977a), bombesin (RIVIER et al. 1978), arginine vasotocin (JOHNSON 1978), neurotensin and substance P (RIVIER et al. 1977b). Each of these substances, however, fails to stimulate prolactin release *in vitro* and these must be eliminated as a candidate for a PRF. On the other hand, the administration of vasoactive intestinal polypeptide (VIP) has been shown to stimulate prolactin release *in vivo* and *in vitro* (RUBERG et al. 1978). Whether VIP functions as a physiological PRF undoubtedly will be the subject of further investigation.

3. Thyrotropin Releasing Hormone (TRH)

The demonstration that TRH directly stimulates prolactin release from rat pituitary tumor cells (TASHJIAN et al. 1971) initiated speculation that this tripeptide may function as a physiological PRF. Although the systemic administration of TRH has little effect on prolactin release in the normal male rat (LU et al. 1972), TRH stimulates prolactin secretion in the proestrus female rat (MUELLER et al. 1973), the estrogen-progesterone treated male rat (RIVIER and VALE 1974), the cow (CONVEY et al. 1973), and in the human (BOWERS et al. 1971; JACOBS et al. 1971). It is believed that the effects of TRH on prolactin release *in vivo* occur at the level of the pituitary.

The observation that prolactin and TSH secretion appear to be independently controlled in a number of circumstances argues against a physiological role for TRH in prolactin regulation. BLAKE (1974) showed that prolactin secretion in rats is increased by nicotine injection or by inhalation of ether during the afternoon of proestrus, whereas TSH secretion was unaffected in the same animals. Stress stimulates prolactin secretion in rodents (MEITES et al. 1972) but has been shown to inhibit TSH release (FORTIER et al. 1970). Exposure of rats to cold temperature causes a sustained rise in plasma TSH levels while eliciting only a transient rise in plasma prolactin levels (JOBIN et al. 1976). In each of these studies prolactin secretion does not parallel TSH release, suggesting that TRH is not a physiological PRF. On the other hand, KOCH et al. (1977) report that passive immunization of either male or diestrous or proestrous female rats with TRH antiserum decreases serum levels of prolactin and TSH by 50% and 70%, respectively. This finding suggests that in this species endogenous TRH may be one of the stimulatory factors for both prolactin and TSH regulation. In most experimental protocols, however, the notion that TRH is a physiological PRF cannot be convincingly demonstrated.

Further, the mechanism by which TRH injection elicits prolactin and TSH secretion may also be different. VALE et al. (1974) showed that somatostatin diminishes TRH-stimulated TSH release in estrogen-progesterone treated male rats without affecting TRH-enhanced prolactin release. Also, CHEN and MEITES (1975) report that L-dopa pretreatment blocks TRH-induced prolactin release without influencing TRH-induced TSH secretion. Since TRH does not compete with dopamine for the same receptor sites on the anterior pituitary (CRONIN et al. 1978) a different mechanism must be involved. Although difficult to interpret, these studies also suggest that the systems controlling prolactin and TSH secretion can usually be dissociated.

4. Estrogens

Reece and Turner (1936) were first to show that the administration of estrogens increases prolactin synthesis in the male rat. With the use of radioimmunoassay it was later shown that estrogens increase both pituitary and serum levels of prolactin in this species (Chen and Meites 1970). Conversely, decreasing estrogen synthesis by ovariectomy causes a fall in prolactin production which can be reversed by injection of estrogens (MacLeod et al. 1969). The proestrus surge in blood prolactin levels is attributed to estrogens, as ovariectomy on the morning of proestrus or passive immunization with antiserum to estradiol prevents the increase in prolactin (Neill et al. 1971).

The mechanism for estrogens action on prolactin secretion may involve interactions with pituitary, hypothalamus or other brain sites. Evidence for a direct action of estrogen on pituitary prolactin release was provided by Nicoll and Meites (1962) who showed that prolactin release in pituitary cell cultures is enhanced by estrogen. Subsequent *in vitro* studies using rat pituitary cells indicate that estradiol blocks the inhibitory influence of dopamine receptor agonists on both basal and TRH-induced prolactin release (Raymond et al. 1978). As estradiol does not compete with catecholamines for binding to pituitary receptor sites (Cronin et al. 1978), the effect of this steroid on pituitary function must involve another mechanism. Estradiol could decrease the sensitivity of the pituitary to dopaminergic inhibition or this steroid could have a direct stimulatory effect on prolactin secretion independent of dopaminergic influences. The latter hypothesis is supported by the finding that estradiol increases prolactin synthesis, secretion, and preprolactin mRNA activity in primary ovine pituitary cell cultures (Vician et al. 1979).

There is also evidence for hypothalamic mediation of the stimulatory effect of estrogens on prolactin release. Stefano and Donoso (1967) found that, in rats, ovariectomy decreased hypothalamic dopamine levels. Subsequently, Fuxe et al. (1969) demonstrated that ovariectomy decreased dopamine turnover in the median eminence; this effect was abolished by hormone replacement with estrogens or testosterone. Since hypothalamic dopamine neurons are thought to be inhibitory to prolactin release, these changes in dopamine metabolism seem paradoxical to the expected effects of ovariectomy and estrogen replacement on prolactin secretion. The demonstration that prolactin itself stimulates dopamine turnover in the median eminence (Hökfelt and Fuxe 1972) and increases dopamine levels in portal blood (Gudelsky and Porter 1980) provides an alternative explanation. That is, the "effect" of estrogens (or ovariectomy) on hypothalamic dopamine turnover could be achieved by direct stimulation of pituitary prolactin secretion by this steroid. On the other hand, Cramer et al. (1979) recently reported that estradiol treatment inhibits dopamine release into hypophyseal portal blood. This finding supports the notion that estrogens can modulate prolactin release by a hypothalamic mechanism in addition to a direct action on the pituitary.

5. Opioid Peptides

The administration of naloxone blocks prolactin secretion following morphine or enkephalin injection (Bruni et al. 1977). Stress-induced prolactin release is also blunted by naloxone, suggesting that endogenous opioid peptides can modulate the release of this hormone (van Vugt et al. 1978). Opioid peptides have no direct effect on pituitary prolactin release (Rivier et al. 1977a) but decrease dopamine turnover in the median eminence (Ferland et al. 1977). Thus, the simulatory effects of opioid peptides on prolactin release may be mediated by an inhibition of dopamine release into hypophyseal portal vessels.

IV. Monoaminergic Control of Prolactin Secretion

Noradrenergic neurons in the brain are reputed to have either stimulatory, inhibitory or no influence on prolactin release depending upon the experimental protocol. This confusion may be attributed to the relatively minor influence of noradrenaline on prolactin regulation, or to the possible lack of specificity of some drugs for noradrenergic neurons. An inhibitory influence of noradrenaline is suggested by the demonstration that phenoxybenzamine increases prolactin release in the cyclic ewe (Davis and Borger 1973) and in the male rat (Martin et al. 1978b). Phentolamine, phenoxybenzamine and propranolol were also reported to increase prolactin release in the rat (Lawson and Gala 1975) and in the crab-eating monkey (Gala et al. 1977). High doses of noradrenaline inhibit prolactin release from rat pituitary *in vitro*, whereas low doses of this catecholamine have the opposite effect (Koch et al. 1970). It is likely that high doses of noradrenaline inhibit pituitary prolactin release by activating dopamine receptors.

In ovariectomized, estrogen-treated rats clonidine, but not isoprenaline, increases prolactin secretion (Lawson and Gala 1975). The administration of the noradrenaline precursor dihydroxyphenylserine to rats depleted of catecholamines by α-methyl-p-tyrosine stimulates prolactin release (Donoso et al. 1971). Similar evidence is provided by Meites and Clemens (1972) who showed that disulfiram, a DBH inhibitor, decreases prolactin secretion in rats. Further, the destruction of noradrenergic neurons in the brain by the intraventricular injection of 6-hydroxydopamine (6-OHDA) suppresses prolactin secretion (Kizer et al. 1975). Also, low doses of phenoxybenzamine block the afternoon prolactin surge in estrogenprimed ovariectomized rats (Subramanian and Gala 1976). Estrogen treatment has been reported to increase the activity of noradrenergic neurons (Carr et al. 1977), providing yet another mechanism for this steroid to stimulate prolactin secretion. Although there are a number of conflicting findings, the bulk of evidence supports the conclusion that noradrenergic neurons have a stimulatory influence on physiological prolaction secretion.

By and large, serotoninergic neurons in the brain are also thought to have a stimulatory influence on prolactin release. Kamberi et al. (1971b) induced prolactin release in rats by injecting 5-HT into the third ventricle, while Lawson and Gala (1975) report increased prolactin secretion following systemic administration of 5-HT. The injection of the 5-HT precursors 5-hydroxytryptophan or tryptophan elicits an increase in blood prolactin levels (Lu and Meites 1973; Mueller et al. 1976b). Pretreatment of rats with fluoxetine, a specific 5-HT reuptake blocker, potentiates prolactin release following 5-hydroxytryptophan as well as prolactin release induced by ether stress or blood withdrawal (Krulich 1975). The 5-hydroxytryptophan-induced prolactin release is also blocked by methysergide (Marchlewska-Koj and Krulich 1975) or metergoline (Cocchi et al. 1977) pretreatment. Quipazine, a 5-HT receptor agonist, was reported to rapidly increase prolactin secretion; this effect is abolished by prior injection of methysergide or bromo-lysergic acid diethylamide (Meltzer et al. 1976). Since 5-HT has no effect on prolactin release from pituitaries *in vitro* (Birge et al. 1970; Lamberts and MacLeod 1978), these agents are thought to modify prolactin secretion through effects on brain neurons.

The depletion of brain 5-HT levels has led to reduced prolactin secretion in some studies. For instance, treatment of rats with p-chlorophenylalanine or p-chloroamphetamine reduces prolactin levels in adult male (Gil-Ad et al. 1976) or in estrogen-primed ovariectomized rats (Caligaris and Taleisnik 1974). Injection of 5,7-dihydroxytryptamine (5,7-DHT), with or without desipramine pretreatment, also reduces prolactin levels in male rats (Gil-Ad et al. 1976), although this effect was not observed by Clemens (1978). Methysergide (Clemens et al. 1977) or 5,7-DHT (Clemens 1978) inhibits the rise in prolactin on the afternoon of estrus, although 5,7-DHT does not affect stress-induced (Wuttke et al. 1978) or the proestrus surge in this hormone (Clemens 1978). These various effects of 5-HT neurotoxins may be related to the restoration of serotoninergic function in these animals due to receptor supersensitivity.

Methysergide also inhibits prolactin secretion induced by suckling (Gallo et al. 1975), estrogen administration (Caligaris and Taleisnik 1974), and stress (Marchlewska-Koj and Krulich 1975). The specificity of methysergide can be questioned, however, as its demethylated metabolite methergine has dopamine receptor agonist properties *in vitro* (Lamberts and MacLeod 1978).

Following complete hypothalamic deafferentation in rats, 5-hydroxytryptophan increases prolactin secretion; this effect is lost in animals with extensive hypothalamic ablation (Ohgo et al. 1976). In summary, these findings support a stimulatory influence of 5-HT on prolactin secretion at the level of the hypothalamus (Ohgo et al. 1976). It has been suggested that serotoninergic neurons in the brain enhance prolactin secretion by stimulating the release of a yet-unidentified prolactin-releasing factor from the hypothalamus (Clemens et al. 1978, Krulich et al. 1980).

V. Conclusions

The release of prolactin by the pituitary is dominated by the inhibitory influence of hypothalamic factors—one of which is dopamine (Fig. 3). In contrast, there is also evidence for a PRF, although such a substance has not been

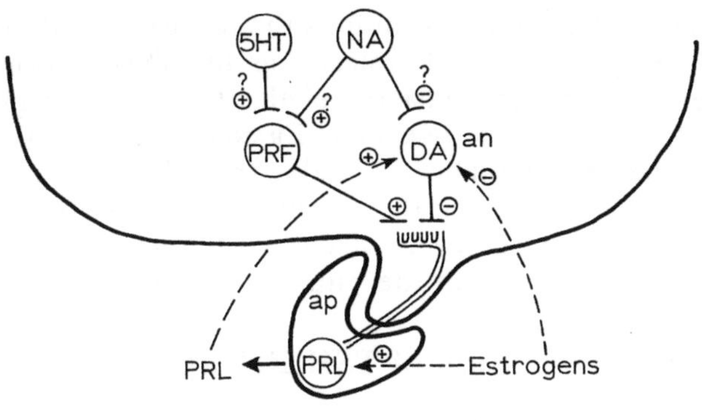

Fig. 3. Diagram of prolactin regulatory system including possible sites at which monoaminergic neurons might interact with hypothalamic releasing factor neurons. Abbreviations: 5-HT, 5-hydroxytryptamine; NA, noradrenaline; DA, dopamine; PRF, prolactin releasing factor; PRL, prolactin; an, arcuate nucleus; ap, anterior pituitary.

chemically identified. TRH is endowed with some prolactin-releasing activity although it fails to qualify fully as a physiological PRF. Estrogens, which are potent stimuli for prolactin synthesis and secretion, act by inhibiting the release of dopamine into hypophyseal portal blood and by direct action on the pituitary. Neuronal release of noradrenaline or 5-HT appears to have a minor influence on prolactin regulation, although it has been proposed that 5-HT stimulates the release of a hypothalamic PRF. Prolactin has been shown to inhibit its own release, probably by stimulating the release of dopamine from the median eminence.

D. Thyrotropin

I. Normal Patterns of Thyrotropin (TSH) Secretion

Thyrotropin, or thyroid-stimulating hormone (TSH), is a glycoprotein synthesized by basophilic cells in the anterior pituitary. TSH is composed of two subunits, α and β, but the biological activity is restricted to the second component (PIERCE 1971). By regulating the release of the thyroid hormones triiodothyronine (T_3) and thyroxine (T_4), TSH is an important factor in mammals for setting the overall metabolic rate.

A nyctohemeral rhythm has been described for TSH in man and experimental animals. Plasma TSH levels in man are highest early in morning (before awakening) and are lowest in the afternoon (PARKER et al. 1976). In rats, however, plasma TSH peaks at midday and falls to a nadir early at night (FUKUDA et al. 1975). At other times of day and night smaller TSH peaks may occur, indicative of episodic release of this hormone (ALFORD et al. 1973).

Acute cold exposure stimulates TSH release in rats (JOBIN et al. 1976) and in human infants (WILBER and BAUM 1970), but does not affect TSH secretion in older human (FISHER and ODELL 1971). In rats hypoglycemia elevates

(LEUNG et al. 1975), whereas stress inhibits the release of TSH (FORTIER et al. 1970). Estrogens may modify TSH release, as the TSH response to TRH is greater in women than in men (ORMSTON et al. 1971), and is highest in women during the late follicular phase (SANCHEZ-FRANCO et al. 1973). In rats, however, basal TSH levels are higher in males than females, and there is no change in hormone levels during the estrous cycle (FUKUDA et al. 1975). In unstressed rats, physiological levels of corticosterone reduce TRH-induced TSH release by a directaction on the pituitary (PAMENTER and HEDGE 1980).

II. Hypothalamic Regulation of TSH Release

The hypothalamus exerts a predominantly stimulatory influence on TSH release, as evidenced by decreased thyroid function in response to pituitary stalk section. Impaired TSH secretion follows destructive lesions of wide areas of the anterior hypothalamic area (AVERILL et al. 1961) and the paraventricular nuclei (MARTIN et al. 1970). On the other hand, increased TSH release results from electrical stimulation of a number of hypothalamic sites, including the preoptic area, paraventricular nucleus, the anterior hypothalamus and the medial-basal hypothalamus (MARTIN and REICHLIN 1972). This wide distribution of hypothalamic sites regulating TSH secretion is consonant with the reported distributaion of TRH (BROWNSTEIN et al. 1974).

Although electrical stimulation of a number of extrahypothalamic sites does not affect TSH release (MARTIN and REICHLIN 1972), there is physiological evidence for this relationship. Hypothalamic deafferentation in rats has been shown to abolish the nyctohemeral TSH rhythm (FUKUDA and GREER 1975) and to block cold-induced TSH release (ANNUNZIATO et al. 1977), although basal TSH levels were not greatly affected in these studies. These findings demonstrate that physiological TSH secretion is dependent on an intact hypothalamus responsive to extrahypothalamic input.

III. Hypothalamic Factors Involved in TSH Regulation

1. TRH

SCHALLY et al. (1966) reported the isolation of a tripeptide from porcine hypothalamus with TSH-releasing properties. This substance subsequently was identified as (pyro)Glu-His-Pro-amide and was shown to stimulate TSH release in the rodent (BOWERS et al. 1970) and in man (HAIGLER et al. 1971). The injection of TRH directly into a pituitary stalk portal vessel also releases TSH in the rat (PORTER et al. 1971), demonstrating a direct action of this tripeptide on the pituitary. In addition to enhancing TSH release, TRH has a stimulatory influence on pituitary TSH synthesis (MITTLER et al. 1969).

The participation of TRH in physiological TSH secretion is supported by studies in which passive immunization of rats with antiserum to TRH suppresses TSH secretion. KOCH et al. (1977) demonstrated that treatment of normal male rats with TRH antiserum reduces plasma TSH levels by over 65%. Cold-induced TSH secretion in the rat is also blunted by the administration of TRH antiserum (SZABO and FROHMAN 1977). In these studies it is assumed that TRH released from neurosecretory cells in the median eminence is important for TSH regulation. Further support for this notion is provided by the finding that hypothalamic TRH level has a diurnal rhythm (KOIVUSALO and LEPPALUOTO 1979) that parallels the diurnal rhythm in blood TSH levels (FUKUDA et al. 1975).

2. Somatostatin

Well-known for its growth hormone inhibiting activity, somatostatin is also a potent inhibitor of TRH-induced TSH release *in vivo* and *in vitro* (VALE et al. 1974). Endogenous somatostatin also inhibits TSH release, as the administration of somatostatin antiserum increases basal TSH levels in conscious and anesthetized rats (FERLAND et al. 1976; CHIHARA et al. 1978b). The increase in blood TSH levels following cold exposure or TRH administration is also enhanced by somatostatin antiserum (FERLAND et al. 1976; ARIMURA and SCHALLY 1976). Pretreating rats with TRH antiserum, however, prevents subsequent somatostatin antiserum treatment from eliciting TSH secretion (CHIHARA et al. 1978b). This finding suggests that somatostatin inhibits TSH release by blocking the response of the thyrotroph to TRH. Alternatively, the inhibitory effect of somatostatin on basal and noradrenaline-stimulated TRH release from cultured rat hypothalamus (HIROOKA et al. 1978) suggests an alternate site of action for somatostatin to inhibit TSH release.

IV. Pituitary-Thyroid Feedback

It is well established that the thyroid hormones T_3 and T_4 participate in the system that controls their secretion by inhibiting pituitary TSH release. Thyroid hormones have a direct inhibitory action on the pituitary (VON EULER and HOLMGREN 1956) that can be blocked by pretreatment with inhibitors of protein synthesis (BOWERS et al. 1968). The pituitary is very sensitive to circulating levels of thyroid hormones; the chronic administration of small quantities of thyroid hormones which do not raise serum T_3 and T_4 levels above the normal range inhibit TRH-induced TSH release (SNYDER and UTIGER 1972). Conversely, the reduced circulating thyroid hormone level in the iodine-deficient rat causes a dramatic increase in TSH secretion which obliterates the diurnal rhythm in blood TSH (FUKUDA et al. 1975). Further control of thyroid hormone interaction with the pituitary is provided by the hypothalamus, whose secretion determines the sensitivity of the pituitary to feedback inhibition (REICHLIN et al. 1972).

There is evidence that thyroid hormones also have effects on hypothalamic TRH and somatostatin. SINHA and MEITES (1965) showed that hypothalamic TRH activity was increased in the hypothyroid rat but was unaltered in T_4-treated rats. The microinjection of small quantities of T_3 into the hypothalamus of hypothyroid monkeys causes a prompt reduction in plasma TSH levels (BELCHETZ et al. 1977). In this study delivery of T_3 to the pituitary by portal blood may have contributed to TSH inhibition. Recently it was shown that hypothalamic somatostatin content is decreased in hypothyroid rats and is restored to euthyroid levels by T_3 treatment (BERELOWITZ et al. 1980). Further, the release of somatostatin from normal rat hypothalamus *in vitro* is unaffected by TRH or TSH but is stimulated by T_3 (BERELOWITZ et al. 1980). These findings support a hypothalamic site of action for thyroid hormones in the regulation of both TSH and GH in this species.

V. Monoaminergic Control of TSH Secretion

The involvement of brain neurons containing noradrenaline, dopamine, or 5-HT in TSH regulation is less well understood than for GH or prolactin. In general, it can be argued that noradrenergic neurons stimulate, whereas dopaminergic neurons inhibit TSH secretion in man and in the rodent. It is difficult to present a unified argument for serotoninergic control of TSH.

In rats the intraventricular injection of either noradrenaline or clonidine elevates blood TSH levels (ANNUNZIATO et al. 1977; HOLAK et al. 1978). Decreasing noradrenaline synthesis with α-methyl-p-tyrosine, FLA-63 (a DBH inhibitor), or diethyldithiocarbamate treatment reduces basal and cold-induced TSH secretion; this effect is reversed by the injection of clonidine (KRULICH et al. 1977; ANNUNZIATO et al. 1977). These studies indicate that in the rat the release of noradrenaline by central neurons may mediate basal and cold-induced TSH release.

In hypothyroid humans the administration of the DBH inhibitor fusaric acid decreases the elevated serum TSH levels (YOSHIMURA et al. 1977). In normal men, however, phentolamine was reported to decrease (NILSSON et al. 1974) or to not affect (WOOLF et al. 1972) TRH-induced TSH release. These few studies suggest a facilitatory influence of noradrenaline neurons on TSH regulation.

The injection of the dopamine receptor agonists apomorphine or piribedil decreases basal TSH levels in rats; this effect is abolished by haloperidol (MUELLER et al. 1976 a). Similarly, RANTA et al. (1977) showed that apomorphine or bromocriptine blocks cold-induced TSH secretion. In hypothyroid humans the administration of bromocriptine (MIYAI et al. 1974) or L-dopa (RAPOPORT et al. 1973) decreases TSH secretion. The intravenous infusion of dopamine was also shown to reduce basal (LEEBAW et al. 1978) and TRH-induced TSH secretion in normal humans (BESSES et al. 1975). These studies indicate that dopamine inhibits TSH release in man and in rodents.

Experiments to delineate the influence of serotoninergic neurons in the brain on TSH secretion have yielded contradictory results. CHEN and MEITES (1975) showed that 5-hydroxytryptophan injection increases while p-chloroamphetamine decreases serum TSH levels. The intraventricular injection of 5-HT into pentobarbital-anesthetized rats also stimulates TSH release; this effect is blocked by cyproheptadine (JORDAN et al. 1978). In this study 5-HT had no effect on TSH release *in vitro*. However, 5-HT did increase TSH release from a pituitary co-incubated with two hypothalami (JORDAN et al. 1978). It was concluded that 5-HT increases TSH secretion by stimulating hypothalamic TRH release. Similarly, the injection of p-chlorophenylalanine or 5,6-dihydroxytryptamine, or the lesioning of the raphe nuclei reduces the midday peak in serum TSH levels in normal male rats (JORDAN et al. 1979).

An inhibitory influence of serotoninergic neurons on TSH secretion is indicated by a study in which tryptophan injection blocked cold-induced TSH release in rats (MUELLER et al. 1976b). The administration of 5-HTP decreases TSH release in hypothyroid but not in euthyroid human patients (YOSHIMURA et al. 1977). Similarly, WOOLF and LEE (1977) report that tryptophan has no effect on TSH secretion in normal human subjects.

The mechanism by which noradrenergic, dopaminergic, or serotoninergic neurons in the brain influence TSH may involve the release of TRH or somatostatin by hypothalamic neuroendocrine transducer cells. GRIMM and REICHLIN (1973) report that dopamine or noradrenaline stimulates TRH release from mouse hypothalamic fragments *in vitro*. In their study disulfiram blocked the effect of dopamine on TRH release, indicating that only by conversion to noradrenaline is dopamine able to elicit TRH release. In other studies noradrenaline was reported to increase (HIROOKA et al. 1978) or to not affect TRH release from hypothalamic preparations *in vitro* (JOSEPH-BRAVO et al. 1979; MAEDA and FROHMAN 1980). It seems plausible that noradrenergic neurons stimulate TSH secretion by increasing TRH release from the hypothalamus.

Dopamine has been variously reported to stimulate (GRIMM and REICHLIN 1973; MAEDA and FROHMAN 1980) or to slightly inhibit (JOSEPH-BRAVO et al. 1979) the release of TRH from hypothalamic preparations *in vitro*. On the other hand, dopamine has consistently been demonstrated to stimulate somatostatin release from median eminence fragments (NEGRO-VILAR et al. 1978), hypothalamic fragments (MAEDA and FROHMAN 1980) and from hypothalamic synaptosomal preparations *in vitro* (WAKABAYASHI et al. 1977). In addition, dopamine recently has been shown to inhibit basal and TRH-induced TSH release from rat pituitary *in vitro* (LEWIS et al. 1980). It is possible that the inhibitory influence of dopamine has two mechanisms, one by increased somatostatin inhibition of TSH release and the other by direct inhibition of the thyrotroph.

The *in vitro* as well as the *in vivo* data for 5-HT regulation of TSH release are contradictory. GRIMM and REICHLIN (1973) report that 5-HT (10^{-4} *M*) decreases TRH release from mouse hypothalamic fragments. Other investigators find that 5-HT ($10^{-7}-10^{-4}$ *M*) has no effect on the release of TRH or somatostatin from hypothalamus *in vitro* (JOSEPH-BRAVO et al. 1979, MAEDA and FROHMAN 1980). It is likely that serotoninergic neurons in the brain have a minor influence on TSH regulation.

VI. Conclusions

The release of thyrotropin by the pituitary is controlled by two hypothalamic hormones, TRH and somatostatin, and by feedback inhibition by the thyroid hormones T_3 and T_4 (Fig. 4). Thyroid hormones are potent inhibitors of TSH release, by a direct action on the thyrotroph, by stimulating hypothalamic somatostatin release and possibly by inhibiting hypothalamic TRH release. Somatostatin inhibits TSH secretion by inhibiting hypothalamic TRH release

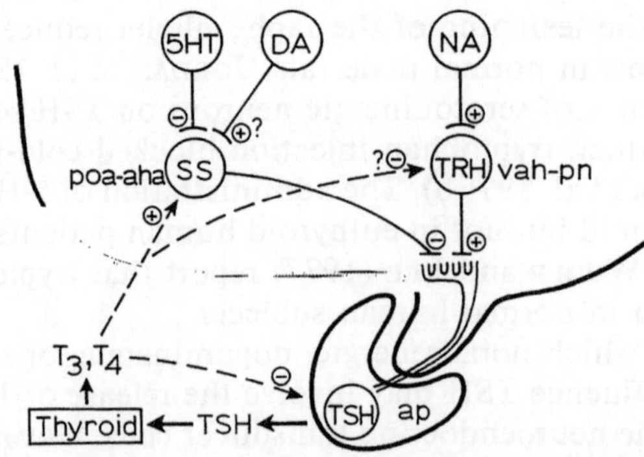

Fig. 4. Diagram of regulation of the hypothalamic-pituitary-thyroid axis including possible sites at which monoaminergic neurons might interact with hypothalamic releasint factor neurons. Abbreviations: 5-HT, 5-hydroxytryptamine; DA, dopamine; NA, noradrenaline; SS, somatostatin; TRH, thyrotropinreleasing hormone; TSH, thyroid stimulating hormone; T_3, triiodothyronine, T_4, thyroxine; poa-aha, preoptic area-anterior hypothalamic area; vah-pn, ventral anterior hypothalamus-paraventricular nucleus; ap, anterior pituitary.

and by blocking the action of TRH on the thyrotroph. Circulating estrogens stimulate, whereas corticosteroids inhibit, TSH secretion. The release of noradrenaline by central neurons increases TSH release probably by stimulating hypothalamic TRH release. Dopamine inhibits TSH secretion by stimulating the release of somatostatin from the hypothalamus and by direct inhibition of the pituitary. The release of 5-HT by brain neurons has a minor influence on TSH release, except perhaps in the maintenance of the diurnal TSH rhythm in the rat.

E. Gonadotropins

I. Normal Patterns of LH and FSH Secretion

Follicle-stimulating hormone (FSH) and luteinizing hormone (LH) are glycoproteins synthesized in basophilic cells in the anterior pituitary. In the testis of the male, testosterone secretion by interstitial cells is stimulated by LH, whereas FSH modulates the growth and maturation of tubule cells. In the female, the development of the ovarian follicle and estrogen production is controlled by FSH and LH, whereas the surge of LH at midcycle is critical for ovulation and for development of the corpus luteum. The gonadal steroids, in turn, feedback on the hypothalamus and the pituitary to regulate gonadotropin secretion. Normally this feedback control is inhibitory, however in the female at midcycle estradiol paradoxically stimulates LH and FSH release, resulting in an ovulatory surge of gonadotropins. This estrogen-gonadotropin feedback loop is common to the rat and to rhesus monkey as treatment of either species with antibodies to estradiol will prevent the midcycle surge in gonadotropin secretion (NEILL et al. 1971; FERIN et al. 1974).

Under basal conditions in men and women gonadotropins are released episodically at approximately hourly intervals (BOYAR et al. 1972). Pulsatile release of LH also occurs in the rat (GAY and SHETH 1972) and in the rhesus monkey (CARMEL et al. 1976) at intervals of approximately 30 minutes in the rat and 1-3 hours in the monkey. Ovariectomy in animals increases blood gonadotropin levels but does not alter the episodic pattern of hormone release.

II. Hypothalamic Regulation of LH and FSH Secretion

Surgical disconnection of the medial basal hypothalamus in rats blocks ovulation (HALASZ and PUPP 1969) without disrupting the tonic pulsatile release of LH (BLAKE and SAWYER 1974). With the use of selective knife cuts it was determined that ovulation depends upon afferent fibers entering the medial basal hypothalamus anteriorly (HALASZ and GORSKI 1967). Discrete lesions of the suprachiasmatic nucleus are particularly effective in inducing an anovulatory persistent estrus (BARRACLOUGH 1963). Lesions of the medial preoptic nucleus also induce persistent estrus (WIEGAND et al. 1980), although lesions placed elsewhere in the medial preoptic area-anterior hypothalamus have no such effect (CLEMENS et al. 1976; WIEGAND et al. 1980). In rats with anterior hypothalamic deafferentation, bilateral electrolytic lesions of the anterior portions of the arcuate nuclei abolish pulsatile LH secretion (SOPER and WEICK 1980). In this study lesions of the arcuate nuclei in intact rats did *not* block pulsatile LH release, however. The authors concluded that episodic LH release in the rat can be generated by two pathways—one involving the arcuate nuclei and another involving afferent fibers entering the medial basal hypothalamus anteriorly from extrahypothalamic structures.

Unlike the rat, the rhesus monkey can maintain ovulatory LH release as well as basal gonadotropin secretion following surgical disconnection of the medial basal hypothalamus (KREY et al. 1975, FERIN et al. 1977). Pituitary stalk section, however, results in immediate and long-lasting inhibition of LH release in monkeys (FERIN 1979). These results indicate that in this species the medial basal hypothalmus contains all the necessary elements for maintaining basal and ovulatory gonadotropin secretion. Little is known about the anatomical organization of gonadotropin regulation in humans.

III. Hypothalamic Luteinizing Hormone-Releasing Hormone

Luteinizing hormone-releasing hormone, or LHRH, was first isolated from porcine hypothalamus and characterized as a linear decapeptide (MATSUO et al. 1971). LHRH stimulates the release of both LH and FSH *in vivo* and *in vitro* (SCHALLY et al. 1971), although the LH secretory response is generally greater than that for FSH (BLACKWELL et al. 1973). There is insufficient evidence at present to establish the existence of a separate FSHRH distinct from LHRH (SCHALLY et al. 1976b).

Evidence for gonadotropin control by endogenous hypothalamic LHRH is provided by studies in which passive immunization of monkeys with antisera to LHRH (MCCORMACK et al. 1977), pituitary stalk section (FERIN 1979), or arcuate lesions (KNOBIL and PLANT 1978) diminish tonic LH secretion. In arcuate-lesioned monkeys constant infusion of LHRH initially increases gonadotropin secretion but after about 24 hours this effect of LHRH is lost (KNOBIL and PLANT 1978). In contrast, the pulsatile administration of LHRH to arcuate-lesioned animals results in long-lasting LH stimulation (KNOBIL and PLANT 1978). Constant infusion of LHRH into normal monkeys (FERIN et al. 1978) or rats (DE KONING et al. 1978) similarly results in an initial stimulation but subsequent decline in blood LH levels. These observations coupled with the demonstration of pulsatile release of LHRH into hypophyseal portal blood of the monkey (CARMEL et al. 1976) argue that tonic gonadotropin secretion is dependent on intermittent rather than constant stimulation by LHRH.

The ovulatory surge in gonadotropins is dependent on stimulation by ovarian estrogens. It is clear that estrogens enhance LH release by increasing pituitary sensitivity to LHRH *in vivo* (ARIMURA and SCHALLY 1971) and *in vitro* (LABRIE et al. 1976). On the other hand, estrogens may also stimulate LH release by increasing hypothalamic LHRH secretion. This notion is supported by the finding of elevated LHRH levels in portal blood of rhesus monkeys (NEILL et al. 1977) and rats (SARKAR et al. 1976) at midcycle. Thus ovulatory gonadotropin secretion may be generated by increased hypothalamic LHRH release in addition to enhanced sensitivity of the gonadotroph to LHRH.

IV. Monoaminergic Control of LH and FSH Secretion

A number of difficulties are encountered in attempting to interpret data from studies involving the neural regulation of gonadotropin secretion. These include: (i) the lack of specificity of many of the drugs used, (ii) the use of surgically altered (e.g. gonadectomized) or anesthetized animal preparations (e.g. pentobarbital-blocked), (iii) the divergency of neural pathways for tonic versus ovulatory LH regulation, (iv) the apparent species difference in gonadotropin regulation (e.g. hypothalamic deafferentation abolishes ovulation in rats, but not monkeys). With these complications in mind it is not surprising that there is much conflicting data and that the formulation of unifying arguments for monoaminergic control of physiological gonadotropin secretion is difficult.

In most experimental protocols, treatments that stimulate central noradrenergic neurons increase gonadotropin secretion in man and animals. The intraventricular infusion of noradrenaline induces ovulation in the rabbit (SAWYER et al. 1949) and stimulates LH release in ovariectomized estrogen and progesterone-primed (KRIEG and SAWYER 1976) and 6-OHDA-treated rats (BACHA and DONOSO 1974); phenoxybenzamine blocks the effect of noradrenaline in 6-OHDA-treated rats. Blocking catecholamine or noradrenaline synthesis with α-methyl-p-tyrosine or diethyldithiocarbamate, respectively, blunts the LH secretory response following electrical stimulation of the preoptic area in anesthetized proestrus rats (KALRA and MCCANN 1973). These synthesis inhibitors also block LH release during proestrus (KALRA and McCANN 1974), but do not affect the LH secretory response following electrical stimulation of the median eminence-arcuate region (KALRA and McCANN 1973). Similarly, infusion of 6-OHDA into the basal midbrain of the rat (MARTINOVIC and McCANN 1977) or into the third ventricle of the rabbit blocks ovulatory LH release (PRZEKOP and DOMANSKI 1976). On the other hand, NICHOLSON et al. (1978) and RABII et al. (1980) report maintenance of estrous cycles and ovulatory LH release in 6-OHDA-treated rats or rabbits. This discrepancy may be related to either the minor role of hypothalamic noradrenergic neurons in gonadotropin regulation or to the development of receptor supersensitivity because of an increase in receptor population following treatment with this neurotoxin. LEPPALUOTO et al. (1976) showed that ovulatory LH but not FSH release in humans is reduced by blocking noradrenaline synthesis with fusaric acid.

The pulsatile release of LH in ovariectomized rats or monkeys is inhibited by the administration of noradrenaline synthesis inhibitors (DROUVA and GALLO 1976; GNODDE and SCHUILING 1976) or α-adrenoceptor antagonists (BHATTACHARYA et al. 1972; WEICK 1978). Conversely, the infusion of noradrenaline into the third ventricle of ovariectomized rats inhibits pulsatile LH secretion (GALLO and DROUVA 1979). It is possible that one effect of intraventricular noradrenaline infusion is the nonspecific activation of an inhibitory dopaminergic influence on hypothalamic LHRH release (GALLO 1980).

Dopaminergic mechanisms have been reported to stimulate, inhibit, or not to affect gonadotropin release. Ovulatory LH and FSH release in the rat are blocked by haloperidol or pimozide injection (DICKERMAN et al. 1974; BEATTIE et al. 1976), suggesting a stimulatory role for dopamine. This notion is supported by the finding that L-dopa stimulates LH secretion in the proestrus rat (WEDIG and GAY 1970).

Pulsatile LH release, however, is reported to be inhibited by dopamine receptor agonists (DROUVA and GALLO 1976; GNODDE and SCHUILING 1976), suggesting an inhibitory role for dopamine in this response. Similarly, pulsatile LH release is inhibited by the intraventricular infusion of dopamine in the intact rats (GALLO and DROUVA 1979), or by the systemic administration of apomorphine in rats with complete deafferentation of the medial basal hypothalamus (ARENDASH and GALLO 1978a). On the other hand, pimozide treatment is reported not to affect (DROUVA and GALLO 1976) or to inhibit LH release in ovariectomized rats (OJEDA et al. 1974). Furthermore, dopamine has also been shown to stimulate the *in vitro* release of LHRH from medial basal hypothalamus of normal male and estradiol-treated ovariectomized rats, but not from ovariectomized rats who have not received estradiol (ROTSZTEJN et al. 1976). Thus, depending upon the exerimental protocol and the steroid environment of the animal preparation, dopaminergic mechanisms can elicit these various effects on gonadotropin secretion.

The administration of 5-HT into the third ventricle of male rats decreases LH and FSH release (KAMBERI et al. 1970a, 1971b). Similarly, the intraventricular injection of 5-HT into pentobarbital-treated proestrus rats decreases LH release following electrical stimulation of the medial preoptic area (CRAMER and Barraclough 1978). Dorsal raphe stimulation blocks pulsatile LH release in unanesthetized, ovariectomized rats; this effect is abolished by pretreating rats with p-chlorophenylalanine or metergoline (Arendash and Gallo 1978b). On the other hand, Hery et al. (1976) report that p-chlorophenylalanine treatment inhibits the cyclic release of LH in ovariectomized, estrogen-primed rats. Since 5-HT has no effect on LH or FSH release from pituitaries *in vitro* (Kamberi et al. 1970b), yet can inhibit the release of LHRH from rat medial basal hypothalamus *in vitro* (Charli et al. 1978), the release of 5-HT from nerve terminals in the hypothalamus may inhibit pulsatile LH secretion by suppressing LHRH release.

The treatment of rats with morphine inhibits (Pang et al. 1977), whereas naloxone stimulate gonadotropin secretion (Bruni et al. 1977). These findings point to a possible role for endogenous opioid peptides in gonadotropin regulation. The mechanism for opioid modulation of gonadotropin release may involve 5-HT, as p-chlorophenylalanine potentiates, whereas 5-hydroxytryptophan attenuates, naloxone-induced LH release (Ieiri et al. 1980). The observation that morphine or β-endorphin injection increases hypothalamic 5-HT turnover (Meites et al. 1979; van Loon and de Souza 1978) supports the hypothesis that endogenous opioid peptides inhibit gonadotropin secretion by stimulating the release of 5-HT by hypothalamic neurons.

V. Conclusions

Under basal conditions the gonadotropins are released episodically, reflective of pulsatile release of a stimulatory hypothalamic releasing factor (LHRH). Normally gonadal steroids inhibit LH and FSH release, however at midcycle estrogens paradoxically have a stimulatory influence resulting in an ovulatory surge of gonadotropins. The release of monoamines by brain neurons can also modulate gonadotropin release, probably by interactions with hypothalamic LHRH neurons. Most likely, noradrenergic neurons are stimulatory, whereas serotoninergic neurons are inhibitory for gonadotropin regulation. Dopamine neurons appear to be stimulatory for ovulatory gonadotropin secretion, although episodic gonadotropin release is inhibited by dopamine neurons.

F. Adrenocorticotropin

I. Normal Patterns of Adrenocorticotropin (ACTH) Secretion

The presence of ACTH in mammals has been clearly documented in the anterior and intermediate lobes of the pituitary as well as throughout the central nervous system (Krieger 1980) and the gut (Larsson 1979). As a hormone released from the pituitary, ACTH functions to stimulate the production and release of adrenal steroids. Adrenal corticosteroids, in turn, feedback on the hypothalamic-pituitary axis to inhibit pituitary ACTH release.

ACTH secretion is characterized as episodic and circadian. In man, blood ACTH levels peak in early morning before awakening (Krieger 1975), whereas in the rat the ACTH peak is observed late in the afternoon or early in the evening (SZAFARCZYK et al. 1979). In each of these mammals the periodicity in blood corticosteroids closely parallels that for ACTH. Stress, in rats (ROSSIER et al. 1977) and in humans (NEWSOME and ROSE 1971), causes prompt increases in ACTH secretion. The site of corticosteroid inhibition of ACTH release is at both hypothalamic and pituitary levels (JONES and HILLHOUSE 1976).

II. Hypothalamic Regulation of ACTH Secretion

Pituitary stalk section in the monkey (KENDALL and ROTH 1964) or relocation of the pituitary in the rat (GREER et al. 1963) results in diminished basal and stress-related ACTH secretion. On the other hand, surgical deafferentation of the medial basal hypothalamus or the median eminence does not abolish ACTH release consequent to histamine injection or insulin-induced hypoglycemia (MAKARA et al. 1970). From these studies it is clear that the brain exerts a stimulatory influence on ACTH release and that the ACTH response to some stressful stimuli requires only an intact medial basal hypothalamus-pituitary unit.

There is also evidence for extrahypothalamic control of ACTH secretion. GANN et al. (1977) have provided a detailed description in the cat of the extrahypothalamic neural pathways involved in the ACTH response to hemorrhage. Further, they implicate the medial-dorsal posterior hypothalamus in the hemorrhage response, and show a stimulatory and an inhibitory influence of this hypothalamic area on ACTH release. KNIGGE (1961) showed that lesions of the amygdala inhibit, whereas lesions of the hippocampus enhance, stress-related ACTH secretion. SZAFARCZYK et al. (1979) showed that lesions of suprachiasmatic nuclei abolish the circadian rhythm of ACTH secretion in rats.

III. Corticotropin-Releasing Factor (CRF)

Three decades ago SCHALLY et al. (1958) reported partial purification of a peptide from posterior pituitary tissue with CRF-activity *in vitro*. Yet only recently has a 41 residue peptide been isolated from ovine hypothalamus which qualifies as a CRF (VALE et al. 1981; SPIESS et al. 1981). Synthetic CRF is a potent stimulus for ACTH release *in vivo* and *in vitro* (VALE et al. 1981; SPIESS et al. 1981; RIVIER et al. 1982a). The notion that endogenous CRF modulates ACTH release is supported by the finding that injection of anti-CRF serum into rats reduces ACTH release (RIVIER et al. 1982b).

Vasopressin has also been postulated to be the elusive CRF. High levels of vasopressin are observed in monkey hypophyseal portal blood (ZIMMERMAN et al. 1973). This nonapeptide stimulates ACTH release *in vitro* (DE WIED et al. 1969) and potentiates the ACTH response to a hypothalamic extract with CRF activity (YATES et al. 1971). Recent studies have demonstrated that vasopressin enhances the ACTH-releasing effect of synthetic CRF on cultured corticotropic cells (VALE et al. 1983). In view of these findings it is possible that vasopressin may contribute to ACTH release *in vivo*.

IV. Monoaminergic Control of ACTH Secretion

The question of control of ACTH release by the catecholamines or 5-HT remains a very controversial issue. While this confusion may reflect differences in experimental protocol, it undoubtedly also reflects the inherent complexity in ACTH secretory physiology.

There is considerable evidence that the release of noradrenaline by brain neurons inhibits ACTH release in the rat, the dog, and in man. The intraventricular administration of noradrenaline or clonidine inhibits stress-induced ACTH secretion in dogs (VAN LOON et al. 1971; GANONG et al. 1976). The intraventricular injection of phenoxybenzamine, but not propranolol, blocks the inhibition of ACTH release by intravenous L-dopa or clonidine (GANONG et al. 1976). On the other hand, BHARGAVA et al. (1972) report that intraventricular noradrenaline or phenylephrine increases plasma cortisol in the dog and the effect of α-adrenoceptor agonists is blocked by yohimbine. Agents that mimic or antagonize the effects of dopamine have no effect on ACTH release in the dog (GANONG et al. 1976).

In the rat, the intraventricular injection of α-methyl-p-tyrosine (VAN LOON et al. 1971) or 6-OHDA (CUELLO et al. 1974) results in increased corticosterone secretion. Similarly, the systemic injection of phentolamine (SCAPAGNINI and PREZIOSI 1973) is associated with increased ACTH secretion. These data suggest that α-adrenoceptors inhibit ACTH release in the rat. Conversely, ABE and HIROSHIGE (1974) find that in the chronically cannulated rat intraventricular noradrenaline or dopamine increase plasma corticosterone levels. In addition they report that 6-OHDA or reserpine have no effect on circadian or stress-induced changes in plasma corticosterone. Recently GIRAUD et al. (1980) showed that haloperidol injection markedly enhances ACTH release in the rat. Most likely, the release of noradrenaline or dopamine by brain neurons inhibits pituitary ACTH release.

Some confusion is noted in interpretation of the effects of catecholamines on *in vitro* release of CRF or ACTH from rat pituitary. VALE et al. (1978) find that noradrenaline stimulates ACTH release from cultured anterior pituitary cells, an effect blocked by phentolamine. VAN LOON and KRAGT (1970) report that dopamine has no effect on ACTH release from intact anterior pituitary. Although BRIAUD et al. (1979) similarly report no effect of catecholamines on anterior lobe ACTH release, they find increases ACTH release by noradrenaline or dopamine from neurointermediate lobe. In their study the effect of noradrenaline or dopamine is blocked by phentolamine or haloperidol, respectively. JONES et al. (1976) find that neither noradrenaline nor dopamine affect CRF release from hypothalamus *in vitro*, although noradrenaline inhibits 5-HT or acetylcholine-stimulated CRF release. Conversely, FEHM et al. (1980) report that noradrenaline or dopamine stimulates CRF release from whole hypothalamus or medial basal hypothalamus. In both studies CRF activity apparently is not due to vasopressin.

In man, the role of catecholamines in ACTH regulation is also not clear. NAKAI et al. (1973) report that the peripheral α-adrenoceptor agonist methoxamine increases ACTH release. Phentolamine blocks the effect of methoxamine (NAKAI et al. 1973) but does not affect the corticosteroid response to insulin-induced hypoglycemia (NAKAGAWA et al. 1971). Conversely, noradrenaline but not dopamine infusion inhibits ACTH release (WILCOX et al. 1975), and guanfacine, an α-adrenoceptor agonist, inhibits ACTH secretion stimulated by insulin-induced hypoglycemia (LANCRANJAN et al. 1979).

The release of ACTH from anterior and neurointermediate lobes of rat pituitary *in vitro* is not affected by 5-HT (BRIAUD et al. 1979). JONES et al. (1976) report that 5-HT releases CRF from rat hypothalamus *in vitro*, although FEHM et al. (1980) did not observe this effect.

Although most data support a facilitatory role for 5-HT in ACTH regulation, there is also evidence for ACTH inhibition by this indoleamine. The intraventricular injection or the implantation of 5-HT into the medial hypothalamus of rats blocks stress-induced corticosterone release (VERMES and TELEGDY 1972). Stimulation of the raphe nuclei inhibits, whereas lesions of the raphe nuclei enhance, stress-induced corticosterone secretion (VERMES et al. 1974).

On the other hand, the injection of the 5-HT precursor 5-hydroxytryptophan, or of fluoxetine, fenfluramine or quipazine are each associated with enhanced corticosterone release in the rat (FULLER and SNODDY 1979). Corticosterone release is also stimulated by 5-hydroxytryptophan in rats with hypothalamic deafferentation, suggesting hypothalamic control of CRF release by 5-HT (POPOVA et al. 1972). Recently, SZAFARCZYK et al. (1979) report that injection of p-chlorophenylalanine into rats reduces plasma ACTH concentrations while abolishing the circadian variations in this hormone. Paradoxically, plasma corticosterone levels in the same animals were actually increased after drug treatment. This finding suggests that 5-HT may have a direct effect on adrenal function, or that adrenal steroid release may occur autonomously. In man, the administration of tryptophan is associated with increases in cortisol (MODLINGER et al. 1979) and ACTH secretion (MODLINGER et al. 1980). Thus in man and in the rat, the release of 5-HT by brain neurons probably stimulates ACTH release through interactions with hypothalamic CRF.

V. Conclusions

The release of ACTH by the pituitary is stimulated by a hypothalamic CRF and possibly by vasopressin (Fig. 5). Corticosteroids are potent inhibitors of ACTH release, by a direct action on the corticotroph and at the level of the hypothalamus. ACTH secretion is described as episodic and circadian, and is rapidly stimulated by stress. There is considerable controversy regarding control of ACTH release by neurotransmitters in the brain. In man and in rats most data indicate that stimulation of central α-adrenoceptors inhibits ACTH release, whereas stimulation of 5-HT receptors enhances ACTH secretion. The stimulation of dopamine receptors probably has little influence on ACTH release in man but may inhibit ACTH secretion in rats.

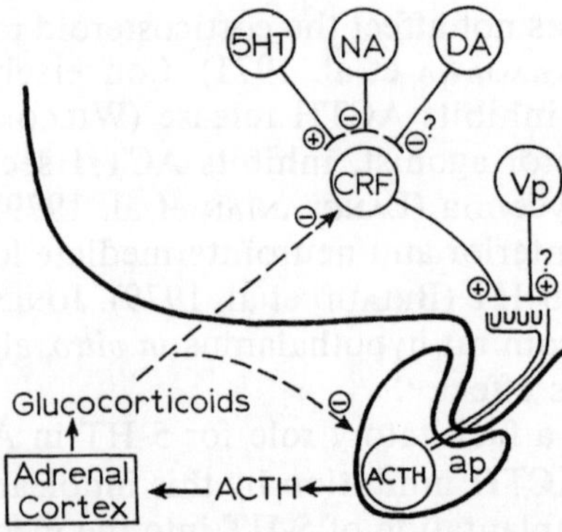

Fig. 5. Diagram of regulation of the hypothalamic-pituitary-adrenal axis including possible sites at which monoaminergic neurons might interact with hypothalamic releasing factor neurons. Abbreviations: 5-HT, 5-hydroxytryptamine; NA, noradrenaline; DA, dopamine; CRF, corticotropin-releasing factor, Vp, vasopressin; ACTH, adrenocorticotropic hormone; ap, anterior pituitary.

G. References

Abe K, Hiroshige T (1974) Changes in plasma corticosterone and hypothalamic CRF levels following intraventricular injection or drug-induced changes of brain biogenic amines in the rat. Neuroendocrinology 14:195–211

Abrams RL, Kaplan S, Grumbach M (1971) The effect of administration of human growth hormone on the plasma GH, cortisol, glucose, and free fatty acid response to insulin: evidence for growth hormone autoregulation in man. J Clin Endocrinol Metab 50:940–950

Alford FP, Baker HWG, Burger HG, De Kretser DM, Hudson B, Johns MW, Masterson JP, Patel YC, Rennie GC (1973) Temporal patterns of integrated plasma hormone levels during sleep and wakefulness. I. Thyroid stimulating hormone, growth hormone and cortisol. J Clin Endocrinol Metab 37:841–847

Amenomori Y, Meites J (1970) Effect of a hypothalamic extract on serum prolactin levels during the estrous cycle and lactation. Proc Soc Exp Biol Med 134:492–495

Amenomori Y, Chen CL, Meites J (1970) Serum prolactin levels in rats during the different reproductive states. Endocrinology 86:506–510

Annunziato L, Di Renzo G, Lombardi G, Scopacasa F, Schettini G, Preziosi P, Scapagnini U (1977) The role of central noradrenergic neurons in the control of thyrotropin secretion in the rat. Endocrinology 100:738–744

Arendash GW, Gallo RV (1978a) Apomorphine-induced inhibition of episodic LH release in ovariectomized rats with complete hypothalamic deafferentation. Proc Soc Exp Biol Med 159:121–125

Arendash GW, Gallo RV (1978b) Serotonin involvement in the inhibition of episodic luteinizing hormone release during electrical stimulation of the midbrain dorsal raphe nucleus in ovariectomized rats. Endocrinology 102:1199–1206

Arimura A, Schally AV (1971) Augmentation of pituitary responsiveness to LH-releasing hormone (LH-RH) by estrogen. Proc Soc Exp Biol Med 136:290–293

Arimura A, Schally AV (1976) Increases in basal and thyrotropin-releasing hormone (TRH)-stimulated secretion of thyrotropin (TSH) by passive immunization with antiserum to somatostatin in rats. Endocrinology 98:1069–1072

Arimura A, Smith WD, Schally AV (1976) Blockade of the stress-induced decrease in blood GH by antisomatostatin serum in rats. Endocrinology 98:540-543
Arnold MA, Fernstrom JD (1978) Serotonin receptor antagonists block a natural, short-term surge in serum growth hormone levels. Endocrinology 103:1159-1163
Arnold MA, Fernstrom JD (1979) Tryptophan injection stimulates growth hormone secretion in rats. Abstract of the Endocrine Society 61st Annual Meeting, p 223
Arnold MA, Fernstrom JD (1980) Administration of antisomatostatin serum to rats reverses the inhibition of pulsatile growth hormone secretion produced by injection of metergoline but not yohimbine. Neuroendocrinology 31:194-199
Averill RLW, Purves HD, Sirett NE (1961) Relation of the hypothalamus to anterior pituitary thyrotropin secretion. Endocrinology 69:735-745
Bacha JC, Donoso AD (1974) Enhanced luteinizing hormone release after noradrenaline treatment in 6-hydroxydopamine treated rats. J Endocrinol 62:169-170
Barraclough CA (1963) Secretion and release of LH and FSH: discussion. In: Advances in Neuroendocrinology Urbana: University of Illinois Press
Bartke A, Croft BT, Dalterio S (1975) Prolactin restores plasma testosterone levels and stimulates testicular growth in hamsters exposed to short day-length. Endocrinology 97:1601-1604
Beattie CW, Gluckman MI, Corbin AA (1976) A comparison of gamma-butyrolactone and pimozide on serum gonadotropins and ovulation in the rat. Proc Soc Exp Biol Med 153:147-150
Belchetz PE, Gredley G, Bird D, Himsworth RL (1977) Regulation of thyrotropin secretion by negative feedback of triiodothyronine on the hypothalamus. J Endocrinol 76:439-448
Ben-Jonathan N, Olivaer C, Weiner HJ, Mical RS, Porter JC (1977) Dopamine in hypophysial portal plasma of the rat during the estrous cycle and throughout pregnancy. Endocrinology 100:452-458
Berelowitz M, Maeda K, Harris S, Frohman LA (1980) The effect of alterations in the pituitary-thyroid axis on hypothalamic content and in vitro release of somatostatin-like immunoreactivity. Endocrinology 107:24-29
Besser GM, Parke L, Edwards CRW, Forsyth IA, McNeilly AS (1972) Galactorrhea: successful treatment with reduction of plasma prolactin by bromo-ergocryptine. Br Med J 3:669-672
Besses GS, Burrow GN, Spaulding SW, Donabedian RK (1975) Dopamine infusion acutely inhibits the TSH and prolactin response to TRH. J Clin Endocrinol Metab 41:985-988
Bhargava KP, Bhargava R, Gupta MB (1972) Central adrenoceptors concerned in the release of adrenocorticotrophic hormone. Br J Pharmacol 45:682-683
Bhattacharya AN, Dierschke DJ, Yamaji T, Knobil E (1972) The pharmacologic blockade of the circhoral mode of LH secretion in the ovariectomized rhesus monkey. Endocrinology 90:778-786
Birge CA, Jacobs LS, Hammer CT, Daughaday WH (1970) Catecholamine inhibition of prolactin secretion by isolated rat adenohypophyses. Endocrinology 86:120-130
Bivens CH, Lebovitz HE, Feldman JM (1973) Inhibition of hypoglycemia-induced growth hormone secretion by the serotonin antagonists cyproheptadine and methysergide. N Engl J Med 289:236-239
Blackard WG, Heidingsfelder SA (1968) Adrenergic receptor control mechanisms for growth hormone secretion. J Clin Invest 47:1407-1414
Blackwell RE, Amoss M Jr, Vale W, Burgus R, Rivier J, Monahan M, Ling N, Guillemin R (1973) Concomitant release of FSH and LH induced by native and synthetic LRF. Am J Physiol 224:170-175

Blake CA (1974) Stimulation of pituitary prolactin and TSH release in lactating and proestrous rats. Endocrinology 94:503–508

Blake CA, Sawyer CH (1974) Effects of hypothalamic deafferentation on the pulsatile rhythm in plasma concentrations of luteinizing hormone in ovariectomized rats. Endocrinology 94:730–736

Bowers CY, Lee KL, Schally AV (1968) A study of the interaction of thyrotropin-releasing factor and L-triiodothyronine: effects of puromycin and cycloheximide. Endocrinology 82:75–82

Bowers CY, Schally AV, Enzmann F, Boler J, Folkers K (1970) Porcine thyrotropin releasing hormone is (pyro)Glu-His-Pro (NH_2). Endocrinology 86:1143–1153

Bowers CY, Friesen HG, Hwang P, Guyda HJ, Folkers K (1971) Prolactin and thyrotropin release in man by synthetic pyroglutamyl-histidyl-prolinamide. Biochem Biophys Res Commun 45:1033–1041

Boyar R, Perlow M, Hellman L, Kapen S, Weitzman E (1972) Twenty-four hour pattern of luteinizing hormone secretion in normal men with sleep stage recording. J Clin Endocrinol Metab 35:73–81

Boyd AE, Lebovitz HE, Pfeiffer JB (1970) Stimulation of growth hormone secretion by L-dopa. N Engl J Med 283:1425–1429

Brazeau P, Vale W, Burgus R, Ling N, Butcher M, Rivier J, Guillemin R (1973) Hypothalamic polypeptide that inhibits the secretion of immunoreactive pituitary growth hormone. Science 179:77–79

Briaud B, Koch B, Lutz-Bucher B, Mialhe C (1979) In vitro regulation of ACTH release from neurointermediate lobe of rat hypophyses. II. Effect of neurotransmitters. Neuroendocrinology 28:377–385

Brown GM, Martin JB (1974) Corticosterone, prolactin and growth hormone responses to handling and new environment in the rat. Psychosomatic Medicine 36:241–247

Brown GM, Schalch DS, Reichlin S (1971) Hypothalamic mediation of growth hormone and adrenal stress response in the squirrel monkey. Endocrinology 89:694–703

Brownstein M, Palkovits M, Saavedra JM, Bassiri RM, Utiger RD (1974) Thyrotropin releasing hormone in specific nuclei of the brain. Science 185:267–269

Bruni JF, van Vugt DA, Marshall S, Meites J (1977) Effect of naloxone, morphine and methionine enkephalin on serum prolactin, luteinizing hormone, thyroid stimulating hormone and growth hormone. Life Sci 21:461–466

Caligaris L, Taleisnik S (1974) Involvement of neurons containing 5-hydroxytryptamine in the mechanism of prolactin release induced by estrogen. J Endocrinol 62:25–33

Carmel PW, Araki S, Ferin M (1976) Pituitary stalk portal blood collection in rhesus monkeys: evidence for pulsatile release of gonadotropin-releasing hormone (GnRH). Endocrinology 99:243–248

Carr LA, Conway PM, Voogt JL (1977) Role of norepinephrine in the release of prolactin induced by suckling and estrogen. Brain Res 133:305–314

Chambers JW, Brown GM (1976) Neurotransmitter regulation of growth hormone and ACTH in the rhesus monkey: effects of biogenic amines. Endocrinology 98:420–428

Chamley WA, Fell LR, Alford FP, Goding JR (1974) Twenty-four hour secretory profiles of ovine prolactin and growth hormone. J Endocrinol 61:165–166

Charli JL, Rotsztejn WH, Pattou E, Kordon C (1978) Effect of neurotransmitters on *in vitro* release of luteinizing hormone releasing hormone from the mediobasal hypothalamus of male rats. Neurosci Lett 10:159–163

Chen CL, Meites J (1970) Effects of estrogen and progesterone on serum and pituitary prolactin levels in ovariectomized rats. Endocrinology 86:503–505

Chen CL, Amenomori Y, Lu KH, Voogt JL, Meites J (1970) Serum prolactin levels in rats with pituitary transplants or hypothalamic lesions. Neuroendocrinology 5:220-227

Chen HJ, Meites J (1975) Effects of biogenic amines and TRH on release of prolactin and TSH in the rat. Endocrinology 96:10-14

Chen HJ, Mueller GP, Meites J (1974) Effects of L-dopa and somatostatin on suckling-induced release of prolactin and GH. Endocrine Res Commun 1:283-291

Cheng KW, Martin JB, Friesen HG (1972) Studies of neurophysin release. Endocrinology 91:177-184

Chihara K, Kato Y, Maeda K, Matsukura S, Imura H (1976) Suppression by cyproheptadine of human growth hormone and cortisol secretion during sleep. J Clin Invest 57:1393-1402

Chihara K, Arimura A, Schally AV (1978a) Immunoreactive somatostatin levels in hypophyseal portal blood of rats anesthetized with urethane, pentobarbital, or Althesin. Fed Proc 37:638

Chihara K, Arimura A, Chihara M, Schally AV (1978b) Studies on the mechanism of growth hormone and thyrotropin responses to somatostatin antiserum in anesthetized rats. Endocrinology 103:1916-1923

Chihara K, Arimura A, Schally AV (1979) Effect of intraventricular injection of dopamine, norepinephrine, and 5-hydroxytryptamine on immunoreactive somatostatin release into rat hypophyseal portal blood. Endocrinology 104:1656-1662

Clemens JA (1978) Effects of serotonin neurotoxins on pituitary hormone release. Ann NY Acad Sci 305:399-410

Clemens JA, Smalstig EB, Sawyer BD (1976) Studies on the role of the preoptic area in the control of reproductive function in the rat. Endocrinology 99:728-735

Clemens JA, Sawyer BD, Cerimele (1977) Further evidence that serotonin is a neurotransmitter involved in the control of prolactin secretion. Endocrinology 100:692-698

Clemens JA, Roush ME, Fuller RW (1978) Evidence that serotonin neurons stimulate secretion of prolactin releasing factor. Life Sci 22:2209-2214

Cocchi D, Gil-Ad L, Panerai AE, Locatelli V, Muller EE (1977) Effect of 5-hydroxytryptophan on prolactin and growth hormone release in the infant rat: evidence for different neurotransmitter mediation. Neuroendocrinology 24:1-13

Collu R, Franschini F, Visconti P, Martini L (1972) Adrenergic and serotonergic control of growth hormone secretion in adult rats. Endocrinology 90:1231-1237

Convey EM, Tucker HA, Smith VG, Zolman J (1973) Bovine prolactin, growth hormone, thyroxine, and corticoid response to thyrotropin-releasing hormone. Endocrinology 92:471-476

Cramer OM, Barraclough CA (1978) The actions of serotonin, norepinephrine, and epinephrine on hypothalamic processes leading to adenohypophyseal luteinizing hormone release. Endocrinology 103:694-703

Cramer OM, Parker CR jr, Porter JC (1979) Estrogen inhibition of dopamine release into hypophysial portal blood. Endocrinology 104:419-422

Cronin MJ, Roberts JM, Weiner RI (1978) Dopamine and dihydroergocryptine binding to the anterior pituitary and other brain areas of the rat and sheep. Endocrinology 103:302-309

Cuello AC, Shoemaker WJ, Ganong WF (1974) Effect of 6-hydroxydopamine on hypothalamic norepinephrine and dopamine content, ultrastructure of the median eminence, and plasma corticosterone. Brain Res 78:57-69

Davis SL, Borger ML (1973) Hypothalamic catecholamine effects on plasma levels of prolactin and growth hormone in sheep. Endocrinology 92:303-309

De Greef WJ, Neill JD (1979) Dopamine levels in hypophysial stalk plasma of the rat during surges of prolactin secretion induced by cervical stimulation. Endocrinology 105:1093-1099

De Koning J, van Dieten JAMJ, van Rees GP (1978) Refractoriness of the pituitary gland after continuous exposure to luteinizing hormone-releasing hormone. J Endocrinology 79:311-318

Devane GW, Siler TM, Yen SSC (1973) Acute suppression of insulin and glucose levels by synthetic somatostatin in normal human subjects. J Clin Endocrinol Metab 38:913-915

De Wied D, Witter A, Versteeg DHG, Muldar AH (1969) Release of ACTH by substances of central nervous system origin. Endocrinology 85:561-569

Dickerman S, Kledzik G, Gelato M, Chen HJ, Meites J (1974) Effects of haloperidol on serum and pituitary prolactin, LH and FSH, and hypothalamic PIF and LRF. Neuroendocrinology 15:10-20

Diefenbach WP, Carmel PW, Frantz AG, Ferin M (1976) Suppression of prolactin secretion by L-dopa in the stalk sectioned rhesus monkey. J Clin Endocrinol Metab 43:638-642

Donoso AO, Bishop W, Fawcett CP, Krulich L, McCann SM (1971) Effects of drugs that modify brain monoamine concentrations on plasma gonadotropin and prolactin levels in the rat. Endocrinology 89:774-784

Donoso AO, Bishop W, McCann SM (1973) The effects of drugs which modify catecholamine synthesis on serum prolactin in rats with median eminence lesions. Proc Soc Exp Biol Med 143:360-363

Donoso AO, Banzan AM, Barcaglioni JC (1974) Further evidence on the direct action of L-dopa on prolactin release. Neuroendocrinology 15:236-239

Drouva SV, Gallo RV (1976) Catecholamine involvement in episodic luteinizing hormone release in adult ovariectomized rats. Endocrinology 99:651-658

Dular R, Labella F, Vivian S, Eddie L (1974) Purification of prolactin-releasing and inhibiting factors from beef. Endocrinology 94:563-567

Durand D, Martin JB, Brazeau P (1977) Evidence for a role of α-adrenergic mechanisms in regulation of episodic growth hormone secretion in the rat. Endocrinology 100:722-728

Enjalbert A, Moos F, Carbonell L, Priam M, Kordon C (1977) Prolactin inhibiting activity of dopamine-free subcellular factors from rat mediobasal hypothalamus. Neuroendocrinol 24:147-161

Epelbaum J, Tapia-Arancibia L, Besson J, Rotsztejn WH, Kordon C (1979) Vasoactive intestinal polypeptide inhibits release of somatostatin from hypothalamus in vitro. Eur J Pharmacol 58:493-495

Ettigi P, Lal S, Martin JB, Friesen H (1975) Effects of sex, oral contraceptives and glucose loading on apomorphine induced growth hormone secretion. J Clin Endocrinol Metab 40:1094-1098

Euker J, Meites J, Reigle G (1973) Serum LH and prolactin following restraint stress in the rat. Physiologist 16:307

Everett JW (1954) Luteotrophic function of autographs of the rat hypophyses. Endocrinology 54:685-690

Fehm HL, Voight KH, Lang RE, Pfeiffer EF (1980) Effects of neurotransmitters on the release of corticotropin releasing hormone (CRH) by hypothalamic tissue in vitro. Exp Brain Res 39:229-234

Ferin M (1979) Central sites controlling pituitary secretion in the rhesus monkey. J Steroid Biochem II:1015-1019

Ferin M, Dyrenfurth I, Cowchock S, Warren M, van de Wiele RL (1974) Active immunization to 17β-estradiol and its effects upon the reproductive cycle of the rhesus monkey. Endocrinology 94:765–776

Ferin M, Antunes JL, Zimmerman EA, Dyrenfurth I, Frantz AG, Robinson A, Carmel PW (1977) Endocrine function in female rhesus monkeys after hypothalamic disconnection. Endocrinology 101:1611–1620

Ferin M, Bogumil J, Drewes J, Dyrenfurth I, Jewelewicz R, van de Wiele RL (1978) Pituitary and ovarian hormonal responses to 48 h gonadotropin releasing hormone (GnRH) infusion in female rhesus monkeys. Acta Endocrinol 89:48–59

Ferland L, Labrie F, Jobin M, Arimura A, Schally AV (1976) Physiological role of somatostatin in the control of growth hormone and thyrotropin secretion. Biochem Biophys Res Commun 68:149–156

Ferland L, Fuxe K, Eneroth P, Gustafsson JA, Skett P (1977) Effects of methionine-enkephalin on prolactin release and catecholamine levels and turnover in the median eminence. Eur J Pharmacol 43:89–90

Finkelstein JW, Roffwarg P, Boyar RM, Kream J, Hellman L (1972) Age-related changes in the twenty-four hour spontaneous secretion of growth hormone. J Clin Endocrinol Metab 35:665–670

Fisher DA, Odell WA (1971) Effect of cold on TSH secretion in man. J Clin Endocrinol Metab 33:859–861

Fortier C, Degaldo A, Ducommun P, Ducommun S, Dupont A, Jobin M, Kraicer J, Macintosh-Hardt B, Marceau H, Miahle P, Miahle-Voloss C, Rerup C, van Rees GP (1970) Functional interrelationships between the adenohypophysis, thyroid, adrenal cortex and gonads. Can Med Assoc J 103:864–874

Frantz AG, Kleinberg DL (1970) Prolactin: evidence that it is separate from growth hormone in human blood. Science 170:745–749

Frantz AG, Kleinberg DL, Noel GL (1972) Studies on prolactin in man. Recent Progress in Hormone Research 28:527–590.

Fraser WM, Tucker HS, Grubb SR, Wigand JP, Blackard WG (1979) Effect of L-tryptophan on growth hormone and prolactin release in normal volunteers and patients with secretory pituitary tumors. Horm Metab Res 11:149–155

Frohman LA, Bernardis LL, Kant K (1968) Hypothalamic stimulation of growth hormone secretion. Science 162:580–582

Frohman LA, Bernardis LL, Burck L, Maran JW, Dhariwal APS (1972) Hypothalamic control of growth hormone secretion in the rat. In: Growth and Growth Hormone. Excerpta Medica, Amsterdam

Fukuda H, Greer MA (1975) The effect of basal hypothalamic deafferentation on the nyctohemeral rhythm of plasma TSH. Endocrinology 97:749–752.

Fukuda H, Greer MA, Roberts L, Allen CF, Critchlow V, Wilson M (1975) Nyctohemeral and sex-related variations in plasma thyrotropin, thyroxine, and triiodothyronine. Endocrinology 97:1424–1431

Fuller RW, Snoddy HD (1979) The effects of metergoline and other serotonin receptor antagonists on serum corticosterone in rats. Endocrinology 105:923–928

Fuxe K, Hökfelt T, Nilsson O (1969) Castration, sex hormones and tuberoinfundibular dopamine neurons. Neuroendocrinology 5:107–120

Gala RR, Subramanian MG, Peters JA, Pieper PR (1977) The effects of serotonergic and adrenergic receptor antagonists on prolactin release in the monkey. Life Sci 20:631–638

Gallo RV (1978) The effect of blockade of dopamine receptors on the inhibition of episodic luteinizing hormone release during electrical stimulation of the arcuate nucleus in ovariectomized rats. Endocrinology 102:1026–1035

Gallo RV (1980) Neuroendocrine regulation of pulsatile luteinizing hormone release in the rat. Neuroendocrinology 30:122–131

Gallo RV, Drouva SV (1979) Effect of intraventricular infusion of catecholamines on luteinizing hormone release in ovariectomized and ovariectomized, steroid-primed rats. Neuroendocrinology 29:149-162
Gallo RV, Rabbi J, Moberg GP (1975) Effect of methysergide, a blocker of serotonin receptors on plasma prolactin levels in lactating and ovariectomized rats. Endocrinology 97:1096-1105
Gann DS, Ward DG, Baertschi AJ, Carlson DE, Maran JW (1977) Neural control of ACTH release in response to hemorrhage. Am NY Acad Sci 297:477-497
Ganong WF, Kramer N, Salmon J, Reid IA, Lovinger R, Scapagnini U, Boryczka AT, Shackelford R (1976) Pharmacological evidence for inhibition of ACTH secretin by a central noradrenergic system in the dog. Neuroscience 1:167-174
Gay VL, Sheth NA (1972) Evidence for periodic release of LH in castrated male and female rats. Endocrinology 90:158-162
Gibbs DM, Neill JD (1978) Dopamine levels in hypophysial stalk blood in the rat are sufficient to inhibit prolactin secretion in vivo. Endocrinology 102:1895-1900
Gil-Ad I, Zambotti F, Carruba MO, Vicentini L, Muller EE (1976) Stimulatory role for brain serotonergic system on prolactin secretion in the male rat. Proc Soc Exp Biol Med 151:512-518
Giraud P, Lissitzky J-C, Conte-Devolx B, Gillioz P, Oliver C (1980) Influence of haloperidol on ACTH and β-endorphin secretion in the rat. Eur J Pharmacol 62:216-217
Glass AR, Schaaf M, Dimond RC (1979) Absent growth hormone response to L-tryptophan in acromegaly. J Clin Endocrinol Metab 48:664-666
Gospodarowicz D, Legault-Demare J (1963) Etude de l'activite biologique *in vitro* des hormones gonadotropes. Acta Endocrinol 42:509-513
Gnodde HP, Schuiling GA (1976) Involvement of catecholaminergic and cholinergic mechanisms in the pulsatile release of LH in the long-term ovariectomized rat. Neuroendocrinology 20:212-223
Greer MA, Kendall JW, Duyck C (1963) Failure of heterotropic rat pituitary transplants to maintain adrenocorticol secretion. Endocrinology 72:499-501
Grimm Y, Reichlin S (1973) Thyrotropin-releasing hormone (TRH): neurotransmitter regulation by mouse hypothalamic tissue *in vitro*. Endocrinology 93:626-631
Grosvenor CE, Turner CW (1960) Pituitary lactogenic hormone concentration during pregnancy in the rat. Endocrinology 66:96-99
Gudelsky GA, Porter JC (1980) Release of dopamine from tuberoinfundibular neurons into pituitary stalk blood after prolactin or haloperidol administration. Endocrinology 106:526-529
Guillemin R, Brazeau P, Bohlen P, Esch F, Ling N, Wehrenberg W (1982) Growth hormone-releasing factor from a human pancreatic tumor that caused acromegaly. Science 218:585-587
Haigler ED jr, Pittman JA jr, Hershman JM, Baugh CM (1971) Direct evaluation of pituitary thyrotropin reserve utilizing synthetic thyrotropin releasing hormone. J Clin Endocrinol Metab 33:573-581
Halasz B, Gorski RA (1967) Gonadotropic hormone secretion in female rats after partial or total interruption of neural afferents to the medial basal hypothalamus. Endocrinology 80:608-622
Halasz B, Pupp L (1969) The endocrine effects of isolation of the hypothalamus from the rest of the brain. In: W. F. Ganong and L. Martini (Eds) Frontiers in Neuroendocrinology. Oxford University Press, New York
Hall R, Besser GM, Schally AV, Coy DH, Evered D, Goldie DJ, Kastin AJ, McNeilly AS, Mortimer CH, Phenekos C, Tunbridge WMG, Weightman D (1973) Action of growth hormone-release inhibitory hormone in healthy men and in acromegaly. Lancent 2:581-586

Harris GW (1948) Electrical stimulation of the hypothalamus and the mechanism of neural control of the adenohypophysis. J Physiol (Lond) 101:418-429

Hery M, Laplante E, Kordon C (1976) Participation of serotonin in the phasic release of LH. I. Evidence from pharmacological experiments. Endocrinology 99:496-503

Hirooka Y, Hollander CS, Sukuki S, Ferdinand P, Juan S-I (1978) Somatostatin inhibits release of thyrotropin releasing factor from organ cultures of rat hypothalamus. Proc Nat Acad Sci USA 75:4509-4513

Hökfelt T (1967) The possible ultrastructural identification of tubero-infundibular dopamine-containing nerve endings in the median eminence of the rat. Brain Res 5:121-123

Hökfelt T, Fuxe K (1972) Effects of prolactin and ergot alkaloids on the tubero-infundibular dopamine (DA) neurons. Neuroendocrinology 9:100-122

Holak H, Baldys A, Jarzab B, Wtstrychowski A, Skrzypek J (1978) Changes in serum TSH level after intraventricular injection of various neuromediators in rats. Acta Endocrinol 87:279-282

Hunter WM, Fonseka CC, Passmore R (1965) The role of growth hormone in mobilization of fuel for muscular exercise. J Exp Physiol 50:406-416

Hwang P, Guyda H, Friesen H (1971) A radioimmunoassay for human prolactin. Proc Natl Acad Sci USA 68:1902-1906

Ieiri T, Chen HT, Meites J (1980) Naloxone stimulation of luteinizing hormone release in prepubertal female rats; role of serotonergic system. Life Sci 26:1269-1274

Imura H, Kato Y, Ikeda M, Morimoto M, Yawata M (1971) Effect of adrenergic blocking or stimulating agents on plasma growth hormone, immunoreactive insulin and blood free fatty acid levels in man. J Clin Invest 50:1069-1071

Imura H, Nakai Y, Yoshimi T (1973) Effect of 5-hydroxytryptophan (5-HTP) on growth hormone and ACTH release in man. J Clin Endocrinol Metab 36:204-206

Jacobs LS, Snyder PJ, Wilber JF, Utiger RD, Daughaday WH (1971) Increased serum prolactin after administration of synthetic thyrotropin releasing hormone (TRH) in man. J Clin Endocrinol Metab 33:996-998

Jacoby JH, Greenstein M, Sassin JF, Weitzman ED (1974) The effect of monoamine precursors on the release of growth hormone in the rhesus monkey. Neuroendocrinology 14:95-102

Jobin M, Ferland L, Labrie F (1976) Effect of pharmacological blockade of ACTH and TSH secretion on the acute stimulation of prolactin release by exposure to cold and ether stress. Endocrinology 99:146-151

Johnson LY (1978) The effect of vasotocin, a pineal peptide, on prolactin secretion in the female rat. Anat Rec 190:433

Jones MT, Hillhouse EW (1976) Structure-activity relationships and the mode of action of corticosteroid feedback on the secretion of corticotropin-releasing factor (corticoliberin). J Ster Biochem 7:1189-1202

Jones MT, Hillhouse E, Burden J (1976) Secretion of corticotropin-releasing hormone in vitro. In: L. Martini and W. F. Ganong and L. Martini (Eds) Frontiers in Neuroendocrinology, vol 4. Raven Press, New York

Jordan D, Poncet C, Mornex R, Ponsin G (1978) Participation of serotonin in thyrotropin release. I. Evidence for the action of serotonin on thyrotropin releasing hormone release. Endocrinology 103:414-419

Jordan D, Pigeon P, McRae-Degueurce A, Pujol JF, Mornex R (1979) Participation of serotonin in thyrotropin release. II. Evidence for the action of serotonin on the phasic release of thyrotropin. Endocrinology 105:975-979

Joseph-Bravo P, Charli JL, Palacios JM, Kordon C (1979) Effect of neurotransmitters on the *in vitro* release of immunoreactive thyrotropin-releasing hormones from rat mediobasal hypothalamus. Endocrinology 104:801-806

Kalra SP, McCann SM (1973) Effect of drugs modifying catecholamine synthesis on LH release induced by preoptic stimulation in the rat. Endocrinology 93:356-362
Kalra SP, McCann SM (1974) Effects of drugs modifying catecholamine synthesis on plasma LH and ovulation in the rat. Neuroendocrinology 15:79-91
Kamberi IA, Mical RS, Porter JC (1970a) Effect of anterior pituitary perfusion and intraventricular injection of catecholamiens and indoleamines on LH release. Endocrinology 87:1-12
Kamberi IA, Mical RS, Porter JC (1970b) Action of dopamine to induce release of FSH-releasing factor (FRF) from hypothalamic tissue in vitro. Endocrinology 86:278-284
Kamberi IA, Mical RS, Porter JC (1971a) Pituitary portal vessel infusion of hypothalamic extract and release of LH, FSH, and prolactin. Endocrinology 88:1294-1299
Kamberi IA, Mical RS, Porter JC (1971b) Effects of melatonin and serotonin on the release of FSH and prolactin. Endocrinology 88:1288-1293
Kanematsu S, Hilliard J, Sawyer CH (1963) Effect of reserpine on pituitary prolactin content and its hypothalamic site of action in the rabbit. Acta Endocrinol (Kbh) 44:467-474
Kansel PC, Buse J, Talbert OR, Buse MG (1972) Effect of L-dopa on plasma growth hormone, insulin, and thyroxine. J Clin Endocrinol Metab 34:99-105
Kendall JW, Roth JG (1964) Adrenocortical function in monkeys after forebrain removal or pituitary stalk section. Endocrinology 84:686-691
Kizer JS, Zwin JA, Jacobowitz DM, Kopin IJ (1975) The nyctohemeral rhythm of plasma prolactin: effects of ganglionectomy, pinealectomy, constant light, constant darkness or 6-OH-dopamine administration. Endocrinology 96:1230-1240
Knigge KM (1961) Adrenocortical response to stress in rats with lesions in hippocampus or amygdala. Proc Soc Exp Biol Med 108:18-21
Knobil E, Plant TM (1978) Neuroendocrine control of gonadotropin secretion in the female rhesus monkey. In: W. F. Ganong and L. Martini (Eds) Frontiers in Neuroendocrinology, vol 5. Raven Press, New York
Koch Y, Lu KH, Meites J (1970) Biphasic effect of catecholamines on pituitary prolactin release *in vitro*. Endocrinology 87:673-675
Koch Y, Goldhaber G, Fireman I, Zor U, Shani J, Tal E (1977) Suppression of prolactin and thyrotropin secretion in the rat by antiserum to thyrotropin-releasing hormone. Endocrinology 100:1476-1478
Koerker DJ, Goodner CJ, Ruch W (1974) Somatostatin action on pancreas. N Engl J Med 291:262-263
Koivusalo F, Leppaluoto J (1979) Brain TRF immunoreactivity during various physiological and stress conditions in the rat. Neuroendocrinology 29:231-236
Koulo M, Lammintausta R (1979) Effect of melatonin on L-tryptophan- and apomorphine stimulated growth hormone secretion in man. J Clin Endocrinol Metab 49:70-72
Krey LC, Butler WR, Knobil E (1975) Surgical disconnection of the medial basal hypothalamus and pituitary function in the rhesus monkey. I. Gonadotropin secretion. Endocrinology 96:1073-1087
Krieg RJ, Sawyer CH (1976) Effects of intraventricular catecholamines on luteinizing hormone release in ovariectomized-steroid-primed rats. Endocrinology 99:411-419
Krieger DT (1975) Rhythms of ACTH and corticosteroid secretion in health and disease, and their experimental modification. J Ster Biochem 6:785-791
Krieger DT (1980) Pituitary hormones in the brain: what is their function? Fed Proc 39:2937-2941
Krieger D, Glick SM (1974) Sleep EEG stages and plasma growth hormone concentra-

tion in states of endogenous or exogenous hypercortisolemia or ACTH elevation. J Clin Endocrinol Metab 39:986-1000
Krulich L (1975) The effect of a serotonin uptake inhibitor (Lilly 110 140) on the secretion of prolactin in the rat. Life Sci 17:1141-1144
Krulich L, Dhariwal APS, McCann SM (1965) Growth hormone-releasing activity of crude ovine hypothalamic extracts. Proc Soc Exp Biol Med 120:180-184
Krulich L, Giachetti A, Marchlewska-Koj A, Hefco E, Jameson HE (1977) On the role of central noradrenergic and dopaminergic systems in the regulation of TSH secretion in the rat. Endocrinology 100:496-505
Krulich L, Coppings RJ, Giachetti A, McCann SM, Mayfield MA (1980) Lack of evidence that the central serotonergic system plays a role in the activation of prolactin secretion following inhibition of dopamine synthesis or blockade of dopamine receptors in the male rat. Neuroendocrinology 30:133-138
Labrie F, Pelletier G, Borgeat P, Drouin J, Ferland L, Belanger A (1976) Mode of action of hypothalamic regulatory hormones in the adenohypophysis. In: L. Martini and W. F. Ganong (Eds) Frontiers in Neuroendocrinology, vol 4. Raven Press, New York
Lal S, De la Vega CE, Sourkes TL, Friesen HG (1973) Effect of apomorphine on growth hormone, prolactin, luteinizing hormone and follicle stimulating hormone levels in human serum. J Clin Endocrinol Metab 37:719-724
Lal S, Tolis G, Martin JB, Brown GM, Guyda H (1975) Effects of clonidine on growth hormone, prolactin, luteinizing hormone, follicle-stimulating hormone and thyroid-stimulating hormone in the serum of normal men. J Clin Endocrinol Metab 41:703-708
Lamberts SWJ, Macleod RM (1978) The interaction of the serotonergic and dopaminergic systems on prolactin secretion in the rat. Endocrinology 103:287-295
Lancranjan I, Wirz-Justice A, Puhringer W, Del Pozo E (1977) Effect of L-5-hydroxytryptophan infusion on growth hormone and prolactin secretion in man. J Clin Endocrinol Metab 45:583-593
Lancranjan I, Ohnhaus E, Girard (1979) The α-adrenoceptor control of adrenocorticotropin secretion in man. J Clin Endocrinol Metab 49:227-230
Larsson L-L (1979) Radioimmunochemical characterization of ACTH-like peptides in the antropyloric mucosa. Life Sci 25:1565-1570
Lawson DM, Gala RR (1975) The influence of adrenergic, dopaminergic, cholinergic, and serotonergic drugs on plasma prolactin levels in ovariectomized, estrogen-treated rats. Endocrinology 96:313-318
Leebaw WF, Lee LA, Woolf PD (1978) Dopamine affects basal and augmented pituitary hormone secretion. J Clin Endocrinol Metab 47:480-487
Leppaluoto J, Mannisto P, Ranta T, Linnoila M (1976) Inhibition of mid-cycle gonadotropin release in healthy women by pimozide and fusaric acid. Acta Endocrinol 81:455-460
Leung Y, Guansing AR, Ajlouni K, Hagen TC, Rosenfeld PS, Barboriak JJ (1975) The effect of hypoglycemia on hypothalamic thyrotropin-releasing hormone (TRH) in the rat. Endocrinology 97:380-384
Lewis M, Scanlon MF, Foord S, Shale DS, McDonald C, Hall R (1980) Dopamine inhibits thyrotropin secretion by direct action on the rat thyrotroph. Int Congr Endocrinol 6:381
Lovinger RD, Connors MH, Kaplan SL, Ganong WF, Grumbach MM (1974) Effect of L-dihydroxyphenylalanine (L-dopa), anesthesia and surgical stress on the secretion of growth hormone in the dog. Endocrinology 95:1317-1321
Lovinger RD, Holland J, Kaplan S, Grumbach M, Boryczka AT, Shackleford R, Sal-

mon J, Reid IA, Ganong WF (1976) Pharmacological evidence for stimulation of growth hormone secretion by a central noradrenergic system in dogs. Neuroscience 1:443–450

Lu KH, Meites J (1971) Inhibition of L-dopa and monoamine oxidase inhibitors on pituitary prolactin release stimulation by methyl-dopa and d-amphetamine. Proc Soc Exp Biol Med 137:480–483

Lu KH, Meites J (1972) Effects of L-dopa on serum prolactin and PIF in intact and hypophysectomized, pituitary grafted rats. Endocrinology 91:868–872

Lu KH, Meites J (1973) Effects of serotonin precursors and melatonin on serum prolactin release in rats. Endocrinology 93:152–155

Lu KH, Amenomori Y, Chen CL, Meites J (1970) Effects of central acting drugs on serum and pituitary prolactin levels in rats. Endocrinology 87:667–670

Lu KH, Schaar CH, Kortright KH, Meites J (1972) Effects of synthetic TRH on *in vitro* and *in vivo* prolactin release in the rat. Endocrinology 91:1540–1544

Macintyre HB, Odell WD (1974) Physiological control of growth hormone in the rabbit. Neuroendocrinology 16:8–21

Macleod RM (1969) Influence of norepinephrine and catecholamine-depleting agents on the synthesis and release of prolactin and growth hormone. Endocrinology 85:916–923

Macleod RM, Lehmeyer JE (1974) Restoration of prolactin synthesis and release by administration of monoaminergic blocking agents to pituitary tumor-bearing rats. Cancer Res 34:345–350

Macleod RM, Abad A, Eidson LL (1969) In vivo effect of sex hormones on the in vitro synthesis of prolactin and growth hormone in normal and pituitary tumor-bearing rats. Endocrinology 84:1475–1483

Maeda K, Frohman LA (1980) Release of somatostatin and thyrotropin-releasing hormone from rat hypothalamic fragments in vitro. Endocrinology 106:1837–1842

Makara GB, Stark E, Palkovits M (1970) Afferent pathways of stressful stimuli: corticotropin release after hypothalamic deafferentation. J Endocrinol 47:411–416

Malarkey WB, Jacobs LS, Daughaday WH (1971) Levodopa suppression of prolactin in nonpuerperal galactorrhea. N Engl J Med 285:1160–1163

Marchlewska-Koj A, Krulich L (1975) The role of central monoamines in the stress-induced prolactin release in the rat. Fed Proc 34:252

Martin JB (1972) Plasma growth hormone (GH) response to hypothalamic or extrahypothalamic electrical stimulation. Endocrinology 91:107–115

Martin JB, Reichlin S (1972) Plasma thyrotropin (TSH) response to hypothalamic stimulation and to injection of synthetic thyrotropin releasing hormone (TRH). Endocrinology 90:1079–1085

Martin JB, Boshans R, Reichlin S (1970) Feedback regulation of TSH secretion in rats with hypothalamic lesions. Endocrinology 87:1032–1040

Martin JB, Kantor J, Mead P (1973) Plasma growth hormone responses to hypothalamic, hippocampal and amygdaloid electrical stimulation: effects of variations in stimulation parameters and treatment with alpha-methyl-p-tyrosine (α-MT). Endocrinology 92:1354–1361

Martin JB, Renaud L, Brazeau P (1974a) Pulsatile growth hormone secretion: suppression by hypothalamic ventromedial lesions and by long-acting somatostatin. Science 186:538–540

Martin JB, Lal S, Tolis G, Friesen HG (1974b) Inhibition by apomorphine of prolactin secretion in patients with elevated serum prolactin. J Clin Endocrinol Metab 39:180–182

Martin JB, Audet J, Saunders A (1975) Effect of somatostatin and hypothalamic ventromedial lesions on GH release induced by morphine. Endocrinology 96:881-889

Martin JB, Brazeau P, Tannenbaum GS, Willoughby JO, Epelbaum J, Terry LC, Durand D (1978a) Neuroendocrine organization of growth hormone secretion. In: S. Reichlin and R. Baldessarini (Eds) The Hypothalamus. Raven Press, New York

Martin JB, Durand D, Gurd W, Faille G, Audet J, Brazeau P (1978b) Neuropharmacologic regulation of episodic growth hormone and prolactin secretion in the rat. Endocrinology 102:106-113

Martinovic JV, McCann SM (1977) Effect of lesions in the ventral noradrenergic tract produced by microinjection of 6-hydroxydopamine on gonadotropin release in the rat. Endocrinology 100:1206-1213

Matsuo H, Baba Y, Nair RMG, Arimura A, Schally AV (1971) Structure of the porcine LH- and FSH-releasing hormone. I. The proposed amino acid sequence. Biochem Biophys Res Commun 43:1334-1339

McCormack JT, Plant TM, Hess DL, Knobil E (1977) The effect of luteinizing hormone releasing hormone (LHRH) antiserum administration on gonadotropin secretion in the rhesus monkey. Endocrinology 100:663-667

Meites J, Clemens JA (1972) Hypothalamic control of prolactin secretion. Vitamins and Hormones 30:165-221

Meites J, Talwalker PK, Nicoll CS (1960) Initiation of lactation in rats with hypothalamic or cerebral tissue. Proc Soc Exp Biol Med 103:298-300

Meites J, Lu KH, Wuttke W, Welsch GW, Nagasawa H, Quadri SK (1972) Recent studies on functions and control of prolactin secretion in rats. Recent Prog Horm Res 28:471-526

Meites J, Bruni JF, van Vugt DA (1979) Effects of endogenous opiate peptides on release of anterior pituitary hormones. In: R. Collu (Ed) Central nervous system effects of hypothalamic hormones and other peptides. Raven Press, New York

Meltzer H, Fang VS (1976) Effect of apomorphine plus 5-hydroxytryptophan on plasma prolactin levels in male rats. Psychopharmacol Commun 2:189-198

Meltzer HY, Fang VS, Paul SM, Kaluskar R (1976) Effect of quipazine on rat plasma prolactin levels. Life Sci 19:1073-1078

Mims RB, Scitt CL, Modebe OM, Bethune JE (1973) Prevention of L-dopa-induced growth hormone stimulation by hyperglycemia. J Clin Endocrinol Metab 40:363-366

Mittler JC, Redding TW, Schally AV (1969) Stimulation of thyrotropin (TSH) by TSH-releasing factor (TRF) in organ cultures of anterior pituitary. Proc Soc Exp Biol 130:406-409

Miyai K, Onishi T, Hosokawa M, Ishibashi K, Kumahara Y (1974) Inhibition of thyrotropin and prolactin secretion in primary hypothyroidism by 2-Br-α-ergocryptine. J Clin Endocrinol Metab 39:391-394

Modlinger RS, Schonmuller JM, Arora SP (1979) Stimulation of aldosterone, renin, and cortisol by tryptophan. J Clin Endocrinol Metab 48:599-603

Modlinger RS, Schonmuller JM, Arora SP (1980) Adrenocorticotropin release by tryptophan in man. J Clin Endocrinol Metab 50:360-363

Mueller GP, Chen HJ, Meites J (1973) *In vivo* stimulation of prolactin release in the rat by synthetic TRH. Proc Soc Exp Biol Med 144:613-617

Mueller GP, Simpkins J, Meites J, Moore KE (1976a) Differential effects of dopamine agonists and haloperidol on release of prolactin, thyroid stimulating hormone, growth hormone and luteinizing hormone in rats. Neuroendocrinol 20:121-135

Mueller GP, Twohy CP, Chen HT, Advis JP, Meites J (1976b) Effects of L-tryptophan and restraint stress on hypothalamic and brain serotonin turnover and pituitary TSH and prolactin release in rats. Life Sci 18:715-724

Muller EE, Giustina G, Miedico D, Cocchi D, Pecile A (1972) Growth and growth hormone. Excerpta Medica, Amsterdam

Muller EE, Brambilla F, Cavagnini F, Peracchi M, Panerai A (1974) Slight effect of L-tryptophan on growth hormone release in normal subjects. J Clin Endocrinol Metab 39:1-5

Nakagawa K, Horiuchi Y, Mashimo K (1971) Further studies on the relation between growth hormone and corticotropin secretion in insulin-induced hypoglycemia. J Clin Endocrinol Metab 32:188-191

Nakai Y, Imura H, Yoshimi T, Matsukura S (1973) Adrenergic control mechanism for ACTH secretion in man. Acta Endocrinol 74:263-270

Nakai Y, Imura H, Sakurai H, Kurahachi H, Yoshimi T (1974) Effect of cyproheptadine on human growth hormone secretion. J Clin Endocrinol Metab 38:446-449

Negro-Vilar A, Saad WA, McCann SM (1977) Evidence for a role of prolactin in prostate and seminal vesicle growth in immature male rats. Endocrinology 100:729-737

Negro-Vilar A, Ojeda SR, Arimura A, McCann SM (1978) Dopamine and norepinephrine stimulate somatostatin release by median eminence fragments *in vitro*. Life Sci 23:1493-1497

Neill JD (1970) Effect of stress on serum prolactin and luteinizing hormone levels during the estrus cycle in the rat. Endocrinology 87:1192-1197

Neill JD (1972a) Comparison of plasma prolactin levels in cannulated and decapitated rats. Endocrinology 90:568-572

Neill JD (1972b) Sexual differences in the hypothalamic regulation of prolactin secretion. Endocrinology 90:1154-1159

Neill JD, Freeman ME, Tillson SA (1971) Control of the proestrus surge of prolactin and luteinizing hormone secretion by estrogens in the rat. Endocrinology 89:1448-1453

Neill JD, Patton JM, Dailey RA, Tsou RC, Tindall GT (1977) Luteinizing hormone releasing hormone (LHRH) in pituitary stalk blood of rhesus monkeys: relationship to level of LH release. Endocrinology 101:430-434

Newsome HH, Rose JC (1971) The response of human adrenocorticotropic hormone and growth hormone to surgical stress. J Clin Endocrinol Metab 33:481-487

Nicholson G, Greeley G, Humm J, Youngblood W, Kizer JS (1978) Lack of effect of noradrenergic denervation of the hypothalamus and medial preoptic area on the feedback regulation of gonadotropin secretion and the estrous cycle in the rat. Endocrinology 103:559-566

Nicoll CS, Meites J (1962) Estrogen stimulation of prolactin production by rat adenohypophysis *in vitro*. Endocrinology 70:272-277

Nicoll CS, Fiorindo RP, McKennee CT, Parsons JA (1970a) Assay of hypothalamic factors which regulate prolactin secretion. In: J. Meites (Ed) Hypophysiotropic hormones of the hypothalamus: assay and chemistry. Williams and Wilkens, Baltimore

Nicoll CS, Yaron Z, Nutt N, Daniels E (1970b) Effects of ergotamine tartrate on prolactin and growth hormone secretion by rat adenohypophysis in vitro. Biol Reprod 5:59-66

Nilsson KO, Thorell JI, Mikflet B (1974) The effect of thyrotropin releasing hormone on the release of thyrotropin and other pituitary hormones in man under basal conditions and following adrenergic blocking agents. Acta Endocrinol 76:24-28

Ohgo S, Kato Y, Chihara K, Imura H, Maeda K (1976) Effect of hypothalamic surgery on prolactin release induced by 5-hydroxytryptophan (5-HTP) in rats. Endocrinol Jap 23:485-491

Ojeda SR, Harms PG, McCann SM (1974) Effect of blockade of dopaminergic receptors on prolactin and LH release: median eminence and pituitary sites of action. Endocrinology 94:1650–1657

Ormston BJ, Garry R, Cryer RJ, Besser GM, Hall R (1971) Thyrotropin-releasing hormone as a thyroid function test. Lancet 2:10–14

Pamenter RW, Hedge GA (1980) Inhibition of thyrotropin secretion by physiological levels of corticosterone. Endocrinology 106:162–166

Pang CN, Zimmerman E, Sawyer CH (1977) Morphine inhibition of the preovulatory surges of plasma luteinizing hormone and follicle stimulating hormone in the rat. Endocrinology 101:1726–1732

Parker DC, Rossman LG (1973) Physiology of human growth hormone release in sleep. In: R. O. Scow (Ed) Endocrinology. Excerpta Medica, Amsterdam

Parker DC, Morishima M, Koerker DJ, Gale CC, Goodner CJ (1972) Pilot study of growth hormone release in sleep of the chair-adapted baboon: potential as model of human sleep release. Endocrinology 91:1462–1467

Parker DC, Pakary AE, Hershman JM (1976) Effect of normal and reversed sleep-wake cycles upon nyctohemeral rhythmicity of plasma thyrotropin: evidence suggestive of an inhibitory influence in sleep. J Clin Endocrinol Metab 43:318–329

Peroutka S, Lebovitz R, Snyder D (1981) Two distinct central serotonin receptors with different physiological functions. Science 212:827–829

Pierce JG (1971) The subunits of pituitary thyrotropin—their relationship to other glycoprotein hormones. Endocrinology 89:1331–1334

Popova NK, Moslova LN, Naumenko EV (1972) Serotonin and the regulation of the pituitary-adrenal system after deafferentation of the hypothalamus. Brain Res 47:61–67

Porter JC, Vale W, Burgus R, Mical RS, Guillemin R (1971) Release of TSH by TRH injected directly into a pituitary stalk portal vessel. Endocrinology 89:1054–1056

Przekop F, Domanski E (1976) Role of catecholamines in release of gonadotrophic hormones in the rabbit. Acta Physiol Pol 27:163–168

Quabbe H-J, Schilling E, Hedge H (1966) Pattern of growth hormone secretion during the 24-hour fast in normal adults. J Clin Endocrinol Metab 26:1173–1177

Quadri SK, Spies HG (1976) Cyclic and diurnal patterns of serum prolactin in the rhesus monkey. Biol Reprod 14:495–501

Rabii J, Ehlers C, Clifton D, Sawyer CH (1980) Effects of intraventricular infusions of 6-hydroxydopamine (6-OHDA) on pituitary LH release and ovulation in the rabbit. Neuroendocrinology 30:362–368

Racagni G, Apud JA, Locatelli V, Cocchi D, Nistico G, Di Giorgio RM, Muller EE (1979) GABA of CNS origin in the rat anterior pituitary inhibits prolactin secretion. Nature 281:575–578

Ranta T, Mannisto P, Tuomisto J (1977) Evidence for dopaminergic control of thyrotropin secretion in the rat. J Endocrinol 72:329–335

Rapoport B, Refetoff S, Fang VS, Friesen HG (1973) Suppression of serum thyrotropin (TSH) by L-DOPA in chronic hypothyroidism: interrelationships in the regulation of TSH and prolactin secretion. J Clin Endocrinol Metab 36:256–262

Raymond V, Beaulieu M, Labrie F, Bossier J (1978) Potent antidopaminergic activity of estradiol at the pituitary level on prolactin release. Science 200:1173–1175

Reece RP, Turner CW (1936) Influence of estrone upon galactin content of the male rat pituitary. Proc Soc Exp Biol Med 34:402–404

Reichlin S (1974) Regulation of somatotrophic hormone. In: E. Knobil and W. H. Sawyer (Eds) Handbook of Physiology-Endocrinology, vol 4, part 2. Williams and Wilkens, Baltimore

Reichlin S, Martin JB, Mitnick M, Boshans RL, Grimm Y, Bollinger J, Gordon J, Mal-

acara J (1972) The hypothalamus in pituitary-thyroid regulation. Recent Progress in Hormone Research 28:229-286
Renaud LP, Brazeau P, Martin JB (1975) Depressant action of TRH, LH-RH, and somatostatin on the activity of central neurons. Nature (London) 255:233-235
Rivier C, Vale W (1974) *In vivo* stimulation of prolactin secretion in the rat by thyrotropin releasing factor, related peptides and hypothalamic extracts. Endocrinology 95:978-983
Rivier C, Vale W, Ling N, Brown M, Guillemin R (1977a) Stimulation *in vivo* of the secretion of prolactin and growth hormone by β-endorphin. Endocrinology 100:238-241
Rivier C, Brown M, Vale W (1977b) Effect of neurotensin, substance P and morphine sulfate on the secretion of prolactin and growth hormone in the rat. Endocrinology 100:751-754
Rivier C, Rivier J, Vale W (1978) The effect of bombesin and related peptides on prolactin and growth hormone secretion in the rat. Endocrinology 102:519-522
Rivier C, Brownstein M, Spiess J, Rivier J, Vale W (1982a) In vivo corticotropin-releasing factor-induced secretion of adrenocorticotropin, β-endorphin and corticosterone. Endocrinology 110:272-278
Rivier C, Rivier J, Vale W (1982b) Inhibition of adrenocorticotropic hormone secretion in the rat by immunoneutralization of corticotropin-releasing factor (CRF). Science 218:377-379
Rossier J, French ED, Riviier C, Ling N, Guillemin R, Bloom FE (1977) Foot-shock induced stress increases β-endorphin levels in blood but not brain. Nature 270:618-620
Rotsztejn WH, Charli JL, Pattou E, Epelbaum J, Kordon C (1976) In vitro release of luteinizing hormone-releasing hormone (LHRH) from rat mediobasal hypothalamus: effects of potassium, calcium and dopamine. Endocrinology 99:1663-1666
Ruberg M, Rotsztejn WH, Arancibia S, Besson J, Enjalbert A (1978) Stimulation of prolactin release by vasoactive intestinal polypeptide (VIP). Eur J Pharmacol 51:319-320
Sanchez-Franco F, Garcia MD, Cacicedo L, Martin-Zurro A, Escobar Del Rey F (1973) Influence of sex phase of the menstrual cycle on thyrotropin (TSH) response to thyrotropin-releasing hormone (TRH). J Clin Endocrinol Metab 37:736-740
Sar M, Meites J (1969) Effects of suckling on pituitary release of prolactin, GH and TSH in postpartum lactating rats. Neuroendocrinology 4:25-31
Sarkar DK, Chiappa SA, Fink G, Sherwood NM (1976) Gonadotropin-releasing hormone surge in proestrus rats. Nature 264:461-463
Sassin JF, Frantz AG, Weitzman ED, Kapen S (1972) Human prolactin; 24-hour pattern with increased release during sleep. Science 177:1205-1207
Saunders A, Terry LC, Audet J, Brazeau P, Martin JB (1976) Dynamic studies of growth hormone and prolactin secretion in the female rat. Neuroendocrinology 21:193-203
Sawyer CH, Markee JE, Townsend BF (1949) Cholinergic and adrenergic components in the neurohumoral control of the release of LH in the rabbit. Endocrinology 44:18-37
Scapagnini U, Preziosi P (1973) Receptor involvement in the central noradrenergic inhibition of ACTH secretion in the rat. Neuropharmacology 12:56-62
Schalch DS, Reichlin S (1968) Stress and growth hormone release. In: A. Pecile and E. E. Muller (Eds) Growth Hormone. Excerpta Medica, Amsterdam
Schally AV, Saffran M, Zimmerman B (1958) A corticotrophin-releasing factor: partial purification and amino acid composition. Biochem J 70:97-103

Schally AV, Bowers CY, Redding TW, Barrett JF (1966) Isolation of thyrotropin releasing factor (TRF) from porcine hypothalamus. Biochem Biophys Res Commun 25:165-169

Schally AV, Arimura A, Baba Y, Nair RMG, Matsuo H, Redding TW, Debeljuk L, White WF (1971) Isolation and properties of the FSH and LH-releasing hormone. Biochem Biophys Res Commun 43:393-399

Schally AV, Dupont A, Arimura A, Takahara J, Redding T, Clemens JA, Shaar CJ (1976a) Purification of a catecholamine-rich fraction with prolactin-release-inhibiting factor (PIF) activity from porcine hypothalami. Acta Endocrinol 82:1-14

Schally AV, Kastin AJ, Coy DH (1976b) LH-releasing hormone and its analogues: recent basic and clinical investigations. Int J Fertil 21:1-30

Schally AV, Redding TW, Arimura A, Dupont A, Linthicum GL (1977) Isolation of gamma-amino butyric acid from pig hypothalami and demonstration of its prolactin release-inhibiting (PIF) activity in vivo and in vitro. Endocrinology 100:681-691

Shaar CJ, Clemens JA (1974) The role of catecholamines in the release of anterior pituitary prolactins *in vitro*. Endocrinology 95:1202-1212

Shin SH (1979) Prolactin secretion in acute stress is controlled by prolactin releasing factor. Life Sci 25:1829-1836

Sinha D, Meites J (1965) Effects of thyroidectomy and thyroxine on hypothalamic concentration of "thyrotropin releasing factor" and pituitary content of thyrotropin in rats. Neuroendocrinology 1:4-14

Smalstig EB, Sawyer BD, Clemens JA (1974) Inhibition of rat prolactin release by apomorphine *in vivo* and *in vitro*. Endocrinology 95:123-129

Smith MS, Freeman ME, Neill JD (1975) The control of progesterone secretion during the estrous cycle and early pseudopregnancy in the rat: prolactin, gonadotropin, and steroid levels associated with rescue of the corpus luteum of pseudopregnancy. Endocrinology 96:219-226

Smythe GA, Lazarus L (1973) Growth hormone regulation by melatonin and serotonin. Nature (London) 244:230-231

Smythe GA, Lazarus L (1974) Suppression of human growth hormone by melatonin and cyproheptadine. J Clin Invest 54:116-121

Snyder PJ, Utiger RD (1972) Inhibition of thyrotropin response to thyrotropin-releasing hormone by small quantities of thyroid hormones. J Clin Invest 51:2077-2084

Soper BD, Weick RF (1980) Hypothalamic and extrahypothalamic mediation of pulsatile discharges of luteinizing hormone in ovariectomized rat. Endocrinology 106:348-355

Spiess J, Rivier J, Rivier C, Vale W (1981) Primary structure of corticotropin-releasing factor from ovine hypothalamus. Proc Natl Acad Sci USA 78:6517-6521

Spiess J, Rivier J, Vale W (1983) Characterization of rat hypothalamic growth hormone-releasing factor. Nature 303:532-535

Stefano FJ, Donoso AO (1967) Norepinephrine levels in the rat hypothalamus during the estrous cycle. Endocrinology 81:1405-1406

Steiner RA, Stewart JK, Barber J, Koerker D, Goodner CJ, Brown A, Illner P, Gale CC (1978) Somatostatin: a physiological role in the regulation of growth hormone secretion in the adolescent male baboon. Endocrinology 102:1587-1594

Subramanian MG, Gala RR (1976) Further studies on the effects of adrenergic serotonergic and cholinergic drugs on the afternoon surge in plasma prolactin in ovariectomized, estrogen-treated rats. Neuroendocrinology 22:240-249

Sussman PM, Tushinski RJ, Bancroft FC (1976) Pregrowth hormone: product of translation *in vitro* of messenger RNA coding for growth hormone. Proc Nat Acad Sci USA 73:29-33

Sutton J, Lazarus L (1974) Effect of adrenergic blocking agents on growth hormone responses to physical exercise. Horm Metab Res 6:428–429

Szabo M, Frohman LA (1977) Suppression of cold-stimulated thyrotropin secretion by antiserum to thyrotropin-releasing hormone. Endocrinology 101:1023–1033

Szafarczyk A, Ixart G, Malaval F, Nouguier-Soule J, Assenmacher I (1979) Effects of lesions of the suprachiasmatic nuclei and of p-chlorophenylalanine on the circadian rhythms of adrenocorticotrophic hormone and corticosterone in the plasma, and on locomotor activity of rats. J Endocrinol 83:1–16

Takahara J, Arimura A, Schally AV (1974) Suppression of prolactin release by a purified porcine PIF preparation and catecholamines infused into a rat hypophyseal portal vessel. Endocrinology 95:462–465

Takahishi T, Kipnis DM, Daughaday WH (1968) Growth hormone secretion during sleep. J Clin Invest 47:2079–2090

Talwalker PK, Ratner A, Meites J (1963) *In vitro* inhibition of pituitary prolactin synthesis and release by hypothalamic extracts. Am J Physiol 205:213–218

Tannenbaum GS, Martin JB (1976) Evidence for an endogenous ultradian rhythm governing growth hormone secretion in the rat. Endocrinology 98:562–570

Tannenbaum GS, Martin JB, Colle E (1976) Ultradian growth hormone rhythm in the rat. Effects of feeding, hyperglycemia, and insulin-induced hypoglycemia. Endocrinology 99:720–727

Tannenbaum GS, Epelbaum J, Colle E, Brazeau P, Martin JB (1978) Antiserum to somatostatin reverses starvation-induced inhibition of growth hormone but not insulin secretion. Endocrinology 102:1909–1914

Tashjian AH Jr, Barowsky NV, Jensen DK (1971) Thyrotropin releasing hormone: direct evidence for stimulation of prolactin production by pituitary cells in culture. Biochem Biophys Res Commun 43:516–523

Terry LC, Willoughby JO, Brazeau P, Martin JB, Patel Y (1976) Antiserum to somatostatin prevents stress-induced inhibition of growth hormone secretion in the rat. Science 192:565–566

Terry LC, Epelbaum J, Brazeau P, Martin JB (1977) Passive immunization with antiserum to somatostatin (AS-SS): effects on the dynamics of growth hormone (GH), prolactin (PRL), and thyroid stimulating hormone (TSH) secretion in cannulated rats. Soc Neurosci 3:359

Toivola PTK, Gale CC (1971) Growth hormone release by microinjection of norepinephrine into hypothalamus of conscious baboon. Fed Proc 30:26

Toivola PTK, Gale CC (1972) Stimulation of growth hormone release by microinjection of norepinephrine into hypothamalus of baboons. Endocrinology 90:895–902

Toivola PTK, Gale CC, Goodner CJ, Werrbach JH (1972) Central alpha-adrenergic regulation of growth hormone and insulin. Hormones 3:193–213

Vale W, Rivier C, Brazeau P, Guillemin R (1974) Effects of somatostatin on the secretion of thyrotropin and prolactin. Endocrinology 95:968–977

Vale W, Rivier C, Yang L, Minick S, Guillemin R (1978) Effects of purified hypothalamic corticotropin-releasing factors and other substances on the secretion of adrenocorticotropin and β-endorphin-like immunoreactivities *in vitro*. Endocrinology 103:1910–1915

Vale W, Spiess J, Rivier C, Rivier J (1981) Characterization of a 41 residue ovine hypothalamic peptide that stimulates the secretion of corticotropin and β-endorphin. Science 213:1394–1397

Vale W, Vaughan J, Smith M, Yamamoto G, Rivier J, Rivier C (1983) Effects of synthetic ovine corticotropin-releasing factor, glucocorticoids, catecholamines, neurohypophysial peptides, and other substances on cultured corticotropic cells. Endocrinology 113:1121–1131

Valverde C, Chieffo V, Reichlin S (1972) Prolactin releasing factor in porcine and rat hypothalamic tissue. Endocrinology 91:982-993

Van Loon GR, De Souza EB (1978) Effects of β-endorphin on brain serotonin metabolism. Life Sci 23:971-978

Van Loon GR, Kragt CL (1970) Effect of dopamine on the biological activity and *in vitro* release of ACTH and FSH. Proc Soc Exp Biol Med 133:1137-1141

Van Loon GR, Scapagnini U, Cohen R, Ganong WF (1971) Effect of the intraventricular administration of adrenergic drugs on the adrenal venous 17-hydroxycorticosteroid response to surgical stress in the dog. Neuroendocrinology 8:257-272

Van Maanen JH, Smelik PG (1968) Induction of pseudopregnancy in rats following local depletion of monoamines in the median eminence of the hypothalamus. Neuroendocrinology 3:177-186

Van Vugt DA, Bruni JF, Meites J (1978) Naloxone inhibition of stress-induced increase in prolactin secretion. Life Sci 22:85-90

Vekemans M, Delvoye P, L'Hermite M, Robyn C (1977) Serum prolactin levels during the menstrual cycle. J Clin Endocrinol Metab 44:989-993

Vermes I, Telegdy G (1972) Effect of intraventricular injection and intrahypothalamic implantation of serotonin on the hypothalamo-hypophyseal adrenal system in the rat. Acta Physiol Acad Sci Hung 42:49-69

Vermes I, Telegdy G, Lissak K (1974) Effect of midbrain raphe lesion on diurnal and stress-induced changes in serotonin content of discrete regions of the limbic system and in adrenal function in the rat. Acta Physiol Acad Sci Hung 45:217-224

Vician L, Shupnik MA, Gorski J (1979) Effects of estrogen on primary ovine pituitary cell cultures: stimulation of prolactin secretion, synthesis, and preprolactin messenger ribonucleic acid activity. Endocrinology 104:736-743

Von Euler C, Holmgren B (1956) The thyroxine receptor of the thyroid-pituitary system. J Physiol (Lond) 131:125-136

Wakabayashi I, Miyazawa Y, Kanda M, Miki N, Demura R, Demura H, Shizume K (1977) Stimulation of immunoreactive somatostatin release from hypothalamic synaptosomes by high (K^+) and dopamine. Endocrinol Jap 24:601-604

Wakabayashi I, Kanda M, Miki N, Miyoshi H, Ohmura E, Demura R, Shizume K (1980) Effects of chlorpromazine and naloxone on growth hormone secretion in rats. Neuroendocrinology 30:319-322

Wedig JH, Gay VL (1970) L-dopa as a stimulus for LH release in the proestrous rat: blockade of its action by pentobarbital anesthesia. Neuroendocrinology 15:99-105

Weick RF (1978) Acute effects of adrenergic receptor blocking drugs and neuroleptic agents on pulsatile discharges of luteinizing hormone in the ovariectomized rat. Neuroendocrinology 26:108-117

Wiegand SJ, Terasawa E, Bridson WE, Goy RW (1980) Effects of discrete lesions of preoptic and suprachiasmatic structures in the female rat. Neuroendocrinology 31:147-157

Wiggins JF, Fernstrom JD (1977) L-dopa inhibits prolactin secretion in proestrous rats. Endocrinology 101:469-474

Wilber JF, Baum D (1970) Elevation of plasma TSH during surgical hypothermia. J Clin Endocrinol Metab 31:372-375

Wilcox CS, Aminoff MJ, Miller JGB, Keenan J, Kremer M (1975) Circulating levels of corticotrophin and cortisol after infusions of L-dopa, dopamine, and noradrenaline in man. Clin Endocrinol 4:191-198

Willoughby JO, Martin JB (1978) Neural structure and neurotransmitters regulating growth hormone and prolactin secretion. In: K. Lederis and W. L. Veale (Eds) Current Studies of Hypothalamic Function. Karger, Basel

Willoughby JO, Brazeau P, Martin JB (1977) Pulsatile growth hormone and prolactin: effects of (+)-butaclamol, a dopamine receptor blocking agent. Endocrinology 101:1298–1303

Wislocki GB, King LS (1936) The permeability of the hypophysis and hypothalamus to vital dyes with a study of the hypophyseal vascular supply. Am J Anat 58:421–472

Woolf P, Lee L (1977) Effect of the serotonin precursor tryptophan on pituitary hormone secretion. J Clin Endocrinol Metab 45:123–133

Woolf PD, Lee LA, Schalch DS (1972) Adrenergic manipulation and thyrotropin-releasing hormone (TRH) induced thyrotropin (TSH) release. J Clin Endocrinol Metab 35:616–618

Wurtman RJ (1973) Role of catecholamines in neuroendocrine function. In: E. Usdin and S. Snyder (Eds) Frontiers in catecholamine research. Pergamon Press, Great Britain

Wuttke W, Hancke JL, Hohn KG, Baumgarten HG (1978) Effect of intraventricular injection of 5,7-dihydroxytryptamine on serum gonadotropins and prolactin. Ann NY Acad Sci 305:423–436

Yates FE, Russell SM, Dallman MF, Hedge GA, McCann SM, Dhariwal APS (1971) Potentiation by vasopressin of corticotropin release induced by corticotropin-releasing factor. Endocrinology 88:3–15

Yen SS, Silver TM, Devan EGW (1974) Effect of somatostatin in patients with acromegaly: suppression of growth hormone, prolactin, insulin, and glucose levels. N Engl J Med 290:935–938

Yoshimura M, Hachiya T, Ochi Y, Magasaka A, Takeda A, Kidaka H, Reffetoff S, Fang SV (1977) Suppression of elevated serum TSH levels in hypothyroidism by fusaric acid. J Clin Endocrinol Metab 45:95–98

Zimmerman EA, Carmel PW, Hussain MK, Ferin M, Tannenbaum M, Frantz AG, Robinson AG (1973) Vasopressin and neurophysin: high concentration in monkey hypophyseal portal blood. Science 182:925–927

CHAPTER 13

Regulation of Catecholamine Development

G. M. JONAKAIT and I. B. BLACK

A. Introduction

How does the mature, fully-differentiated nervous system develop from its embryonic, undifferentiated precursors? This question, posed a century ago by classical embryologists, remains the central issue of contemporary developmental neurobiology. Even partial answers would begin to define the mechanisms which govern normal growth, development and aging of the nervous system. Moreover, such information may help to define the molecular pathogenesis of developmental and degenerative neurological disorders and the events which trigger transformation of normal neurons into their neoplastic counterparts, including pheochromocytomas and neuroblastomas.

The immensity of the underlying question, however, has necessitated a narrower focus. Development of catecholaminergic neurons has constituted one focus of recent intense study which promises to yield insights relevant to the nervous system as a whole. This chapter describes these efforts, articulates recent observations, and attempts to define critical problems for future investigation. A number of recent reviews address related subjects (BRONNER-FRASER and COHEN 1980; LE DOUARIN 1980, 1982, 1984, 1986; LE DOUARIN et al. 1980; BLACK 1982; Black et al. 1984, 1987; WESTON 1982).

This volume attests to the fact that available techniques for study of the catecholamine neuron are among the most sophisticated in neurobiology. These neurobiological techniques have been complemented by a host of procedures drawn from classical embryology and tissue culture to approach some of the central issues in catecholamine development: What are the embryonic origins of catecholaminergic neurons? When does initial expression of noradrenergic characteristics occur? Is a precursor cell committed to become catecholaminergic before the first detectable appearance of that phenotype? What factors control such commitment? When—if ever—is a cell irrevocably committed to the expression of the catecholaminergic phenotype? Are all noradrenergic phenotypic characters regulated as a single genetic unit during development? What events regulate the development of neuropeptides in catecholaminergic neurons? Once established in its definitive site, what factors support and maintain the catecholaminergic phenotype? How does a catecholamine neuron define its field of innervation and make specific synaptic connections? Do maternal factors influence normal and abnormal catecholamine development?

Some answers exist. In general terms, it is known that both the peripheral catecholamine neuron and the adrenal chromaffin cell originate from the neural crest, a temporary cluster of cells in the embryo. The crest population migrates widely throughout the embryo, and its derivatives display remarkable phenotypic variety, encompassing both neuronal and non-neuronal cell types. The initial expression of catecholaminergic phenotypic characters appears to be influenced by critical interactions with factors in the migratory microenvironment. Subsequent development and survival may be regulated by other factors such as nerve growth factor (NGF), steroid hormones and ortho- and retrograde trans-synaptic interactions.

Recent studies of catecholamine development have fostered a number of new concepts concerning neuronal function. For example, the developing catecholamine neuron is remarkably labile with respect to transmitter phenotypic expression, and may alter expression under appropriate conditions. Recent experiments suggest that transplantation of embryonic catecholamine cells to appropriate sites elicits cholinergic expression. Some embryonic cells which express noradrenergic characteristics during one phase of development normally lose those traits later during ontogeny *in vivo*, or add other transmitter characters. Moreover, sympathetic ganglion cells express cholinergic and peptidergic properties *in vitro*. Finally, phenotypic plasticity may persist well beyond the normal, "developmental" period remaining a property of adult ganglia. Consequently, the capacity for change and phenotypic adaptation may not be confined to the "developing" neuron.

These observations are necessitating re-evaluation of the phenotypic potential of catecholamine neurons, and re-examination of the neurobiological meaning of the term "differentiated." In turn, these considerations have generated a number of new questions: Is phenotypic plasticity shared by the entire nervous system? When—if ever—does a neuron lose its potential for such change? What factor(s) trigger these changes? To what degree are neurons restricted in the phenotypic "choices" they can make? How many transmitters normally coexist in neurons? How does the manipulation of one neurotransmitter affect its co-localized partners?

In this chapter we approach these questions and relevant observations sequentially, by following catecholamine development in the embryo, fetus and neonate. We concentrate on peripheral catecholamine systems, since so much information is available in this area. After discussing the induction and migration of the neural crest, transmitter phenotypic expression is examined in detail. The roles of NGF, glucocorticoids and other growth factors are defined. Phenotypic expression during the postnatal period is subsequently characterized by examining the role of intercellular interactions as manifested by trans-synaptic regulation and regulation of transmitter expression *in vitro*.

B. Prenatal Development

I. The Neural Crest

1. Phenotypic Heterogeneity of Neural Crest Derivatives

Early studies definitively identified the neural crest as the precursor population for neurons of sympathetic ganglia (HAMMOND and YNTEMA 1947; see HORSTADIUS 1950 for review) and catecholamine-containing chromaffin tissue of the adrenal medulla and pre-aortic areas (HAMMOND and YNTEMA 1947; YNTEMA and HAMMOND 1954). In addition to these catecholaminergic populations, however, neural crest gives rise to sensory neurons of the dorsal root ganglia (TENNYSON 1965), parasympathetic neurons (VAN CAMPENHOUT 1946; NARAYANAN and NARAYANAN 1978), intrinsic neurons of the enteric plexuses (YNTEMA and HAMMOND 1945; ANDREW 1969, 1970, 1971; LE DOUARIN and TEILLET 1973), and sub-populations within cranial ganglia V, VII, IX and X (JOHNSTON 1966; NODEN 1975, 1978a, b; NARAYANAN and NARAYANAN 1980; AYER-LE LIÈVRE and LE DOUARIN 1982). It is apparent, of course, that these different neuronal populations use a variety of neurotransmitters and, consequently, express diverse phenotypic characters. In addition to neuronal derivatives, the crest also generates *non-neuronal* cell lines, including melanocytes (DORRIS 1938), cartilage of the upper face and jaw (HAMMOND and YNTEMA 1964; JOHNSTON 1966; LE LIÈVRE and LE DOUARIN 1975; NODEN 1975; LE LIÈVRE 1978), and glial and supporting elements of the peripheral nervous system (COULOMBRE et al. 1974).

The variety of cellular and neurotransmitter phenotypes derived from the neural crest raises questions concerning the heterogeneity of the crest population itself. Is each premigratory neural crest cell committed to a specific phenotypic fate before migration; and does migration then proceed along predetermined pathways to definitive sites of terminal differentiation? Alternatively, is the crest an homogeneous population of pluripotent cells which awaits cues from its migratory or definitive environments to select among a variety of phenotypic options?

2. Premigratory Neural Crest

To approach these questions studies have focused on induction of the neural crest from ectodermal precursors, and the nature of the premigratory crest cells.

The crest is an ectodermal derivative of the neural plate, neural folds and neural tube and appears as a wedge of cells at the edges of the neural folds during the neurula stage of development. Since only induced ectoderm yields cells with neuronal phenotypes (DUPRAT et al. 1985), a fundamental change in the developmental potential of these cells accompanies induction. While the mechanism of induction is unknown, it has been suggested that lateral sectors of the archenteron roof induce crest formation (RAVEN and KLOOS 1945). More recent studies have postulated a two-step induction process, requiring a series of critical ion concentrations (BARTH and BARTH 1969; MESSENGER and

WARNER 1976). Additionally, cyclic nucleotides can induce amphibian epidermis to differentiate into cells with neural crest potential (WAHN et al. 1975).

The precise consequences of crest induction are poorly defined, although the developmental potential obviously differs from that of the surrounding tissue. Recent studies have noted positive acetylcholinesterase staining in premigrating chick neural crest (SMITH et al. 1980; COCHARD and COLTEY 1983). More recently a variety of monoclonal antibodies that recognize neural crest-specific epitopes has been generated. These antibodies, designated HNK-1 (ABBO and BALCH 1981; VINCENT et al. 1983), NC-1 (LIPINSKI et al. 1983; VINCENT and THIERY 1984), EC/8 (CIMENT and WESTON 1982), A2B5 (GIRDLESTONE and WESTON 1985) and GLN1 (BARBU et al. 1986) allow early detection of migrating crest because that are among the earliest manifestation of neural crest induction.

It is more difficult to distinguish one neural crest cell from its sister. Cells are indistinguishable from one another at the ultrastructural level (BANCROFT and BELLAIRS 1976), and lectin-binding studies performed *in vitro* reveal an undifferentiated pattern of surface carbohydrate moieties (SIEBER-BLUM and COHEN 1978). However, using monoclonal antibodies to neuron-specific gangliosides, subpopulations of crest cells have been identified (GIRDLESTONE and WESTON 1985), suggesting, that some crest cells may be determined or differentiated prior to migration. Furthermore, studies of the discrete pigment patterning in adult allophenic mice, produced from artificial composites of genetically different blastomeres (MINTZ 1974), have suggested that the developmental potential of pigment cells, at least, is completely restricted prior to neural tube closure and subsequent migration (MINTZ 1967). Whether commitment to the catecholaminergic phenotype has occurred at this early stage remains an issue (ZILLER et al. 1983). However, it is clear that catecholaminergic precursors are profoundly influenced by the embryonic microenvironment, and depend on intercellular interactions for normal phenotypic expression. The interaction of intrinsic cellular information and environment may differ among crest sub-populations.

3. The Migratory Environment

Immediately following closure of the neural tube, crest cells begin extensive migration (NODEN 1975; BRONNER-FRASER and COHEN 1980; VINCENT et al. 1983; WESTON et al. 1984; LE DOUARIN et al. 1984e; LE DOUARIN 1986). Studies utilizing cell-marking techniques have traced this movement dorsolaterally beneath the ectoderm (TEILLET 1971) and ventrolaterally toward the mesoderm (WESTON 1963, 1970; JOHNSTON 1966; NODEN 1975; VINCENT et al. 1983). Cells destined to populate the sympathetic ganglia follow a path between somites to reach paraaortic sites (THIÈRY et al 1982a). Dorsal root ganglia arise primarily from cells opposite the anterior half of each somite (RICKMANN et al. 1986; KALCHEIM and LE DOUARIN 1987; LORING and ERICKSON 1987). The bulk of migration occurs opposite the anterior portion of the somite (TEILLET et al. 1987). Cells located in more rostral areas of the embryo leave the neural folds first, and subsequent migration proceeds in a rostro-caudal wave (WESTON

1970). Changes in cell-cell adhesion (HOLTFRETER 1939; NEWGREEN et al. 1982; THIÈRY et al. 1982b) or contact avoidance may initiate migration (TWITTY and NIU 1948; ABERCROMBIE 1970; EPPERLEIN 1974; NEWGREEN et al. 1979), but definitive evidence is lacking. Scanning and transmission electron microscopic studies have defined a number of morphological rearrangements that precede migration (BANCROFT and BELLAIRS 1976; LÖFBERG 1976; LÖFBERG et al. 1980). However, the factors that cause neural crest cells to begin their journey remain unknown. Nonetheless, the crest is endowed with remarkable migratory aggressiveness, and can move through tissue spaces even in foreign environments (ERICKSON et al. 1980).

Cells begin normal migration by extending filopodia into a cell-free matrix (EBENDAL 1977; TOSNEY 1978) filled with microfibrils (LÖFBERG 1976; EBENDAL 1977; LÖFBERG et al. 1980). Since migrating cells make transient contacts with this fibrillar matrix, it has been suggested that the alignment of fibrils provides some degree of stabilization and/or guidance (EBENDAL 1977; LÖFBERG et al. 1980; TOSNEY 1982). Ventrally-directed migration appears to be imposed by the environment shortly after the initiation of movement (ERICKSON et al. 1980). The extracellular field of migration is rich in glycosaminoglycans (GAGS; DERBY et al. 1976; DERBY 1978; PINTAR 1978; TOSNEY 1978), specifically, hyaluronate and sulfated GAGS (PRATT et al. 1975). These compounds are easily hydrated and may, by expanding the extracellular space, facilitate migration (TOSNEY 1978). Cessation of migration and aggregation into ganglionic clusters is accompanied by a fall in GAG concentration (DERBY 1978; PINTAR 1978), suggesting that environmental GAG synthesis and cell movement are correlated. It is of particular interest that the GAG matrix appears to be synthesized not only by migratory pathway structures but by migrating crest cells themselves (GREENBERG and PRATT 1977; MANASEK and COHEN 1977). Crest cells may, therefore, possess the capacity to condition their own migratory environment.

The path of neural crest cell migration is also rich in the glycoprotein fibronectin (NEWGREEN and THIÈRY 1980; DUBAND and THIÈRY 1982), which has been shown to be a preferred substrate for neural crest migration *in vitro* (GREENBERG et al. 1981; NEWGREEN et al. 1982; ERICKSON and TURLEY 1983; ROVASIO et al. 1983). Antibodies against fibronectin disrupt crest migration (BOUCAUT et al. 1984) arguing for a key role. Like GAGs, fibronectin disappears from the extracellular matrix when neural crest cells aggregate to form ganglia (THIÈRY et al. 1982a). Moreover, neural crest cells possess on their surfaces a 140kD glycoprotein complex (BRONNER-FRASER 1986; DUBAND et al. 1986), identified as a possible fibronectin receptor (CHEN et al. 1985).

The environment, therefore, plays a major role in crest migration. Moreover, "young" premigratory crest cells grafted into progressively older hosts fail to migrate and contribute normally to sympathetic ganglia in the host (WESTON and BUTLER 1966). Conversely, "older" crest cells transplanted to a younger somitic environment re-initiate migration and contribute to the normal range of neural crest derivatives. Consequently, age-related characteristics of the embryonic microenvironment are critical for crest migration and are necessary for full expression of the crest cell phenotypic potential. These studies imply that premigratory crest cells are not phenotypically predetermined, but rather are pluripotent, depending on age-dependent environmental "signals" for final localization and differentiation.

The "young" migratory environment even supports migration of differentiated crest cells. Cloned melanocytes (Cohen and Konigsberg 1975) introduced to a premigratory environment *in vivo* undergo ventral migration and populate "foreign locales" (Bronner and Cohen 1979; Bronner-Fraser and Cohen 1979, 1980). Ventral migration is severely restricted, however, if cloned melanocytes are introduced to rostrocaudal levels where migration is already well underway (Bronner-Fraser and Cohen 1979). Furthermore, neither the young nor the old environment supports migration of *fibroblasts* from somite, limb bud, lateral plate or heart (Erickson et al. 1980), suggesting that these migratory pathways specifically regulate crest mobility.

4. The Migratory Environment and Phenotypic Expression

Increasing evidence, therefore, supports the contention that the migratory and definitive environments critically influence neuronal and non-neuronal fates of crest derivatives. Using a cell-marking technique which relies on species differences in nuclear chromatin condensation (Le Douarin 1973), neural crest from quail has been grafted into chick heterotopically (i.e. to an alien level of the neuraxis) to ascertain whether specificity lies in the crest population or the environment. Heterotopic exchange of premigratory crest within the cephalic region results in normal neuronal and mesenchymal development (Noden 1975, 1978 a,b). Similarly, grafts of cephalic crest to trunk levels results in the normal formation of thoracolumbar sympathetic ganglia and adrenal medulla (Le Douarin and Teillet 1974; Teillet and Le Douarin 1974). These observations suggest that premigratory crest cells possess considerable phenotypic flexibility, and that undefined environmental signals influence phenotypic expression. However, it is possible that the embryonic environment allows survival of only selected cellular populations, which are phenotypically pre-determined (Le Douarin 1984; Girdlestone and Weston 1985). The observation that the progeny of single, cloned crest cells express multiple phenotypes (Cohen and Konigsberg 1975; Sieber-Blum and Cohen 1980) is provocative. Further studies with this system may distinguish between the alternative explanations.

Even differentiated crest derivatives retain an extraordinary degree of plasticity in response to their environment. Cells from sensory ganglia, for example, explanted *in vitro* soon after ganglion formation, develop catecholamine fluorescence (Newgreen and Jones 1975; Xue et al. 1985), or become pigmented (Cowell and Weston 1970; Nichols and Weston 1977). The apparent conversion to melanin production is dependent upon the presence of serum, and is encouraged by culture conditions which disrupt associations between cells (Nichols et al. 1977), suggesting that serum factors and cellular interactions remain critical *in vitro*. Indeed, the importance of serum constituents (Ludueña 1973a, b; Greenberg and Schrier 1977; Ziller et al. 1979, 1983, 1987; Derby and Newgreen 1982; Iacovitti et al 1982; Wolinsky and

Patterson 1985) and culture substrates in affecting neuronal differentiation has been extensively documented (Ludueña 1973a, b; Maxwell 1976; Sieber-Blum and Cohen 1978; Loring et al. 1979, 1982; Hawrot 1980; Adler and Black 1986; Acheson et al. 1986). The discovery that a neural tube-derived factor promotes the survival and differentiation of sensory ganglia *in vivo* (Teillet and Le Douarin 1983; Lindsay et al. 1985; Kalcheim and Le Douarin 1986; Kalcheim et al. 1987) highlights the growing ability to identify specific environmental molecules capable of affecting cell differentiation.

While interactions between cell and environment may guide and influence development, the environment is certainly not instructive in all cases. Experiments performed in both amphibia (Chibon 1966) and birds (Le Douarin and Teillet 1974; Le Lièvre and Le Douarin 1975; Noden 1978a, b) have shown that truncal neural crest, grafted to cephalic levels migrates appropriately to visceral arches, but does not form mesenchymal derivatives of the face and jaw. Similarly, in the chick, transplanted cells from the trunk do not interact normally with placodal tissue to form trigeminal and ciliary ganglia (Noden 1978b). Conversely, cephalic neural crest forms normal cranial chondrocytes *in vitro* (Hall and Tremaine 1979) and at the thoracic level after transplantation *in vivo* (Le Douarin and Teillet 1974). It is unclear whether the apparent pre-determination is due to crest cell numbers involved, inherent differences in crest cell populations, precocious signaling from rostral areas of the embryo, or other factor(s).

In summary, the embryonic neural crest gives rise to extraordinarily heterogeneous cellular populations. While the migratory environment plays a critical role in crest cell phenotypic expression, it does not appear to be determinative in all cases. In fact, the central issue of the relationship of intrinsic cellular information to environmental influences in regulating phenotypic expression remains to be resolved.

II. Initial Appearance of Catecholaminergic Characteristics

When do catecholaminergic phenotypic characters initially appear in neural crest derivatives, and what is the pattern of appearance of different catecholamine traits? There is general agreement that, while cholinergic traits may appear in migrating mesencephalic crest cells (Smith et al 1980; Cochard and Coltey 1983), a number of catecholaminergic characters appears in close temporal proximity at about the time that cells aggregate to form the sympathetic ganglion primordia.

These issues have been examined in a number of species (Enemar et al. 1965; de Champlain et al. 1970; Fernholm 1972; Papka 1972; Cochard et al. 1978, 1979; Teitelman et al. 1979; Pearson et al. 1980). However, studies in the rat embryo have defined the simultaneous appearance of catecholamine fluorescence and immunoreactivity to the noradrenergic biosynthetic enzymes, TH and DBH (Cochard et al. 1978, 1979; Teitelman et al. 1979). Catecholamine characters are undetectable in neural crest or migrating crest cells (Enemar et al. 1965; de Champlain et al. 1970; Allan and Newgreen 1977; Cochard et al. 1978, 1979), and initially appear as cells aggregate to form the ganglion anlage (COCHARD et al. 1978, 1979; TEITELMAN et al. 1979; see also DE CHAMPLAIN et al. 1970). Immunoreactivity to TH, the rate-limiting enzyme in catecholamine biosynthesis, and to DBH, which converts dopamine to noradrenaline, and catecholamine histofluorescence all appear at 11.5 days of gestation (E11.5) in primitive ganglia (COCHARD et al. 1978, 1979; TEITELMAN et al. 1979). Moreover *in situ* hypridization studies reveal the presence of TH-specific mRNA at E11.5 (JONAKAIT et al. 1988). In the chick, catecholamine fluorescence appears at an analogous developmental stage, at 3.5 days of incubation, in primary sympathetic ganglia (ENEMAR et al. 1965; KIRBY and GILMORE 1976; ALLAN and NEWGREEN 1977).

TH, DBH, catecholamines, and TH-specific mRNA also appear simultaneously at E11.5 in a population of cells in the embryonic rat gut (COCHARD et al. 1978, 1979; TEITELMAN et al. 1979; JONAKAIT et al. 1979, 1980, 1988). Remarkably, this population expresses noradrenergic characters only transiently. By E12.5, gut cells exhibiting catecholamine characters have increased in number, and also possess the specific, high-affinity uptake system for noradrenaline (JONAKAIT et al. 1979, 1985; GERSHON et al. 1984). However, by 13.5 days immunoreactivity to TH and DBH and catecholamine fluorescence have disappeared, while uptake persists, allowing identification of the cells (JONAKAIT et al. 1979, 1985).

The simultaneous appearance and disappearance of TH, DBH and catecholamines suggests that a number of catecholaminergic characters may be subject to coordinate regulation during development. Alternatively, of course, the different phenotypic characters may be individually regulated by different intra- or extra-cellular signals developing simultaneously. However, all characters do not disappear synchronously, since the uptake system persists in the gut cells. Moreover, previous studies have indicated that uptake mechanisms appear before endogenous catecholamine histofluorescence in chick ganglia (ROTHMAN et al. 1978), in terminals of chick spinal cord (SINGER et al. 1980), and in rat central noradrenergic neurons (COYLE and AXELROD 1971) and after the appearance of TH in transiently catecholaminergic gut cells (JONAKAIT et al. 1985). Moreover, a variety of non-neuronal cell types in the embryo also take up catecholamines (KIRBY and GILMORE 1972; NEWGREEN et al. 1981). Finally, DOPA decarboxylase (L-aromatic amino acid decarboxylase), a ubiquitous, non-specific, amino acid-metabolizing enzyme, appears prior to catecholamine fluorescence (PEARSE 1969; KIRBY and GILMORE, 1976; ALLAN and NEWGREEN 1977). Consequently, different catecholaminergic characters may appear at different times and may, therefore, be subject to different ontogenetic mechanisms.

The transiently noradrenergic population of the gut is of particular interest, since it indicates that cells may normally exhibit phenotypic plasticity during development *in vivo*. The selective loss of TH, DBH and endogenous catecholamines with the retention of uptake mechanisms, suggests that plasticity extends to different characters of the same transmitter. It is not yet clear whether the gut cells acquire new transmitter phenotypic characters while, or after, losing catecholaminergic traits. In the developing rat pancreas, transiently dopaminergic cells acquire glucagon prior to losing catecholamine traits (TEITELMAN et al. 1981b). Recent studies have suggested further that phenotypic plasticity is also a characteristic of the developing cholinergic sympathetic innervation of eccrine sweat glands, since catecholamine fluorescence, evident in the gland at postnatal week one, fades after three weeks (LANDIS and KEEFE 1983; YODLOWSKI et al. 1984). Phenotypic transformation may be a generalized phenomenon in the nervous system during development: Beyond its detection in cells of gut and pancreas, transient expression of TH has now been found in a variety of populations including cranial sensory ganglia (JONAKAIT et al. 1984) and ventral neural tube (TEITELMAN et al. 1981a; JONAKAIT et al. 1985).

1. Cell Cycle and Initial Expression

The relationship of phenotypic expression and enzyme inducibility to stages of the cell cycle has long been of central interest in developmental and regulatory biology. Elucidation of this problem may lead to insights regarding the role of genomic and post-transcriptional events in phenotypic expression. In developing catecholamine systems, work has thus far focused on the temporal relationship of mitosis and phenotypic expression. In early studies, COHEN demonstrated that catecholamine histofluorescence appears in sympathetic ganglion cells undergoing mitosis (COHEN 1974) and that, consequently, transmitter differentiation is not simply a post-mitotic event. These observations have been subsequently confirmed and extended (HENDRY 1977a; ROTHMAN et al. 1978, 1980; TEITELMAN et al. 1981a; ROHRER and THOENEN 1987), and it is now apparent that withdrawal from the cell cycle is not a prerequisite for expression of noradrenergic characters in sympathetic ganglia or in gut cells. However, it is not clear that these observations are applicable to central populations, since cessation of mitosis precedes expression in central catecholamine neurons (ROTHMAN et al. 1980).

2. Embryonic Environment and Catecholaminergic Expression

The apparent role of the embryonic microenvironment in guiding crest derivatives to their divergent phenotypic fates has been discussed in general terms in previous sections. We now consider environmental influences in greater detail, focusing on catecholamine expression. However, the same caveat applies: It is extremely difficult from *in vivo* studies alone to distinguish between environmental selection of phenotypically pre-determined cells, and direct environmental influences on phenotypic expression in each cell.

Cephalic or "vagal" (somites 1-7) pre-migratory crest, normally destined to populate non-catecholaminergic enteric ganglia (LE DOUARIN and TEILLET 1973), transplanted to "adrenomedullary" levels (somites 18-24), populates sympathetic ganglia and adrenal medullae and exhibits catecholamine fluorescence (LE DOUARIN and TEILLET 1974). Conversely, cells from the "adrenomedullary" region, heterotopically transplanted to cephalic levels, invade the enteric ganglia, produce morphologically normal enteric plexuses, and fail to exhibit catecholamine fluorescence (LE DOUARIN and TEILLET 1974; LE DOUarin et al. 1975). Furthermore, association of "adrenomedullary" crest with the aneural embryonic gut *in vitro*, results in the formation of normal, non-adrenergic enteric ganglia (Le Douarin and Teillet 1974; Le Douarin 1977; Le Douarin et al. 1977; Smith et al. 1977; Le Douarin et al. 1980). Moreover, transiently catecholaminergic cells in the embryonic gut show abrupt cessation of TH mRNA accumulation upon entering the gut microenvironment (Jonakait et al. 1988). These observations raise the possibility that the environment directly influences phenotypic expression, although selection of predetermined populations is possible. In the following sections we examine structures and molecules which may participate in environmental regulation, regardless of ultimate mechanism.

a. The Notochord and Somites

Several studies suggest that interactions among neural crest, somite and notochord can promote expression of noradrenergic characteristics. Contact with somitic mesenchyme elicits catecholamine appearance in crest cells *in vitro* (Cohen 1972; Norr 1973). However, the ventral neural tube area (presumably the notochord) must first "prime" the somitic mesenchyme (Cohen 1972; Norr 1973; Le Douarin 1977). The ventral neural tube exerts this influence across a membrane (Norr 1973), suggesting that the activating substance is diffusible. Since the presence of the notochord allows even aneural gut (splanchinc mesenchyme) to suport noradrenergic expression in cells of the enteric ganglia (Le Douarin et al. 1977), the specificity of the mesenchyme appears to be less important that the influence of the notochord (Le Douarin 1977).

The dorsal region of the embryo in the "adrenomedullary" area may elicit catecholamine expression even in undifferentiated precursor cells from non-sympathetic ganglia. Transplantation of the ganglion of Remak (Le Douarin et al. 1977), the ciliary ganglion (Le Douarin et al. 1978; Ziller et al. 1979; Dupin 1984), or sensory ganglia (Le Lièvre et al. 1980; Ayer-Le Lièvre and Le Douarin 1982; Schweizer et al. 1983) into the dorsal area of the embryo, results in disaggregation, re-migration and invasion of sympathetic ganglia by donor cells which exhibit catecholamine histofluorescence.

In cultures of neural crest, growth on a substrate conditioned by somitic fibroblasts or the addition of fibronectin significantly increases the proportion of catecholamine-containing cells (Sieber-Blum et al. 1981; Loring et al. 1982). Since neural tube, notochord, and somites synthesize fibronectin, it has been suggested that this molecule is involved in the induction of catecholaminergic characteristics by these structures (Sieber-Blum et al. 1981; Loring et al. 1982).

In contrast to the foregoing observations, however, are the studies of Bjerre (1973), Cohen (1977), and Kahn et al. (1980) which indicate that catecholaminergic traits may appear in crest cells cultured in the absence of somite and neural tube. Still other studies find that culture medium conditioned by notochord and somite will not support adrenergic differentiation in the absence of neural tube-derived factor(s) (Howard and Bronner-Fraser 1985). Although factors in the culture medium or substrate may have substituted for somite and notochord, it is also possible that these structures are not strictly determinative in eliciting phenotypic expression *de novo*.

b. Nerve Growth Factor (NGF)

NGF is a well-characterized protein (LEVI-MONTALCINI and ANGELETTI 1968; GREENE and SHOOTER 1980; THOENEN and BARDE 1980; THOENEN et al. 1985; YANKNER and SHOOTER 1982), which is present in a number of sympathetic targets (HENDRY 1972; HENDRY and IVERSEN 1973b; EBENDAL et al. 1980, 1983; SHELTON and REICHARDT 1984; KORSCHING and THOENEN 1983a, 1988; DAVIES et al. 1987; BANDTLOW et al. 1987), synthesized in high titers in the male mouse salivary gland (BUEKER et al. 1960; ISHII and SHOOTER 1975), and subject to retrograde axonal transport by postnatal sensory and sympathetic neurons (HENDRY et al. 1974a, b; STÖCKEL et al. 1975a,b; HENDRY 1977b; JOHNSON et al. 1978; BRUNSO-BECHTOLD and HAMBURGER 1979; SCHWAB et al. 1982; KORSCHING and THOENEN 1983b). It has long been recognized that the protein stimulates development and enhances survival of *postnatal* sympathetic neurons *in vivo* and *in vitro* (see subsequent section for discussion of postnatal effects), and that antiserum to NGF (anti-NGF) prevents normal postnatal development (LEVI-MONTALCINI and ANGELETTI 1963; COHEN et al. 1964; ANGELETTI et al. 1972), and is necessary for postnatal catecholaminergic maturation. More recently, however, it has become apparent that NGF may be important for development of *embryonic* noradrenergic neuroblasts as well (KLINGMAN 1966; GORIN and JOHNSON 1979, 1980; KESSLER et al. 1979; KESSLER and BLACK 1980). Sympathetic neuroblasts *respond* to NGF *in vivo* and *in vitro* shortly after ganglia coalesce in the embryo, and become increasingly dependent on the protein with age (KLINGMAN 1966; COUGHLIN et al. 1977, 1978; KESSLER and BLACK 1980; COUGHLIN and COLLINS 1985). Moreover, transuterine injection of embryos with NGF on E11.5 results in the prolongation of TH activity and immunoreactivity as well as catecholamine histofluorescence in transiently catecholaminergic cells of the gut (KESSLER et al. 1979; JONAKAIT et al. 1981b). Since NGF is known to enhance survival of cultured, migrating crest cells which exhibit catecholamine fluorescence (NORR 1973), local concentrations of NGF in the embryonic microenvironment *in vivo* may allow survival of cells, which, in turn, permits catecholamine expression.

Although neuroblasts *respond* to NGF early in gestation, the factor does not appear to be *necessary* for survival and development at this stage. Ganglion explants as well as dissociated cells from E13-14 mice elaborate neurites and exhibit normal TH development in culture in the absence of added NGF or in the presence of anti-NGF (BLACK and COUGHLIN 1977; COUGHLIN et al. 1977; BLOOM and BLACK 1979; COUGHLIN and COLLINS 1985). Nevertheless, addition of high titers of NGF to the medium results in enhanced neurite elaboration and elevated TH activity (COUGHLIN et al. 1977, 1978). The ganglia appear to develop an absolute dependence on NGF *in vitro* at 16-17 days of gestation (COUGHLIN et al. 1977, 1978). These *in vitro* studies are complemented by *in vivo* experiments which demonstrate that anti-NGF administration has an increasingly inhibitory effect on ganglion development with gestational age

(KLINGMAN 1966; KESSLER and BLACK 1980). Nevertheless, even at the earliest stages, anti-NGF has a small deleterious effect, suggesting that a small subpopulation of ganglion neuroblasts may already be dependent on NGF at this stage. In summary, then, NGF plays a significant role in catecholaminergic neuron development prenatally as well as postnatally.

Although NGF is apparently necessary for catecholamine neuronal survival and development, until recently it has been unclear whether the molecule could directly alter phenotypic expression (CHUN and PATTERSON 1977a,b,c). However, NGF may affect phenotypic expression. Addition of NGF to cultured, dissociated adrenomedullary chromaffin cells, for example, results in *de novo* neurite elaboration and increased TH activity (UNSICKER et al. 1978; TISCHLER and GREENE 1975; DOUPE et al. 1985a). A similar extension of neurites from embryonic adrenal chromaffin cells accompanies NGF treatment *in vivo* (ALOE and LEVI-MONTALCINI 1979). Recent studies have also described neurite elaboration and loss of fluorescence intensity in cultures of small, intensely fluorescent (SIF) cells grown with NGF (DOUPE et al. 1985b). Consequently, certain catecholaminergic populations do respond to NGF by altering phenotypic expression.

While it is clear that NGF is capable of eliciting multiple responses in receptive catecholamine cells, the molecular basis of action is undefined. One model system that has been of value in studying the mechanisms of NGF action is the PC12 pheochromocytoma clonal cell line (see GREENE and SHOOTER 1980 for review). In the absence of NGF, PC12 cells survive and exhibit a variety of noradrenergic traits, including catecholamine fluorescence, biosynthetic enzymes, appropriate vesicular morphology and noradrenaline uptake and release (GREENE and TISCHLER 1976; GREENE and REIN 1977a, b, 1978; TISCHLER and GREENE 1978). The clone also exhibits cholinergic properties (GREENE and TISCHLER 1976; GREENE and REIN 1977c; SCHUBERT et al. 1977) as well as gabaergic traits (HATANAKA et al. 1980). PC12 cells possess NGF receptors (HERRUP and THOENEN 1979), and NGF elicits cessation of mitosis, extension of neurites (TISCHLER and GREENE 1975, 1978; GREENE and TISCHLER 1976), development of action potentials (TISCHLER et al. 1976; DICHTER et al. 1977), increased choline acetyltransferase (CAT) activity (GREENE and REIN 1977c; SCHUBERT et al. 1977), and rapid activation of TH via CA^{2+}-independent phosphorylation of the TH molecule (GOODMAN and HERSCHMAN 1978; HALEGOUA and PATRICK 1980; GREENE et al. 1984; LEE et al. 1985). Since so many of these changes involve membrane phenomena, it is of interest that NGF treatment also results in the increased synthesis of a membrane glycoprotein, the NGF-Inducible Large External (NILE) glycoprotein (MCGUIRE et al. 1978). The effects of NGF are reversible; withdrawal results in the disappearance of neurites and resumption of cell division (GREENE and TISCHLER 1976).

c. Glucocorticoid Hormones

It is well documented that glucocorticoid hormones regulate PNMT, the adrenaline-synthesizing enzyme, in the adrenal medulla both *in vivo* (COU-

PLAND and MacDougall 1966; Wurtman and Axelrod 1966; Ciaranello 1978) and *in vitro* (Doupe et al. 1985a), and TH in sympathetic ganglia (Hanbauer et al. 1975; Otten and Thoenen 1975, 1976a,b, 1977) postnatally. Recent studies indicate that glucocorticoids also influence catecholaminergic phenotypic expression during embryonic development.

Adrenomedullary precursors initially migrate to caudal thoracic ganglion-primordia where TH and DBH, but not PNMT, are expressed at approximately E13.0–13.5 (Cochard et al. 1979; Teitelman et al. 1979; Bohn et al. 1981). Subsequently, TH and DBH-containing cells migrate to the adrenal anlage, but PNMT does not appear until E17.5 (Teitelman et al. 1979; Bohn et al. 1981) suggesting that glucocorticoids might elicit PNMT appearance (Chevallier 1972; Unsicker et al. 1978). However, treatment of mothers or fetuses with dexamethasone, cortisol or ACTH does not cause precocious appearance of PNMT (Bohn et al. 1981). Moreover, steroids do not elicit precocious PNMT appearance *in vitro* (Teitelman et al. 1982). Conversely, inhibition of glucocorticoid production does not prevent the initial appearance of PNMT at E17.5 (Bohn et al. 1981). However, these treatments prevent the normal developmental increase in PNMT, once it has appeared (Bohn et al. 1981; Teitelman et al. 1982). Consequently, in the adrenal medulla, glucocorticoids appear to regulate development of the adrenergic phenotype after steroid-independent initial expression has occurred. It may be concluded that *initial expression* and subsequent *maintenance* of the adrenergic phenotype are separate processes (Bohn et al. 1981).

Glucocorticoids also influence adrenergic expression in sympathetic ganglia. Treatment of rats either *in utero* or during the first postnatal week results in radiometric (Ciaranello et al. 1973; Phillipson and Moore 1975; Liuzzi et al. 1977; Bohn et al. 1982; Eränkö et al. 1982) and immunocytochemical (Bohn et al. 1982; Eränkö et al. 1982) detection of PNMT in a ganglion population corresponding to the SIF cell (Eränkö et al. 1972, 1973, 1982; Bohn et al. 1982, 1984; Doupe et al. 1985b). Moreover, in the case of the ganglion, as in the adrenal medulla, this represents an increase in PNMT activity already present and not *de novo* appearance of enzyme (Bohn et al. 1982).

In addition to influencing *adrenergic* expression *in vivo*, glucocorticoids also affect *noradrenergic* expression. Treatment of pregnant rats with agents that increase endogenous glucocorticoids, or with the steroids themselves, delays the normal disappearance of noradrenergic characters in cells of the embryonic gut (Jonakait et al. 1980, 1981a, 1988). Thus, maternal glucocorticoids may influence development of the noradrenergic phenotype in the embryo.

In cultures of postnatal sympathetic neurons or adrenal medullary cells, which have expressed cholinergic traits (see below), dexamethasone decreases CAT activity and increases TH activity (McLennan et al. 1980; Fukada 1981; Doupe et al. 1985a,b). Similarly in PC12 cells (see above), dexamethasone increases TH activity and decreases both baseline and NGF-induced activities of CAT (Edgar and Thoenen 1978; Schubert et al. 1980). The increase in TH activity is blocked by actinomycin D (Edgar and Thoenen 1978) suggesting that dexamethasone acts at the transcriptional level, a contention supported by the report of increased levels of TH-specific mRNA in PC12 cells exposed to dexamethasone (Baetge et al. 1981; Lewis et al. 1983). Moreover, transiently catecholaminergic cells of the embryonic gut show increased accumulation of TH mRNA in response to increases in maternal glucocorticoids (Jonakait et al. 1988).

In summary, glucocorticoids preferentially select for noradrenergic expression in a number of *in vitro* systems, and effects may be exerted at the transcriptional level. However, the basis of this selectivity remains to be elucidated.

d. Growth Factors in Culture

Extensive evidence indicates that environmental factors regulate development and phenotypic expression in catecholamine neurons. These factors are currently being analyzed *in vitro* by studying the effects of medium conditioned by different cellular populations on neuronal growth and development. A number of new concepts is emerging. First, and most generally, families of factors, perhaps analogous to NGF, appear to guide receptive neurons during development. Second, during development catecholamine neurons, at least, can be induced to express more than one transmitter phenotype by appropriate factors. Finally, such transmitter phenotypic plasticity may extend well beyond the conventional developmental period, perhaps throughout the life of the neuron. In this section of the chapter we briefly summarize some of the work on which these conclusions are based.

Studies from a number of different laboratories indicate that neurons are capable of expressing more than one transmitter phenotype. If grown in a nutrient medium which permits the proliferation of ganglionic non-neuronal cells, cultures of sympathetic neurons, initially predominantly noradrenergic, acquire the ability to synthesize acetylcholine (ACh; Patterson and Chun 1974; Johnson et al. 1976) and the neuropeptide substance P (SP; Kessler et al. 1981; Kessler 1984; Roach et al. 1987). Therefore, the culture environment plays a critical role in the acquisition of cholinergic and peptidergic characteristics. However, neuronal depolarization with the attendant influx of Ca^{2+} reduces the ability of cells to respond to the cholinergic factor in the medium (Walicke et al. 1977; Walicke and Patterson 1981; Kessler et al. 1981; Roach et al. 1987). While cells in these cultures exhibit specific uptake and Ca^{2+}-dependent release of ^{3}H-noradrenaline (Patterson et al. 1975; Burton and Bunge 1975; Wakshull et al. 1978; Greene and Rein 1978), and retain morphological characteristics of noradrenergic cells (Claude 1973; Rees and Bunge 1974), transmission at many of the synapses formed, particularly in older cultures, is cholinergic since it (1) is blocked by curare and hexamethonium, (2) is inhibited by high doses of atropine, and (3) can be mimicked by iontophoresis of ACh (O'Lague et al. 1974; Ko et al. 1976b; Higgins et al. 1981; Nurse 1981). Moreover, vesicular morphology becomes increasingly characteristic of cholinergic terminals (Johnson et al. 1976). Membrane glycoproteins and glycolipids, measured both morphologically and biochemically, also show significant alterations as noradrenergic cells acquire cholinergic traits in culture (SCHWAB and LANDIS 1981; BRAUN et al. 1981).

A variety of other cell types and culture conditions induces similar changes in dissociated sympathetic neurons indicating that the effect is not restricted to ganglion non-neuronal cells. Sympathetic neurons and even adrenal medullary cells express cholinergic and peptidergic traits when cultured in the presence of heart (O'LAGUE et al. 1975; PATTERSON et al. 1977; DOUPE et al. 1985a,b), skeletal myotubes (NURSE and O'LAGUE 1975; NURSE 1981), spinal cord explants (BUNGE et al. 1974; KO et al. 1976a, b), sympathetic targets (KESSLER et al. 1984), in medium supported by chick embryo extract and/or serum (KO et al. 1976a, b; IACOVITTI et al. 1981, 1987; OGAWA et al. 1984; WOLINSKY and PATTERSON 1985), or in medium conditioned by a variety of tissues including rat heart, blood vessel, skeletal muscle, liver, brain or embryonic fibroblasts (PATTERSON et al. 1975; PATTERSON and CHUN 1977a, b; PATTERSON et al. 1977). Partial purification of a protein differentiation factor from conditioned medium has been accomplished (FUKADA 1985; WEBER et al. 1985).

However, while ample evidence exists for the existence of a diffusible, conditioned-medium factor, a membrane-bound factor plays a similar role. Increases in cell density with attendant increases in cell-to-cell interaction and/or aggregation also causes an increase in CAT and SP (ADLER and BLACK 1985, 1986) and TH (ACHESON and THOENEN 1983). Addition of membranes from a variety of sources including DRG and brain (ADLER and BLACK 1986; KESSLER et al. 1986) or adrenal medulla (SAADAT and THOENEN 1986) mimics the effects of increased cell density. Partial purification of this membrane-bound factor has also been accomplished (WONG and KESSLER 1987).

Studies on *single* sympathetic neurons grown on heart monolayers in culture, support the contention that "dual function", noradrenergic-cholinergic neurons do, indeed, exist. In these cultures most synapses formed on cardiac myocytes exhibit cholinergic, noradrenergic and even purinergic properties electrophysiologically (FURSHPAN et al. 1976, 1986a,b; O'LAGUE et al. 1978; POTTER et al. 1980, 1986). Moreover, electrophysiologically-identified dual-function neurons contain small vesicles, some of which appear granular following permanganate fixation, suggesting dual function on a morphological basis as well (LANDIS 1976, 1980). Electrophysiologically identified cholinergic neurons, in addition, exhibit TH immunoreactivity (HIGGINS et al. 1981), and take up 5-hydroxydopamine, a substrate for the noradrenergic uptake system, into their vesicles (LANDIS 1976). At a time when cultures exhibit significant levels of CAT, more than 95% of neurons exhibit immunoreactivity to TH (IACOVITTI et al. 1981). Moreover, neurons are simultaneously capable of ACh synthesis and noradrenaline uptake (REICHARDT and PATTERSON 1977; WAKSHULL et al. 1978; SCHERMAN and WEBER 1986). Others, however, report a loss of TH activity (WOLINSKY and PATTERSON 1983; SWERTS et al. 1983) suggesting a conversion from one phenotype to another. These studies clearly indicate that developing neonatal neurons are capable of expressing multiple neurotransmitter characteristics in response to appropriate environmental cues in culture.

Since cultured sensory and ciliary neurons will, under certain conditions, express TH immunoreactivity (PRICE and MUDGE 1983; IACOVITTI et al. 1985; TEITELMAN et al. 1985), the capacity to respond to environmental cues by acquiring or altering neurotransmitter phenotypes may be a phenomenon shared by many neuronal types (see JONAKAIT and BLACK 1986).

Moreover, the period of neuronal phenotypic plasticity may extend well beyond the accepted "developmental period." *Adult* sympathetic neurons in culture also express cholinergic (WAKSHULL et al. 1979 a,b) and peptidergic (ADLER and BLACK 1984; POTTER et al. 1986) properties. That some adult sympathetic neurons release both noradrenaline and ACh under normal circumstances has long been proposed (BURN and RAND 1965; BURNSTOCK 1976, 1978), and other studies have established the existence of noradrenaline uptake in the cholinergic sympathetic neurons which innervate eccrine sweat glands (LANDIS and KEEFE 1983). Furthermore, these observations are consistent with a rapidly expanding body of evidence indicating the coexistence of neuromodulatory peptides and "classical" neurotransmitter molecules in the same neuron (see HÖKFELT et al. 1986). The potential for expressing dual function, therefore, may be a widespread phenomenon, not restricted to the *developing* nervous system.

C. Postnatal Development

During postnatal development the sympathetic neuron establishes communication with the afferent pre-synaptic (cholinergic) terminal as well as appropriate target organs in the periphery. These connections not only promote survival and maturation but also govern critical regulatory relationships which persist throughout adulthood.

I. Anterograde Trans-synaptic Regulation of Ganglionic Neurons

Avian and mammalian superior cervical ganglia (SCG) exhibit a dramatic postnatal increase in CAT activity (BLACK et al. 1971 a; DOLEZALOVA et al. 1974). Since this rise is accompanied by a parallel increase in ganglionic synapse formation (BLACK et al. 1971 a; ROSS et al. 1977; SMOLEN and RAISMAN 1980), the development of CAT activity in the SCG is an appropriate index of preganglionic synapse formation (BLACK et al. 1971 a).

Preganglionic synapse formation, in turn, is followed by an increase in TH and DBH activity within sympathetic ganglion cells (BLACK et al. 1971 a,b; THOENEN et al. 1972 a; FAIRMAN et al. 1976). Immunotitration with specific antibodies to TH has revealed that the developmental rise in TH is not due to the activation of pre-existent enzyme molecules, but is instead entirely attributable to increased numbers of TH molecules (BLACK et al. 1974). Furthermore, transection of the preganglionic cholinergic trunk in neonatal mice or rats prevents the normal developmental increase in ganglionic TH activity (BLACK et al. 1971 a, 1972 b, 1974; THOENEN et al. 1972 c). Ganglionic blocking agents, which compete with ACh for nicotinic cholinoceptors, mimic the effects of decentralization (BLACK 1973; HENDRY 1973; BLACK and GEEN 1974; FAIRMAN et al. 1977). However, muscarinic antagonists do not affect postsynaptic maturation (BLACK et al. 1972b; BLACK and GEEN 1974). Conversely, stimulation of ganglionic transmission in developing rats produces a long-lasting elevation of TH and DBH activities (BARTOLOMÉ and SLOTKIN 1976). These observations suggest that presynaptic terminals regulate ganglionic ontogeny through the action of ACh on nicotinic receptors. Since ACh cannot by itself replace presynaptic terminals *in vivo* (BLACK et al. 1972b), ACh may be a necessary, but not sufficient factor in anterograde trans-synaptic regulation. In addition to ACh, cholinergic terminals may release additional molecules subserving this particular developmental function, as has been suggested for neurotrophic regulation at the neuromuscular junction (JESSELL et al. 1979; CONNOLLY et al. 1982). Exposure of dissociated sympathetic ganglion cell cultures to depolarizing concentrations of K^+ increases neuronal survival (WAKADE et al. 1983; WAKADE and THOENEN 1984) and produces increased intraneuronal levels of both noradrenaline (MACKAY and IVERSEN 1972; OTTEN and THOENEN 1976c; WALICKE et al. 1977; WALICKE and PATTERSON 1981) and TH (HEFTI et al. 1982), suggesting that depolarization itself may be the relevant stimulus.

Anterograde trans-synaptic influences affect target organ innervation as well as ganglionic maturation. Deafferentation of neonatal rat SCG produces a decrease of ^{3}H-noradrenaline uptake, innervation density, ground plexus ramification and fluorescence intensity in irides innervated by decentralized ganglia (BLACK and MYTILINEOU 1976a), suggesting that normal presynaptic connections are required for functional target organ innervation.

NGF does not appear to be involved in orthograde trans-synaptic regulation. NGF cannot prevent or reverse the effects of ganglionic decentralization or blockade (BLACK et al. 1972b; THOENEN et al. 1972c; HENDRY 1973; BLACK and MYTILINEOU 1976b). Furthermore, NGF and anti-NGF exert their effects prenatally (KESSLER and BLACK 1980a; GORIN and JOHNSON 1979, 1980), postnatally (BLACK et al. 1972b; HENDRY 1973) and *in vitro* (JOHNSON et al. 1972; CHUN and PATTERSON 1977a,b,c; GREENE 1977) in the absence of intact preganglionic innervation.

II. CNS Influence on Anterograde Trans-synaptic Regulation

Presynaptic cholinergic neurons located in the intermediolateral cell column of the mammalian spinal cord receive afferents descending from suprasegmental levels, including direct projections from the hypothalamus (HANCOCK 1976; SAPER et al. 1976) and ventral medulla (DAHLSTROM and FUXE 1965). Interruption of these descending fibers by spinal cord transection prevents the normal biochemical and morphological development of presynaptic cholinergic terminals as well as the ontogenetic increase of TH and AADC in ganglionic perikarya distal to the site of the lesion (BLACK et al. 1976; HAMILL et al. 1977; LAWRENCE et al. 1981). A similar phenomenon has been described following spinal cord transection in chick (SMOLEN and ROSS 1978; ROSS and COSIO 1979). This suggests that anterograde trans-synaptic regulation of peripheral sympathetic development depends not only on the integrity of the presynaptic cholinergic neuron, but on central neurons which regulate intermediolateral cell column development as well. Since sympathetic ganglion cell number is unaffected by spinal cord transection (LAWRENCE et al. 1981), it appears that while enzymatic maturation is dependent upon intact presynaptic innervation, ganglion cell survival is not. The effect is not, apparently, secondary to surgical injury to intermediolateral cell neurons, since histologic examination does not reveal necrosis, reactive gliosis, or inflammation in the spinal cord (HAMILL et al. 1977). Moreover, *morphologic* development of intermediolateral cell column neurons proceeds normally (HAMILL et al. 1977; ROSS and COSIO 1979). However, the number of terminal synaptic densities present in the ganglion is greatly reduced following spinal cord transection (ROSS and COSIO 1979; LAWRENCE et al. 1981), suggesting that ganglionic innervation by cholinergic terminals is disrupted.

III. Retrograde Trans-synaptic Regulation of Ganglia

While presynaptic cholinergic terminals regulate the development of sympathetic ganglion cells through an *anterograde* process, developing ganglion cells regulate presynaptic terminal maturation through a reciprocal *retrograde* process. Selective destruction of noradrenergic ganglion cells surgically by target extirpation (DIBNER et al. 1977) or axotomy (HENDRY 1975a; PURVES and NJÅ 1976), pharmacologically with 6-hydroxydopamine (BLACK et al. 1972a) or immunologically with anti-NGF (BLACK et al. 1972a; GOEDERT et al. 1978; NJÅ and PURVES 1978; GORIN and JOHNSON 1979, 1980) prevents the normal developmental rise in presynaptic CAT activity, and results in synaptic depression within the ganglion. As a corollary, abnormal *increases* in noradrenergic cell growth induced by NGF administration (THOENEN et al. 1972b; OPPENHEIM et al. 1982) or increased target organ size (DIBNER and BLACK 1978) result in supranormal development of presynaptic CAT activity.

It may be concluded that normal postnatal survival and maturation of the peripheral catecholamine neurons require the establishment and maintenance of both afferent and efferent synaptic connections. Consequently, an interneuronal symbiotic relationship at multiple synaptic sites is necessary for normal catecholamine neuron development.

IV. Target Organ Regulation of Ganglionic Development

Cell death is characteristic of normal development in a variety of neuronal systems (see HAMBURGER and OPPENHEIM 1982 for review), including sympathetic ganglia (HENDRY and CAMPBELL 1975). Removal of targets increases developmental cell death in the SCG, and also results in a failure of enzyme development (HENDRY and IVERSEN 1973a; HENDRY and THOENEN 1974; HENDRY 1975c; DIBNER and BLACK 1976; DIBNER et al. 1977; BANKS and WALTER 1977). Conversely, target enlargement increases survival of noradrenergic neurons of the sympathetic ganglia (DIBNER and BLACK 1978).

The importance of target organs for the growth and development of sympathetic neurons has been examined in greater detail *in vitro*. In the presence of vas deferens, for example, sympathetic neurons in culture increase neuritic outgrowth, cell body and nuclear size and fluorescence intensity (CHAMLEY et al. 1973). Similarly, biochemical and morphological differentiation of mouse submandibular ganglion (COUGHLIN 1975) as well as SCG (COUGHLIN et al. 1978; BLACK et al. 1979) are profoundly stimulated by co-culture with salivary glands. In addition, a variety of tissue explants, including heart, kidney, and colon, elicit increased fiber outgrowth from co-cultured sensory and sympathetic ganglia (COUGHLIN et al. 1978; EBENDAL 1979).

These findings have prompted investigations of the active target organ factors which may be responsible for end-organ regulation of neuronal maturation.

1. NGF and Target Organ Regulation

A vast literature indicates that NGF stimulates the postnatal development of catecholamine neurons and suggests that the protein may be necessary for normal postnatal development (See above p.147). Repetitive treatment of neonates with NGF results in a variety of hypertrophic effects on developing SCG, including increases in ganglionic volume (LEVI-MONTALCINI and BOOKER 1960a; ANGELETTI et al. 1972; HENDRY and CAMPBELL 1975), ganglion cell number (BANKS et al. 1975; HENDRY 1977b), elevated catecholamine-specific enzymes, TH and DBH (HENDRY and IVERSEN 1971; THOENEN et al. 1971; 1972a,c; OTTEN et al. 1977) as well as increased concentrations of catecholamines (CRAIN and WIEGAND 1961). Conversely, treatment of neonates with anti-sera directed against NGF (anti-NGF) results in irreversible destruction of sympathetic ganglia (LEVI-MONTALCINI and BOOKER 1960b), which is accompanied by a profound depression of catecholamine biosynthetic enzymes (HENDRY and IVERSEN 1971; ANGELETTI et al. 1972; OTTEN et al. 1978; GOEDERT et al. 1978; GORIN and JOHNSON 1979, 1980), as well as decreased catecholamine concentration (KLINGMAN 1965; KLINGMAN and KLINGMAN 1967), fluorescence intensity (BJERRE et al. 1975), and noradrenaline uptake in peripheral sympathetic terminals (IVERSEN et al. 1966). Since these effects are present even in complement-deficient animals, they do not appear to be caused by a complement-mediated cytotoxicity (GOEDERT et al. 1980). Consequently, neonatal neurons respond to NGF and appear to require the protein for survival and normal development.

Sympathetic ganglion cell death accompanies surgical or pharmacological treatments which disrupt connections between ganglionic perikarya and targets (see above). Since treatment with NGF attenuates the biochemical and cytological destruction caused by postganglionic axotomy (HENDRY 1975b; HENDRY and CAMPBELL 1975; PURVES and NJÅ 1976; BANKS and WALTER 1977), 6-hydroxydopamine (ALOE et al. 1975; TIFFANY-CASTIGLIONI and PEREZ-POLO 1979), guanethidine (JOHNSON and ALOE 1974; JOHNSON 1978), and vinblastine (CHEN et al. 1977; JOHNSON et al. 1979), survival and maintenance of sympathetic neurons is probably dependent upon target organs as peripheral depots of NGF (Thoenen and Stöckel 1975; Johnson 1978; Thoenen et al. 1978). The detection of NGF (Schwab et al. 1976; Gresik et al. 1980; Hofmann and Drenckhahn 1981; Ebendal et al. 1983; Korsching and Thoenen 1983a, 1988) and NGF mRNA (Shelton and Reichardt 1984; Heumann et al. 1984; Davies et al. 1987; Bandtlow et al. 1987) in target tissues and the discovery that irides placed in culture elaborate NGF (Ebendal et al. 1980; Barth et al, 1984) and NGF mRNA (Shelton and Reichardt 1986) gives more direct confirmation of this allegation. The demonstration that both exogenous (Stöckel et al. 1973; Hendry et al. 1974a; Paravicini et al. 1975; Iversen et al. 1975; Schwab and Thoenen 1977) and endogenous (Korsching and Thoenen 1983b) NGF is subject to retrograde transport from target to ganglion cell body also supports this contention. Moreover, unilateral ocular injection of NGF results in elevated TH levels, enlarged cells, and increased levels of ornithine decarboxylase in the ipsilateral SCG only (Paravicini et al. 1975; Stöckel and Thoenen 1975; Hendry 1977b; Hendry and Bonyhady 1980). The NGF transport process is saturable (Hendry et al. 1974b) and shows considerable specificity for NGF (Stöckel et al. 1974).

The many actions of NGF on responsive neurons are initiated by interaction with a specific membrane-bound receptor (Banerjee et al. 1973, 1975; Frazier et al. 1974a,b; Herrup et al. 1974; Snyder et al. 1974; Costrini and Bradshaw 1979; Kim et al. 1979). Kinetic analyses of the NGF receptor have indicated that there are both high- and low-affinity forms of the receptor molecule (Sutter et al. 1984). The receptor has been purified (Puma et al. 1983) and the gene encoding the NGF receptor molecule(s) isolated (Chao et al 1986).

One active area of current research focuses on the early intracellular events that mediate the variety of NGF effects. Since early actions of NGF are blocked by inhibitors of methyltransferase (Greene et al. 1984; Seeley et al. 1984; Landreth and Rieser 1985; Acheson et al. 1986; Acheson and Thoenen 1987), protein methylation may be one early consequence of NGF activation. Moreover, since NGF causes phosphorylation and dephosphorylation of specific proteins (Halegoua and Patrick 1980; Yu et al. 1980; Por and Huttner 1984; Landreth and Rieser 1985; Lee et al. 1985), including TH (Halegoua and Patrick 1980; Lee et al. 1985), the involvement of cAMP-dependent protein kinases has been widely investigated. While some membrane-permeant cAMP analogs mimic the actions of NGF in promoting survival, neurite outgrowth, and increased TH activity (Mackay and Iverson 1972b) in cultured sympathetic neurons, cAMP antagonists inhibit cAMP, but not NGF-induced survival and differentiation (Rydel and Greene 1988), suggesting the existence of parallel but independent neurotrophic pathways.

In addition to serving a survival and developmental function, NGF may also exert chemotactic effects. Sympathetic terminals are apparently attracted to a capillary tube source of NGF over distances of 2 mm *in vitro* (Charlwood et al. 1972; Le Tourneau 1978; Gunderson and Barrett 1979), and growth cones will change directional growth over periods of seconds.

In addition to playing a role in postnatal maturation, NGF is necessary for catecholamine integrity throughout life. Treatment of adult rats with anti-

NGF, axotomy, sialectomy or placement of a postganglionic colchicine cuff, which prevents retrograde axonal transport of NGF, results in decreased catecholamine histofluorescence and reduced TH activity in sympathetic ganglia (HENDRY and THOENEN 1974; BJERRE et al. 1975; OTTEN et al. 1977; KESSLER and BLACK 1979). Moreover, these effects are prevented by NGF administration. A developmental increase in the NGF receptor (YAN and JOHNSON 1987) and NGF receptor mRNA (BUCK et al. 1987) argues for a continuing role for NGF throughout life.

D. Concluding Observations

The developmental process begins to emerge, *not* as a distinct era in the life of an organism, circumscribed by a beginning moment and a concluding event, but rather as a series of events containing the earliest inductive processes of the embryo as well as the neuronal regulatory mechanisms operating in the aging adult. While the details of the processes may differ, trans-synaptic regulation, NGF-induced events, hormone interactions, target organ and environmental influences operate in the adult organism in ways similar to their actions in immature animals. What is more intriguing, however, is the fact that the remarkable phenotypic plasticity exhibited by developing peripheral neurons can be reinitiated in the adult, suggesting that a certain developmental potential remains throughout the life of the organism. Determination of the factors that restrict that potential may reveal the fundamental processes involved in the maintenance and/or aging of the adult nervous system. Furthermore, awareness of the factors that reinitiate developmental processes in the adult may alert us to the specific ways in which the adult nervous system can be destabilized (as in disease processes) or reshaped (following injury).

E. References

Abbo T, Balch CM (1981) A differentiation antigen of human NK and K cells identified by a monoclonal antibody (HNK-1). J Immunol 127:1024–1029

Abercrombie M (1970) Contact inhibition in tissue culture. In Vitro 6:128–142

Acheson A, Thoenen H (1987) Both short- and long-term effects of nerve growth factor on tyrosine hydroxylase in calf adrenal chromaffin cells are blocked by S-adenosylhomocysteine hydrolase inhibitors. J Neurochem 48:1416–1424

Acheson A, Vogl W, Huttner WB, Thoenen H (1986) Methyltransferase inhibitors block NGF-regulated survival and protein phosphorylation in sympathetic neurons. EMBO J 5:2799–2803

Acheson AL, Thoenen H (1983) Cell contact-mediated regulation of tyrosine hydroxylase synthesis in cultured bovine adrenal chromaffin cells. J Cell Biol 97:925–928

Acheson AL, Edgar D, Timpl R, Thoenen H (1986) Laminin increases both levels and activity of tyrosine hydroxylase in calf adrenal chromaffin cell. J Cell Biol 102:151–159

Adler, JE, Black IB (1984) Plasticity of substance P in mature and aged sympathetic neurons in culture. Science 225:1499–1500

Adler JE, Black IB (1985) Sympathetic neuron density differentially regulates transmitter phenotypic expression in culture. Proc Natl Acad Sci USA 82:4296-4300
Adler JE, Black IB (1986) Membrane contact regulates transmitter phenotypic expression. Dev Brain Res 30:237-241
Allan IJ, Newgreen DF (1977) Catecholamine accumulation in neural crest cells and the primary sympathetic chain. Amer J Anat 149:413-421
Aloe L, Levi-Montalcini R (1979) Nerve Growth Factor-induced transformation of immature chromaffin cells *in vivo* into sympathetic neurons—Effect of antiserum to Nerve Growth Factor. Proc Natl Acad Sci USA 76:1246-1250
Aloe L, Mugnaini L, Levi-Montalcini R (1975) Light and electron microscopic studies on the excessive growth of sympathetic ganglia in rats injected daily from birth with 6-hydroxydopamine and nerve growth factor. Arch italienne de Biol 113:326-353
Andrew A (1969) The origin of intramural ganglia. II. The trunk neural crest as a source of enteric ganglion cells. J Anat 105:89-101
Andrew A (1970) The origin of intramural ganglia. III. The 'vagal' source of enteric ganglion cells. J Anat 107:327-336
Andrew A (1971) The origin of intramural ganglia. IV. A critical review and discussion of the present stage of the problem. J Anat 108:169-184
Angeletti PU, Levi-Montalcini R, Kettler R, Thoenen H (1972) Comparative studies on the effect of the Nerve Growth Factor on sympathetic ganglia and adrenal medulla in newborn rats. Brain Res 44:197-206
Ayer-LeLièvre CS, LeDouarin NM (1982) The early development of cranial sensory ganglia and the potentialities of their component cells studies in quail-chick chimeras. Devel Biol 94:291-310
Baetge E, Kaplan BB, Reis D, Joh T (1981) Translation of tyrosine hydroxylase from poly (A) mRNA in pheochromocytoma cells (PC 12) is enhanced by dexamethasone. Proc Natl Acad Sci USA 78:1269-1273
Bandtlow CE, Heumann R, Schwab ME, Thoenen H (1987) Cellular localization of nerve growth factor synthesis by in situ hybridization. EMBO J 6:891-899
Bancroft M, Bellairs R (1976) The neural crest cells of the trunk region of the chick embryo studied by scanning electron microscopy and transmission electron microscopy. ZOON 4:73-85
Banerjee SP, Snyder SH, Cuatrecasas P, Greene LA (1973) Binding of Nerve Growth Factor receptor in sympathetic ganglia. Proc Natl Acad Sci USA 70:2519-2523
Banerjee SP, Cuatrecasas P, Snyder SH (1975) Nerve Growth Factor receptor binding—influence of enzymes, ions and protein reagents. J Biol Chem 250:1427-1433
Banks BE, Walter SJ (1977) The effects of postganglionic axotomy and nerve growth factor on the superior cervical ganglia of developing mice. J Neurocytol 6:287-297
Banks BEC, Charlwood KA, Edwards DC, Vernon CA, Walter SJ (1975) Effects of Nerve Growth Factor from salivary glands and snake venom on the sympathetic ganglia of neonatal and developing mice. J Physiol London 247:289-298
Barbu M, Ziller C, Rong RM, LeDouarin NM (1986) Heterogeneity in migrating neural crest cells revealed by a monoclonal antibiody. J Neurosci 6:2215-2225
Barth EM, Korsching S, Thoenen H (1984) Regulation of nerve growth factor synthesis and release in organ cultures of rat iris. J Cell Biol 99:839-843
Barth LG, Barth LJ (1969) The sodium dependence of embryonic induction. Devel Biol 20:236-262
Bartolomé J, Slotkin TA (1976) Effects of postnatal reserpine administration on sympathoadrenal development in the rat. Biochem Pharmacol 25:1513-1519
Bjerre B (1973) The production of catecholamine-containing cells *in vitro* by young

chick embryos studied by the histochemical fluorescence method. J Anat (Lond) 115:119-131

Bjerre B, Wiklund L, Edwards DC (1975) A study of the de- and regenerative changes in the sympathetic nervous system of the adult mouse after treatment with the antiserum to Nerve Growth Factor. Brain Res 92:257-278

Black IB (1973) Development of adrenergic neurons *in vivo*: inhibition by ganglionic blockade. J Neurochem 20:1265-1267

Black IB (1982) Stages of neurotransmitter development in autonomic neurons. Science 215:1198-1204

Black IB, Adler JE, Dreyfus CF, Friedman WF, LaGamma EF, Roach AH (1987) Biochemistry of information storage in the nervous system. Science 236:1263-1268

Black IB, Adler JE, Dreyfus CF, Jonakait GM, Katz DM, LaGamma EF, Markey KM (1984) Neurotransmitter plasticity at the molecular level. Science 225:1266-1270

Black IB, Coughlin M (1977) Ontogeny of an embryonic mouse sympathetic ganglion *in vivo* and *in vitro*. In: Vernadakis A, Giacobini E, Filogamo G (Eds) Maturation of Neurotransmission. Karger, Basel pp 65-75

Black IB, Geen SC (1974) Inhibition of the biochemical and morphological maturation of adrenergic neurons by nicotinic receptor blockade. J Neurochem 22:301-306

Black IB, Mytilineou C (1976a) Trans-synaptic regulation of the development of end organ innervation by sympathetic neurons. Brain Res 101:503-521

Black IB, Mytilineou C (1976b) The interaction of Nerve Growth Factor and trans-synaptic regulation in the development of target organ innervation by sympathetic neurons. Brain Res 108:199-204

Black IB, Hendry IA, Iversen LL (1971a) Trans-synaptic regulation of growth and development of adrenergic neurons in a mouse sympathetic ganglion. Brain Res 34:229-240

Black IB, Hendry IA, Iversen LL (1971b) Differences in the regulation of tyrosine hydroxylase and DOPA-decarboxylase in sympathetic ganglia and adrenal. Nature 231:27-29

Black IB, Hendry IA, Iversen LL (1972a) The role of post-synaptic neurons in the biochemical maturation of presynaptic cholinergic nerve terminals in a mouse sympathetic ganglion. J Physiol London 221:149-159

Black IB, Hendry IA, Iversen LL (1972b) Effects of surgical decentralization and Nerve Growth Factor on the maturation of adrenergic neurons in a mouse sympathetic ganglion. J Neurochem 19:1367-1377

Black IB, Joh TH, Reis DJ (1974) Accumulation of tyrosine hydroxylase molecules during growth and development of the superior cervical ganglion. Brain Res 75:133-144

Black IB, Bloom EM, Hamill RW (1976) Central regulation of sympathetic neuron development. Proc Natl Acad Sci USA 73:3575-3578

Black IB, Coughlin MD, Cochard P (1979) Factors regulating neuronal differentiation. Soc Neurosci Symp 4:184-207

Bloom EM, Black IB (1979) Metabolic requirements for differentiation of embryonic sympathetic ganglia cultured in the absence of exogenous nerve growth factor. Devel Biol 68:568-578

Bohn MC, Bloom E, Goldstein M, Black IB (1984) Glucocorticoid regulation of phenylethanolamine N-methyltransferase (PNMT) in organ culture of superior cervical ganglia. Dev Biol 105:130-136

Bohn MC, Goldstein M, Black IB (1981) Role of glucocorticoids in expression of the adrenergic phenotype in rat embryonic adrenal gland. Devel Biol 82:1-10

Bohn MC, Goldstein M, Black IB (1982) Expression of phenylethanolamine N-methy-transferase in rat sympathetic ganglia and extra adrenal chromaffin tissue. Devel Biol 89:299–308

Boucaut JC, Darribere T, Poole TJ, Aoyama H, Yamada KM, Thiery JP (1984) Biological active synthetic peptides as probes of embryonic development: A competitive peptide inhibitor of fibronectin function inhibits gastrulation in amphibian embryos and neural crest cell migration in avian embryos. J Cell Biol 99:1822–1830

Bronner-Fraser M (1986) An antibody to a receptor for fibronectin and laminin perturbs cranial neural crest development in vivo. Dev Biol 117:528–536

Bronner ME, Cohen AM (1979) Migratory patterns of cloned neural crest melanocytes injected into host chicken embryo. Proc Natl Acad Sci USA 76:1843–1847

Bronner-Fraser ME, Cohen AM (1979) Analysis of the neural crest ventral pathway using injected tracer cells. Devel Biol 77:130–141

Bronner-Fraser ME, Cohen AM (1980) The Neural Crest: What can it tell us about cell migration and determination? Current Topics in Devel Biol 15:1–25

Brunso-Bechtold JK, Hamburger V (1979) Retrograde transport of nerve growth factor in chicken embryo. Proc Natl Acad Sci USA 76:1494–1496

Buck CR, Martinez HJ, Black IB, Chao MV (1987) Developmentally regulated expression of the nerve growth factor receptor gene in the periphery and brain. Proc Natl Acad Sci USA 84:3060–3063

Bueker ED, Scheinkein I, Barre JL (1960) Distribution of Nerve Growth Factor specific for spinal and sympathetic ganglia. Cancer Res 20:1220–1227

Bunge RP, Rees R, Wood P, Burton H, Ko C-P (1974) Anatomical and physiological observations on synapses formed on isolated autonomic neurons in tissue culture. Brian Res 66:401–412

Burn JH, Rand MJ (1965) Acetylcholine in adrenergic transmission. Ann Rev Pharmacol 5:163–182

Burnstock G (1976) some nerve cells release more than one transmitter? Neurosci 1:239–248

Burnstock G (1978) Do some sympathetic neurones synthesize and release both noradrenaline and acetylcholine? Prog Neurobiol 11:205–222

Burton H, Bunge RP (1975) A comparison of the uptake and release of ^{3}H-Norepinephrine in rat autonomic and sensory ganglia in tissue culture. Brain Res 97:157–162

Chamley JH, Campbell GR, Burnstock G (1973) An analysis of the interactions between sympathetic nerve fibers and smooth muscle cells in tissue culture. Devel Biol 33:344–361

Chao MV, Bothwell MA, Ross AH, Koprowski H, Lanahan T, Buck CR, Sehgal A (1986) Gene transfer and molecular cloning of the human NGF receptor. Science 232:518–521

Charlwood KA, Lamont DM, Banks B (1972) Apparent orienting effects produced by nerve growth factor. In: Zaimis E, Knight I (Eds) Nerve Growth Factor Athlone Press, London, pp 102–107

Chen MGM, Chen JS, Calissano P, Levi-Montalcini R (1977) Nerve growth factor prevents vinblastine destructive effects on sympathetic ganglia in newborn mice. Proc Natl Acad Sci USA 74:5559–5563

Chen WT, Greve JM, Gottlieb DI, Singer SJ (1985) Immunocytological localization of 140 kd cell adhesion molecules in cultured chicken fibroblasts, and in chicken smooth muscle and intestinal epithelial tissues. J Histochem Cytochem 33:576–586

Chevallier A (1972) Localisation et durée des potentialites médullo-surrénaliennes des cretes neurales chez l'embryon de Poulet. J Embryol exp Morph 27:603–614

Chibon P (1966) Étude autoradiographique après marquage par la thymidine tritiée des dérivés de la crête neurale troncale chez l'amphibian urodèle Pleurodeles waltlii Michah. CR Acad Sci 261:5645-5648

Chun LLY, Patterson PH (1977a) Role of nerve growth factor in the development of rat sympathetic neurons *in vivo*. I. Survival, growth and differentiation of catecholamine production. J Cell Biol 75:694-704

Chun LLY, Patterson PH (1977b) Role of nerve growth factor in the development of rat sympathetic neurons *in vivo*. II. Developmental studies. J Cell Biol 75:705-711

Chun LLY, Patterson PH (1977c) Role of nerve growth factor in the development of rat sympathetic neurons *in vivo*. III. Effect on acetylcholine production. J Cell Biol 75:712-718

Ciaranello RD (1978) Regulation of phenylethanolamine N-methyltransferase synthesis and degradation. I. Regulation by rat adrenal glucocorticoids. Mol Pharmacol 14:478-489

Ciaranello RD, Jacobowitz D, Axelrod J (1973) Effect of dexamethasone on phenylethanolamine-N-methyltransferase in chromaffin tissue of the neonatal rat. J Neurochem 799-805

Ciment G, Weston JA (1982) Early appearance in neural crest and crest-derived cells of an antigenic determinant present in avian neurons. Dev Biol 93:355-367

Claude P (1973) Electron microscopy of dissociated rat sympathetic neurons *in vitro*. J Cell Biol 59:57a

Cochard P, Coltey P (1983) Cholinergic traits in the neural crest: acetylcholinesterase in crest cells of the chick embryo. Dev Biol 98:221-238

Cochard P, Goldstein M, Black IB (1978) Ontogenetic appearance and disappearance of tyrosine hydroxylase and catecholamines in the rat embryo. Proc Natl Acad Sci USA 75:2986-2990

Cochard P, Goldstein M, Black IB (1979) Initial development of the noradrenergic phenotype in autonomic neuroblasts of the rat embryo. Devel Biol 71:100-114

Cohen AM (1972) Factors directing the expression of sympathetic nerve traits in cells of neural crest origin. J Exp Zool 179:167-182

Cohen AM (1974) DNA synthesis and cell division in differentiating avian adrenergic neuroblasts. In: Fuxe K, Olson L, Zotterman Y (Eds), Dynamics of degeneration and growth in neurons, Pergamon Press, New York, pp 359-370

Cohen AM (1977) Independent expression of the adrenergic phenotype by neural crest cells *in vitro*. Proc Natl Acad Sci USA 74:2899-2903

Cohen AM, Konigsberg IR (1975) A clonal approach to the problem of neural crest determination. Devel Biol 46:262-280

Cohen A, Nicol EC, Richter W (1964) Nerve growth factor requirement for development of dissociated embryonic sensory and sympathetic ganglia in culture. Proc Soc Exp Biol Med 116:784-789

Connolly JA, St. John PA, Fischbach GD (1982) Extracts of electric lobe and electric organ from *Torpedo californica* increase the total number as well as the number of aggregates of chick myotube acetylcholine receptors. J Neurosci 2:1207-1213

Costrini NV, Bradshaw RA (1979) Binding characteristics and apparent molecular size of detergent-solubilized nerve growth factor receptors of sympathetic ganglia. Proc Natl Acad Sci USA 76:3242-3245

Coughlin MD (1975) Target organ stimulation of parasympathetic nerve growth in the developing mouse submandibular gland. Devel Biol 43:140-158

Coughlin MD, Boyer DM, Black IB (1977) Embryologic development of a mouse sympathetic ganglion *in vivo* and *in vitro*. Proc Natl Acad Sci USA 74:3438-3442

Coughlin MD, Collins MB (1985) Nerve growth factor-independent development of

embryonic mouse sympathetic neurons in dissociated cell culture. Dev Biol 110:392-401
Dahlstrom A, Fuxe K (1965) Evidence for the existence of monoamine-containing neurons in the central nervous system. II. Experimentally induced changes in the intraneuronal amine levels of bulbospinal neuron system. Acta Physiol Scand Suppl 232 62:1-53
Coughlin MD, Dibner MD, Boyer DM, Black IB (1978) Factors regulating development of an embronic mouse sympathetic ganglion. Devel Biol 66:513-528
Coulombre AJ, Johnston MC, Weston JA (1974) Conference on neural crest in normal and abnormal embryogenesis. Devel Biol 36:f1-5
Coupland RE, Macdougal JDB (1966) Adrenaline formation in noradrenaline-storing chromaffin cells *in vitro* induced by corticosterone. J Endocrin 36:317-324
Cowell LC, Weston JA (1970) An analysis of melanogenesis in cultured chick embryo spinal ganglia. Devel Biol 22:670-697
Coyle JT, Axelrod J (1971) Development of the uptake and storage of L-[^{3}H] norepinephrine in the rat brain. J Neurochem 18:2061-2075
Crain SM, Wiegand RC (1961) Catecholamine levels of mouse sympathetic ganglia following hypertrophy produced by salivary nerve growth factor. Proc Soc Exp Biol Med 107:6637
Davies AM, Bandtlow C, Heumann R, Korsching S, Rohrer H, Thoenen H (1987) Timing and site of nerve growth factor synthesis in developing skin in relation to innervation and expression of the receptor. Nature 326:353-358
DeChamplain J, Malmfors T, Olson L, Sachs C (1970) Ontogenesis of peripheral adrenergic neurons in the rat: pre- and post-natal observations. Acta Physiol Scand 80:276-288
Derby MA (1978) Analysis of glykosaminoglycans within the extracellular environments encountered by migrating neural crest cells. Devel Biol 66:321-336
Derby MA, Newgreen DF (1982) Differentiation of avian neural crest cells in vitro: Absence of a developmental bias toward melanogenesis. Cell Tiss Res 225:365-378
Derby MA, Pintar JE, Weston JA (1976) Glycosaminoglycans and the development of the trunk neural crest. J Gen Physiol 68:4a
Dibner MD, Black IB (1976) The effect of target organ removal on the development of sympathetic neurons. Brain Res 103:93-102
Dibner MD, Black IB (1978) Biochemical and morphological effects of testosterone treatment on developing sympathetic neurons. J Neurochem 30:1479-1483
Dibner MD, Mytilineou C, Black IB (1977) Target organ regulation of sympathetic neurons development. Brain Res 123:301-310
Dichter MA, Tischler AS, Greene LA (1977) Nerve growth factor-induced increase in electrical excitability and acetylcholine sensitivity of a rat pheochromocytoma cell line. Nature 268:501-504
Dolezalova H, Giacobini E, Giacobini G, Rossi A, Toschi G (1974) Developmental variations of choline acetylase, dopamine B-hydroxylase, and monoamine oxidase in chick sympathetic ganglia. Brain Res 73:309-320
Dorris F (1938) The production of pigment *in vitro* by chick neural crest. Roux Arch 138:323-334
Doupe AJ, Landis SC, Patterson PH (1985a) Environmental influences in the development of neural crest derivatives: Glucocorticoids, growth factors, and chromaffin cell plasticity. J Neurosci 5:2119-2142
Doupe AJ, Patterson PH, Landis SC (1985b) Small intensely fluorescent cells in culture: Role of glucocorticoids and growth factors in their development and interconversions with other neural crest derivatives. J Neurosci 5:2143-2160

Duband JL, Thiery JP (1982) Distribution of fibronectin in the early phase of avian cephalic neural crest cell migration. Dev Biol 93:308–323
Duband JL, Rocher S, Chen WT, Yamada KM, Thiery JP (1986) Cell adhesion and migration in the early vertebrate embryo: location and possible role of the putative fibronectin-receptor complex. J Cell Biol 102:160–178
Dupin E (1984) Cell division in the ciliary ganglion of quail embryos in situ and after back-transplantation into the neural crest migration pathway of chick embryos. Dev Biol 105:288–299
Duprat AM, Kan P, Foulquier F, Weber M (1985) *In vitro* differentiation of neuronal precursor cells from amphibian late gastrulae: morphological, immunocytochemical studies; biosynthesis, accumulation and uptake of neurotransmitters. J Embryol Exp Morph 86:71–87
Ebendal T (1977) Extracellular matrix fibrils and cell contacts in the chick embryo. Possible roles in orientation of cell migration and axon extension. Cell Tissue Res 175:439–458
Ebendal T (1979) Stage-dependent stimulation of neurite outgrowth exerted by nerve growth factor and chick heart in cultured embryonic ganglia. Devel Biol 72:276–290
Ebendal T, Olson L, Seiger A, Hedlund K-O (1980) Nerve growth factors in the rat iris. Nature 286:25–28
Ebendal T, Olson L, Seiger A (1983) The level of nerve growth factor (NGF) as a function of innervation. Exp Cell Res 148:311–317
Edgar DH, Thoenen H (1978) Selective enzyme induction in a nerve growth factor-responsive *pheochromocytoma* cell line (PC 12). Brain Res 154:186–190
Enemar A, Falck B, Hakanson R (1965) Observations on the appearance of norepinephrine in the sympathetic nervous system of the chick embryo. Devel Biol 11:268–283
Epperlein HH (1974) The ectomesenchymal-endodermal interaction system (EEIS) or *Triturus Alpestris* in tissue culture. I. Observations on attachment, migration and differentiation of neural crest cells. Differentiation 2:157–168
Eränkö O, Eränkö L, Hill CE, Burnstock G (1972) Hydrocortisone-induced increase in the number of small intensely fluorescent cells and their histochemically demonstrable catecholamine content on cultures of sympathetic ganglia of the newborn rat. Histochem J 4:49–58
Eränkö O, Eränkö L, Hervonen H (1973) Cultures of sympathetic ganglia and the effect of glucocorticoids on SIF cells. In: Eranko O (Ed) SIF cells. Structure and function of the small, intensely fluorescent sympathetic cells. US Govt Printing Office Washington DC, pp 196–214
Eränkö O, Pickel VM, Harkonen M, Eränkö L, Joh TH, Reis DJ (1982) Effect of hydrocortisone on catecholamines and the enzymes synthesizing them in the developing sympathetic ganglion. Histochem J 14:461–478
Erickson CA, Tosney KW, Weston JA (1980) Analysis of migratory behavior of neural crest and fibroblastic cells in embryonic tissue. Devel Biol 77:142–156
Erickson CA, Turley EA (1983) Substrata formed by combinations of extracellular matrix components alter neural crest cell motility in vitro. J Cell Sci 61:299–323
Fairman K, Giacobini E, Chiappinelli V (1976) Developmental variations of tyrosine hydroxylase and acetylcholinesterase in embryonic and post-hatching chicken sympathetic ganglia. Brain Res 102:301–312
Fairman K, Chiappinelli V, Giacobini E, Yurkewicz L (1977) The effect of a single dose of reserpine administered prior to incubation on the development of tyrosine hydroxylase activity in chick sympathetic ganglia. Brain Res 122:503–512

Fernholm M (1972) On the appearance of monoamines in the sympathetic systems and the chromaffin tissue in the mouse embryo. Z Anat Entw Gesch 135:350–361

Frazier WA, Boyd LF, Bradshaw RA (1974a) Properties of specific binding of ^{125}I-nerve growth factor to responsive peripheral neurons. J Biol Chem 249:5513–5519

Frazier WA, Boyd LF, Szutowicz A, Pulliam MW, Bradshaw RA (1974b) Specific binding-sites for ^{125}I-nerve growth factor in peripheral tissues and brain. Biochem Biophys Res Commun 57:1096–1103

Fukada K (1981) Hormonal control of neurotransmitter choice in sympathetic neuron cultures. Nature 287:553–554

Fukada K (1985) Purification and partial characterization of a cholinergic neuronal differentiation factor. Proc Natl Acad Sci USA 82:8795–8799

Furshpan EJ, Landis SC, Matsumoto SG, Potter DD (1986a) Synaptic functions in rat sympathetic neurons in microcultures. I. Secretion of norepinephrine and acetylcholine. J Neurosci 6:1061–1079

Furshpan EJ, Macleish PR, O'Lague PH, Potter DD (1976) Chemical transmission between rat sympathetic neurons and cardiac myocytes developing in microcultures: Evidence for cholinergic, adrenergic and dual-function neurons. Proc Natl Acad Sci USA 73:4225–4229

Furshpan EJ, Potter DD, Matsumoto SG (1986b) Synaptic functions in rat sympathetic neurons in microcultures. III. A purinergic effect on cardiac myocytes. J Neurosci 6:1099–1107

Gershon MD, Rothman TP, Joh TH, Teitelman GN (1984) Transient and differential expression of aspects of the catecholaminergic phenotype during development of the fetal bowel of rats and mice. J Neurosci 4:2269

Girdlestone J, Weston JA (1985) Identification of early neuronal subpopulations in avian neural crest cell cultures. Develop Biol 109:274–287

Goedert M, Otten U, Thoenen H (1978) Biochemical effects of antibodies against nerve growth factor on developing and differentiated sympathetic ganglia. Brain Res 148:264–268

Goedert M, Otten U, Schäfer Th, Schwab M, Thoenen H (1980) Immunosympathectomy: lack of evidence for a complement-mediated cytotoxic mechanism. Brain Res 201:399–409

Goodman R, Herschman H (1978) Nerve growth factor-mediated induction of tyrosine hydoxylase in a clonal pheochromocytoma cell line. Proc Natl Acad Sci USA 75:4587–4590

Gorin PD, Johnson EM (1979) Experimental autoimmune model of nerve growth factor deprivation: Effects on developing peripheral sympathetic and sensory neurons. Proc Natl Acad Sci USA 76:5387–5391

Gorin PD, Johnson EM (1980) Effects of exposure to nerve growth factor antibodies on the developing nervous system of the rat: an experimental autoimmune approach. Devel Biol 80:313–323

Greenberg JH, Pratt RM (1977) Glycosaminoglycan and glycoprotein synthesis by cranial neural crest cells *in vitro*. Cell Differ 6:119–132

Greenberg JH, Schrier BK (1977) Development of choline acetyltransferase activity in chick cranial neural crest cells in culture. Devel Biol 61:86–93

Greenberg JH, Seppä S, Seppä H, Hewitt AT (1981) Role of collagen and fibronectin in neural crest cell adhesion and migration. Devel Biol 87:259–266

Greene LA (1977) Quantitative *in vitro* studies on the nerve growth factor requirement of neurons. I. Sympathetic neurons. Devel Biol 58:96–105

Greene LA, Rein G (1977a) Release of (^{3}H)norepinephrine from a clonal line of *pheochromocytoma* cells (PC 12) by nicotinic cholinergic stimulation. Brain Res 138:521–528

Greene LA, Rein G (1977b) Release, storage and uptake of catecholamines by a clonal cell line of nerve growth factor (NGF) responsive pheochromocytoma cells. Brain Res 129:247–263
Greene LA, Rein G (1977c) Synthesis, storage and release of acetylcholine by a noradrenergic pheochromocytoma cell line. Nature (Lond) 268:349–351
Greene LA, Rein G (1978) Release of (^{3}H)norepinephrine from a clonal line of pheochromocytoma cells (PC 12) by nicotinic cholinergic stimulation. Brain Res 138:521–528
Greene LA, Seeley PJ, Rukenstein A, di Piazza M, Howard A (1984) Rapid activation of tyrosine hydroxylase in response to nerve growth factor. J Neurochem 42:1728–1734
Greene LA, Shooter E (1980) The Nerve Growth Factor: Biochemistry, synthesis and mechanism of action. Ann Rev Neurosci 3:353–402
Greene KA, Tischler AS (1976) Establishment of a noradrenergic clonal line of rat adrenal pheochromocytoma cells which respond to nerve growth factor. Proc Natl Acad Sci USA 73:2424–2428
Gresik EW, Chung KW, Barka T, Schenkein I (1980) Immunocytochemical localization of nerve growth factor, epidermal growth factor, renin and protease A in the submandibular glands of Tfm/Y mice. Am J Anat 158:247–250
Gunderson RW, Barrett JN (1979) Neuronal chemotaxis: chick dorsal root axons turn towards high concentrations of nerve growth factor. Science 206:1079–1080
Halegoua S, Patrick J (1980) Nerve growth factor mediates phosphorylation of specific proteins. Cell 22:571–581
Hall BK, Tremaine R (1979) Ability of neural crest cells from the embyronic chick to differentiate into cartilage before their migration away from the neural tube. Anat Rec 194:469–476
Hamburger V, Oppenheim RW (1982) Naturally occurring neuronal death in vertebrates. Neurosci Comm 1:39–55
Hamill RW, Bloom EM, Black IB (1977) The effect of spinal cord transection on the development of cholinergic and adrenergic sympathetic neurons. Brain Res 134:269–278
Hammond WS, Yntema CL (1947) Depletion in the thoracolumbar sympathetic system following removal of neural crest in the chick. J Comp Neurol 85:237–265
Hammond WS, Yntema CL (1964) Depletions of pharyngeal arch cartilage following extirpation of cranial neural crest in chick embryos. Acta Anat 56:21–34
Hanbauer I, Lovenberg W, Guidotti A, Costa E (1975) Role of cholinergic and glucocorticoid receptors in the tyrosine hydroxylase induction elicited by reserpine in superior cervical ganglion. Brain Res 96:197–200
Hancock MB (1976) Cells of origin of hypothalamo-spinal projections in the rat. Neurosci Lett 3:179–184
Hatanaka H, Tanaka M, Amano T (1980) A clonal rate pheochromocytoma cell line possesses synthesizing ability of gamma-amino-butyric acid together with catecholamine and acetylcholine. Brain Res 183:490–493
Hawrot E (1980) Cultured sympathetic neurons: effects of cell-derived and synthetic substrata on survival and development. Devel Biol 74:136–151
Hefti F, Gnahn H, Schwab ME, Thoenen H (1982) Induction of tyrosine hydroxylase by nerve growth factor and by elevated K^+ concentrations in cultures of dissociated sympathetic neurons. J Neurosci 2:1554–1566
Hendry IA (1972) Developmental changes in tissue and plasma concentrations of the biologically active species of Nerve Growth Factor in the mouse, by using a two site radioimmunoassay. Biochem J 128:1265–1272

Hendry IA (1973) Trans-synaptic regulation of tyrosine hydroxylase activity in a developing mouse sympathetic ganglion: effects of Nerve Growth Factor (NGF), NGF-antiserum and pempidine. Brain Res 56:313-320

Hendry IA (1975a) The retrograde trans-synaptic control of the development of cholinergic terminals in sympathetic ganglia. Brain Res 86:483-487

Hendry IA (1975b) The response of adrenergic neurons to axotomy and Nerve Growth Factor. Brain Res 94:87-97

Hendry IA (1975c) The effects of axotomy on the development of the rat superior cervical ganglion. Brain Res 90:235-244

Hendry IA (1977a) Cell division in the developing sympathetic nervous system. J Neurocytol 6:299-309

Hendry IA (1977b) The effect of the retrograde axonal transport of nerve growth factor on the morphology of adrenergic neurones. Brain Res 134:213-223

Hendry IA, Bonyhady R (1980) Retrogradely transported nerve growth factor increases ornithine decarboxylase activity in rat superior cervical ganglia. Brain Res 200:39-45

Hendry IA, Campbell J (1975) Morphometric analysis of rat superior cervical ganglion after axotomy and Nerve Growth Factor treatment. J Neurocytol 5:351-360

Hendry IA, Iversen LL (1971) Effect of Nerve Growth Factor and its antiserum on tyrosine hydroxylase activity in the mouse superior cervical ganglion. Brain Res 29:159-162

Hendry IA, Iversen LL (1973a) Changes in tissue and plasma concentrations of Nerve Growth Factor following removal of the submaxillary glands in adult mice and their effects on the sympathetic nervous system. Nature 243:500-504

Hendry IA, Iversen LL (1973b) Reduction in the concentration of nerve growth factor in mice after sialectomy and castration. Nature 243:550-554

Hendry IA, Thoenen H (1974) Changes of enzyme pattern in the sympathetic nervous system of adult mice after submaxillary gland removal: response to exogenous Nerve Growth Factor. J Neurochem 22:999-1004

Hendry IA, Stach R, Herrup K (1974a) Characteristics of the retrograde axonal transport system for Nerve Growth Factor in the sympathetic nervous system. Brain Res 82:117-128

Hendry IA, Stöckel K, Thoenen H, Iversen LL (1974b) The retrograde axonal transport of Nerve Growth Factor. Brain Res 68:103-121

Herrup K, Thoenen H (1979) Properties of the nerve growth factor receptor of a clonal line of rat pheochromocytoma (PC 12) cells. Exp Cell Res 121:71-78

Herrup K, Stickgold R, Shooter EM (1974) The role of nerve growth factor in the development of sensory and sympathetic ganglia. Ann NY Acad Sci 228:381-392

Heumann R, Korsching S, Scott J, Thoenen H (1984) Relationship between levels of nerve growth factor (NGF) and its messenger RNA in sympathetic ganglia and peripheral target tissues. EMBO J 3:3183-3189

Higgins D, Iacovitti L, Joh TH, Burton H (1981) The immunocytochemical localization of tyrosine hydroxylase within rat sympathetic neurons that release acetylcholine in culture. J Neurosci 1:126-131

Hofmann H-D, Drenckhahn D (1981) Distribution of nerve growth factor in the submandibular gland of the male and female mouse. A re-examination by use of an improved immunohistochemical procedure. Cell Tiss Res 221:77-83

Hökfelt T, Holets VR, Staines W, Meister B, Melander T, Schalling M, Schultzberg M, Freedman J, Bjorklund A, Olson L, Lindh B, Elfvin L-G, Lundberg JM, Lindgren JA, Samuelsson B, Pernow B, Terenius L, Post C, Everitt B, Goldstein M (1986) Coexistence of neuronal messengers—an overview. Prog. in Brain Research 68:33-70

Holtfreter J (1939) Gewebeaffinität, ein Mittel der embryonalen Formbildung. Arch exp Zellforsch 23:169-209
Horstadius S (1950) The neural crest. Oxford University Press, London
Howard MJ, Bronner-Fraser M (1985) The influence of neural tube derived factors on differentiation of neural crest cells in vitro: I. Histochemical study on the appearance of adrenergic cells. J Neurosci 5:3302-3309
Iacovitti L, Joh TH, Park DH, Bunge RP (1981) Dual expression of neurotransmitter synthesis in cultured autonomic neurons. J Neurosci 1:685-690
Iacovitti L, Joh TH, Albert VR, Park DH, Reis DJ, Teitelman G (1985) Partial expression of catecholaminergic traits in cholinergic chick ciliary ganglia: Studies *in vivo* and *in vitro*. Develop Biol 110:402-412
Iacovitti L, Johnson MI, Joh TH, Bunge RP (1982) Biochemical and morphological characterization of sympathetic neurons grown in a chemically defined medium. Neuroscience 7:2225-2239
Iacovitti L, Teitelman G, Joh TH, Reis DJ (1987) Chick eye extract promotes expression of a cholinergic enzyme in sympathetic ganglia in culture. Brain Res 430:59-65
Ishii DN, Shooter EM (1975) Regulation of nerve growth factor synthesis in mouse submaxillary glands by testosterone. J Neurochem 25:843-851
Iversen LL, Glowinski J, Axelrod J (1966) The physiologic disposition and metabolism of norepinephrine in immunosympathectomized animals. J Pharmacol Exp Ther 151:273-284
Iversen LL, Stöckel K, Thoenen H (1975) Autoradiographic studies of the retrograde axonal transport of nerve growth factor in mouse sympathetic neurones. Brain Res 88:37-43
Jessell TM, Siegel RE, Fischbach GD (1979) Induction of acetylcholine receptors on cultured skeletal muscle by a factor extracted from brain and spinal cord. Proc Natl Acad Sci USA 76:5397-5401
Johnson DG, Silberstein SD, Hanbauer I, Kopin IJ (1972) The role of Nerve Growth Factor in the ramification of sympathetic nerve fibres into the rat iris in organ culture. J Neurochem 19:2025-2029
Johnson EM, jr (1978) Destruction of the sympathetic nervous system in neonatal rats and hamsters by vinblastine: prevention by concommitant administration of NGF. Brain Res 141:105-118
Johnson EM, Aloe L (1974) Suppression of the *in vitro* and *in vivo* cytotoxic effects of guanethidine in sympathetic neurons by Nerve Growth Factor. Brain Res 81:519-532
Johnson EM, jr, Andres RY, Bradshaw RA (1978) Characterization of the retrograde transport of nerve growth factor (NGF) using high specific activity (^{125}I) NGF. Brain Res 150:319-331
Johnson EM, jr, Macia RA, Andres RY, Bradshaw RA (1979) The effects of drugs which destroy the sympathetic nervous system on the retrograde transport of nerve growth factor. Brain Res 171:461-472
Johnson M, Ross D, Meyers M, Rees R, Bunge R, Wakshull E, Burton H (1976) Synaptic vesicle cytochemistry changes when cultured sympathetic neurons develop cholinergic interactions. Nature (Lond) 262:308-310
Johnston MC (1966) A radioautographic study of the migration and fate of cranial neural crest cells in the chick embryo. Anat Rec 156:143-156
Jonakait GM, Black IB (1986) Neurotransmitter phenotypic plasticity in the mammalian embryo. In: Current Topics in Developmental Biology, vol 20 pp 165-174
Jonakait GM, Wolf J, Cochard P, Goldstein M, Black IB (1979) Selective loss of norad-

renergic phenotypic characters in neuroblasts of the rat embryo. Proc Natl Acad Sci USA 76:4683-4686
Jonakait GM, Bohn MC, Black IB (1980) Maternal glucocorticoid hormones influence neurotransmitter phenotypic expression in embryos. Science 210:551-553
Jonakait GM, Bohn MC, Goldstein M, Black IB (1981a) Elevation of maternal glucocorticoid hormones alters neurotransmitter phenotypic expression in embryos. Devel Biol 88:288-296.
Jonakait GM, Kessler JA, Goldstein M, Markey K, Black IB (1981b) Characterization of transiently noradrenergic cells of embryonic gut *in vivo* and *in vitro.* Abstract Soc Neurosci 7:289
Jonakait GM, Markey KA, Goldstein M, Black IB (1984) Transient expression of selected catecholaminergic traits in cranial sensory and dorsal root ganglia of the embryonic rat. Devel Biol 101:51-60
Jonakait GM, Markey KA, Goldstein M, Dreyfus CF, Black IB (1985) Selective expression of high-affinity uptake of catecholamines by transiently catecholaminergic cells of the rat embryo: studies *in vivo* and *in vitro.* Devel Biol 108:6-17
Jonakait GM, Rosenthal M, Morrell JI (1988) Regulation of tyrosine hydroxylase mRNA in catecholaminergic cells of the embryonic rat: Analysis by *in situ* hybridization. J Histochem Cytochem, in press.
Kahn CR, Coyle JT, Cohen AM (1980) Head and trunk neural crest *in vitro*: Autonomic neuron differentiation. Devel Biol 77:340-348
Kalcheim C, Barde Y-A, Thoenen H, LeDouarin NM (1987) In vivo effect of brain-derived neurotrophic factor on the survival of developing dorsal root ganglion cells. EMBO J 6:2871-2873
Kalcheim C, LeDouarin NM (1986) Requirement of a neural tube signal for the differentiation of neural crest cells into dorsal root ganglia. Dev Biol 116:451-466
Kalcheim C, LeDouarin NM (1987) Formation of the dorsal root ganglia in the avian embryo: segmental origin and migratory behavior of neural crest progenitor cells. Dev Biol 120:329-347
Kessler JA (1984) Non-neuronal cell conditioned medium stimulates peptidergic expression in sympathetic and sensory neurons in vitro. Dev Biol 106:61-69
Kessler JA, Adler JE, Jonakait GM, Black IB (1984) Target organ regulation of substance P in sympathetic neurons in culture. Dev Biol 103:71-79
Kessler JA, Black IB (1979) The role of axonal transport in the regulation of enzyme activity in sympathetic ganglia of *adult* rats. Brain Res 171:415-424
Kessler JA, Cochard P, Black IB (1979) Nerve growth factor alters the fate of embryonic neuroblasts. Nature (Lond) 280:141-142
Kessler JA, Conn G, Hatcher VB (1986) Isolated plasma membranes regulate neurotransmitter expression and facilitate effects of a soluble brain cholinergic factor. Proc Natl Acad Sci USA 83:3528-3532
Kessler JA, Black IB (1980) The effects of nerve growth factor (NGF) and antiserum to NGF on the development of embryonic sympathetic neurons *in vivo.* Brain Res. 189:157-168
Kessler JA, Adler JE, Bohn MC, Black IB (1981) Substance P in principal sympathetic neurons: regulation by impulse activity. Science 214:335-336
Kim SU, Hogue-Angeletti R, Gonatas NK (1979) Localization of nerve growth factor receptors in sympathetic neurons cultured *in vitro.* Brain Res 168:602-608
Kirby ML, Gilmore SA (1972) A fluorescence study on the ability of the notochord to synthesize and store catecholamines in early chick embryos. Anat Rec 173:469-478
Kirby ML, Gilmore SA (1976) A correlative histofluorescence and light microscopic

study of the formation of sympathetic trunks in chick embryos. Anat Rec 186:437-450
Klingman GI (1965) Catecholamine levels and DOPA-decarboxylase activity in peripheral organs and adrenergic tissue in the rat after immunosympathectomy. J Pharmacol Exp Ther 148:14-21
Klingman GI (1966) *In utero* immunosympathectomy of mice. Int J Neuropharmacol 5:163-170
Klingman GI, Klingman JD (1967) Prenatal and postnatal treatment of mice with antiserum to Nerve Growth Factor. Int J Neuropharmacol 6:501-508
Ko C-P, Burton H, Bunge RP (1976a) Synaptic transmission between rat spinal cord explants and dissociated superior cervical ganglion neurons in tissue culture. Brain Res 117:437-460
Ko C-P, Burton H, Johnson M, Bunge RP (1976b) Synaptic transmission between rat superior cervical ganglion neurons in dissociated cell cultures. Brain Res 117:461-485
Korsching S, Thoenen H (1983a) Nerve growth factor in sympathetic ganglia and corresponding target organs of the rat: correlation with density of sympathetic innervation. Proc Natl Acad Sci USA 80:3513-3516
Korsching S, Thoenen H (1983b) Quantitative demonstration of the retrograde axonal transport of endogenous nerve growth factor. Neurosci Lett 39:1-4
Korsching S, Thoenen H (1988) Developmental changes of nerve growth factor levels in sympathetic ganglia and their target organs. Dev Biol 126:40-46
Landis SC (1976) Rat sympathetic neurons and cardiac myocytes developing in microcultures: correlation of the fine structure of endings with neurotransmitter function in single neurons. Proc Natl Acad Sci USA 73:4220-4224
Landis SC (1980) Developmental changes in the neurotransmitter properties of dissociated sympathetic neurons; A cytochemical study of the effects of medium. Devel Biol 77:349-361
Landis SC, Keefe D (1983) Evidence for neurotransmitter plasticity *in vivo*: Developmental changes in properties of cholinergic sympathetic neurons. Devel Biol 98:349-372
Landreth GE, Rieser GD (1985) Nerve growth factor- and epidermal growth factor-stimulated phosphorylation of a PC 12 cytoskeletally associated protein in situ. J Cell Biol 100:677-683
Lawrence JM, Hamill RW, Cochard P, Raisman G, Black IB (1981) Effects of spinal cord transection on synapse numbers and biochemical maturation in rat lumbar sympathetic ganglia. Brain Res 212:83-88
Le Douarin N (1973) A biological cell labelling technique and its use in experimental embryology. Devel Biol 30:217-222
Le Douarin N (1977) The differentiation of the ganglioblasts of the autonomic nervous system studied in chimeric avian embryos. In: Karkinen-Jaaskelainen M, Saxen L, Weiss L (Eds) Cell interaction and differentiation. Academic Press, London pp 171-190
Le Douarin NM (1980) The ontogeny of the neural crest in avian embryo chimaeras. Nature (Lond) 286:663-669
Le Douarin NM (1982) The Neural Crest. Cambridge, Cambridge Univ. Press
Le Douarin NM (1984) A model for cell-line divergence in the ontogeny of the peripheral nervous system. In: Black IB (Ed) Cellular and Molecular Biology of Neuronal Development. Plenum Press, New York, pp 3-28
Le Douarin NM (1986) Cell line segregation during peripheral nervous system ontogeny. Science 231:1515-1522

Le Douarin N, Teillet MA (1973) The migration of neural crest cells to the wall of the digestive tract in avian embryo. J Embryol Exp Morph 30:31-48

Le Douarin N, Teillet MA (1974) Experimental analysis of the migration and differentiation of neuroblasts of the autonomic nervous system and of neurectodermal mesenchymal derivatives, using a biological cell marking technique. Devel Biol 41:162-184

Le Douarin NM, Renaud D, Teillet MA, LeDouarin GH (1975) Cholinergic differentiation of presumptive adrenergic neuroblasts in interspecific chimeras after heterotopic transplantations. Proc Natl Acad Sci USA 72:728-732

Le Douarin NM, Teillet MA, Lelièvre C (1977) Influence of the tissue environment on the differentiation of neural crest cells. In: Lash JW, Burger MM (Eds) Cell and tissue interactions. Raven Press, New York, pp 11-27

Le Douarin NM, Teillet MA, Ziller C, Smith J (1978) Adrenergic differentiation of cells of the cholinergic ciliary and Remak ganglia in avian embryo after *in vivo* transplantation. Proc Natl Acad Sci USA 75:2030-2034

Le Douarin NM, Smith J, Teillet MA, Lelièvre CS, Ziller C (1980) The neural crest and its developmental analysis in avian embryo chimaeras. Trends in Neurosci 3:39-42

Le Douarin NM, Cochard P, Vincent M, Duband J-L, Tucker GC, Teillet M-A, Thiery J-P (1984) Nuclear, cytoplasmic, and membrane markers to follow neural crest cell migration: A comparative study. In: The Role of Extracellular Matrix in Development, New York, Alan R Liss, Inc, pp 373-398

Lee K, Seeley PJ, Muller TH, Helmer-Matyjek E, Sabban E, Goldstein M, Greene LA (1985) Regulation of tyrosine hydroxylase phosphorylation in PC 12 pheochromocytoma cells by elevated K^+ and nerve growth factor. Evidence for different mechanisms of action. Mol Pharmacol 28:220-228

Le Lièvre CS (1978) Participation of neural crest-derived cells in the genesis of the skull in birds. J Embryol exp Morph 47:17-37

Le Lièvre CS, LeDouarin NM (1975) Mesenchymal derivatives of the neural crest: analysis of chimaeric quail and chick embryos. J Embryol Exp Morph 34:125-154

Le Lièvre CS, Schweizer GG, Ziller CM, LeDouarin NM (1980) Restrictions of developmental capabilities in neural crest derivatives as tested by *in vivo* transplantation experiments. Devel Biol 77:362-380

Le Tourneau PC (1978) Chemotactic response of nerve fiber elongation to nerve growth factor. Devel Biol 66:183-196

Levi-Montalcini R, Angeletti PU (1963) Essential role of nerve growth factor in the survival and maintenance of dissociated sensory and sympathetic embryonic nerve cells *in vitro*. Devel Biol 7:653-659

Levi-Montalcini R, Angeletti PU (1968) Nerve growth factor. Physiol Rev 48: 534-569

Levi-Montalcini R, Booker B (1960a) Excessive growth of the sympathetic ganglia evoked by a protein isolated from mouse salivary glands. Proc Natl Acad Sci USA 46:373-383

Levi-Montalcini R, Booker B (1960b) Destruction of the sympathetic ganglia in mammals by antiserum to a Nerve Growth Factor. Proc Natl Acad Sci USA 46:384-391

Lewis EJ, Tank AW, Weiner N, Chikaraishi DM (1983) Regulation of tyrosine hydroxylase mRNA by glucocorticoid and cyclic AMP in a rat pheochromocytoma cell line. Isolation of a cDNA clone for tyrosine hydroxylase mRNA. J Biol Chem 258:14632-14637

Lindsay RM, Thoenen H, Barde Y-A (1985) Placode and neural crest-derived sensory neurons are responsive at early developmental stages to brain-derived neurotrophic factor. Dev Biol 112:319-328

Lipinski M, Graham K, Caillaud JM, Crlu C, Tursz T (1983) HNK-1 antibody detects an antigen expressed on neuroectodermal cells. J Exp Med 158:1775-1780
Liuzzi A, Foppen FH, Saavedra JM, Jacobowitz D, Kopin IJ (1977) Effect of NGF and dexamethasone on phenylethanolamine-N-methyltransferase (PNMT) activity in neonatal rat superior cervical ganglia. J Neurochem 28:1215-1220
Löfberg J (1976) Scanning and transmission electron microscopy of early neural crest migration and extracellular fiber systems of the amphibian embryo. J Ultrastruct Res 54:484a
Löfberg J, Ahlfors K, Fallstrom C (1980) Neural crest cell migration in relation to extracellular matrix organization in the embryonic axolotl trunk. Devel Biol 75:148-167
Loring JF, Erickson CA (1987) Neural crest cell migratory pathways in the trunk of the chick embryo. Dev Biol 121:220-236
Loring J, Glimelius B, Weston J (1979) Substrata of extracellular matrix (EMC) components affect the choice and expression of neural crest phenotypes. J Cell Biol 83:41a
Loring J, Glimelius B, Weston JA (1982) Extracellular matrix materials influence quail neural crest cell differentiation *in vitro*. Devel Biol 90:165-174
Ludueña MA (1973a) Nerve cell differentiation *in vitro*. Devel Biol 33:268-284
Ludueña MA (1973b) The growth of spinal ganglion neurons in serum free medium. Devel Biol 33:470-476
Mackay AVP, Iversen LL (1972a) Transsynaptic regulation of tyrosine hydroxylase activity in adrenergic neurons: Effects of potassium concentration on cultured sympathetic ganglia. Naunyn-Schmiedeberg's Arch Exp Path Pharmacol 272:225-229
Mackay A, Iverson L (1972b) Increased tyrosine hydroxylase activity of sympathetioc ganglia cultured in the presence of dibutyrl cyclic AMP. Brain Res 48:424-426
Manasek FJ, Cohen AM (1977) Anionic glycopeptides and glycosaminoglycans synthesized by embryonic neural tube and neural crest. Proc Natl Acad Sci USA 74:1057-1061
Maxwell GE (1976) Substrate dependence of cell migration from explanted neural tubes *in vitro*. Cell Tiss Res 172:325-330
McGuire JC, Greene LA, Furano AV (1978) NGF stimulates incorporation of fucose or glucosamine into an external glycoprotein in cultured rat PC 12 pheochromocytoma cells. Cell 15:357-365
McLennan IS, Hill CE, Hendry IA (1980) Glucocorticosteroids modulate transmitter choice in developing superior cervical ganglion. Nature 283:206-207
Messenger EA, Warner AE (1976) The effect of inhibiting the sodium pump on the differentiation of nerve cells. J Physiol (Lond) 263:P211-212
Mintz B (1967) Gene control of mammalian pigmentation differentiation. I. Clonal origin of melanocytes. Proc Natl Acad Sci USA 58:344-351
Mintz B (1974) Gene control of mammalian differentiation. Ann Rev Genetics 4:411-470
Narayanan CH, Narayanan Y (1978) On the origin of the ciliary ganglion in birds studied by the method of interspecific transplantation of embryonic brain regions between quail and chick. J Embryol Exp Morph 47:137-148
Narayanan CH, Narayanan Y (1980) Neural crest and placodal contributions in the development of the glossopharyngealvagal complex in the chick. Anat Rec 196:71-82
Newgreen DF, Jones RO (1975) Differentiation *in vitro* of sympathetic cells from chick embryo sensory ganglia. J Embryol Exp Morph 33:43-56
Newgreen DF, Ritterman M, Peters EA (1979) Morphology and behavior of neural crest cells of chick embryo *in vitro*. Cell and Tiss Res 203:115-140

Newgreen DF, Thièry J-P (1980) Fibronectin in early avian embryos: synthesis and distribution along the migration pathways of neural crest cells. Cell Tiss Res 211:269-291
Newgreen DF, Allan IJ, Young HM, Southwell BR (1981) Accumulation of exogenous catecholamines in the neural tube and non-neural tissues of the early fowl embryo. Correlation with morphogenetic movements. Wilhelm Roux's Arch 190:320-330
Newgreen DF, Gibbins IL, Sauter J, Wallenfels B, Wütz R (1982) Ultrastructural and tissue-culture studies on the role of fibronectin collage and glycosaminoglycans in the migration of neural crest cells in the fowl embryo. Cell Tiss Res 221:521-549
Nichols DH, Kaplan RA, Weston JA (1977) Melanogenesis in cultures of peripheral nervous system. II. Environmental factors determining the fate of pigment-forming cells. Devel Biol 60:226-237
Nichols DH, Weston JA (1977) Melanogenesis in cultures of peripheral nervous tissue. I. The origin of prospective fate of cells giving rise to melanocytes. Devel Biol 60:217-225
Njå A, Purves D (1978) The effects of nerve growth factor and its antiserum on synapses in the superior cervical ganglion of the guinea pig. J Physiol (Lond) 277:53-75
Noden D (1975) An analysis of the migratory behavior of avian cephalic neural crest cells. Devel Biol 42:106-130
Noden DM (1978a) The control of avian cephalic neural crest cytodifferentiation. I. Skeletal and connective tissues. Devel Biol 67:296-312
Noden DM (1978b) The control of avian cephalic neural crest cytodifferentiation. II. Neural tissues. Devel Biol 67:313-329
Norr SC (1973) *In vitro* analysis of sympathetic neuron differentiation with chick neural crest cells. Devel Biol 34:16-38
Nurse CA (1981) Interactions between dissociated rat sympathetic neurons and skeleton muscle cells developing in cell culture. I. cholinergic transmission. Devel Biol 88:55-70
Nurse CA, O'Lague PH (1975) Formation of cholinergic synapses between dissociated sympathetic neurons and skeletal myotubes of the rat in the cell culture. Proc Natl Acad Sci USA 72:1955-1959
Ogawa M, Isikawa T, Irimajiri A (1984) Adrenal chromaffin cells from functional cholinergic synapses in culture. Nature 307:66-68
O'Lague PH, Obata K, Claude P, Furshpan EJ, Potter DD (1974) Evidence for cholinergic synapses between dissociated rat sympathetic neurons in cell culture. Proc Natl Acad Sci USA 71:3602-3606
O'Lague PH, MacLeish PR, Nurse CA, Claude P, Furshpan EJ, Potter DD (1975) Physiological and morphological studies on developing sympathetic neurons in dissociated cell culture. Cold Spring Harbor Symp 40:399-407
O'Lague PH, Potter DD, Furshpan EJ (1978) Studies on rat sympathetic neurons developing in cell culture. III. Cholinergic transmission. Devel Biol 67:424-443
Oppenheim RW, Maderdrut JL, Wells DJ (1982) Reduction of naturally-occurring cell death in the thoraco-lumbar preganglionic cell column of the chick embryo by nerve growth factor and hemicholinium-3. Devel Brain Res 3:134-139
Otten U, Thoenen H (1975) Circadian rhythm of tyrosine hydroxylase induction by shrot-term cold stress: modulatory action of glucocorticoids in newborn and adult rats. Proc Natl Acad Sci USA 72:1415-1419
Otten U, Thoenen H (1976a) Modulatory role of glucocorticoids on NGF-mediated enzyme induction in organ cultures of sympathetic ganglia. Brain Res 111:438-441
Otten U, Thoenen H (1976b) Selective induction of tyrosine hydroxylase and dopa-

mine-beta-hydroxylase in sympathetic ganglia in organ culture: role of glucocorticoids as modulators. Mol Pharmacol 12:353-361
Otten U, Thoenen H (1976c) Mechanism of tyrosine hydroxylase and dopamine-beta-hydroxylase induction in organ cultures of rat sympathetic ganglia by potassium depolarization and cholinomimetics. Naunyn-Schmiedeberg's Arch Pharmacol 292:153-159
Otten U, Thoenen H (1977) Effect of glucocorticoids on nerve growth factor-mediated enzyme induction in organ cultures of rat sympathetic ganglia: enhanced response and reduced time requirement to initiate enzyme induction. J Neurochem 29:69-75
Otten U, Schwab M, Gagnon C, Thoenen H (1977) Selective induction of tyrosine hydroxylase and dopamine beta-hydroxylase by nerve growth factor: comparison between adrenal medulla and sympathetic ganglia of adult and newborn rats. Brain Res 133:291-303
Papka RE (1972) Ultrastructural and fluorescence histochemical studies of developing sympathetic ganglia in the rabbit. Am J Anat 134:337-364
Paravicini U, Stöckel K, Thoenen H (1975) Biological importance of retrograde axonal transport of Nerve Growth Factor in adrenergic neurons. Brain Res 84:279-291
Patterson PH, Chun LLY (1974) The influence of nonneuronal cells on catecholamine and acetylcholine synthesis and accumulation in cultures of dissociated sympathetic neurons. Proc Natl Acad Sci USA 71:3607-3610
Patterson PH, Chun LLY (1977a) The induction of acetylcholine synthesis in primary cultures of dissociated rat sympathetic neurons. I. Effects of conditioned medium. Devel Biol 56:263-280
Patterson PH, Chun LLY (1977b) The induction of acetylcholine synthesis in primary cultures of dissociated rat sympathetic neurons. II. Developmental aspects. Devel Biol 60:473-481
Patterson PH, Reichardt LF, Chun LLY (1975) Biochemical studies on the development of primary sympathetic neurons in cell culture. Cold Spring Harbor Symp Quant Biol 40:389-397
Patterson PH, Chun LLY, Reichardt LF (1977) The role of nonneuronal cells in the development of sympathetically derived neurons. Prog Clin Biol Res 15:95-103
Pearse AGE (1969) The cytochemistry and ultrastructure of polypeptide hormone-producing cells (the APUD series) and the embryologic, physiologic and pathologic implications of the concept. J Histochem Cytochem 17:303-313
Pearson J, Brandeis L, Goldstein M (1980) Appearance of tyrosine hydroxylase immunoreactivity in the human embryo. Devel Neurosci 3:140-150
Phillipson OT, Moore KE (1975) Effects of dexamethasone and nerve growth factor on phenylethanolamine-N-methyltransferase and adrenaline in organ cultures of newborn rat superior cervical ganglion. J Neurochem 25:295-298
Pintar JE (1978) Distribution and synthesis of glycosaminoglycans during quail neural crest morphogenesis. Devel Biol 67:444-464
Por SB, Huttner WB (1984) A M_r 70,000 phosphoprotein of sympathetic neurons regulated by nerve growth factor and by depolarization. J Biol Chem 259:6526-6533
Potter DD, Landis SC, Furshpan EJ (1980) Dual function during development of rat sympathetic neurons in culture. J exp Biol 89:57-71
Potter DD, Landis SC, Matsumoto SG, Furshpan EJ (1986) Synaptic functions in rat sympathetic neurons in microcultures. II. Adrenergic/cholinergic dual status and plasticity. J Neurosci 6:1080-1098
Pratt RM, Larsen MA, Johnston MC (1975) Migration of cranial neural crest cells in a cell-free hyaluronate-rich matrix. Devel Biol 44:298-305

Price J, Mudge AW (1983) A subpopulation of rat dorsal root ganglion neurones is catecholaminergic. Nature 301:241-243

Puma P, Buseser SE, Watson L, Kelleher DJ, Johnson GL (1983) Purification of the receptor for nerve growth factor from A875 melanoma cells by affinity chromatography. J Biol Chem 258:3370-3375

Purves D, Njå A (1976) Effect of nerve growth factor on synaptic depression after axotomy. Nature 260:535-536

Raven CP, Kloos J (1945) Induction by medial and lateral pieces after archenteron roof, with special reference to the determination of the neural crest. Acta Neerl Morphol Norm Pathol 5:348-362

Rees R, Bunge RP (1974) Morphological and cytochemical studies of synapses formed in culture between isolated rat superior cervical ganglion neurons. J Comp Neurol 157:1-12

Reichardt LF, Patterson PH (1977) Neurotransmitter synthesis and uptake by individual rat sympathetic neurons developing in microcultures. Nature (Lond) 270:147-151

Rickmann M, Fawcett JW, Keynes RJ (1986) Interactions between neurites and somite cells: inhibition and stimulation of nerve growth in the chick embryo. J Embryol exp Morphol 91:209-226

Roach A, Adler JE, Black IB (1987) Depolarizing influences regulate preprotachykinin mRNA in sympathetic neurons. Proc Natl Acad Sci USA 84:5078-5081

Rohrer H, Thoenen H (1987) Relationship between differentiation and terminal mitosis: chick sensory and ciliary neurons differentiate after terminal mitosis of precursor cells, whereas sympathetic neurons continue to divide after differentiation. J Neurosci 7:3739-3748

Ross D, Johnson M, Bunge R (1977) Evidence that development of cholinergic characteristics in adrenergic neurons is age-dependent. Nature (Lond) 267:536-539

Ross LL, Cosio L (1979) A morphometric study of synapse formation in the chick sympathetic ganglion following spinal cord transection. Soc for Neurosci Abst 5:176

Rothman TP, Gershon MD, Holtzer H (1978) The relationship of cell division to the acquisition of adrenergic characteristics by developing sympathetic ganglion cell precursors. Devel Biol 65:322-341

Rothman T, Specht LA, Gershon MD, Joh TH, Teitelman G, Pickel VM, Reis DJ (1980) Catecholamine biosynthetic enzymes are expressed in replicating cells of the peripheral but not central nervous system. Proc Natl Acad Sci USA 77:6221-6225

Rovasio RA, DeLouvee A, Yamada KM, Timpl R, Thiery JP (1983) Neural crest cell migration: Requirement for exogenous fibronectin and high cell density. J Cell Biol 96:462-473

Rydel RE, Greene LA (1988) cAMP analogs promote survival and neurite outgrowth in cultures of rat sympathetic and sensory neurons independently of nerve growth factor. Proc Natl Acad Sci USA 85:1257-1261

Saadat S, Thoenen H (1986) Selective induction of tyrosine hydroxylase by cell-cell contact in bovine adrenal chromaffin cells is mimicked by plasma membranes. J Cell Biol 103:1991-1997

Saadat S, Stehle AD, Lamouroux A, Mallet J, Thoenen H (1987) Influence of cell-cell contact on levels of tyrosine hydroxylase in cultured bovine adrenal chromaffin cells. J Biol Chem 262:13007-13014

Saper CB, Loewy AD, Swanson LW, Cowan WM (1976) Direct hypothalamo-autonomic connection. Brain Res 117:305-312

Schubert D, Heinemann S, Kidokoro Y (1977) Cholinergic metabolism and synapse formation by a rat nerve cell line. Proc Natl Sci USA 74:2579-2583

Scherman D, Weber MJ (1986) Characterization of the vesicular monoamine transporter in cultured rat sympathetic neurons: persistence upon induction of cholinergic phenotypic traits. Dev Biol 119:68-74

Schubert D, Lacorbiere M, Klier FG, Steinbach JH (1980) The modulation of neurotransmitter synthesis by steroid hormones and insulin. Brain Res 190:67-79

Schwab ME, Heumann R, Thoenen H (1982) Communication between target organs and nerve cells: retrograde axonal transport and site of action of nerve growth factor. Cold Spring Harbor Symp Quant Biol 46:125-154

Schwab M, Landis S (1981) Membrane properties of cultured rat sympathetic neurons: morphological studies of adrenergic and cholinergic differentiation. Devel Biol 84:67-78

Schwab M, Thoenen H (1977) Selective trans-synaptic migration of tetanus toxin after retrograde axonal transport in peripheral sympathetic nerves: a comparison with nerve growth factor. Brain Res 122:459-474

Schwab M, Stöckel K, Thoenen H (1976) Immunocytochemical localization of nerve growth factor (NGF) in the submandibular gland of adult mice by light and electron microscopy. Cell Tiss Res 169:289

Schweizer G, Ayer-Le Lievre C, LeDouarin NM (1983) Restrictions of developmental capacities in the dorsal root ganglia during the course of development. Cell Differ 13:191-200

Seeley PJ, Rukenstein A, Connolly JL, Greene LA (1984) Differential inhibition of nerve growth factor and epidermal growth factor effects on the PC 12 pheochromocytoma line. J Cell Biol 98:417-426

Shelton DL, Reichardt LF (1984) Expression of B-nerve growth factor gene correlates with the density of sympathetic innervation in effector organs. Proc Natl Acad Sci USA 81:7951-7955

Shelton DL, Reichardt L (1986) Studies on the regulation of beta-nerve growth factor gene expression in the rat iris: the level of mRNA-encoding nerve growth factor is increased in irises placed in explant cultures in vitro, but not in irises deprived of sensory or sympathetic innervation in vivo. J Cell Biol 102:1040-1048

Sieber-Blum M, Cohen AM (1978) Lectin binding to neural crest cells. J Cell Biol 76:628-638

Sieber-Blum M, Cohen AM (1980) Clonal analysis of quail neural crest cells: They are pluripotent and differentiate *in vitro* in the absence of noncrest cells. Devel Biol 80:96-106

Sieber-Blum M, Sieber F, Yamada KM (1981) Cellular fibronectin promotes adrenergic differentiation of quail neural crest cells *in vitro*. Exp Cell Res 133:285-295

Singer HS, Coyle JT, Vernon N, Kallman CH, Price DL (1980) The development of catecholaminergic innervation in chick spinal cord. Brain Res 191:417-428

Smith J, Cochard P, LeDouarin NM (1977) Development of choline acetyltransferase and cholinesterase activities in enteric ganglia derived from presumptive adrenergic and cholinergic levels of the neural crest. Cell Different 6:199-216

Smith J, Fauquet M, Ziller C, LeDouarin N (1980) Acetylcholine synthesis by mesencephalic neural crest cells in the process of migration *in vivo*. Nature (Lond) 282:853-855

Smolen AJ, Raisman G (1980) Synapse formation in the rat superior cervical ganglion during normal development and after neonatal deafferentation. Brain Res 181:315-324

Smolen AJ, Ross LL (1978) The bulbospinal monoaminergic system of the chick: degeneration in the sympathetic nucleus following surgical and chemical lesions. Brain Res 139:153-159

Snyder SH, Banerjee SP, Cuatrecasas P, Greene LA (1974) The nerve growth factor receptor: demonstration of specific binding in sympathetic ganglia. In: Fuxe K, Olson L, Zotterman Y (Eds) Dynamics of degeneration and growth in neurons, Pergamon Press, Oxford, pp 347-358

Stöckel K, Hendry IA, Thoenen H (1973) Retrograde axonal transport of nerve growth factor. Experientia 29:767

Stöckel K, Paravicini U, Thoenen H (1974) Specificity of the retrograde axonal transport of Nerve Growth Factor. Brain Res 76:413-422

Stöckel K, Schwab M, Thoenen H (1975a) Specificity of retrograde transport of nerve growth factor (NGF) in sensory neurons: a biochemical and morphological study. Brain Res 89:1-14

Stöckel K, Schwab M, Thoenen H (1975b) Comparison between the retrograde axonal transport of nerve growth factor and tetanus toxin in motor, sensory and adrenergic neurons. Brain Res 99:1-16

Sutter A, Hosang M, Vale RD, Shooter EM (1984) The interaction of nerve growth factor with its specific receptors. In: Black IB (Ed) Cellular and Molecular Biology of Neuronal Development. Plenum Press, New York, p 201-214

Swerts JP, Thai ALV, Vigny A, Weber MJ (1983) Regulation of enzymes responsible for neurotransmitter synthesis and degradation in cultured rat sympathetic neurons. I. Effects of muscle-conditioned medium. Dev Biol 100:1-11

Teillet MA (1971) Recherches sur le mode de migration et la differenciation des melanoblastes cutanes chez l'embryon d'Oiseau: etude experimentale par la methode des greffes heterospecifiques entre embryons de Caille et de Poulet. Ann Embryol Morphol 4:95-109

Teillet MA, Le Douarin N (1974) Determination par la methode des greffes heterospecifiques d'ebauches neurales de Caille sur l'embryon de Poulet, du niveau du neuraxe dont derivent les cellules medullo-surrenaliennes. Arch Anat Micr Morph Exp 63:57-62

Teillet M-A, Kalcheim C, LeDouarin NM (1987) Formation of the dorsal root ganglia in the avian embryo: segmental origin and migratory behavior of neural crest progenitor cells. Dev Biol 120:329-347

Teillet M-A, LeDouarin NM (1983) Consequences of neural tube and notochord excision on the development of the peripheral nervous system in the chick embryo. Dev Biol 98:192-211

Teitelman G, Baker H, Joh TH, Reis DJ (1979) Appearance of catecholamine-synthesizing enzymes during development of rat sympathetic nervous system: possible role of tissue environment. Proc Natl Acad Sci USA 76:509-513

Teitelman G, Gershon MD, Rothman TP, Joh TH, Reis DJ (1981a) Proliferation and distribution of cells that transiently express a catecholaminergic phenotype during development in mice and rats. Devel Biol 86:348-355

Teitelman G, Joh TH, Reis DJ (1981b) Transformation of catecholaminergic precursors into glucagon(A) cells of mouse embryonic pancreas. Proc Natl Acad Sci USA 78:5225-5229

Teitelman G, Joh TH, Park D, Brodsky M, New M, Reis DJ (1982) Expression of the adrenergic phenotype in cultured fetal adrenal medullary cells: Role of intrinsic and extrinsic factors. Devel Biol 89:450-459

Teitelman G, Joh TH, Grayson L, Park DH, Reis DJ, Iacovitti L (1985) Cholinergic neurons of the chick ciliary ganglia express adrenergic traits *in vivo* and *in vitro*. J Neurosci 5:29-39

Tennyson V (1965) Electron microscopic study of the developing neuroblasts of the dorsal root ganglion of the rabbit embryo. J Comp Neurol 124:267-317

Thiery J-P, Duband J-L, Delouvee A (1982a) Pathways and mechanisms of avian trunk neural crest cell migration and localization. Develop Biol 93:324-343

Thiery J-P, Duband J-L, Rutishauser U, Edelman GM (1982b) Cell adhesion molecules in early chick embryogenesis. Proc Natl Acad Sci USA 79:6737-6741

Thoenen H, Barde Y-A (1980) Physiology of nerve growth factor. Physiol Rev 60:1284-1335

Thoenen H, Korsching S, Heumann R, Acheson A (1985) Nerve growth factor. In: Growth Factors in Biology and Medicine (CIBA Foundation Symposium 116) 113-128 Pitman, London

Thoenen H, Stöckel K (1975) Ortho-and retro-grade axonal transport: importance for the function of adrenergic neurones. Clin Exp Pharmacol Physiol Suppl 2:1-5

Thoenen H, Angeletti PU, Levi-Montalcini R, Kettler R (1971) Selective induction by nerve growth factor of tyrosine hydroxylase and dopamine-beta-hydroxylase in the rat superior cervical ganglia. Proc Natl Acad Sci USA 68:1598-1602

Thoenen H, Kettler R, Saner A (1972a) Time course of the development of enzymes involved in the synthesis of norepinephrine in the superior cervical ganglion of the rat from birth to adult life. Brain Res 40:459-468

Thoenen H, Saner A, Angeletti PU, Levi-Montalcini R (1972b) Increased activity of choline acetyltransferase in sympathetic ganglia after prolonged administration of Nerve Growth Factor. Nature New Biol 236:26-28

Thoenen H, Saner A, Kettler R, Angeletti PU (1972c) Nerve growth factor and preganglionic cholinergic nerves: their relative importance to the development of the terminal adrenergic neuron. Brain Res 44:593-602

Thoenen H, Schwab M, Otten U (1978) Nerve growth factor as a mediator of information between effector organs and innervating neurons. Symp Soc Devel Biol 35:101-118

Tiffany-Castiglioni E, Perez-Polo JR (1979) The role of nerve growth factor *in vitro* in cell resistance to 6-hydroxydopamine toxicity. Exp Cell Res 121:179-189

Tischler AS, Greene LA (1975) Nerve growth factor-induced process formation by cultured rat pheochromocytoma cells. Nature (Lond) 258:341-342

Tischler AS, Greene LA (1978) Morphologic and cytochemical properties of a clonal line of rat adrenal phenochromocytoma cells which respond to nerve growth factor. Lab Invest 39:77-89

Tischler AS, Dichter MA, Biales B, Delellis RA, Wolfe HW (1976) Neural properties of cultured human endocrine tumor cells of proposed neural crest origin. Science 192:902-904

Tosney KW (1978) The early migration of neural crest cells in the trunk region of the avian embryo: an electron microscopic study. Devel Biol 62:317-333

Tosney KW (1982) The segregation and early migration of cranial neural crest cells in the avian embryo. Devel Biol 89:13-24

Twitty VC, Niu MC (1948) Causal analysis of chromatophore migration. J Exp Zool 108:405-437

Unsicker KB, Krisch V, Otten U, Thoenen H (1978) Induced fiber outgrowth from isolated rat adrenal chromaffin cells: impairment by glucocorticoids. Proc Natl Acad Sci USA 75:3498-3502

Van Campenhout E (1946) The epithelioneural bodies. Q Rev Biol 21:327-347

Vincent M, Duband JL, Thiery JP (1983) A cell surface determinant expressed early on migrating avian neural crest cells. Dev Biol 98:235-238

Vincent M, Thiery JP (1984) A cell surface marker for neural crest and placodal cells: Further evolution in peripheral and central nervous system. Dev Biol 103:468-481

Wahn HL, Lightbody LE, Tchen TT, Taylor JD (1975) Induction of neural differentiation in cultures of amphibian undetermined presumptive epidermis by cyclic AMP derivatives. Science 88:366-369

Wakade AR, Edgar D, Thoenen H (1983) Both nerve growth factor and high K^+ concentrations support the survival of chick embryo sympathetic neurons. Evidence for a common mechanism of action. Exp Cell Res 144:377-384

Wakade AR, Thoenen H (1984) Interchangeability of nerve growth factor and high potassium in the long-term survival of chick sympathetic neurons in serum-free culture medium. Neurosci Lett 45:71-74

Wakshull E, Johnson MI, Burton H (1978) Persistence of an amine uptake system in cultured rat sympathetic neurons which use acetylcholine as their transmitter. J Cell Biol 79:121-131

Wakshull E, Johnson MI, Burton H (1979a) Postnatal rat sympathetic neurons in culture. I. A comparison with embryonic neurons. J Neurophysiol 42:1410-1425

Wakshull E, Johnson MI, Burton H (1979b) Postnatal rat sympathetic neurons in culture. II. Synaptic transmission by postnatal neurons. J Neurophysiol 42:1426-1436

Walicke P, Patterson PH (1981) On the role of Ca^{2+} in the transmitter choice made by cultured sympathetic neurons. J Neurosci 1:343-350

Walicke PA, Campenot PB, Patterson PH (1977) Determination of transmitter function by neuronal activity. Proc Natl Acad Sci USA 74:5767-5771

Weber MJ, Raynaud B, Delteil C (1985) Molecular properties of a cholinergic differentiation factor from muscle-conditioned medium. J Neurochem 45:1541-1547

Weston JA (1963) A radioautographic analysis of the migration and localization of trunk neural crest cells in the chick. Devel Biol 6:279-310

Weston JA (1970) The migration and differentiation of neural crest cells. Advanc Morphogenes 8:41-114

Weston JA (1982) Motile and social behavior of neural crest cells. In: Bellairs R, Curtis A, Dunn G (Eds) Cell Behavior, Cambridge Univ Press p 429-469

Weston JA , Butler SL (1966) Temporal factors affecting localization of neural crest cells in the chicken embryo. Devel Biol 14:246-266

Weston JA, Ciment G, Girdlestone J (1984) The role of extracellular matrix in neural crest development: A reevaluation. In: Alan R. Liss, New York (Eds) The Role of Extracellular Matrix in Development. pp 433-460

Wolinsky E, Patterson PH (1983) Tyrosine hydroxylase activity decreases with induction of cholinergic properties in cultured sympathetic neurons. J Neurosci 3:1495-1500

Wolinsky E, Patterson PH (1985) Rat serum contains a developmentally regulated cholinergic inducing activity. J Neurosci 5:1509-1512

Wong V, Kessler JA (1987) Solubilization of a membrane factor that stimulates levels of substance P and choline acetyltransferase in sympathetic neurons. Proc Natl Acad Sci USA 84:8726-8729

Wurtman RJ, Axelrod J (1966) Control of enzymatic synthesis of adrenaline in the adrenal medulla by adrenal cortical steroids. J Biol Chem 241:2301-2305

Xue Z-G, Smith J, LeDouarin NM (1985) Differentiation of catecholaminergic cells in cultures of embryonic avian sensory ganglia. Proc Natl Acad Sci USA 82:8800-8804

Yan Q, Johnson EM (1987) A quantitative study of the developmental expression of nerve growth factor (NGF) receptor in rats. Dev Biol 121:139-148

Yankner BA, Shooter EM (1982) The biology and mechanism of action of nerve growth factor. Ann Rev Biochem 51:845-868

Yntema CL, Hammond WS (1945) Depletions and abnormalities in the cervical sympathetic system of the chick following extirpation of the neural crest. J Exp Zool 100:237-263

Yntema CL, Hammond WS (1954) The origin of intrinsic ganglia of trunk viscera from vagal neural crest in the chick. J Comp Neurol 101:515-542

Yodlowski M, Fredieu JR, Landis SC (1984) Neonatal 6-hydroxydopamine treatment eliminates cholinergic sympathetic innervation and induces sensory sprouting in rat sweat glands. J Neurosci 4:1535-1548

Yu MW, Tolson NW, Guroff G (1980) Increased phosphorylation of specific nuclear proteins in superior cervical ganglia and PC 12 cells in response to nerve growth factor. J Biol Chem 255:10481-10492

Ziller C, Fauquet M, Kalcheim C, Smith J, LeDouarin NM (1987) Cell lineages in peripheral nervous system ontogeny: Medium-induced modulation of neuronal phenotypic expression in neural crest cell cultures. Dev Biol 120:101-111

Ziller C, Smith J, Fauquet M, Ledouarin NM (1979) Environmentally directed nerve cell differentiation: *In vivo* and *in vitro* studies. In: Development and chemical specificity of neurons, progress in brain research, (Eds Cuenod M, Kreutzberg GW, and B Bloom FE), vol 51, Amsterdam: Elsevier, pp 59-74

Ziller C, Dupin E, Brazeau P, Paulin D, LeDouarin NM (1983) Early segregation of a neuronal precursor cell line in the neural crest as revealed by culture in a chemically defined medium. Cell 32:627-638

CHAPTER 14

β-Phenylethylamine, Phenylethanolamine, Tyramine and Octopamine

J. M. SAAVEDRA

A. Introduction

Phenylethylamine derivatives (Fig. 1) were first synthesized and tested as sympathomimetic compounds (BARGER and DALE 1910). Tyramine and octopamine were classified as indirectly acting sympathomimetic amines (TRENDELENBURG 1972). Octopamine was considered a "false" neurotransmitter, meaning that the amine was taken up, stored in, and released from catecholaminergic terminals upon nerve stimulation, but its release resulted in failure of transmission, since it exerted only a weak postsynaptic effect (KOPIN 1968).

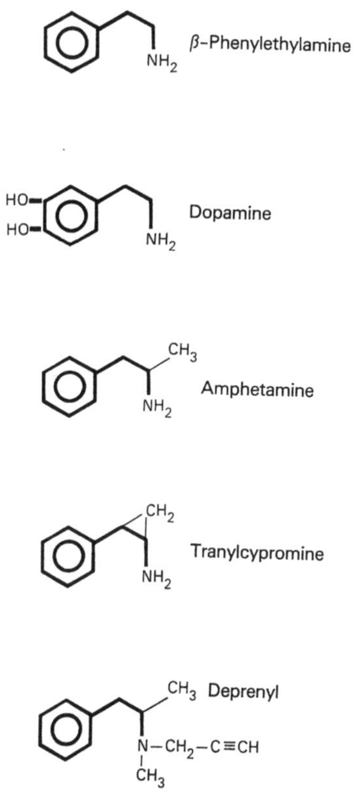

Fig. 1. Structures of β-phenylethylamine and some phenylethylamine derivatives

Interest in these amines has been stimulated by their structural, metabolic, physiological and pharmacological relationship with the catecholamines. With the development of specific and sensitive methods, β-phenylethylamine, phenylethanolamine, tyramine and octopamine were found to be widely distributed in invertebrate and vertebrate tissues and should therefore be considered as biogenic amines (MOLINOFF and AXELROD 1969; SAAVEDRA and AXELROD 1973; SAAVEDRA 1974a,b; PHILIPS et al. 1974; TALLMAN et al. 1976a).

There is mounting evidence of the existence of specific octopaminergic cells in invertebrates, where octopamine may play a role of its own as a neurohormone, neuromodulator, or neurotransmitter (EVANS 1984a; NATHANSON 1985a). In vertebrates, these amines are partially localized in sympathetic nerves and are present in brain. This suggests that they could play a role in sympathetic neurotransmission (MOLINOFF and AXELROD 1972; SAAVEDRA and AXELROD 1973; SAAVEDRA 1974a, b). The proposed role of β-phenylethylamine, tyramine and octopamine in human pathology is, however, a matter of controversy (SAAVEDRA 1978; BOULTON and JUORIO 1979).

The distribution, physiology, pharmacology and biological importance of β-phenylethylamine, phenylethanolamine, tyramine and octopamine have been reviewed elsewhere (AXELROD and SAAVEDRA 1977; SAAVEDRA 1977, 1978; BOULTON 1978, 1979; BOULTON and JUORIO 1979; SAAVEDRA and AXELROD 1976).

B. Methods

I. Enzymatic Radioisotopic Methods

β-Phenylethylamine, phenylethanolamine, tyramine and octopamine can be quantified by incubation with partially purified PNMT in the presence of the methyl donor [^{3}H]-methyl-S-adenosyl-L-methionine (Ado-met). β-phenylethylamine and tyramine are first transformed into phenylethanolamine and octopamine by incubation in the presence of partially purified DBH (MOLINOFF et al. 1969; SAAVEDRA 1974a, b; SAAVEDRA and AXELROD 1973; TALLMAN et al. 1976a).

II. Mass Spectrographic Methods

These amines are measured by mass spectrographic procedures, based on the analysis of dansyl or other amine derivatives with or without gas chromatography (DURDEN et al. 1973; WILLNER et al. 1974; PHILIPS et al. 1974, 1975).

III. Other Methods

Other accepted methods utilize gas chromatography with electron capture detection (EDWARDS and BLAU 1972a; SCHWEITZER et al. 1975) and high performance liquid chromatography (FLATMARK et al. 1978).

IV. Choice of Method

Radioenzymatic and mass-spectrographic methods are specific, have high sensitivity, and are the methods of choice. Mass spectrographic methods are used for the identification of meta- and ortho-hydroxylated tyramine and octopamine isomers (PHILIPS et al. 1975; WILLIAMS and COUCH 1978). Gas chromatographic techniques are used for the analysis of large quantities of amines, such as those present in urine (EDWARDS and BLAU 1972b). The methodology for β-phenylethylamine and related amines has been critically reviewed elsewhere (SCHWEITZER and FRIEDHOFF 1978; SAAVEDRA 1978; BOULTON 1979).

C. Distribution

β-Phenylethylamine, tyramine, octopamine, and to a lesser extent phenylethanolamine are widely distributed throughout the animal kingdom. Of special interest are the existence of specific octopaminergic cells in invertebrates and the very low endogenous levels of all four amines in vertebrate tissues.

I. Invertebrates

1. Octopaminergic Cells

Very high levels of tyramine and octopamine are present in octopus and other invertebrates (JUORIO and PHILIPS 1975; DAVID and COULON 1985). In some invertebrate species, the octopamine/noradrenaline ratio is greater than unity (JUORIO and MOLINOFF 1974).

There are specific octopaminergic cells in several invertebrate species. In lobster, there is a specific and distinct distribution of dopamine-forming and octopamine-forming cells (WALLACE 1976; BARKER et al. 1979). Octopamine concentration is especially high in the thoracic nerve cord, and octopamine-containing neurosecretory cells have been characterized (WALLACE et al. 1974; EVANS et al. 1975). These cells are able to form octopamine but not noradrenaline, and do not contain dopamine or any other identified biogenic amine (BARKER et al. 1972, 1979).

Octopamine, but not noradrenaline, occurs in a number of cell bodies from single, identified neurons of *Aplysia* (SAAVEDRA et al. 1974a) or in the corresponding ganglionic neuropil area where the neuronal terminals are concentrated (FARNHAM et al. 1978). The ability of *Aplysia* and lobster nerves to synthesize dopamine and octopamine but not noradrenaline can best be explained by the occurrence of two types of nerve cells: octopaminergic cells that contain AADC and DBH but not TH, and dopaminergic cells containing TH and AADC but not DBH. In some invertebrate species, the catecholamine synthesizing enzymes may not coexist or may not be phenotypically expressed in the same cell. Several other specific octopaminergic cells have been characterized in invertebrates (DAVID and COULON 1985).

II. Vertebrates

1. Peripheral Tissues

β-Phenylethylamine, phenylethanolamine, tyramine and octopamine are present in low concentrations in most mammalian tissues (TALLMAN et al. 1976a; MOLINOFF et al. 1969; SAAVEDRA and AXELROD 1973; DURDEN et al. 1973; WILLNER et al. 1974; SAAVEDRA 1974a, b, 1978; PHILIPS and JUORIO 1978). Octopamine occurs predominantly in visceral sympathetically innervated organs, such as the heart, spleen, vas deferens, and salivary glands (MOLINOFF et al. 1969), and in sympathetic ganglia (BUCK et al. 1977). The adrenal gland contains high levels of m- and o-octopamine isomers (WILLIAMS and COUCH 1978).

2. Body Fluids

Human plasma contains a relatively high β-phenylethylamine level, comparable to the concentrations of catecholamines (WILLNER et al. 1974; SAAVEDRA 1978). The major β-phenylethylamine metabolite, phenylacetic acid, is present in human plasma and cerebrospinal fluid (SANDLER et al. 1982).

β-Phenylethylamine, octopamine, tyramine, and the m- and o-tyramine isomers are excreted in the urine of humans and other mammals (JEPSON et al. 1960; PISANO et al. 1961; KAKIMOTO and ARMSTRONG 1962a; BOULTON and DYCK 1974; KING et al. 1974). β-Phenylethylamine is excreted normally in the human urine at low concentrations, but with large day-to-day fluctuations, and its secretion is dependent on age, sex, drug intake, urine volume and pH, and probably diet (SCHWEITZER et al. 1975; SLINGSBY and BOULTON 1976; HUEBER and BOULTON 1979).

3. Brain

Very low levels of β-phenylethylamine, phenylethanolamine, tyramine and octopamine occur in mammalian brain, where they represent less than 5 per cent of the total endogenous levels of catecholamines (MOLINOFF et al. 1969; SAAVEDRA and AXELROD 1973; SAAVEDRA 1974a; BOULTON et al. 1975; JUORIO 1977). The amines are heterogeneously distributed in mammalian brain with highest levels in hypothalamus and extrapyramidal structures (MOLINOFF and AXELROD 1972; SAAVEDRA 1974b; WILLNER et al. 1974; WU and BOULTON 1975; BOULTON et al. 1976; TALLMAN et al. 1976a; PHILIPS et al. 1978). Tyramine and dopamine have been reported to coexist within the same nigro-striatal pathway (JUORIO and JONES 1981).

Phenylethanolamine and octopamine are present in higher concentrations in the fetal rat brain at 16 to 17 days of gestation. Their levels decline sharply on day 18 of gestation to approximately those of the adult. This decrease coincides with an increase in the activity of MAO and the noradrenaline synthesizing enzymes, and with the appearance of a saturable, active uptake mechanism for noradrenaline, suggesting a relation between octopamine and catecholamine metabolism in the brain (SAAVEDRA et al. 1974b; DAVID 1984).

Whereas octopamine is present in noradrenaline-containing nerve endings of the rat brain (MOLINOFF and BUCK 1976), only a fraction of rat brain β-phenylethylamine and tyramine is localized in synaptosomes (BOULTON and BAKER 1975); β-phenylethylamine is formed in brain *in situ*, but since it readily crosses all membranes, its levels in brain are influenced directly by its concentration in the periphery, determined by, among other factors, the activities of its main degrading enzyme MAO (YANG and NEFF 1973) and of its main biosynthetic enzyme AADC (LOVENBERG et al. 1962) and the blood and tissue levels of its immediate precursor amino acid L-phenylalanine (DAVID et al. 1974).

In addition, the mammalian brain containts m-octopamine (DANIELSON et al. 1977 a), m-tyramine (BOULTON et al. 1976; JUORIO 1977; PHILIPS et al. 1978) and the tyramine metabolites p-hydroxyphenylacetic acid, p-hydroxymandelic acid and m-hydroxyphenylacetic acid (KAROUM et al. 1975). Phenylacetic acid, the major metabolite of β-phenylethylamine, occurs in rat brain at a level 20 times higher than that of its parent amine (DURDEN and BOULTON 1982).

D. Uptake, Storage and Release

I. Invertebrates

Insect neural tissues contain a high affinity uptake mechanism for octopamine. The system is specific for β-hydroxylated compounds and has a higher affinity for phenolic amines than catecholamines (EVANS 1978). An identified group of cells in the lobster nervous system can release endogenous octopamine into the haemolymph by a Ca^{2+}-dependent neurosecretory process after depolarization by high K^{+} (EVANS et al. 1975, 1976a).

II. Vertebrates

1. Peripheral Tissues

In mammals, both p- and m-octopamine are localized within sympathetic nerve endings, in close association with catecholamines (MOLINOFF et al. 1969; BUCK et al. 1977; IBRAHIM et al. 1985). It is possible that octopamine is released from nerves as a co-transmitter together with noradrenaline (MOLINOFF and BUCK 1976).

At least part of the phenylethanolamine formed in peripheral tissues is stored within sympathetic nerves (SAAVEDRA and AXELROD 1973). In contrast, there are no specific uptake or storage mechanisms for β-phenylethylamine (ROSS and RENYI 1971; WU and BOULTON 1975).

2. Brain

Octopamine can be taken up by nerve endings of the rat brain by a temperature, glucose and sodium-dependent process, which is inhibited by drugs that prevent the uptake of catecholamines (BALDESSARINI and VOGT 1971, 1972). Octopamine can displace noradrenaline previously accumulated in central nerve terminals. Similarly, exogenous noradrenaline displaced previously stored octopamine (BALDESSARINI 1971). Both noradrenaline and octopamine are released *in vitro* from slices of rat brain by electrical or ionic depolarizing stimuli. Intrathecal 6-hydroxydopamine, as well as reserpine, decreased the ability of brain slices to store and release both noradrenaline and dopamine (MOLINOFF and BUCK 1976). From these experiments it can be concluded that octopamine enters central, noradrenaline-containing nerve endings, is stored in presynaptic vesicles, and is released by nerve stimulation.

There is a high affinity uptake of tyramine into mouse brain homogenates (UNGAR et al. 1978) and into rat brain synaptosomes (LENTZEN and PHILIPPU 1977). Both tyramine and its m-isomer are concentrated and released from slices of the rat caudate and hypothalamus in a manner analogous to that of dopamine (BALDESSARINI and VOGT 1972; BOULTON 1978). In contrast to tyramine and octopamine, the *in vitro* uptake of β-phenylethylamine into rat brain synaptosomes occurs by passive diffusion (BALDESSARINI and VOGT 1971; ROSS and RENYI 1971).

E. Metabolism

β-Phenylethylamine, phenylethanolamine, tyramine and octopamine are metabolically related to catecholamines. They share common amino acid precursors, formation and degradation pathways, and can be interconverted with catecholamines by ring(de)hydroxylation (Figs. 2–4) (BOULTON 1979).

I. Formation

1. Decarboxylation and β-Hydroxylation

The main pathways for the formation of β-phenylethylamine and tyramine are decarboxylation of L-phenylalanine and L-tyrosine, respectively, by AADC (Figs. 2–4). The most active β-phenylethylamine formation occurs in kidney, an organ with high AADC activity (LOVENBERG et al. 1962).

Decarboxylation is a minor metabolic pahtway of phenylalanine and tyrosine metabolism (BOULTON and DYCK 1974; TALLMAN et al. 1976b). The concentration of phenylalanine in blood, however, is about two orders of magnitude lower than its Michaelis constant for decarboxylation (LOVENBERG et al. 1962). Thus, increased blood phenylalanine concentrations result in increased decarboxylation rates and increased formation of β-phenylethylamine (DAVID et al. 1974; SAAVEDRA 1974a). A similar phenomenon occurs when blood L-tyrosine levels are increased, with increased formation of tyramine (DAVID et

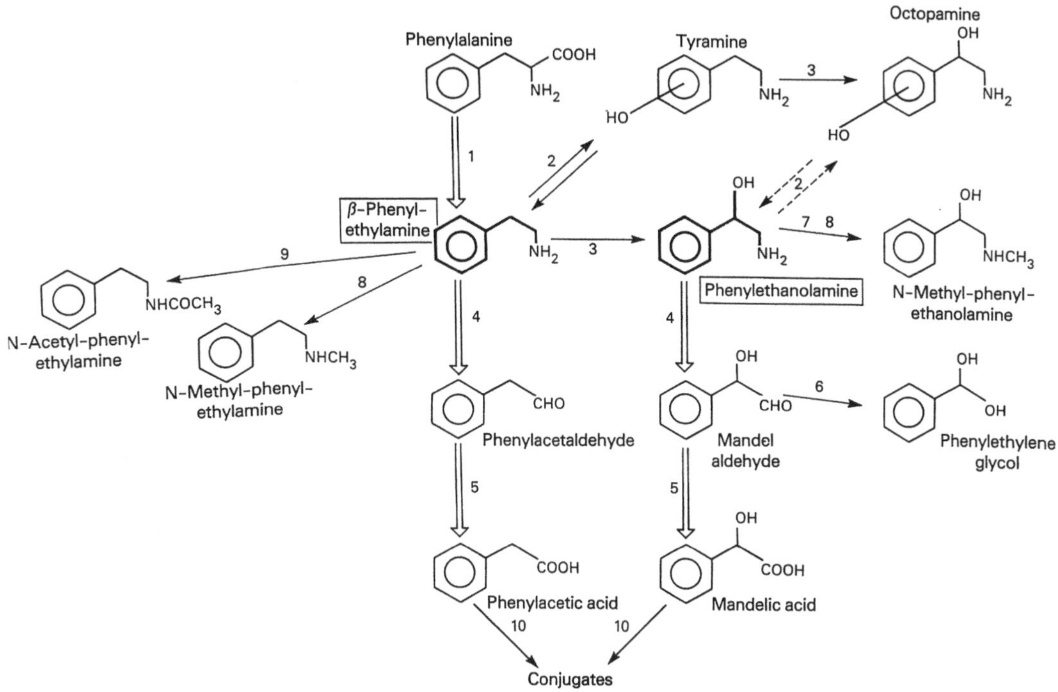

Fig. 2. Metabolic pathways for β-phenylethylamine and phenylethanolamine. Full arrows correspond to reactions which have been unequivocally identified in animal tissues. Double arrows indicate main metabolic pathways. Broken arrows correspond to postulated mechanisms. 1. AADC (EC 4.1.1.28); 2. Ring (de) hydroxylase; 3. DBH (EC 1.14.17.1); 4. MAO (EC 1.14.3.4); 5. Alcohol dehydrogenase (EC 1.1.1.1); 6. Aldehyde dehydrogenase (EC 1.2.1.3); 7. PNMT (EC 2.1.1.28); 8. Nonspecific N-methyltransferase; 9. Arylamine acetyltransferase (EC 2.3.1.5); 10. Conjugases

al. 1974). Peripheral inhibition of AADC increases the formation of brain tyramines, by increasing the brain concentration of the parent aminoacids (Juorio 1983).

The main pathway for the biosynthesis of phenylethanolamine and octopamine proceeds by β-hydroxylation of β-phenylethylamine and tyramine, respectively, a reaction catalyzed by DBH in adrenergic granules (Figs. 2-4) (Molinoff and Axelrod 1969; Creveling et al. 1962a; Saavedra and Axelrod 1973). Tyramine hydroxylation to octopamine is one of the major routes for tyramine metabolism (Wu and Boulton 1974; Tallman et al. 1976b). The tyramine isomers m-tyramine and o-tyramine are readily formed by decarboxylation of m-tyrosine and o-tyrosine (Tong et al. 1979) and m-tyramine can be hydroxylated to m-octopamine in mammalian brain (Danielson et al. 1977a) (Figs. 3 and 4).

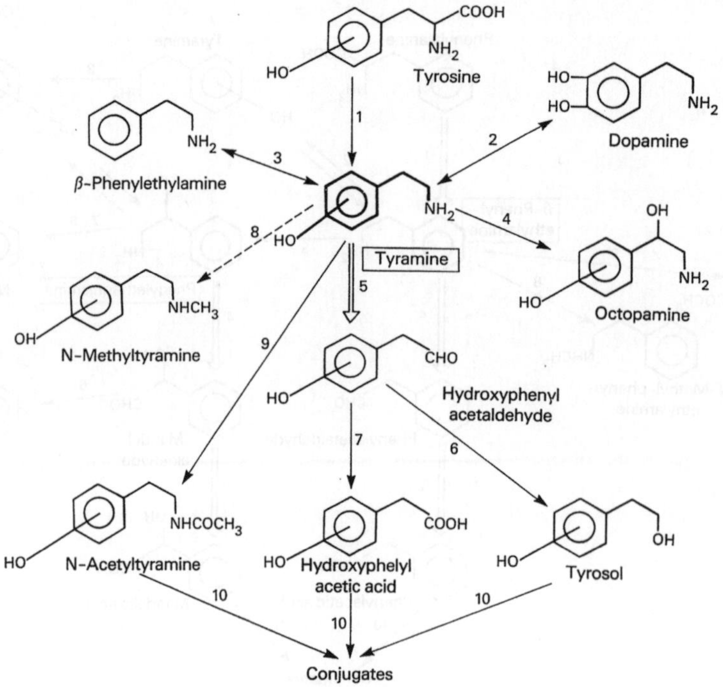

Fig. 3. Metabolic pathways for tyramine. Full arrows correspond to reactions which have been unequivocally identified in animal tissues. Double arrows indicate main metabolic pathways. Broken arrows show postulated mechanisms. 1. AADC; 2 and 3. Ring (de) hydroxylase; 4. DBH; 5. MAO; 6. Aldehyde dehydrogenase; 7. Alcohol dehydrogenase; 8. Nonspecific N-methyltransferase; 9. Arylamine N-acetyltransferase; 10. Conjugases

2. Ring (de) Hydroxylation

Tyramine, octopamine, β-phenylethylamine, phenylethanolamine and catecholamines can be interconverted by ring (de) hydroxylation. Ring (de) hydroxylation of catecholamines results in tyramine and octopamine formation; in turn, the ring (de) hydroxylation of tyramine forms β-phenylethylamine, and that of octopamine yields phenylethanolamine (Figs. 2–4) (BRANDAU and AXELROD 1972, 1973; BOULTON and WU 1972, 1973). Dehydroxylation of acidic catecholic metabolites leads to hydroxyphenylacetic acids, p-dehydroxylation being more active than m-dehydroxylation (SCHELINE et al. 1960; SMITH et al. 1964). Ring hydroxylation of β-phenylethylamine yields tyramine (BOULTON et al. 1974; JONSSON et al. 1975), m-tyramine (PHILIPS et al. 1975) and dopamine (LEMBERGER 1966). Octopamine can be formed by ring hydroxylation of phenylethanolamine (BOULTON and WU 1972) (Fig. 2). Catecholamines can be formed by ring hydroxylation of tyramine and octopamine (Figs. 3 and 4) (CREVELING et al. 1962b; AXELROD 1963; BOULTON 1979).

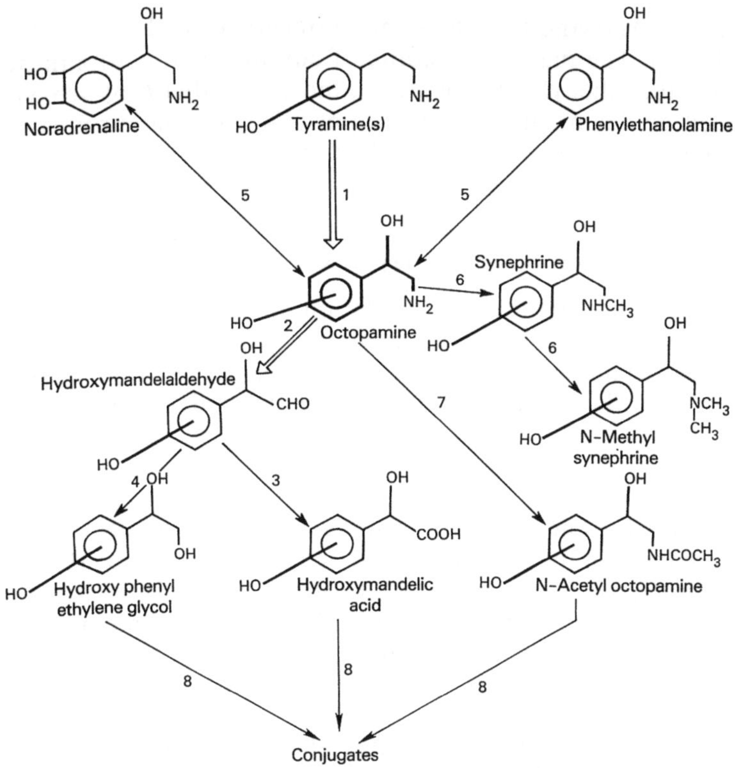

Fig. 4. Metabolic pathways for octopamine. Full arrows correspond to reactions which have been unequivocally identified in animal tissues. Double arrows indicate main metabolic pathways. 1. DBH; 2. MAO; 3. Aldehyde dehydrogenase; 4. Alcohol dehydrogenase; 5. Ring (de) hydroxylase; 6. PNMT; 7. Arylamine acetyltransferase; 8. Conjugases

II. Degradation and Biotransformation

1. Oxidative Deamination

Oxidative deamination by MAO is the principal metabolic route for β-phenylethylamine (EDWARDS and BLAU 1972b), phenylethanolamine (MOLINOFF et al. 1969; SAAVEDRA and AXELROD 1973), tyramine (TALLMAN et al. 1976b; YANG and NEFF 1973), and octopamine (MOLINOFF et al. 1969), both in mammals and in some, but not all, invertebrate species (BARKER et al. 1972). At physiological concentrations, β-phenylethylamine is a preferred substrate for MAO B (YANG and NEFF 1973). Further metabolism after MAO proceeds via aldehyde oxidation to form phenylacetic acid in the case of β-phenylethylamine (WU and BOULTON 1975), whereas the deaminated aldehyde of phenylethanolamine yields mandelic acid and phenylethyleneglycol via alcohol dehydrogenase and aldehyde dehydrogenase, respectively (LOO et al. 1976; EDWARDS and RIZK 1979) (Fig. 2). Oxidative deamination of tyramine yields hydroxyphenylacetic acid (Fig. 3) and that of octopamine, hydroxymandelic acid

(KAKIMOTO and ARMSTRONG 1962a, b; LEMBERGER et al. 1966; BOULTON et al. 1976) (Fig. 4). The major metabolite of administered octopamine in brain tissue is 4-hydroxyphenylethylene glycol, indicating that β-hydroxylated aldehydes formed by deamination of phenol or phenylethanolamines in cerebral tissues are enzymatically reduced to glycol derivatives (RUTLEDGE and JONASON 1967).

2. β-Hydroxylation and Ring (de) Hydroxylation

β-Hydroxylation and ring (de) hydroxylation are alternative metabolic routes for β-phenylethylamine, phenylethanolamine, tyramine and octopamine, analogous to the metabolism of catecholamines (see above) (Figs. 2-4). Their relative importance when compared to monoamine oxidation is not clear.

3. N-Methylation

Phenylethanolamine and octopamine are N-methylated by PNMT, an enzyme localized predominantly in the adrenal medulla (AXELROD 1962b) and to much smaller degree in the heart (AXELROD 1962b) and specific brain areas (SAAVEDRA et al. 1974c). Small amounts of endogenous N-methyl octopamine (synephrine) have been identified in human urine (PISANO et al. 1961), mammalian adrenal gland (MIDGLEY et al. 1980) and rat brain (BOULTON and WU 1972; BRANDAU and AXELROD 1973).

β-Phenylethylamine can be N-methylated by a non-specific N-methyltransferase, predominantly localized in rabbit lung (AXELROD 1962a) but also present in mammalian brain (SAAVEDRA et al. 1973). Exogenous β-phenylethylamine is concentrated in lung, an organ which may contribute to its inactivation (BAKHLE and YOUDIM 1979). N-methyl-phenylethylamine is apparently devoid of substantial pharmacological activity (MCEWEN et al. 1969).

4. N-Acetylation

N-acetylation of β-phenylethylamine and octopamine was demonstrated in mammalian tissues, including brain (YANG and NEFF 1976; WU and BOULTON 1979). Free and conjugated N-acetyltyramine appears in rat urine after injection of labelled tyramine (TACKER et al. 1970) (Figs. 2-4). N-acetylation is a major metabolic route for tyramine and octopamine in some invertebrates (MIR and VAUGHAN 1981).

5. Conjugation

Substantial proportions of urinary β-phenylethylamine, tyramine, octopamine, and their deaminated metabolites are excreted in the human urine as glucuronic and other conjugates (TACKER et al. 1970; JAMES and SMITH 1973; SMITH and MITCHELL 1974; EDWARDS and RIZK 1979).

III. Turnover

The turnover rates of pharmacologically active compounds may be a more significant index of their importance than their endogenous concentrations. Although levels of β-phenylethylamine, phenylethanolamine, tyramine and octopamine are very low in vertebrate tissues, their turnover rates are very fast, both in peripheral tissues and brain of mammals (SAAVEDRA and AXELROD 1973; SAAVEDRA 1974a, 1978; BOULTON et al. 1975; WU and BOULTON 1975; TALLMAN et al. 1976b).

Octopamine turnover, estimated either by measuring the decrease in its level after synthesis inhibition, or by following the rate of disappearance of accumulated ^{3}H-octopamine, was considerably faster than that reported for noradrenaline both in the rat heart or brain (Fig. 5). Thus, steady-state concentration ratios of noradrenaline to octopamine may not reflect the exact relative contribution of each amine to the total intravesicular amine pool. The synthesis of octopamine may constitute as much as one-third of the biogenic amines synthesized in the adrenergic neurons in the rat heart and probably other tissues (MOLINOFF and AXELROD 1972; BRANDAU and AXELROD 1972).

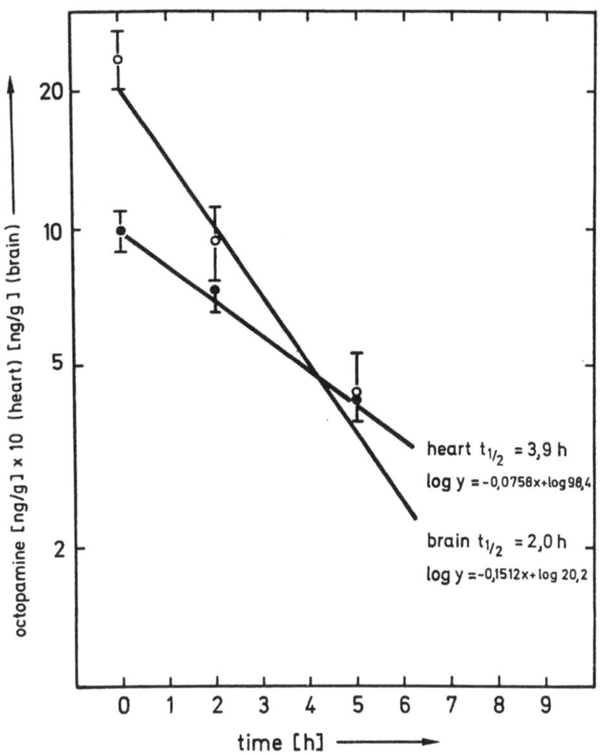

Fig. 5. Rate of disappearance of endogenous octopamine in rat brain and heart after dopamine-β-hydroxylase blockade with FLA 63 (50 mg, i.p.). Each value represents the mean ± SEM of six animals. Half-life times for octopamine turnover were calculated from the slopes of these plots (regression lines). (From BRANDAU and AXELROD 1972)

F. Effects of Drugs on Tissue Levels

I. Metabolic Precursors and Inhibitors of Degradation

Administration of L-phenylalanine and blockade of phenylalanine hydroxylase (EC 1.14.16.1) with p-chlorophenylalanine results in large increments in β-phenylethylamine and phenylethanolamine levels in rat tissues, indicating that the metabolism of L-phenylalanine can be diverted to the decarboxylation and β-hydroxylation pathway (SAAVEDRA and AXELROD 1973; SAAVEDRA 1974a; BOULTON 1979).

Conversely, administration of the precursor amino acids L-phenylalanine and L-tyrosine produces a decrease, rather than an increase, in brain tyramine levels (TALLMAN et al. 1976b). The inhibition of TH by α-methyl-p-tyrosine does not result in an increase in tissue tyramine content. A decrease in octopamine levels occurs after a tyrosine load or TH inhibition (BRANDAU and AXELROD 1972). These results are difficult to interpret in terms of simple enzyme inhibition or precursor load. Several additional phenomena may be involved, among those an increase in the release of the non-catechol biogenic amines followed by rapid metabolism (TALLMAN et al. 1976b).

β-Phenylethylamine administration results in a substantial elevation of phenylethanolamine levels (MOLINOFF et al. 1969) and tyramine administration increases octopamine in rat tissues (BUCK et al. 1977). In both cases, the changes can be prevented by prior DBH inhibition (MOLINOFF and AXELROD 1972; SAAVEDRA and AXELROD 1973). Such a tretment produces a decrease in endogenous octopamine in invertebrates and in mammals (BRANDAU and AXELROD 1972; MOLINOFF and AXELROD 1972), and results in increased levels of the octopamine precursor tyramine (BOULTON 1976).

Increases in rat brain octopamine occur after injection of β-phenylethylamine and after inhibition of phenylalanine hydroxylase. Administration of MAO inhibitors results in large increases in the levels of β-phenylethylamine, tyramine, phenylethanolamine and octopamine and potentiates the accumulation of amines produced after administration of precursor aminoacids (MOLINOFF et al. 1969; BRANDAU and AXELROD, 1972; SAAVEDRA and AXELROD 1973; SAAVEDRA 1974a; BOULTON et al. 1975; JUORIO 1976; TALLMAN et al. 1976b; BUCK et al. 1977; BOULTON 1979).

II. Psychotropic and Other Drugs

Tyramine and octopamine levels are decreased in rat brain after systemic administration of reserpine or central administration of 6-hydroxydopamine (TALLMAN et al. 1976b; BOULTON et al. 1977; BUCK et al. 1977).

Amphetamine treatment decreases brain tyramine (DANIELSON et al. 1977b) and octopamine (JUORIO and DANIELSON 1978) without affecting β-phenylethylamine concentrations (DANIELSON et al. 1976).

Antipsychotic drugs decrease tyramine (JUORIO 1977; JUORIO and DANIELSON 1978), but increase the levels of its m-isomer (JUORIO 1977) in mammalian brain. Trycyclic antidepressants increase octopamine levels in rat brain

(ROTH and GILLIS 1974). The administration of the broad spectrum antibiotic neomycin does not significantly change tissue levels of octopamine and does not interfere with the increase of octopamine after MAO inhibitors, indicating that in normal conditions the intestinal tract is not an important source of tissue octopamine (BUCK et al. 1977).

G. Pharmacological Actions and Receptors

I. Invertebrates

Several well characterized octopaminergic systems have been described in invertebrates (DAVID and COULON 1985).

1. Aplysia Californica Ganglion Cells

In *Aplysia californica*, iontophoretic application of octopamine increases a specific hyperpolarizing conductance, localized in the cerebral ganglia neuropil where functional synapses are present (CARPENTER and GAUBATZ 1974). In addition, octopamine causes a considerable elevation in cyclic AMP and a long lasting increase in protein phosphorylation in abdominal ganglia of *Aplysia*, effects localized postsynaptically (LEVITAN and BARONDES 1974). Specific ionic responses, associated with increased membrane conductance, are also produced in *Aplysia* ganglia by phenylethanolamine (SAAVEDRA et al. 1977).

2. The Firefly Lantern

Octopamine, but not catecholamines, is present throughout the luminiscence-producing pathway of the firefly (CHRISTENSEN et al. 1983). The firefly lantern contains an octopamine-sensitive adenylate cyclase (NATHANSON and HUNNICUTT 1979). The relative potency of octopamine and related amines in stimulating adenylate cyclase in cell-free homogenates of the lantern correlates well with the known effects of these compounds in affecting light production. Octopamine and N-methyl-octopamine are the most potent compounds in eliciting firefly lantern luminiscence (CARLSON 1969) and are equipotent in stimulating adenylate cyclase (25-fold increase over basal levels), while dopamine produced only a 4-fould increase in cyclic AMP formation (NATHANSON 1979).

3. Lobster Neurosecretory Neurons

The octopaminergic cells in lobster have their cell bodies located in the proximal regions of the second roots of the thoracic ganglia, and send their axons to the pericardial neurohaemal organ (EVANS et al. 1975, 1976a). Both the cell body area and the neurosecretory plexus of the heart contain large quantities of octopamine (WALLACE et al. 1974).

Octopamine is selectively released by potassium depolarization from the neurosecretory cells into the haemolymph, a process dependent on Ca^{2+} and occurring at two distinct sites (EVANS et al. 1975; 1976a). One site is close to the cell bodies and results in release of octopamine to the ventral blood sinus and from there to the general circulation, where it may be involved in the regulation of the contractile responses of exoskeletal muscles (EVANS et al. 1975). In addition, octopamine released from the terminals adjacent to the pericardial cells is directed to the heart (EVANS et al. 1976a; KONISHI and KRAVITZ 1978). Octopamine increases the lobster heart rate at a very low concentration and enhances the contractile response to nerve stimulation (FLOREY and RATHMAYER 1978). The persistence of the excitatory actions of octopamine, both in the heart and at the neuro-muscular synapses, suggest that octopamine could function as a neurohormone.

In addition, octopamine produces a characteristic pattern of nerve firing, elevates cyclic AMP in lobster ganglia, and is probably released synaptically (HARRIS-WARRICK and KRAVITZ 1984) to act as a central neurotransmitter.

4. Locust DUMETi Neurons

DUM (dorsal, unpaired, median) neurons from the thoracic ganglia of the locust make neurosecretory connections to the leg muscle (extensor tibia) (HOYLE et al. 1980) (and are therefore called DUMETi, i.e. neurons projecting to the extensor tibia). Their stimulation inhibits the intrinsic, slow rhythmic contractions made by the extensor tibia. These effects are mimicked by a low concentration of octopamine (HOYLE 1975). The DUMETi cells contain octopamine and synthesize it from tyrosine (EVANS and O'SHEA 1977). The effects of octopamine on this system are of long duration, far outlasting the application of the amine (O'SHEA and EVANS 1979). An additional role for octopamine in the modulation of muscular activity in the visceral system has been proposed (ORCHARD and LANGE 1985; LANGE and ORCHARD 1986).

5. Limulus Visual System

Octopamine has been proposed as a neurotransmitter candidate in the visual system of the horseshoe crab, Limulus polyphemus (BATELLE et al. 1982) and probably regulates the dark-adaptation processes through adenylate cyclase stimulation (KAUPP et al. 1982) as well as retinal sensitivity (KASS and BARLOW 1984).

6. Other Systems

In addition, octopamine has been recently proposed to play a role in other invertebrate systems, such as the neuromuscular transmission in moths (KLAASEN and KAMMER 1985), the crayfish's lateral giant escape reaction (GLANZMAN and KRASNE 1983), the presynaptic activity-dependent effects at crayfish neuromuscular junctions (BREEN and ATWOOD 1983) and in the modulation of the membrane potential of the Schwann cell of the squid giant nerve fiber (REALE et al. 1986).

7. Octopamine Receptors

Octopamine-sensitive adenylate cyclases were described in *Aplysia*, firefly lantern, lobster, and in the cockroach nervous system (NATHANSON and GREENGARD 1973). Stimulation of octopamine receptors in insect nervous system results in phosphorylase activation, glycogenolysis, and modulation of potassium permeability (SCHOFIELD and TREHERNE 1985). Guanylyl nucleotides stimulate basal and octopamine-sensitive adenylate cyclase activities in insects (BODMARYK 1979) and other invertebrate species (BATTELLE and KRAVITZ 1978).

High affinity binding of [^{3}H]-octopamine, probably corresponding to specific octopamine receptors, was recently reported in Drosophila (DUDAI and ZVI 1984).

Three clases of octopamine receptors have been defined in the locust, on the basis of pharmacological criteria and physiological studies, and designated as octopamine-1, octopamine-2 A and 2 B (EVANS 1981). Octopamine-2-receptors appear to be present primarily in invertebrates, and many of the effects of octopamine at these receptors are mediated by adenylate cyclase stimulation (NATHANSON 1985a; LAFON-CAZAL and BOCKAERT 1985).

The role of cyclic nucleotides in the mediation of the effects of octopamine of the locust skeletal muscle has been extensively investigated (EVANS 1984a,b; LAFON-CAZAL and BOCKAERT 1985). Cyclic AMP mediates the activation of octopamine receptors that modulate neuromuscular transmission and muscle contraction in the extensor tibiae muscle of the locust hind leg, probably through octopamine 2 A presynaptic and octopamine 2 B postsynaptic receptors, but not through octopamine class 1 receptors (EVANS 1984a,b).

In the gregarious form of *Locusta migratoria* the octopamine-sensitive adenylate cyclase is more responsive, the octopamine level is lower, and the increase in cyclic AMP in cell-free homogenates is greater than in the solitary stage (FUZEAU-BRAESCH et al. 1979). This suggests that there is a relationship between the level of transmitter and the sensitivity of receptor not unlike that seen, for example, in the mammalian brain for noradrenaline and noradrenaline-sensitive adenylate cyclase. In addition, the presence of endogenous octopamine and of an octopamine-sensitive adenylate cyclase has been documented in the central nervous system of the locust (COULON et al. 1984; MORTON 1984; HIRIPI and ROZSA 1984).

The invertebrate octopamine-sensitive adenylate cyclase has been demonstrated to be a site of action for pesticides (EVANS and GEE 1980; NATHANSON and HUNNICUTT 1981). Recently developed, highly potent and selective octopamine-2 agonists, substituted phenyliminoimidazolines, have been proposed as insect toxins with low toxicity in vertebrates (NATHANSON 1985b).

II. Vertebrates

1. Peripheral Effects

BARGER and DALE (1910) first described the peripheral sympathomimetic effects of large doses of β-phenylethylamine and tyramine. These amines, together with octopamine, were later classified as indirectly acting sympathomimetic amines (TRENDELENBURG, 1972; BÖNISCH and TRENDELENBURG, Chap. 5, Vol. I).

Octopamine can be taken up and stored in adrenergic vesicles and released upon nerve stimulation, being two orders of magnitude less potent than noradrenaline in the test organs studied (TRENDELENBURG 1972). For these reasons octopamine was included among the "false neurotransmitters" (KOPIN 1968). Such a mechanism could play a role in the regulation of sympathetic activity and in the response to various drugs. Octopamine occurs normally in sympathetic nerves, and the noradrenaline-octopamine ratio decreases substantially after inhibition of MAO (MOLINOFF and AXELROD 1972). The adrenal gland of mammals contains m-octopamine (WILLIAMS and COUCH 1978) which exhibits substantial α-adrenergic agonist activity and has a pressor action (FREGLY et al. 1979).

2. Central Effects

The central, stimulatory effects (Fig. 1) of β-phenylethylamine include increased locomotor activity (MANTEGAZZA and RIVA 1963; NAKAJIMA et al. 1964), anorexia, hyperthermia and differences in lethality between isolated and aggregated mice (MANTEGAZZA and RIVA 1963), altered conditioned behavior (HUANG and HO 1974) and induction of stereotyped movements in rats, similar to those produced by amphetamine; aggressive behavior is not elicited (BRAESTRUP and RANDRUP 1978).

β-Phenylethylamine stimulatory effects only occur after large doses of the amine are injected, or after smaller doses are given to animals pretreated with MAO inhibitors. A good correlation exists between the brain levels of β-phenylethylamine and its stimulant properties (NAKAJIMA et al. 1964). However, significant behavioral changes occur only after attainment of brain concentrations in the order of 40 to 90 nanomoles per gram of tissue, which are three orders of magnitude higher than the normal endogenous brain concentrations (about 0.012 nanomoles per gram of brain tissue) (NAKAJIMA et al. 1964; SAAVEDRA 1974a), and several times higher than the maximum endogenous β-phenylethylamine concentrations obtained after administration of precursors and stimultaneous blockade of degradation (EDWARDS and BLAU 1973; SAAVEDRA 1974a; BOULTON 1979; PHILIPS and BOULTON 1979). Furthermore, the administration of the specific type B MAO inhibitor deprenyl (Fig. 1), which results in a selective increase in brain β-phenylethylamine, does not result in behavioral activation (BRAESTRUP and RANDRUP 1978).

The central effects of β-phenylethylamine are explained by an indirect mechanism through catecholamine release (JONSSON et al. 1966; FUXE et al. 1967; BALDESSARINI and VOGT 1971; JACKSON and SMYTHE 1973; BRAESTRUP et al. 1975). Massive doses of β-phenylethylamine, however, result in an initial direct effect on postsynaptic dopamine receptors (FUXE et al. 1967). β-Phenylethylamine inhibits the electrically evoked release of acetylcholine from rat striatal slices, an effect dependent on MAO inhibition and on the integrity for the dopaminergic nerve terminals (BAUD et al. 1985). The direct actions of β-phenylethylamine are not likely to have a physiological correlate and can be considered as a pharmacological toxicological effect.

Phenylethanolamine acts as an indirectly acting amine when administeed locally to brain structures. Iontophoretic application of phenylethanolamine depresses the spontaneous discharge of cortical rat neurons and the glutamate-induced excitation of caudate rat neurons, an effect similar to that of dopamine and noradrenaline (HENWOOD et al. 1979). When applied directly to the nucleus accumbens of rats treated with MAO inhibitors, phenylethanolamine causes increased locomotor activity, an effect blocked by prior inhibition of catecholamine synthesis (JACKSON et al. 1975).

A similar indirect mechanism of action can be postulated for tyramine in the central nervous system of mammals. The actions of tyramine are qualitatively similar to those of catecholamines, and include hyperpolarization of cortical and cerebellar Purkinje cells (SASTRY and PHILIS 1977) and depression of caudate neurons (BOAKES et al. 1976). Both m- and p-tyramine can release noradrenaline and dopamine from brain synaptosomes (RAITERI et al. 1977).

When administered intraventricularly, tyramine (especially its ortho isomer) can increase motor activity (STOOF et al. 1976). These data strongly suggest that the mechanism for the central action of tyramine is similar to its indirect peripheral effects through release of endogenous catecholamines from presynaptic elements.

The mechanisms of the central effects of octopamine are controversial. This amine has been reported to have similar (HENWOOD et al. 1979) or opposite (HICKS and MCLENNAN 1978) effects to those of catecholamines when applied iontophoretically to specific areas of the rat brain.

No specific receptor for β-phenylethylamine, phenylethanolamine, tyramine or octopamine has been described so far in mammalian brain. Tyramine is a poor ligand for the β-adrenoceptors in the central nervous system (U'PRICHARD et al. 1977), and fails to bind to β-adrenoceptors in erythrocyte membranes (SCHRAMM et al. 1972). In addition, tyramine does not activate or stimulate cyclic AMP formation in brain homogenates or in brain slices (WOODRUFF 1978) and it does not stimulate Na^+-K^+ ATPase (SCHAEFER et al. 1972). No specific receptor blockers for octopamine have been descriebed to date in mammalian systems.

H. Function in Pathological States

β-Phenylethylamine, phenylethanolamine, tyramine or octopamine have been implicated in phenylketonuria, hepatic coma, depression, schizophrenia, aggression, stress, migraine, epilepsy, Parkinson's disease, and hypertensive crises. To date, a role for these amines in human pathology remains controversial, with the exception of their possible involvement in phenylketonuria and hepatic coma.

I. Phenylketonuria

High levels of L-phenylalanine in phenylketonuria result in an increased formation, excretion and metabolism of β-phenylethylamine, phenylethanolamine and the major β-phenylethylamine metabolite, phenylacetic acid (JEPSON et al. 1960; PERRY 1962; OATES et al. 1963; EDWARDS and RIZK 1979). This is quantitatively the most striking disturbance of amine metabolism in phenylketonuria found to date. A similar biochemical profile, which includes increased octopamine levels, occurs in rats treated with combinations of phenylalanine hydroxylase inhibitors and phenylalanine, to simulate phenylketonuria (EDWARDS and BLAU 1973; SAAVEDRA and AXELROD 1973; SAAVEDRA et al. 1974a; LOO et al. 1976). These abnormalities could play a role in the mental impairment associated with this disease through formation of neurotoxic compounds such as phenylacetate (EDWARDS and BLAU 1972b; LOO et al. 1980), or alterations in neurotransmitter function (FISCHER and BALDESSARINI 1971; MCKEAN 1972; SAAVEDRA 1974a).

II. Hepatic Encephalopathy

The neurological and cardiovascular symptoms of hepatic failure may result from the accumulaton of octopamine and probably tyramine in central and peripheral sympathetic nerves (FISCHER and BALDESSARINI 1971). A substantial percent of urinary octopamine originates in the intestinal flora (PERRY et al. 1966) and this or other amines might not be fully metabolized by a malfunctioning liver, resulting in higher blood levels of amines of intestinal origin. Increased peripheral octopamine could be taken up by sympathetic nerves and, by a "false neurotransmitter" mechanism, interfere with peripheral noradrenergic transmission.

A role for increased octopamine of intestinal origin in the central symptoms of hepatic coma is more difficult to explain. Octopamine could enter the brain in the form of a precursor amino acid, such as tyrosine. Brain levels of tyrosine correlate well with brain octopamine levels and with the degree of encephalopathy (JAMES et al. 1975). A high brain tyrosine concentration could also result in increased tyramine levels (DAVID et al. 1974) raising the possibility that several amines might contribute to the pathogenesis of the nervous system alterations in this disease.

Clinical improvement in patients with hepatic coma has been obtained after intestinal sterilization with antibiotics to reduce body octopamine levels (LAM et al. 1973), after administration of directly acting sympathomimetic amines (FISCHER and JAMES 1972), and after L-dopa administration, which results in displacement of octopamine from noradrenergic nerves (JAMES and FISCHER 1975).

III. Hypertension

Tyramine and related amines can participate in cardiovascular regulation through its indirect sympathomimetic effects and interactions with the metabolism of catecholamines.

The intake of food rich in tyramine by patients receiving MAO inhibitors leads to hypertensive crises (HORWITZ et al. 1964). After MAO inhibiton, an ingested load of tyramine, which is more slowly metabolized, will displace catecholamines from synaptic vesicles, increasing their release and resulting in acute hypertension. The central effects of tyramine my be opposite to its perpheral effects since in DOCA (deoxycorticosterone acetate)-saline treated animals the high blood pressure is lowered after treatment with tyramine precursors (SHALITA and DIKSTEIN 1977). This effect, like that on peripheral organs, may be mediated indirectly through noradrenaline release.

IV. Schizophrenia and Affective Disorders

β-Phenylethylamine has long been implicated in the symptomatology or pathogenesis of schizophrenia. This hypothesis was based on its pharmacological effects, similar to those of amphetamine, and in reports that schizophrenics presented alterations in β-phenylethylamine excretion or metabolism and in the activity of its main catabolizing enzyme, MAO (SANDLER et al. 1978a, b; POTKIN et al. 1979). Many of the clinical findings, however, have not been substantiated (SAAVEDRA 1978; DAVIS et al. 1982). Alterations in β-phenylethylamine have been reported in affective disorders (SANDLER et al. 1979; MOISES et al. 1985). The therapeutic effect of the β-phenylethylamine derivative deprenyl (Fig. 1), a selective MAO B inhibitor, in depressed patients, could be explained by *in vivo* formation of methamphetamine from deprenyl (REYNOLDS et al. 1978; MANN and GERSHON 1980). The role of β-phenylethylamine in psychiatric disorders remains highly controversial (SCHWEITZER et al. 1975; SLINGSBY and BOULTON 1976; SAAVEDRA 1978).

V. Miscellanea

The evidence for involvement of β-phenylethylamine, tyramine and octopamine in Parkinson's disease, migraine, and other diseases, is indirect and subject to controversy (NAKAJIMA et al. 1964; MOFFETT et al. 1972; SANDLER et al. 1976; REYNOLDS et al. 1978).

J. Conclusions

The presence of β-phenylethylamine, phenylethanolamine, tyramine and octopamine in animal tissues has been established beyond controversy. As phenylethylamine derivatives, these amines are structurally related to catecholamines and to therapeutically active drugs (Fig. 1). They are metabolically and physiologically related to the catecholamines, since they share the same precursors and metabolic enzymes, and can be interconverted to catecholamines by ring (de) hydroxylation.

β-Phenylethylamine administration results in an amphetamine-like stimulation syndrome when its brain levels reach a concentation three orders of magnitude higher than its endogenous levels. These changes are mainly indirect and mediated through release of endogenous catecholamines. There are no storage and release mechanisms or receptors specific for β-phenylethylamine, and the amine is uniformly distributed across membranes by a mechanism of passive diffusion. The physiological role of the metabolic relationship beween β-phenylethylamine and other biogenic amines, including catecholamines, has not yet been determined.

Octopamine is an important neurohormone and neurotransmitter in invertebrates. In mammals, it is likely that endogenous octopamine functions as a cotransmitter in the peripheral nervous system. Even though the endogenous octopamine concentrations are low, and its pharmacological effects in mammals are one order of magnitude weaker than those of noradrenaline, the very rapid turnover demonstrated for octopamine indicates that it can represent a significant proportion of the total number of amine molecules produced at a given time in the adrenergic nerves, and that changes in octopamine metabolism could substantially modify the organ response to adrenergic stimulation.

As in the case with octopamine, phenylethanolamine is partially localized in sympathetic nerves of mammls, but its physiological role has not been studied in detail.

Tyramine has some, but not all, the properties necessary to be considered a neurotransmitter. Its mechanism of action is primarily related to its ability to release catecholamines. The physiological importance of the metabolic interactions between tyramine, related amines, and catecholamines has not yet been determined.

On the basis of their low concentrations in mammalian brain, octopamine and related biogenic amines have been classified as "micro", "minor" or "trace" amines. These classifiction are too restrictive, since the concentrations of the different amines vary with the species, the organs and the physiological conditions. Their turnover rates are high even in mammals, and the amines may play significant but different role in different organs and animal species. Octopamine, for example, acts as a hormone and/or as a neurotransmitter in some invertebrates and could act as a co-transmitter in some special cases in mammalian systems. A more appropriate classification for the present group of amines would be within the general and expanding group of the biogenic amines, with specification of their actions as co-transmitters, neuromodulators, neurotransmitters, or neurohormones.

K. References

Axelrod J (1962a) The enzymatic N-methylation of serotonin and other amines. J Pharmacol Exp Ther 138:28-33

Axelrod J (1962b) Purification and properties of phenylethanolamine-N-methyl transferase. J Biol Chem 237:1657-1660

Axelrod J (1963) Enzymatic formation of adrenaline and other catechols from monophenols. Science 140:499-500

Axelrod J, Saavedra JM (1977) Octopamine. Nature 265:501-504

Bakhle YS, Youdim MBH (1979) The metabolism of 5-hydroxytryptamine and β-phenylethylamine in perfused rat lung and *in vitro*. Br J Pharmacol 65:147-154

Baldessarini RJ (1971) Release of aromatic amines from brain tissues of the rat *in vitro*. J Neurochem 18:2509-2518

Baldessarini RJ, Vogt M (1971) The uptake and subcellular distribution of aromatic amines in the brain of the rat. J Neurochem 18:2519-2533

Baldessarini RJ, Vogt M (1972) Regional release of aromatic amines from tissues of the rat brain *in vitro*. J Neurochem 19:755-761

Barger A, Dale HH (1910) Chemical structure and sympathomimetic action of amines. J Physiol (Lond) 41:19-59

Barker DL, Molinoff PB, Kravitz EA (1972) Octopamine in the lobster nervous system. Nature New Biol 236:61-63

Barker DL, Kushner PD, Hooper NK (1979) Synthesis of dopamine and octopamine in the crustacean stomatogastric nervous system. Brain Res 161:99-113

Battelle BA, Kravitz EA (1978) Targets of octopamine action in the lobster: cyclic nucleotide changes and physiological effects in hemolymph, heart and exoskeletal muscle. J Pharmacol Exp Ther 205:438-448

Battelle BA, Evans JA, Chamberlain SC (1982) Efferent fibers to Limulus eyes synthesize and release octopamine. Science 216:1250-1252

Baud P, Arbilla S, Cantrill RC, Scatton B, Langer SZ (1985) Trace amines inhibit the electrically evoked release of [^{3}H]acetylcholine from slices of rat striatum in the presence of pargyline: similarities between β-phenylethylamine and amphetamine. J Pharmacol Exp Ther 235:220-229

Boakes RJ, Dua PR, Baker, GB (1976) Actions of microiontophoretically applied p- and m-tyramine on caudate neurones. Proc Meeting Eur Soc Neurochem Bath UK Abst 59C

Bodnaryk RP (1979) Basal dopamine- and octopamine stimulated adenylate cyclase activity in the brain of the moth, *Mamestra configurata*, during its metamorphosis. J Neurochem 33:275-282

Boulton AA (1976) Cerebral aryl alkyl aminergic mechanisms. In: Usdin E, Sandler M (Eds) Trace Amines and the Brain. Marcell Dekker, New York, pp 21-39

Boulton AA (1978) The tyramines: functionally significant biogenic amines or metabolic accidents? Life Sci. 23:659-672

Boulton AA (1979) Trace amines in the central nervous system. In: Tipon KF (Ed) International Review of Biochemistry. Physiological and Pharmacological Biochemistry vol 26. University Park Press, Baltimore, pp 179-206

Boulton AA, Baker GB (1975) The subcellular distribution of β-phenylethylamine, p-tyramine and tryptamine in rat brain. J Neurochem 25:477-481

Boulton AA, Dyck LE (1974) Biosynthesis and excretion of meta and para tyramine in the rat. Life Sci 14:2497-2506

Boulton AA, Juorio AV (1979) The tyramines: are they involved in the psychosis? Biol Psychiat 14:413-419

Boulton AA, Wu PH (1972) Biosynthesis of cerebral phenolic amines. I *In vivo* formation of p-tyramine, octopamine and synephrine. Can J Biochem 50:261-267

Boulton AA, Wu PH (1973) Biosynthesis of cerebral phenolic amines. II *In vivo* regional formation of p-tyramine and octopamine from tyrosine and dopamine. Can J Biochem 51:428-435

Boulton AA, Dyck LE, Durden DA (1974) Hydroxylation of β-phenylethylamine in the rat. Life Sci 15:1673-1683

Boulton AA, Juorio AV, Philips SR, WU PH (1975) Some arylalkylamines in rabbit brain. Brain Res. 96:212-216

Boulton AA, Philips SR, Durden DA, Davis BA, Baker GB (1976) The tissue and cerebral subcellular distribution of some arylalkylamines in the rat and the effect of certain drug treatments on those distributions. Adv Mass Spectrom Biochem Med 1:193-205

Boulton AA, Juorio AV, Philips SR, Wu PA (1977) The effects of reserpine and 6-hydroxydopamine on the concentrations of some arylalkylamines in rat brain. Br J Pharmacol 59:209-214

Braestrup C, Randrup A (1978) Stereotyped behavior in rats induced by phenylethylamine, dependence on dopamine and noradrenaline, and possible relation to psychoses. In Mosnaim AD, Wolf ME (Eds) Noncatecholic Phenylethylamines part 1. Phenylethylamine: Biological Mechanisms and Clinical Aspects. Marcell Dekker New York pp 245-269

Braestrup C, Andersen H, Randrup A (1975) The monoamine oxidase B inhibitor deprenyl potentiates phenylethylamine behavior in rats without inhibition of catecholamine metabolite formation. Eur J Pharmacol 34:181-187

Brandau K, Axelrod J (1972) The biosynthesis of octopamine. Naunyn-Schmiedeberg's Arch Pharmacol 273:123-133

Brandau K, Axelrod J (1973) Ring dehydroxylation and N-methylation of noradrenaline and dopamine in the intact rat brain. In: Usdin E, Snyder S (Eds) Frontiers in Catecholamine Research. Pergamon Press, New York, pp 129-131

Breen CA, Atwood HL (1983) Octopamine—a neurohormone with presynaptic activity dependent effects at crayfish neuromuscular junctions. Nature 303:716-718

Buck SH, Murphy RC, Molinoff PB (1977) The normal occurrence of octopamine in the central nervous system of the rat. Brain Res. 122:281-297

Carlson AD (1969) Neural control of firefly luminescence. Adv Insect Physiol 6:51-96

Carpenter DO, Gaubatz GL (1974) Octopamine receptors on *Aplysia* neurones mediate hyperpolarization by increasing membrane conductance. Nature 252:483-485

Christensen TA, Sherman TG, McCaman RE, Carlson AD (1983) Presence of octopamine in firefly photomotor neurons. Neuroscience 9:183-189

Coulon JF, Lafon-Cazal M, David JC (1984) In vitro occurrence of m-octopamine in the cultured cephalic ganglion of Locusta migratoria L. after L-dopa administration. Comp Biochem Physiol 78C:77-80

Creveling CR, Daly JW, Witkop B, Udenfriend S (1962a) Substrates and inhibitors of dopamine-β-oxidase. Biochem Biophys Acta 64:125-134

Creveling CR, Levitt M, Udenfriend S (1962b) An alternative route for the biosynthesis of norepinephrine. Life Sci 10:523-526

Danielson TJ, Wishart TB, Boulton AA (1976) Effect of acute and chronic injection of amphetamine on intracranial self-stimulation (ICS) and some arylalkylamines in rat brain. Life Sci 18:1237-1243

Danielson TJ, Boulton AA, Robertson HA (1977a) m-Octopamine, p-octopamine and phenylethanolamine in rat brain: a sensitive, specific assay and the effects of some drugs. J Neurochem 29:1131-1135

Danielson TJ, Wishart TB, Robertson HA, Boulton AA (1977b) Effect of acute and chronic injections of amphetamine on intracranial self stimulation: amphetamine levels and effects upon some arylalkylamines in rat brain. Prog Neuropsychopharmacol 1:279-284

David JC (1984) Relationship between phenolamines and catecholamines during rat brain embryonic development in vivo and in vitro. J Neurochem 43:668-674

David JC, Coulon JF (1985) Octopamine in invertebrates and vertebrates. A Review. Progress in Neurobiology 24:141-185

David JC, Dairman W, Udenfriend S (1974) On the importance of decarboxylase in the metabolism of phenylalanine, tyrosine and tryptophan. Arch Biochem Biophys 160:561-568

Davis BA, Yu EH, Carson K, O'Sullivan K, Boulton AA (1982) Plasma levels of phenylacetic acid, m- and p-hydroxyphenyl acetic acid and platelet monoamine oxidase activity in schizophrenic and other patients. Psychiat Res 6:97-105

Dudai Y, Zvi S (1984) High affinity [^{3}H]octopamine binding sites in Drosophila melanogaster: interaction with ligands and relationship to octopamine receptors. Comp Biochem Physiol 772:145-151

Durden DA, Boulton AA (1982) Identification and distribution of phenylacetic acid in the brain of the rat. J Neurochem 38:1532-1536

Durden DA, Philips SR, Boulton AA (1973) Identification and distribution of β-phenylethylamine in the rat. Can J Biochem 51:995-1002

Edwards DJ, Blau K (1972a) Analysis of phenylethylamines in biological tissues by gas-liquid chromatography with electrocapture detection. Anal Biochem 45:387-402

Edwards DJ, Blau K (1972b) Aromatic acids derived from phenylalanine in the tissues of rats with experimentally induced phenylketonuria-like characteristics. Biochem J 130:495-503

Edwards DJ, Blau K (1973) Phenethylamines in brain and liver of rats with experimentally induced phenylketonuria-like characteristics. Biochem J 132:95-100

Edwards DJ, Rizk M (1979) Identification and quantification of phenylethylene glycol in human and rat urine, and its elevation in phenylketonuria. Clin Chim Acta 95:1-10

Evans PD (1978) Octopamine: a high-affinity uptake mechanism in the nervous system of the cockroach. J Neurochem 30:1015-1022

Evans PD (1981) Multiple receptor types for octopamine in the locust. J Physiol (Lond) 318:99-122

Evans PD (1984a) The role of cyclic nucleotides and calcium in the mediation of the modulatory effects of octopamine on locust skeletal muscle. J Physiol (Lond) 348:325-340

Evans PD (1984b) Studies on the mode of action of octopamine, 5-hydroxytryptamine and proctolin on a myogenic rhythm in the locus. J Exp Biol 110:231-251

Evans PD, Gee JD (1980) Action of formamidine pesticides on octopamine receptors. Nature 287:60-62

Evans PD, O'Shea M (1977) An octopaminergic neurone modulates neuromuscular transmission in the locust. Nature 270:257-259

Evans PD, Talamo BR, Kravitz EA (1975) Octopamine neurons: morphology, release of octopamine and possible physiological role. Brain Res 90:340-347

Evans PD, Kravitz EA, Talamo BR (1976) Octopamine release at two points along lobster nerve trunks. J Physiol (Lond) 262:71-89

Evans PD, Kravitz EA, Talamo BR, Wallace BG (1976) The association of octopamine with specific neurons along lobster nerve trunks. J Physiol (Lond) 262:57-70

Farnham PJ, Novak RA, McAdoo DJ (1978) A re-examination of the distributions of octopamine and phenylethanolamine in the *aplysia* nervous system. J Neurochem 30:1173–1176

Fischer JE, Baldessarini RJ (1971) False neurotransmitters and hepatic failure. Lancet ii 75–80

Fischer JE, James JH (1972) Treatment of hepatic coma and hepatorenal syndrome. Am J Surg 123:222–230

Flatmark T, Skotland T, Jones T, Ingebretsen OC (1978) Fluorimetric detection of octopamine in high-performance liquid chromatography and its application to the assay of dopamine *beta*-monooxygenase in human serum. J Chromatogr 146:433–438

Florey E, Rathmayer M (1978) The effects of octopamine and other amines on the heart and on neuromuscular transmission in decapod crustaceans: further evidence for a role as neurohormone. Comp Biochem Physiol [C] 61c: 229–237

Fregly MJ, Kelleher DL, Williams CM (1979) Adrenergic activity of ortho-, meta-, and para-octopamine. Pharmacology 18:180–187

Fuxe K, Grobecker H, Jonsson J (1967) The effect of β-phenylethylamine on central and peripheral monoamine-containing neurons. Eur J Pharmacol 2:202–207

Fuzeau-Braesch S, Coulon JF, David JC (1979) Octopamine levels during the moult cycle and adult development in the migratory locust, *Locusta migratoria*. Experientia 15:1349–1350

Glanzman DL, Krasne FB (1983) Serotonin and octopamine have opposite modulatory effects on the crayfish's lateral giant escape reaction. J Neurosci 3:2263–2269

Harris-Warrick RM, Kravitz EA (1984) Cellular mechanisms for modulation of posture by octopamine and serotonin in the lobster. J Neuroscience 4:1976–1993

Henwood RW, Boulton AA, Phillis JW (1979) Iontophoretic studies of some trace amines in the mammalian CNS. Brain Res. 164:347–351

Hicks TP, McLennan H (1978) Comparison of the actions of octopamine and catecholamines on single neurones of the rat cerebral cortex. Br J Pharmacol 64:485–491

Hiripi L, Rozsa KS (1984) Octopamine- and dopamine-sensitive adenylate cyclase in the brain of Locusta migratoria during its development. Cell Molec Neurobiol 4:199–206

Horwitz D, Lovenberg W, Engelman K, Sjoerdsma A (1964) Monoamine oxidase inhibitors, tyramine and cheese. J Am Med Assoc 190:1133–1136

Hoyle G (1975) Evidence that insect dorsal unpaired median (DUM) neurons are octopaminergic. J Exp Zool 193:425–431

Hoyle G, Colquhoun W, Williams M (1980) Fine structure of an octopaminergic neuron and its terminals. J Neurobiol 11:103–126

Huang JT, Ho BT (1974) The effect of pretreatment with iproniazid on the behavioral activities of β-phenylethylamine in rats. Psychopharmacologia 35:71–81

Hueber ND, Boulton AA (1979) Longitudinal urinary trace amine excretion in a human male. J Chromatogr. 162:169–176

Ibrahim KE, Couch MW, Williams CM, Fregly MJ, Midgley JM (1985) m-Octopamine: normal occurrence with p-octopamine in mammalian sympathetic nerves. J Neurochem 44:1862–1867

Jackson DM, Smythe DB (1973) The distribution of β-phenylethylamine in discrete regions of the rat brain and its effect on brain noradrenaline, dopamine and 5-hydroxytryptamine levels. Neuropharmacology 12:663–668

Jackson DM, Andén NE, Dahlström A (1975) A functional effect of dopamine in the nucleus accumbens and in some other dopamine-rich parts of the rat brain. Psychopharmacologia 45:139–149

James JH, Fischer JE (1975) Release of octopamine and α-methyl octopamine by L-DOPA. Biochem Pharmacol 24:1099–1101

James JH, Hodgman JM, Funovics JM, Fischer JE (1975) Alterations in brain octopamine and brain tyrosine following portacaval anastomosis in rats. J Neurochem 27:223-227

James MO, Smith RL (1973) The conjugation of phenylacetic acid in phenylketonurics. Eur J Clin Pharmacol 5:243-246

Jepson B, Lovenberg W, Zaltman P, Oates S, Sjoerdsma A, Udenfriend S (1960) Amine metabolism studied in normal and phenylketonuric humans by monoamineoxidase inhibition. Biochem J 74:5P

Jonsson J, Grobecker H, Holtz P (1966) Effect of β-phenylethylamine on content and subcellular distribution of norepinephrine in rat heart and brain. Life Sci 5:2235-2246

Jonsson J, Lindecke B, Cho AK (1975) Oxidation of phenylethylamine yielding tyramine by rat liver microsomes. Acta Pharmacol Toxicol Kbh 37:352-360

Juorio AV (1976) Presence and metabolism of β-phenylethylamine, p-tyramine, m-tyramine and tryptamine in the brain of the domestic fowl. Brain Res 111:442-445

Juorio AV (1977) Effects of d-amphetamine and antipsychotic drug administration on striatal tyramine levels in the mouse. Brain Res 126:181-184

Juorio AV (1983) The effect of some decarboxylase inhibitors on striatal tyramines in the mouse. Neuropharmacology 22:71-73

Juorio AV, Danielson TJ (1978) Effect of haloperidol and d-amphetamine on cerebral tyramine and octopamine levels. Eur J Pharmacol 50:79-82

Juorio AV, Jones RS (1981) The effect of mesencephalic lesions on tyramine and dopamine in the caudate nucleus of the rat. J Neurochem 36:1898-1903

Juorio AV, Molinoff PB (1974) The normal occurrence of octopamine in neural tissues of the *Octopus* and other cephalopods. J Neurochem 22:271-280

Juorio AV, Philips SR (1975) Tyramines in octopus nerves. Brain Res 83:180-184

Kakimoto Y, Armstrong MD (1962a) The phenolic amines of human urine. J Biol Chem 237:208-214

Kakimoto Y, Armstrong MD (1962b) On the identification of octopamine in mammals. J Biol Chem 237:422-427

Karoum F, Gillin JC, Wyatt RJ (1975) Mass fragmentographic determination of some acidic and alcoholic metabolites of biogenic amines in the rat brain. J Neurochem 25:653-658

Kass L, Barlow RB Jr (1984) Efferent neurotransmission of circadian rhythms in Limulus lateral eye. J Neurosc 4:908-917

Kaupp UB, Malbon CC, Batelle BA, Brown JE (1982) Octopamine stimulated rise of cAMP in Limulus ventral photoreceptors. Vision Res 22:1503-1506

King GS, Goodwin BL, Ruthven CRJ, Sandler M (1974) Urinary excretion of o-tyramine. Clin Chim Acta 51:105-107

Klaasen LW, Kammer AE (1985) Octopamine enhances neuromuscular transmission in developing and adult moths, Manduca sexta. J Neurobiol 16:227-243

Konishi S, Kravitz EA (1978) The physiological properties of amine-containing neurones in the lobster nervous system. J Physiol (Lond) 279:215-229

Kopin IJ (1968) False adrenergic transmitters. Annu Rev Pharmacol 8:377-394

Lafon-Cazal M, Bockaert J (1985) Pharmacological characterization of octopamine-sensitive adenylate cyclase in the flight muscle of Locusta migratoria L. Eur J Pharmacol 119:53-59

Lam KC, Tall AR, Goldstein GB, Mistilis SP (1973) Role of a false neurotransmitter, octopamine, in the pathogenesis of hepatic and renal encephalopathy. Scand J Gastroenterol 8:465-472

Lange AB, Orchard I (1986) Identified octopaminergic neurons modulate contractions

of locust visceral muscle via adenosine 3′, 5′-monophosphate (cyclic AMP). Brain Res 363:340-349
Lemberger L, Klutch A, Kuntzman R (1966) The metabolism of tyramine in rabbits. J Pharmacol Exp Ther 153:183-190
Lentzen H, Philippu A (1977) Uptake of tyramine into synaptic vesicles of the caudate nucleus. Naunyn Schmiedeberg's Arch Pharmacol 300:25-30
Levitan IB, Barondes SH (1974) Octopamine- and serotonin-stimulated phosphorylation of specific protein in the abdominal ganglion of *Aplysia californica.* Proc Natl Acad Sci USA 71:1145-1148
Loo YH, Scotto L, Horning MG (1976) Gas chromatographic determination of aromatic metabolites of phenylalanine in brain. Anal Biochem 76:111-118
Loo YH, Fulton T, Miller K, Wisniewski HM (1980) Phenylacetate and brain dysfunction in experimental phenylketonuria: synaptic development. Life Sci 27:1283-1290
Lovenberg W, Weissbach H, Udenfriend S (1962) Aromatic L-amino acid decarboxylase. J Biol Chem 237:89-93
Mann J, Gershon S (1980) L-deprenyl, a selective monoamine oxidase type-B inhibitor in endogenous depression. Life Sci 17:877-882
Mantegazza P, Riva M (1963) Amphetamine-like activity of β-phenylethylamine after a monoamine oxidase inhibitor *in vivo*. J Pharm Pharmacol 15:472-478
McEwen F, Hume A, Hutchinson M, Holland W (1969) The effects of N-methylation on the pharmacological activity of phenethylamine. Arch Int Pharmacodyn 179:86-93
McKean CM (1972) The effects of high phenylalanine concentration on serotonin and catecholamine metabolism in the human brain. Brain Res 47:469-476
Midgley JM, Couch MW, Crowley JR, Williams CM (1980) m-Synephrine: normal occurrence in adrenal gland. J Neurochem 34:1225-1230
Mir AK, Vaughan PFT (1981) Biosynthesis of N-acetyldopamine and N-acetyloctopamine by schistocerca gregaria nervous tissue. J Neurochem 36:441-446
Moffett A, Swash M, Scott DF (1972) Effect of tyramine in migraine: a double blind study. J Neurol Neurosurg Psychiat 35:496-499
Moises HW, Waldmeier P, Beckmann H (1985) Phenylethylamine and personality. In: Boulton AA, Maitre L, Bieck PR, Riederer P, (Eds) Neuropsychopharmacology of the trace amines. Humana Press, Clifton, NJ USA pp 387-394
Molinoff PB, Axelrod J (1969) Octopamine: normal occurrence in sympathetic nerves of rats. Science 164:428-429
Molinoff PB, Axelrod J (1972) Distribution and turnover of octopamine in tissues. J Neurochem 19:157-163
Molinoff PB, Buck SH (1976) Octopamine: normal occurrence in neuronal tissues of rats and other species. In: Sandler M, Usdin E (Eds) Trace Amines in the Brain. Marcel Dekker, New York pp 131-160
Molinoff PB, Landsberg L, Axelrod J (1969) An enzymatic assay for octopamine and other β-hydroxylated phenylethylamines. J Pharmacol Exp Ther 170:253-261
Morton DB (1984) Pharmacology of the octopamine-stimulated adenylate cyclase of the locust and tick CNS. Comp Biochem Physiol 78C:153-158
Nakajima T, Kakimoto Y, Sano I (1964) Formation of β-phenylethylamine in mammalian tissue and its effect on motor activity in the mouse. J Pharmacol Exp Ther 143:319-325
Nathanson JA (1979) Octopamine receptors, adenosine 3′, 5′-monophosphate, and neural control of firefly flashing. Science 203:65-68
Nathanson JA (1985a) Characterization of octopamine-sensitive adenylate cyclase:

elucidation of a class of potent and selective octopamine-2 receptor agonists with toxic effects in insects. Proc Natl Acad Sci USA 82:599-603
Nathanson JA (1985b) Phenyliminoimidazolidines. Characterization of a class of potent agonists of octopamine-sensitive adenylate cyclase and their use in understanding the pharmacology of octopamine receptors. Mol Pharmacol 28:254-268
Nathanson JA, Greengard P (1973) Octopamine-sensitive adenylate cyclase: evidence for a biological role of octopamine in nervous tissue. Science 189:308-310
Nathanson JA, Hunnicutt EJ (1979) Neural control of light emission in *Photuris larvae*: identification of octopamine-sensitive adenylate cyclase (1). J Exp Zool 208:255-262
Nathanson JA, Hunnicutt EJ (1981) N-demethylchlordimeform: a potent partial agonist of octopamine-sensitive adenylate cyclase. Mol Pharmacol 20:68-75
Oates JA, Nirenberg PZ, Jepson JB, Sjoerdsma A, Udenfriend S (1963) Conversion of phenylalanine to phenylethylamine in patients with phenylketonuria. Proc Soc Exp Biol 112:1078-1081
Orchard I, Lange AB (1985) Evidence for octopamine modulation of an insect visceral muscle. J Neurobiol 16:171-181
O'Shea M, Evans PD (1979) Potentiation of neuromuscular transmission by an octopaminergic neurone in the locust. J Exp Biol 79:169-190
Perry TL (1962) Urinary excretion of amines in phenylketonuria and mongolism. Science 136:879-880
Perry TL, Hestrin M, MacDougall L, Hansen S (1966) Urinary amines of intestinal bacterial origin. Clin Chim Acta 14:116-123
Philips SR, Boulton AA (1979) The effect of monoamine oxidase inhibitors on some arylalkylamines in rat striatum. J Neurochem 33:159-167
Philips SR, Juorio AV (1978) Arylalkylamines in the adrenal medulla. Can J Biochem 56:1058-1060
Philips SR, Durden DA, Boulton AA (1974) Identification and distribution of p-tyramine in the rat. Can J Biochem 52:366-373
Philips SR, Davis BA, Durden DA, Boulton AA (1975) Identification and distribution of m-tyramine in the rat. Can J Biochem 53:65-69
Philips SR, Rozdilsky B, Boulton AA (1978) Evidence for the presence of m-tyramine, tryptamine, and phenylethylamine in the rat brain and several areas of the human brain. Biol Psychiat 13:51-57
Pisano JJ, Oates JA, Karmen A, Sjoerdsma A, Udenfriend S (1961) Identification of p-hydroxy-α-(methylaminomethyl) benzyl alcohol (synephrine) in human urine. J Biol Chem 236:898-901
Potkin SG, Karoum F, Chuang LW, Cannon-Spoor HE, Phillips I, Wyatt RJ (1979) Phenylethylamine in paranoid chronic schizophrenia. Science 206:470-471
Raiteri M, Bertollini A, Levi G (1977) Effect of sympathomimetic amines on the synaptosomal transport of noradrenaline, dopamine and 5-hydroxytryptamine. Eur J Pharmacol 41:133-143
Reale V, Evans PD, Villegas J (1986) Octopaminergic modulation of the membrane potential of the Schwann cell of the squid giant nerve fibre. J Exp Biol 121:421-443
Reynolds GP, Riederer P, Sandler M, Jellinger K, Seemann D (1978) Amphetamine and 2-phenylethylamine in post-mortem Parkinsonian brain after (−) deprenyl administration. J Neural Transm 43:271-277
Ross SB, Renyi AL (1971) Uptake and metabolism of β-phenethylamine and tyramine in mouse brain and heart slices. J Pharm Pharmacol 23:276-279

Roth JA, Gillis CN (1974) Deamination of β-phenylethylamine by monoamineoxidase, inhibition by imipramine. Biochem Pharmacol 23:2537-2545

Rutledge CO, Jonason J (1967) Metabolic pathways of dopamine and norepinephrine in rabbit brain *in vitro.* J Pharmacol Exp Ther 157:493-502

Saavedra JM (1974a) Enzymatic isotopic assay for and presence of β-phenylethylamine in brain. J Neurochem 22:211-216

Saavedra JM (1974b) Enzymatic-isotopic method for octopamine at the picogram level. Anal Biochem 59:628-633

Saavedra JM (1977) Microassay of biogenic amines in neurons of *Aplysia*: the coexistence of more than one transmitter molecule in a neuron. In: Osborne N (Ed) Biochemistry of Characterized Neurons. Pergamon Press, Elmsford NY, USA pp 217-238

Saavedra JM (1978) β-Phenylethylamine: is this biogenic amine related to neuropsychiatric diseases? In: Moshaim AD, Wolf ME (Eds) Modern Pharmacology-Toxicology, Vol. 12, Noncatecholic Phenylethylamines. Part 1. Phenylethylamine: Biological Mechanisms and Clinical Aspects. Marcel Dekker, New York, pp 139-157

Saavedra JM, Axelrod J (1973) Demonstration and distribution of phenylethanolamine in brain and other tissues. Proc Natl Acad Sci USA 70:769-772

Saavedra JM, Axelrod J (1976) Octopamine as a putative neurotransmitter. In: Costa E, Giacobini E, Paoletti R (Eds) Advances in Biochemical Psychopharmacology, Vol. 15 Raven Press, New York, pp 95-110

Saavedra JM, Coyle JT, Axelrod J (1973) The distribution and properties of the nonspecific N-methyltransferase in brain. J Neurochem 20:743-752

Saavedra JM, Brownstein MJ, Carpenter DO, Axelrod J (1974a) Octopamine: presence in single neurons of *Aplysia* suggests neurotransmitter function. Science 185:364-365

Saavedra JM, Coyle JT, Axelrod J (1974b) Developmental characteristics of phenylethanolamine and octopamine in the rat brain. J Neurochem 23:511-515

Saavedra JM, Palkovits M, Brownstein MJ, Axelrod J (1974c) Localization of phenylethanolamine N-methyl transferase in the rat brain nuclei. Nature 248:695-696

Saavedra JM, Ribas J, Swann J, Carpenter DO (1977) Phenylethanolamine: a new putative transmitter in *Aplysia.* Science 195:1004-1006

Sandler M, Bonham Carter S, Goodwin BL, Ruthven CRJ (1976) Trace amine metabolism in man. In: Usdin E, Sandler M (Eds) Trace Amines and the Brain. Marcel Dekker, New York, pp 233-281

Sandler M, Ruthven CR, Goodwin GL, Field H, Matthews R (1978a) Phenylethylamine overproduction in aggressive psychopaths. Lancet 2:1269-1270

Sandler M, Ruthven CR, Goodwin BL, King GS, Pettit BR, Renyolds GP, Tyrer SP, Weller MP, Hirsch SR (1978b) Raised cerebrospinal fluid phenylacetic acid concentration: preliminary support for the phenylethylamine hypothesis of schizophrenia? Commun Psychopharmacol 2:199-202

Sandler M, Ruthven CR, Goodwin BL, Coppen A (1979) Decreased cerebrospinal fluid concentration of free phenylacetic acid in depressive illness. Clin Chim Acta 93:169-171

Sandler M, Ruthven CR, Goodwin BL, Lees A, Stern GM (1982) Phenylacetic acid in human body fluids: high correlation between plasma and cerebrospinal fluid concentration values. J Neurol Neurosurg Psychiat 45:366-368

Sastry BSR, Philis JW (1977) Antagonism of biogenic amine-induced depression of cerebral cortical neurons by Na^+-K^+-ATPase inhibitors. Can J Physiol Pharmacol 55:170-179

Schaefer A, Unyi G, Pfeifer AK (1972) The effects of a soluble factor and of catechol-

amines on the activity of adenosine triphosphatase in subcellar fractions of rat brain. Biochem Pharmacol 21:2289-2294
Scheline R, Williams R, Wit J (1960) Biological dehydroxylation. Nature 188:849-850
Schofield PK, Treherne JE (1985) Octopamine reduces potassium permeability of the glia that form the insect blood-brain barrier. Brain Res 360:344-348
Schramm M, Feinstein H, Naim E, Lang M, Lasser M (1972) Epinephrine binding to the catecholamine receptor and activation of the adenylate cyclase in erythrocyte membranes. Proc Natl Acad Sci USA 69:523-527
Schweitzer JW, Friedhoff AJ (1978) A critique of current methods for the analysis of phenethylamine in biological media. In: Mosnaim AD, Wolf ME (Eds) Noncatechol Phenylethylamines, part 1. Phenylethylamine: Biological Mechanisms and Clinical Aspects. Marcel Dekker, New York, pp 475-488
Schweitzer JW, Friedhoff AJ, Schwartz R (1975) Phenethylamine in normal urine: failure to verify high values. Biol Psychiat 10:277-285
Shalita B, Dikstein S (1977) Central tyramine prevents hypertension in uninephrectomized DOCA-saline treated rats. Experientia 33:1430-1431
Slingsby JM, Boulton AA (1976) Separation and quantitation of some urinary arylalkylamines. J Chromatogr 123:51-56
Smith AA, Fabrykant M, Kaplan M, Gavitt J (1964) Dehydroxylation of some catecholamines and their products. Biochim Biophys Acta 86:429-437
Smith I, Mitchell PD (1974) The effect or oral inorganic sulphate on the metabolism of 4-hydroxyphenethylamine (tyramine) in man: tyramine o-sulphate measurement in human urine. Biochem J 142:189-191
Stoof JC, Liem AL, Mulder AH (1976) Release and receptor stimulating properties of p-tyramines in rat brain. Arch Int Pharmacodyn 220:62-71
Tacker M, McIsaac WM, Creaven PJ (1970) Metabolism of tyramine-1-^{14}C by the rat. Biochem Pharmacol 19:2763-2773
Tallman JF, Saavedra JM, Axelrod J (1976a) A sensitive enzymatic-isotopic method for the analysis of tyramine in brain and other tissues. J Neurochem 27:465-469
Tallman JF, Saavedra JM, Axelrod J (1976b) Biosynthesis and metabolism of endogenous tyramine and its normal presence in sympathetic nerves. J Pharmacol Exp Ther 199:216-221
Tong JH, Smyth RG, D'Iorio A (1979) Metabolism of m-tyrosine in the rat. Biochem Pharmacol 28:1029-1036
Trendelenburg U (1972) Classification of sympathomimetic amines. In: Blaschko H, Muscholl E (Eds) Catecholamines (Handbook of Experimental Pharmacology, vol 33) Springer-Verlag, Berlin, Heidelberg, New York pp 336-362
Ungar F, Mosnaim AD, Ungar B, Wolf ME (1978) Preliminary studies of the sodium borohydride stabilizable binding of phenylethylamine and tyramine to brain preparations. Res Commun Chem Pathol Pharmacol 19:427-434
U'Prichard DC, Greenberg DA, Snyder SH (1977) Binding characteristics of a radiolabelled agonist and antagonist at central nervous system alpha noradrenergic receptors. Mol Pharmacol 13:454-473
Wallace BG (1976) The biosynthesis of octopamine-characterization of lobster tyramine β-hydroxylase. J Neurochem 26:761-770
Wallace BG, Talamo BR, Evans PD, Kravitz EA (1974) Octopamine: selective association with specific neurons in the lobster nerve system. Brain Res 74:349-355
Williams CM, Couch MW (1978) Identification of ortho-octopamine and meta-octopamine in mammalian adrenal and salivary gland. Life Sci 22:2213-2120
Willner J, Lefevre H, Costa E (1974) Assay by multiple ion detection of β-phenylethylamine and phenylethanolamine in rat brain. J Neurochem 23:857-859

Woodruff GN (1978) Biochemical and pharmacological studies on dopamine receptors. Adv Biochem Psychopharmacol 19:89–118
Wu PH, Boulton AA (1974) Distribution, metabolism and disappearance of intraventricularly injected p-tyramine in the rat. Can J Biochem 52:374–381
Wu PH, Boulton AA (1975) Metabolism, distribution and disappearance of injected β-phenylethylamine in the rat. Can J Biochem 53:42–50
Wu PH, Boulton AA (1979) N-Acylation of tyramines: purification and characterization of an arylamine N-acetyltransferase from rat brain and liver. Can J Biochem 57:1204–1209
Yang HYT, Neff NH (1973) β-Phenylethylamine: a specific substrate for type B monoamineoxidase in brain. J Pharmacol Exp Ther 187:365–371
Yang HYT, Neff NH (1976) Brain N-acetyltransferase: substrate specificity, distribution and comparison with enzyme activity from other tissues. Neuropharmacology 15:561–564

CHAPTER 15

Plasma Levels of Catecholamines and Dopamine-β-Hydroxylase

I. J. KOPIN

A. Introduction

Shortly after EULER (1948) demonstrated that noradrenaline was the catecholamine present in mammalian sympathetic nerves, PEART (1949) showed that sympathetic nerve stimulation evoked release of this catecholamine into the effluent blood from a perfused bovine spleen. At that time, biological assays were used to assay the catecholamines present in extracts or in perfusates (GADDUM 1959), but generally these methods were inadequante for routine assay of levels of catecholamines normally found in plasma. It was not until the development of biochemical methods for assay that alterations in levels of human plasma catecholamines could be examined.

B. Measurement of Plasma Catecholamines

Although many methods for analysis of catecholamines have been developed, only a few have been consistently used by more than one or two laboratories. The fluorimetric methods dominated the field for about twenty years, until the introduction of more convenient, more sensitive, and highly specific radioenzymatic methods. The most recent development is use of electrochemical detection combined with high pressure liquid chromatography (HPLC). Other methods based on radioimmunoassay, thin-layer chromatography, or gas chromatography coupled with mass spectrometry are less sensitive, require special reagents or expensive instruments, or are used only occasionally.

I. Fluorimetric Method

The first methods which approached the sensitivity required to assay the minute quantities of catecholamines in normal human plasma were fluorimetric methods described by LUND (1950) and WEIL-MALHERBE and BONE (1953). In both methods, adrenaline and noradrenaline are separated from serum or plasma by adsorption on alumina, eluted with acid, and converted to fluorescent derivatives which are estimated fluorimetrically. The method described by LUND (1950), usually referred to as the trihydroxyindole method, involves oxidation of the catecholamines to their corresponding chrome derivatives followed by conversion to the highly fluorescent trihydroxyindole using strong alkali in the presence of an antioxidant. WEIL-MALHERBE and BONE (1953) used ethylenediamine to convert the catecholamines to fluorescent derivatives. Although the ethylenediamine condensation method is more sensitive, it is also less specific and the values obtained were several-fold greater than those obtained using the trihydroxyindole method; the ethylenediamine method was not widely used. Modifications of the details of the procedures for extraction and derivatization of the catecholamines enhanced the sensitivity and specificity of the assay (see EULER 1959) so that it was possible to measure the levels of plasma noradrenaline and adrenaline with considerable accuracy, although as much as 30 ml of blood was required. Normal human venous plasma noradrenaline was about 2 pmol/ml and adrenaline, 1.2 pmol/ml (COHEN and GOLDENBERG 1957). Methods which do not have a sensitivity below 0.3 pmol/ml plasma are useless for most studies.

Subsequent modifications (ANTON and SAYRE 1962; HÄGGENDAL 1963; RENZINI et al. 1970) of the trihydroxyindole method allowed use of smaller volumes of blood, but since the introduction of less tedious, more highly sensitive and specific methods, most investigators have abandoned the fluorimetric methods.

II. Radioenzymatic Methods

The fluorimetric techniques fall short of the requirements for measurements of catecholamines in small fragments of tissue or small volumes of plasma (e.g. rat bood). Radioisotopic techniques employ isotopically labelled reagents

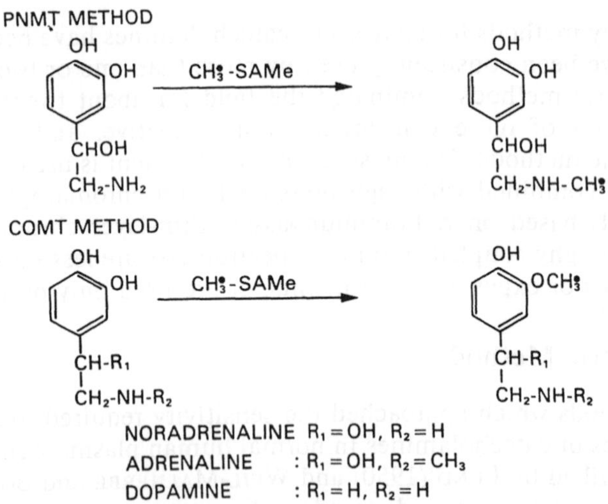

Fig. 1. Radioenzymatic methods used to measure plasma levels of catecholamines. The PNMT method measures only noradrenaline by converting the catecholamine to radioactively labelled adrenaline which is isolated by adsorption on alumina. In the COMT method the catecholamines are O-methylated, separated from the labelled precursor by extraction into an organic solvent and then separated from each other by a chromatographic procedure and the radioactivity assayed by liquid scintillation spectrometry

with high specific activity to convert the catecholamines to labelled derivatives which are separated and assayed. These methods have extended the limits of sensitivity of the assays so that catecholamines in relatively small volumes of plasma can be measured precisely. The specificity of the methods and efficiency of isotope utilization are enhanced by using enzymes to effect the derivatization. Both of the most widely used radioenzymatic methods involve transfer of radio-labelled methyl groups from S-adenosylmethionine to the catecholamines to be measured (Fig. 1). SAELENS et al. (1967) used phenylethanolamine-N-methyl transferase (PNMT) to catalyze formation of labelled adrenaline from labelled S-adenosylmethionine and noradrenaline in small segments of mouse brain, whereas ENGELMAN and PORTNOY (1970) used catechol-O-methyltransferase (COMT) to convert the catecholamines in plasma to the corresponding O-methylated derivatives. Subsequently these methods have been modified to provide convenient assays for plasma catecholamines.

1. PNMT Method

In this method, PNMT, partially purified from adrenal medullae, is used to enzymatically N-methylate noradrenaline with radiolabelled methyl groups from S-adenosylmethionine. The product, adrenaline, is isolated and the isotope content assayed by liquid scintillation spectrometry. Substitution of bovine for rabbit PNMT and use of tritium- rather than ^{14}C-labelled S-adenosylmethionine markedly enhances the sensitivity of this method (IVERSEN and JARROT 1970). Subsequent modifications (HENRY et al. 1975; LAKE et al. 1976) of the PNMT method have provided a relatively convenient and rapid method for assay of plasma noradrenaline levels. Since PNMT is not entirely specific for noradrenaline, octopamine, metanephrine, and adrenaline are also partly N-methylated. After N-methylation the labelled catechols are absorbed on alumina so that octopamine and metanephrine do not interfere. The PNMT method, however, yields slightly greater noradrenaline levels than the COMT method, possibly because some adrenaline, either present in the plasma or formed from noradrenaline, is N-methylated (to a dimethyl derivative) and contributes to the radioisotope measured with the tritium labelled adrenaline.

2. COMT Method

ENGLEMAN and PORTNOY (1970) introduced the use of COMT prepared from rat liver for assay of plasma catecholamines. Because the extent of O-methylation varied, they developed a double-isotope method. Tritiated adrenaline and noradrenaline were added to the sample to be assayed, the catecholamines extracted and eluted from an ion exchange resin and the eluate freeze-dried. The catecholamines in the dried residue were then converted to the O-methylated analogues using ^{14}C-S-adenosylmethionine and COMT. The radiolabelled products were separated from the incubation mixture and oxidized to vanillin using alkaline periodate. The ratio of $^{14}C/^{3}H$ in the vanillin was used to es-

Table 1. Plasma levels of catecholamines in reclining humans measured by a COMT method

	Dopamine	Noradrenaline	Adrenaline
CALLINGHAM and BARRAND (1976)	1.31 ± 0.05 (10)	1.72 ± 0.30	0.93 ± 0.11
CHRISTENSEN et al. (1976)	2.16 ± 0.39 (6)	0.95 ± 0.06 (14)	0.22 ± 0.05
DA PRADA and ZÜRCHER (1976)	0.78 ± 0.13 (13)	1.12 ± 0.14	0.36 ± 0.05
WISE and KOPIN (1976)	0.22 ± 0.04 (13)	1.23 ± 0.10	0.37 ± 0.05
FRANCO-MORSELLI et al. (1977)	0.30 ± 0.04 (11)	1.46 ± 0.33	0.22 ± 0.03
PEULER and JOHNSON (1977)	0.22 ± 0.05 (15)	1.65 ± 0.26	0.12 ± 0.03
BÜHLER et al. (1978)	0.64 ± 0.13 (11)	1.20 ± 0.06	0.36 ± 0.03

Results are expressed as pmol/ml and are mean values (± SEM) for the number of subjects shown in parentheses. In the study by CHRISTENSEN et al. (1976) only six of the fourteen subjects had dopamine determined.

timate the level of total catecholamines (expressed as noradrenaline equivalents). If noradrenaline and adrenaline were to be assayed separately, the labelled metanephrines were extracted by ion exchange from the incubation mixture and were separated by thin-layer chromatography prior to vanillin formation.

In 1973, PASSON and PEULER used high specific activity ^{3}H-methyl-S-adenosylmethionine to increase the sensitivity of the COMT method. The extent of O-methylation varied somewhat so that this modification lacked the precision of the double-isotope method. There was good agreement, however, between the results obtained with the double-isotope (ENGELMAN and PORTNOY 1970; PEDERSEN and CHRISTENSEN 1975) and single-isotope methods (PASSON and PEULER 1973; CRYER et al. 1974). Improvements in the derivatization procedure (WEISE and KOPIN 1976) and in extraction and separation of the methylated products by thin-layer chromatography (DA PRADA and ZÜRCHER 1976; PEULER and JOHNSON 1977) or by high performance liquid chromatography (KLANIECKI 1977; ENDERT 1979; MULLER et al. 1979) have diminished further the variability and increased the precision and sensitivity of the single isotope COMT method. When the single and double isotope methods were compared in the same laboratory (EVANS et al. 1978) they were found to yield virtually identical values and to be equally reliable, but the single isotope method required one-half as much time and only 1/200th the sample size the double-isotope method.

The COMT methods have the further advantage that dopamine is readily assayed along with noradrenaline and adrenaline. After enzymatic O-methylation, 3-methoxy-tyramine, the O-methylated derivative of dopamine, is separated from the other labelled derivatives and its content of labelled methyl groups determined. Dopamine levels in plasma are usually quite low (probably less than 0.7 pmol/ml) but large amounts of conjugated dopamine are present (Buu and Kuchel 1977). Precautions are necessary to prevent hydrolysis and to prevent decarboxylation of endogenous dopa if the enzyme preparation used is contaminated with AADC (Peuler and Johnson 1977; Sole and Hussain 1977).

When measured by the radioenzymatic methods, plasma catecholamines are not affected significantly by moderate degrees of hemolysis (Smith 1980). Furthermore, catecholamines appear to be stable when stored in plasma or whole blood, even without added antioxidants (Pettersson et al. 1980). In some blood samples kept at room temperature for a few hours, high levels may be obtained perhaps because conjugates hydrolyze. Values for plasma catecholamines obtained by use of the COMT method are shown in Table 1.

III. High Performance Liquid Chromatography with Electro-Chemical Detection (HPLC-ED)

The most recent innovation in measurement of catecholamines has been the introduction of electrochemical detectors with sufficiently high sensitivity to measure the amounts of catecholamines in the eluate from HPLC columns (Kissinger et al. 1973). The specificity of the methods is governed by the procedures used to separate the compounds which are detected. Catecholamines are adsorbed on alumina, eluted with acid, and separated from each other and from other readily oxidizable substance by HPLC. The effluent from the column flows through an electrochemical detector which consists of an electrode set at a predetermined potential to oxidize catechols to quinones. The current (pA) generated by the oxidation reaction is measured with a potentiostat and the output fed into a strip chart recorder. The details of construction of the electrodes and typical electrical configurations have been described (Kissinger et al. 1973), but now several models are commercially available. The application of this technique for analysis of brain catecholamines (Refshauge et al. 1974) was modified and made sufficiently sensitive for adaptation to plasma catecholamines (Hallman et al. 1978; Goldstein et al. 1981). The HPLC-ED method presents a number of advantages over the radioenzymatic methods, particularly in studies with human plasma, but generally it is less sensitive than the radioenzymaytic methods (see below).

IV. Gas Chromatography-Mass Spectroscopy (GC-MS)

Although a variety of chromatographic procedures, including paper, thin-layer, open liquid column, and gas-liquid chromatography, satisfactorily separate the catecholamines, lack of specificity and sensitivity of detection methods have usually limited application of these to biological samples containing much higher levels of the catecholamines than are noramlly found in plasma. An exception to this has been gas-liquid chromatography coupled with mass spectrometry (GC-MS). The fragmentation pattern obtained with mass spectrometric analysis combined with the differing retention times on the gas chromatography column of the various derivatives of the catecholamines provide the most definitive method for identification of the compound being measured. Use of deuterium-labelled or unnatural catecholamines (e.g. N-propylnoradrenaline) as internal standards ensures exact determination of recoveries. Thus the GC-MS methods provide a standard against which other methods may be evaluated. While GC-MS can be used for assay of plasma catecholamines (Ehrhardt and Schwartz 1978), the required equipment is expensive, suitable internal standards must be available, the procedures for isolation and derivatization are as exacting and tedious as for any other of the available methods, and the time required for the sequential injections into the gas chromatograph columns limits the number of samples that can be assayed conveniently. Because of these disadvantages, GC-MS methods are reserved for validation of methods which can be more readily adapted for routine use.

V. Other Methods

The most commonly used methods for assay of plasma noradrenaline in studies published during the last ten years have been the trihydroxyindole method as modified by Renzini et al. (1970), and the two radioenzymatic methods. In general, there is quite good agreement among the methods. Using any of the methods there is a considerable variation among "normal" resting values. The variations are reduced somewhat if care is taken to standardize the conditions under which blood is withdrawn. Usually, the investigators carefully specified that the blood samples were obtained via indwelling needles with the subjects reclining for at least 20 minutes before the blood was obtained for analysis. Although there were no significant differences in the low levels of adrenaline found in plasma, the fluorimetric methods appear to yield slightly lower and the PNMT method somewhat higher noradrenaline levels compared to the results obtained by GC-MS and the COMT method. Hjemdahl et al. (1979b) found good agreement between the COMT and HPLC methods in plasma from a stressed baboon, and Goldstein et al. (1981) confirmed this for physiological levels of the catecholamines in human plasma. In the latter study the mean plasma catecholamine levels obtained by the two methods were virtually identical for noradrenaline—(1.84 vs 1.78 pmol/ml) and reasonably similar for adrenaline (0.32 vs. 0.20 pmol/ml) with HPLC-ED and COMT methods, respectively. There was a highly significant correlation (0.99) between the two methods. The coefficients of variation indicated that the COMT method was somewhat more precise than the HPLC-ED method, but the differences were not important unless values of less than 0.60 pmol/ml were being determined. The HPLC-ED method require larges volumes of blood that the COMT method (1 ml vs. 50-100 μl plasma) and may require as long as 15-20 minutes for clearance of the columns before the next injection, but the method is inexpensive, yields results rapidly, and does not involve use of radioisotopes, enzymes, or a liquid scintillation spectrometer. HPLC-ED provides an extremely useful method for routine use in measurement of human plasma catecholamines and their metabolites, but the COMT method is usually still required for accurate measurement of noradrenaline and adrenaline in the small samples of blood or tissues obtained from laboratory animals. Using either radioenzymatic or HPLC methods, however, good standardization and careful quality control are necessary (HJEMDAHL et al. 1984).

C. Plasma Catecholamine Levels

Plasma levels of catecholamines are maintained by their secretion from the adrenal medulla (mostly adrenaline) and the overflow of noradrenaline from the catecholamine released at sympathetic neuro-effector junctions. The origin of dopamine is uncertain; some may be derived from the adrenal medulla, some from sympathetic nerves, and perhaps some from specialized organs e.g. carotid body. The levels of all catecholamines are subject to rapid fluctuations and may vary with the method and site of blood collection, the time of the day, and a host of other factors. Before discussing variations of plasma catecholamines in pathological states or after drug administration, physiological variations in plasma catecholamines must be considered.

I. Basal Levels

In most studies in which plasma catecholamines have been measured in humans, blood was obtained from the antecubital vein from supine subjects. The stress of blood collection in animals and in humans is sufficient to raise the levels of circulating catecholamines (see CALLINGHAM 1975; POPPER et al. 1977; KVETNANSKY et al. 1978). To avoid stress-induced increases in plasma catecholamines it is necessary to introduce a catheter or needle into a blood vessel and wait sufficient time (20–30 min) for the effects of the procedure to dissipate. In animals, a surgically implanted indwelling catheter is usually required.

As indicated above, the mean levels of catecholamines in human plasma obtained from resting subjects are 1.0–1.5 pmol/ml for noradrenaline and about 0.25 pmol/ml for adrenaline and dopamine, with slight variations depending on the method used for assay. In large experimental animals plasma catecholamine levels are similar to those in humans (CALLINGHAM 1975; BÜHLER et al. 1978). In early studies, the values reported in rat blood obtained after decapitation were extraordinarily high, but when suitable precautions were taken to avoid stress, low values, similar to those in humans, are obtained (POPPER et al. 1977; CHIUEH and KOPIN 1978a; KVETNANSKY et al. 1978; BENEDICT et al. 1978; BÜHLER et al. 1978; MICALIZZI and PALS 1979). Decapitation elevates levels of adrenaline most (hundred-fold increases or more) and dopamine least (KVETNANSKY et al. 1978; BÜHLER et al. 1978). Plasma adrenaline becomes undetectable in adrenalectomized rats, but dopamine levels are not altered (KVETNANSKY et al. 1978). Small amounts of dopamine are, however, released from the acetylcholine-stimulated rat adrenal (BEN-JONATHAN and PORTER 1976). Basal plasma catecholamine levels differ with the site from which blood is obtained. Plasma from arterial blood contains slightly lower noradrenaline and somewhat higher adrenaline levels than does venous blood, indicating net release of noradrenaline and uptake of adrenaline during passage of the blood through the tissues (PRICE and PRICE 1957; VENDSALU 1960; STJÄRNE et al. 1975; MIURA et al. 1976; KIM et al. 1979; HALTER et al. 1980). Hepatic venous blood has the lowest levels (VENDSALU 1960; HÄGGENDAL 1963; JONES et al. 1979; REID et al. 1980) reflecting hepatic clearance (75-80 percent) of catecholamines in hepatic artery and portal vein blood (VELASQUEZ and ALEXANDER 1979); left renal vein blood has the highest catecholamine levels because it contains blood from the left adrenal vein. In 10 patients with cirrhosis, arterial-hepatic venous extraction averaged 43 percent, whereas the right kidney appeared to release noradrenaline at a rate sufficient to increase renal venous blood levels by about 34 percent over arterial levels (RING-LARSEN et al. 1982). Plasma from human adrenal vein blood contains about 15 pmol/ml dopamine (PLANZ and PLANZ 1979), whereas the dopamine content of venous blood from the iliac vein is about 1.3 pmol/ml.

The net rate of entry of noradrenaline from a tissue into the plasma can be determined from the arterial and venous concentrations of the catecholamine and the blood flow. If the uptake of radiolabelled noradrenaline is determined simultaneously, the overflow of released noradrenaline can be calculated. ESLER et al. (1984) estimated that at rest, the kidneys and skeletal muscle each contribute about 20% of the total overflow of noradrenaline into plasma.

Removal of noradrenaline from the circulation is the result of its metabolism, uptake into the sympathetic nerve terminals ($uptake_1$) or uptake into other tissues. Since isoprenaline is not taken up by sympathetic neurones ($uptake_1$), the difference in the decrements of venous versus arterial concentrations of ^{3}H-isoprenaline and of ^{3}H-noradrenaline (during their infusion at a constant rate) can be used to estimate in the forearm of humans, the role of $uptake_1$ in removing noradrenaline (GOLDSTEIN et al. 1985). This difference is completely eliminated after treatment with desipramine. From such studies it appears that only about 15% of the arterio-venous differences in plasma levels of ^{3}H-noradrenaline can be attributed to neuronal uptake.

In humans, about 25 percent of exogenous ^{3}H-noradrenaline is removed during passage of blood through the lungs (GILLIS et al. 1972). In perfused rat lungs, noradrenaline is removed by uptake into the endothelium (uptake into sympathetic nerves is negligible) where it is metabolized rapidly (NICHOLAS et al. 1974). Removal of noradrenaline by perfused rabbit lung is reduced significantly by cocaine, phenoxybenzamine, and imipramine (IWASAWA and GILLIS 1974), but cocaine does not appear to diminish pulmonary noradrenaline clearance *in vivo* (CATRAVAS and GILLIS 1980). When the pulmonary capillary endothelium is poisoned with monocrotaline, inactivation of noradrenaline by the perfused lungs is strikingly reduced (GILLIS et al. 1978). There are, however, sympathetic nerves in the pulmonary vasculature (KADOWITZ and HYMAN 1973) which might release varying amounts of noradrenaline into the blood; this compensates, at least in part, for removal of the catecholamine by the endothelium. ESLER et al. (1984) estimated that about one third of the noradrenaline which enters the blood is derived from the lungs. There is, however, a net fall in concentrations of the catecholamine because more is removed than is added during passage of the blood through the lungs. SOLE et al. (1979) found about 25 (±3) percent pulmonary extraction of endogenous noradrenaline in patients with normal pulmonary vascular resistance, but no extraction in patients with elevated resistances. In contrast to this, smaller, nonsignificant differences in plasma noradrenaline levels in mixed-venous versus arterial blood were found in studies from three other laboratories (O'NEILL et al. 1978; VECHT et al. 1978; KIM et al. 1979). It is likely that net differences in mixed-venous and arterial levels of plasma catecholamines are statistically significant, but the magnitude of differences may depend upon permeability factors, flow rates (cardiac output), integrity of pulmonary epithelium and pulmonary sympathetic activity.

The blood may be viewed as a compartment through which noradrenaline is carried from tissues in which sympathetic nerves are active to the liver, lungs, and other tissues in which there is a net loss of the catecholamine. At rest, the concentration of plasma noradrenaline in venous blood from most peripheral tissues (heart, muscle, kidney, skin, etc) is about 25 percent greater than in plasma of arterial blood (Fig. 2). The liver and lungs remove noradrenaline so that concentrations of noradrealine in hepatic venous blood are much lower than in portal venous blood and, as indicated above, arterial plasma contains less noradrenaline than does plasma of mixed venous blood.

Plasma catecholamines measured in antecubital vein blood represent the levels in mixed-venous blood and the net changes which result from uptake and release of the catecholamine in the lungs and tissues of the arm. This site

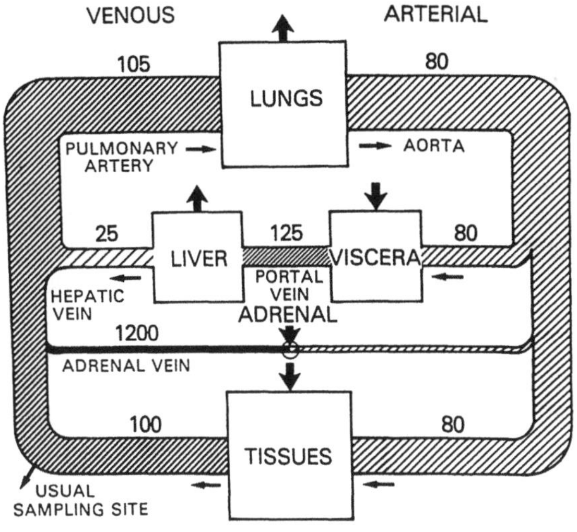

Fig. 2. Relative concentrations of noradrenaline in plasma at various sites in the circulation. There is generally an increase in plasma noradrenaline levels in venous blood from the various tissues because removal of noradrenaline is less than its overflow from sympathetic nerve terminal synapses. In the lungs and liver, however, there ist usually net removal of noradrenaline. Hepatic venous blood contains the lowest levels of noradrenaline because of efficient removal of the catecholamine by the liver

of sampling is convenient, the values for basal plasma catecholamine levels are not greatly distorted from those of mixed venous blood (WATSON et al. 1979), and correlate well with sympathetic nerve activity measured in the peroneal nerve (WALLIN et al. 1981). Under some circumstances, however, forearm venous plasma noradrenaline levels may be misleadingly high (see below).

II. Determinants of Basal Plasma Catecholamine Levels

In most studies of plasma catecholamine kinetics, the plasma is treated as a single compartment with rapid mixing. Under these conditions the mean steady-state basal plasma levels of catecholamines are determined by three parameters: their rate of entry into the circulation; the apparent volumes of their distribution (which includes fluid compartments and binding sites with which plasma catecholamines equilibriate, as well as the plasma volume); and the rate constants for their removal from the circulation (by excretion, metabolism, or uptake into the tissues). COHEN et al. (1959) showed that during constant rates of infusion of adrenaline or noradrenaline, within 10 minutes relatively stable levels of the infused catecholamine were attained. The levels appeared to be directly proportional to the infusion rates, suggesting that the sum of removal rates was equivalent to an overall first order process which was not saturated at the plasma levels (0.75–35.0 pmol/ml) attained during the infusions. They estimated plasma clearance rates of about 110 ml/kg/min for noradrenaline and 160 ml/kg/min for adrenaline. Subsequent studies using the same fluorimetric method (VENDSALU 1960), newer radioenzymatic assays (SILVERBERG et al. 1978; FITZGERALD et al. 1979; YOUNG et al. 1980) or infusion of (−)noradrenaline$-7-{}^3$H (GHIONE et al. 1978; ESLER et al. 1979; ESLER et al. 1980) at slower infusion rates have yielded somewhat lower estimates of plasma clearance rates for noradrenaline (35–50 ml/kg/min). It has been pointed out by CHRISTENSEN (1982) that these measurements of clearance rates, which are equivalent to cardiac output, are erroneously high. "Spillover" rates calculated from the specific activity of plasma noradrenaline during infusion of the labelled catecholamine may be excessively high because of the entry of unlabelled noradrenaline during passage of blood through the tissues. Clearance rates based on increases in noradrenaline concentration are erroneous because of net removal of the amine in the lungs and tissues at the higher noradrenaline concentrations. CLUTTER et al. (1980) found that at lower infusion rates of adrenaline, with steady-state levels (0.13–0.41 pmol/ml) overlapping the normal basal level of the catecholamines, the apparent plasma clearance rates (about 50 ml/kg/min) were significantly lower than those found (about 90 ml/kg/min) at higher levels (0.50–5.5 pmol/ml) of the catecholamines. The acceleration of plasma clearance of catecholamines at higher, physiologically significant levels may account for the discrepancy in estimates of secretion rates by COHEN et al. (1959) and those of subsequent studies noted above. This may partly explain why ESLER et al. (1980) found that patients with idiopathic peripheral autonomic insufficiency (who have low plasma noradrenaline levels) had diminished apparent rates of plasma noradrenaline clearance. It has not been determined whether differences in blood flow, uptake, or metabolism account for the enhanced apparent plasma clearance of higher levels of the catecholamines but stimulation of β-adrenoceptors appears to be involved (CRYER et al. 1980). Blocking β- but not α-adrenoceptors reduces adrenaline stimulation of its own clearance and also reduces clearance of endogenous noradrenaline.

After cessation of infusion of noradrenaline, plasma levels fall precipitously with a half-time of 2.0 to 2.5 minutes (VENDSALU 1960; SILVERBERG et al. 1978; ESLER et al. 1979). When tritium-labelled noradrenaline is infused a second, slower exponential phase of decline of the isotope becomes apparent (GITLOW et al. 1964; ESLER et al. 1979). The half-life of this slower component was 4 hours after infusion of $(\pm)^3$H-noradrenaline (GITLOW et al. 1964) and 30 minutes after $(-)^3$H-noradrenaline (ESLER et al. 1979). LABROSSE et al. (1961) described a similar biphasic decline in plasma levels of $(\pm)^3$H-adrenaline with half-time components of 1.2 min and 75 minutes. Since the first phase is prolonged by desipramine (ESLER et al. 1981), it is likely that uptake into sympathetic neurones accounts for a major fraction of this component. The physiological significance of the second, slower phase of decline in labelled catecholamines, is unclear, but appears to reflect tissue stores of the catecholamines, in sympathetic nerve terminals or muscle, which slowly equilibrate with the plasma. Comparision of the rates of decline in plasma levels of ^{3}H-noradrenaline and ^{3}H-isoprenaline (which is not taken up by sympathetic neurons) provides a means of assessing the role of neuronal uptake in clearing catecholamines from the plasma (GOLDSTEIN et al. 1983).

In normal humans the total rate of urinary excretion of catecholamine metabolites is in excess of 30 μmol/day whereas the apparent noradrenaline secretion rate is about 8.5 nmol/day. Thus, most of the noradrenaline produced in the body appears to be metabolized without entering the general circulation. This is probably not the case for adrenaline, which is produced mainly in the adrenal medulla and is secreted directly into the blood. Several factors favor metabolism of noradrenaline before it reaches the circulation. Although noradrenaline is stored in vesicles in nerve endings where it is protected from destruction, some may "leak" into the cytoplasm and be destroyed by mitochondrial monoamine oxidase. After release into the synapse from sympathetic nerve endings, a variable fraction of the released noradrenaline is actively transported back into the nerve ending where it may be metabolized or again stored in vesicles. The relative importance of reuptake depends at least in part on the width of the neuroeffector junctions (BEVAN 1977). At wide neuroeffector junctions less noradrenaline is taken up into the nerve ending. It has been estimated that at vascular neuroeffector junctions about 2/3 or the released noradrenaline is inactivated by reuptake (KOPIN et al. 1984). After diffusing away from the junction, another portion of the catecholamine is taken up into non-neuronal tissue where it is O-methylated. Finally, the efficient removal of catecholamines from the portal venous blood during passage through the liver (see above) markedly diminishes the contribition of plasma noradrenaline of sympathetic nerves in the visceral organs (e.g. spleen, small intestine, and pancreas).

Significant genetic influences on basal plasma noradrenaline levels were found in a study of mono- and dizygotic twins (MILLER et al. 1980), but the mechanisms (release, clearance, blood distribution) which may be involved were not examined.

III. Physiological Significance of Plasma Catecholamine Levels

Plasma catecholamine levels are usually assumed to be valid indices of sympathetic nerve and adrenal medullary activity. This has been tested in experimental animals. Stimulation of the sympathetic outflow from the spinal cord of pithed rats, as described by GILLESPIE and MUIR (1967), provided a means for controlling sympathetic activity and the newer radioenzymatic methods have made possible measurement of plasma catecholamine responses during such stimulation (YAMAGUCHI and KOPIN 1979). When sympathetic outflow is stimulated at physiologic frequencies (<5 impulses/sec) there is an immediate rise in blood pressure which is sustained at a relatively constant level for at least 5 minutes. Plasma levels of adrenaline rise to a peak within 30 seconds, decline slightly, and then gradually increase (Fig. 3). Noradrenaline levels rise more slowly, attaining a constant, steady-state level in 3–4 minutes. In pithed adrenal-demedullated rats, the plasma levels of noradrenaline attained at 1 minute are about half those reached at steady-state, consistent with the half time for the decline of the elevated levels when stimulation stops. These results suggest that the entry of noradrenaline into the circulation during sympathetic stimulation is prompt and relatively constant. The levels attained after one minute of stimulation are directly proportional to the rate of stimulation, whereas the pressor responses are proportional to the logarithm of the increment in noradrenaline concentration (YAMAGUCHI and KOPIN 1979). Bretylium blocks release of noradrenaline from the sympathetic nerves but does not block adrenal catecholamine release. Treatment with this drug blocks almost completely the pressor response to stimulation; plasma levels of adrenaline are increased to about the same level as in untreated animals, whereas noradrenaline levels are diminished by about 35 percent (YAMAGUCHI and KOPIN 1979). Thus, in pithed rats the pressor response appears to be mediated almost wholly by noradrenaline released at sympathetic nerve endings and the plasma levels of noradrenaline reflect overflow of the released catecholamine. Adrenaline and noradrenaline released into the blood from the adrenal medulla play only a small, insignificant role in the pressor response.

In humans, SILVERBERG et al. (1978) found that hemodynamic changes were produced only by infusions of noradrenaline which were sufficient to elevate plasma levels of noradrenaline to 11 pmol/ml, which is over five times the basal levels of the catecholamine. Such levels are attained during severe stress (see below). In a similar study with adrenaline CLUTTER et al. (1980) found that plasma adrenaline thresholds were 0.28–0.56 pmol/ml for increments in heart rate, 0.41–0.69 pmol/ml for increments in plasma glycerol and systolic pressure, and 0.84–1.12 pmol/ml for increments in diastolic pressure, plasma glucose, lactic acid, and β-hydroxybutyrate. In anesthetized dogs, plasma adrenaline had to be increased from a basal level of 2.5 to 5 pmol/ml

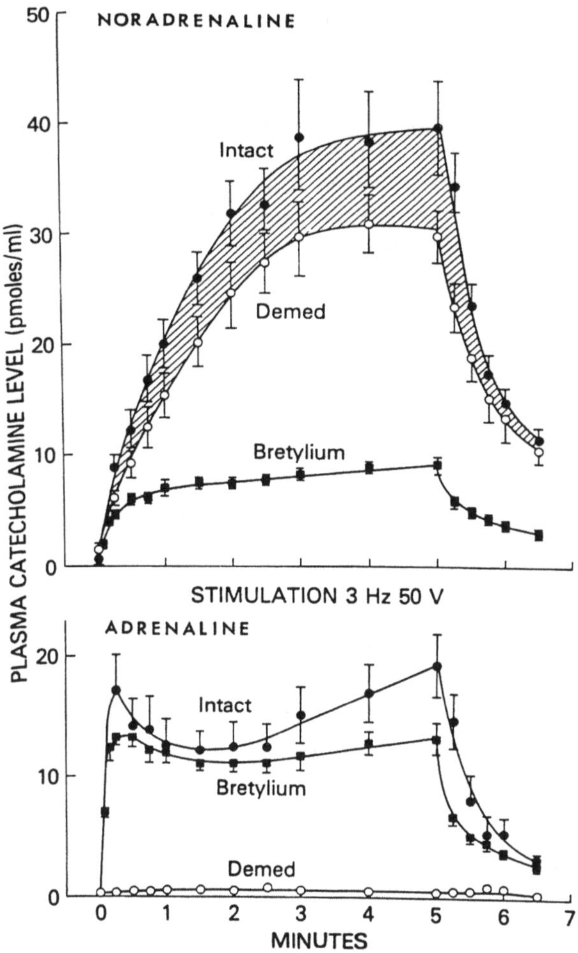

Fig. 3. Plasma catecholamine levels during and after 5 minutes of sustained stimulation of sympathetic outflow of intact, adrenal demedullated (Demed) and bretylium pretreated pithed rats. The difference between adrenal demedulated and intact rats in plasma noradrenaline levels (shaded area) can be attributed to noradrenaline derived from the adrenal medulla. Bretylium blocks release of noradrenaline from sympathetic nerves, but has no effect on adrenal release of catecholamines. From YAMAGUCHI and KOPIN (1979)

to evoke increases in plasma cyclic AMP levels and to 15 pmol/ml to increase plasma glucose or glycerol. To increase plasma glycerol, cyclic AMP, and glucose, noradrenaline levels had to be elevated to 5, 15, and 65 pmol/ml, respectively (HJEMDAHL et al. 1979a). Vasoconstriction was elicited in isolated adipose tissue by 5 pmol/ml adrenaline without affecting blood flow to skeletal muscle, whereas noradrenaline (5 pmol/ml) caused vasoconstriction in smooth muscle without affecting adipose tissue blood flow. The authors suggested the response patterns were related to preferential stimulation of

α_2-adrenoceptors by adrenaline and of α_1-adrenoceptors by noradrenaline. The results in humans and dogs are consistent with the view that adrenaline is an active hormone, whereas noradrenaline reflects primarily sympathetic nerve activity and only under special circumstances does adrenaline reach levels in plasma sufficiently high to be of physiologic importance.

Velasquez and Alexander (1979) showed that, in anesthetized rabbits, carotid artery occlusion resulted in increases in noradrenaline content of the intestinal lymph and superior mesenteric vein blood. Although they claimed that the noradrenaline in arterial plasma did not increase significantly (6.92 ± 1.78 control vs. 9.53 ± 3.02 pmol/ml during carotid occlusion), the increment ($+2.66 \pm 0.47$ pmol/ml) calculated from the matched pairs in their data was highly significant ($p < 0.001$). This increment was significantly smaller, however, than the increment in noradrenaline in superior mesenteric vein plasma (5.20 ± 0.76 pmol/ml), consistent with their conclusion that changes in regional sympathetic activity may occur without significant alteration in the arterial plasma noradrenaline levels. Because of the important role of the liver in removing noradrenaline from the plasma, any event (drug, procedure, or response) which diminishes splanchnic blood flow would be expected to diminish the clearance and prolong the disapparance of plasma catecholamines. As discussed earlier, noradrenaline released at narrow neuroeffector junctions and in areas of the body which contribute venous blood to the hepatic portal vein will be of lesser importance in determining plasma noradrenline levels. There is some selectivity in activation of portions of the sympathetic outflow to meet particular physiologic needs, rather than uniform increases in sympathetic outflow. Selective activation in regions which are poor contributors to the circulating pool of noradrenaline (such as the viscera) will not cause as great an increase in plasma noradrenaline as those areas with wide neuroeffector junctions and venous drainage into the superior or inferior vena cava (e.g. peripheral vasculature). In view of the large bulk of the vasculature and the anatomical structure of the vascular neuroeffector junctions, it appears likely that these are a major source of circulating noradrenaline.

As indicated above, plasma noradrenaline levels found in blood obtained from the antecubital vein may not reflect precisely total body sympathetic neural activity. The levels obtained are the result of the addition of a portion of the noradrenaline which is released from the sympathetic nerve terminals in the forearm and the fraction of the arterial plasma noradrenaline which is removed by neuronal and extraneuronal uptake and metabolism of the catecholamine (Goldstein et al. 1983; Chang et al. 1985). Furthermore, the rate of forearm blood flow has a significant influence on the percent noradrenaline removed (Brick et al. 1967). In a recent study (Goldstein et al., 1987) it was found that during mental challenge, changes in cardiac output on systolic blood pressure were significantly correlated with total body spillover of noradrenaline (determined as described above), whereas there were no significant relationships to venous noradrenaline levels. The influence of forearm blood flow on removal of catecholamine from the blood suggested in earlier studies (Brick et al., 1967) was clearly evident. Although increases in forearm blood flow were attended by net inreases in noradrenaline removal, the net rate of removal failed to keep pace with the net rate of noradrenaline presentation. This resulted in a decrease in the percent extraction of noradrenaline so that the noradrenaline concentration in the venous blood more closely approximates that in arterial blood.

The relationship between the concentration of noradrenaline in plasma and that at vascular neuroeffector junctions has been examined in pithed adrenalmedullectomized rats pretreated with an α_2-adrenoceptor blocking agent (Kopin et al. 1984). In this model system, during constant intravenous infusion of noradrenaline or during constant stimulation of sympathetic outflow from the spinal cord, stable levels of plasma noradrenaline are attained. During infusion of the catecholamine, the plasma levels of noradrenaline associated with a particular level of pressor response are higher than those at the neuroeffector junction because of intervening metabolism and uptake by the tissues and sympathetic nerves (Fig. 4). Because of these same processes, the plasma noradrenaline levels attained during stimulation of sympathetic outflow are lower than those at the vascular neuroeffector juntion. Because it may be assumed that the intervening removal processes are equally effective at removing endogenous or exogenous noradrenaline, the gradients in concentration of noradrenaline between plasma and synaptic cleft may be presumed to be equal and opposite—thus being symmetrical—during sympathetic stimulation or noradrenaline infusion (Fig. 5). The levels of noradenaline at the vascular neuroeffector junction during a given pressor response can be estimated as equal to the geometric mean of plasma noradrenaline concentrations dur-

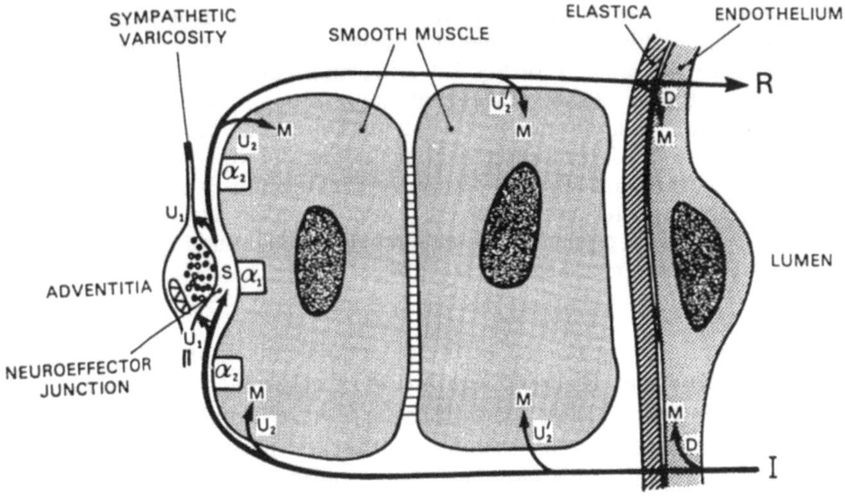

Fig. 4. Schematic representation of the cross-section of a blood vessel and the processes involved in generating the gradient in noradrenaline concentration between the neuroeffector junction and plasma in the lumen of the blood vessel.
U_1, Uptake$_1$; U_2, uptake into non-neural cells adjacent to synapse; U_2', uptake into non-neural cells at a distance from the synapse; I, infusion of noradrenaline; R, release of noradrenaline; M, metabolism; D, diffusion; α_1, α_1-adrenoceptor; α_2, α_2-adrenoceptor. From Kopin et al. 1984

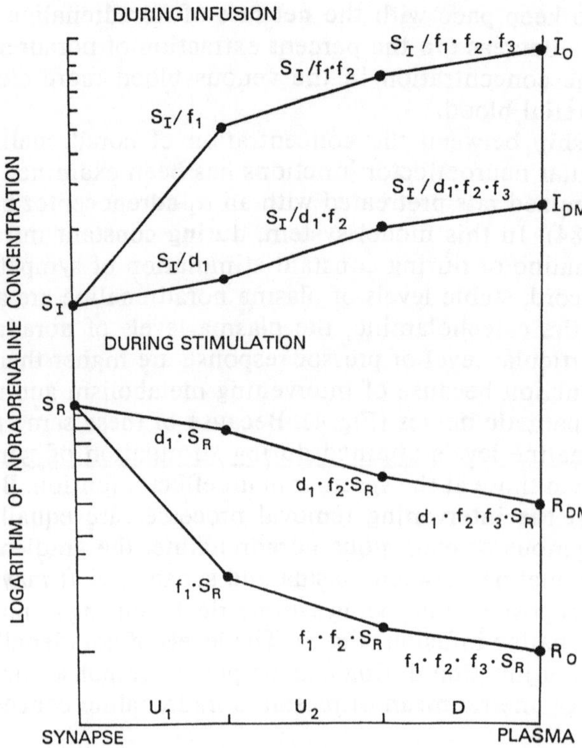

Fig. 5. Hypothesized gradients in noradrenaline concentration between the synaptic cleft and the plasma during noradrenaline infusion or during sympathetic stimulation. Abbreviations are the same as for Fig. 4. During stimulated release, S_R, synaptic noradrenaline concentration; R_0, plasma concentration without desipramine; R_{DMI}, plasma concentration after desipramine pretreatment. During noradrenaline infusion, I_0, plasma noradrenaline concentration without desipramine S_1, synaptic concentration; I_{DMI}, plasma concentration after desipramine pretreatment. f_1 is the proportion of noradrenaline remaining after removal by U_1, f_2 after U_2 (combined U_2 and U_2' of Fig. 4), and f_3 after D; d_1 is the proportion of noradrenaline remaining after inhibition of uptake$_1$ by desipramine. From KOPIN et al. 1984

ing infusion and during release of the catecholamine found during that pressor response (KOPIN et al. 1984). Inhibition of neuronal noradrenaline uptake with desipramine diminishes the difference in noradrenaline plasma levels associated with a given pressor response so that the plasma noradrenaline concentrations approach more closely the neuroeffector junctional level of the catecholamine (KOPIN et al. 1984). The results of such studies indicate that at vascular neuroeffector junctions about two-thirds of released noradrenaline is removed by reuptake into the sympathetic nerves. During the steady-state attained during sympathetic stimulation, the plasma noradrenaline levels in the neuroeffector junction are three-fold greater than in the plasma, whereas during noradrenaline infusion the plasma noradrenaline levels are three-fold greater than in the junction.

A similar procedure has been applied to studies in the forearm of humans during combined ganglionic, α_2-adrenoceptor and uptake blockade. Drug treatment resulted in a five-fold difference in the concentration of infused noradrenaline necessary to elicit a 20 mm Hg pressor response (20.2 pmol/ml before drug administration versus 3.7 pmol/ml after the antagonists). During yohimbine-induced release of endogenous noradrenaline after desipramine treatment, the arterial plasma noradrenaline level attending a 20 mm Hg in blood presure was 2.58 pmol/ml. From this it was concluded that in humans, a pressor response of 20 mm Hg requires a vascular neuromuscular junction noradrenaline mean concentration of about 3 pmol/ml (GOLDSTEIN et al. 1986).

IV. Plasma Catecholamine Variations in Normal and Stressful Situations

The importance of the sympathetic nerves and adrenal medulla in defense of the organism against external threats as well as maintaining homeostasis was summarized in the classical works by CANNON (1929) and CANNON and ROSENBLUETH (1937). The increases in catecholamine levels evoked by postural changes, muscular work, hypoglycemia, environmental temperature changes, hypoxia and physical or mental stresses were apparent using bioassay or fluorimetric methods for measurements of catecholamines in plasma of humans or large animals (CANNON and ROSENBLUETH 1937; VENDSALU 1960; CALLINGHAM 1975). During the last 10 years, availability of the more convenient, sensitive, and precise radioenzymatic methods capable of assaying catecholamines in 50–100 µl of plasma has made possible studies in small animals as well as confirmation and extension of older observations with refinements not possible previously.

1. Time-Dependent Variations in Plasma Catecholamines

Basal plasma catecholamines vary diurnally with peak levels of both noradrenaline and adrenaline in the late morning or early afternoon and a nadir in the early hours of the morning, during sleep (TURTON and DEEGAN 1974; PRINZ et al. 1979; KATO et al. 1980; BARNES et al. 1980; MULLEN et al. 1981). The levels decrease during sleep, but do not appear to differ with stages of sleep (MALING et al. 1979). YOUNG et al. (1980) have reported that an oral glucose meal (200 ml of 50 percent glucose) causes an increase in plasma noradrenaline which in elderly subjects was higher (79 percent vs. 32 percent) and peaked later (120 min vs. 60 min after ingestion) than in the young. SANTIAGO et al. (1980) found small but significant increases in both catecholamines in response to lowering glucose levels from 200 to 100 mg/dl, in normals as well as diabetics. Lowering glucose levels from 95 to 60 mg/dl evoked greater responses. The subjects received constant infusions of insulin throughout and glucose levels were controlled by glucose infusion. Thus sudden changes in insulin levels or alteration of glucose concentrations even within the physiological range can evoke increases in plasma catecholamines.

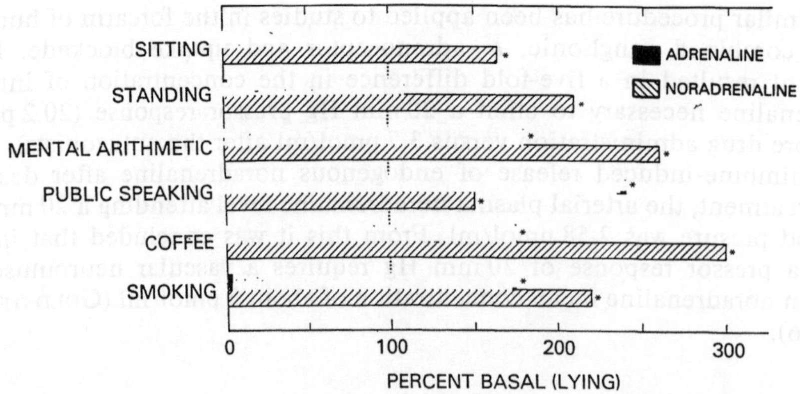

Fig. 6. Effects of usual activities on plasma levels of noradrenaline and adrenaline in plasma during reclining, under standardized conditions are taken as 100 percent (see text)

Caffeine (Fig. 6) increases plasma levels of both adrenaline and noradrenaline, by 200 percent and 70 percent, respectively (ROBERTSON et al. 1979a). Cigarette smoking (Fig. 6) evokes increases in plasma noradrenaline and adrenaline (CRYER et al. 1976; QUIGLEY, et al. 1979). The increases in sympathetic activity are presumably mediated by central nervous system effects of caffeine and nicotine, although nicotine may also have ganglion-stimulating effects. These factors must be considered when clinical studies of catecholamine levels in plasma are planned.

2. Age, Sex and Race

In normotensive humans, plasma levels of noradrenaline increase with age (CALLINGHAM 1975; CHRISTENSEN 1973; ZIEGLER et al. 1976a) although the correlation is not strong nor always apparent (see discussion of plasma noradrenaline in hypertension). The increments in plasma noradrenaline levels do not appear to be due to differences either in metabolic clearance rates or in half-times for disappearance of the catecholamine (YOUNG et al. 1980; RUBIN et al. 1982), indicating that the age-related increase is due to enhanced noradrenaline release into plasma. ESLER et al. (1982) have reported, however, that the mean clearance rates of noradrenaline in normal subjects over 50 years of age was lower than in subjects aged 20–29 years. Higher release rates of noradrenaline with advancing age, however, is consistent with the greater sympathetic nerve activity noted in older subjects (SUNDLÖF and WALLIN 1978; WALLIN and SUNDLÖF 1979). The responses of plasma catecholamines to a variety of stimuli also increase with age (ZIEGLER et al. 1976a; PALMER et al. 1978; YOUNG et al. 1980; BARNES et al. 1982).

In pregnant ewes, fetal sheep plasma catecholamine levels have been measured in blood obtained via a chronically implanted catheter in the jugular veins of the fetus (JONES and ROBINSON 1975; ARTAL et al. 1979). The levels in the plasma of the fetus are similar to those in the ewe. When the ewe is subjected to anoxia (breathing 9 percent O_2) or hemorrhage of about 30 percent of its blood volume, catecholamine levels in the fetus are elevated to a greater extent than in the mother, with a greater increase in adrenaline than noradrenaline. Since the placenta does not transport catecholamines, they must be derived from stimulation of the fetal adrenal medulla. The maximal levels attained are inversely proportional to the pH of the blood. The half-life of ^{3}H-(−)-adrenaline is similar in the fetus and ewe, indicating that the mechanisms for disposition of catecholamines are well developed in the fetal sheep.

Human newborn plasma catecholamine levels are markedly elevated (YOUNG et al. 1979; ELIOT et al., 1980), presumably as consequence of the stress of delivery. The level in blood from the umbilical artery is 4-5-fold greater than in the umbilical vein, indicating that the liver of the newborn is as effective at removing catecholamines as is that of the adult. Fetal peripheral venous and umbilical arterial blood contain about the same levels of both adrenaline and noradrenaline (ELIOT et al. 1980). The levels of noradrenaline remain elevated (4.5-6.0 pmol/ml) during the first three hours after birth, but at 12-45 hours reach normal adult ranges (1.5-2.0 pmol/ml). Adrenaline levels at birth are high (10-fold or higher than adult) but 3 hours after birth are decreased to about 0.5 pmol/ml and are at normal adult levels between 12 and 48 hours (ELIOT et al. 1980). There do not appear to be any significant differences between infants delivered vaginally or by cesarean section, nor does the anesthetic (spinal vs. general) influence the catecholamine levels in the newborn infant (ELIOT et al. 1980).

In most studies, there have not been sex differences in plasma catecholamine levels (e.g. ENGELMAN and PORTNOY 1970); PEDERSEN and CHRISTENSEN 1975; ZIEGLER et al. 1976a; HENRY et al. 1980) although in one relatively large study, 67 white women had a significantly higher mean level of noradrenaline (3.13 ± 0.30 pmol/ml) than 60 white men (2.30 ± 0.24 pmol/ml) and 22 black women had higher levels than 8 black men (JONES et al. 1978). SEVER et al. (1978, 1979) failed to find any significant race differences in plasma noradrenaline levels.

3. Postural Changes

On assuming upright posture, pooling of blood in the lower portion of the body reduces venous return, diminishes cardiac output, and the resultant tendency to orthostatic fall in blood pressure activates homeostatic reflexes which usually prevent a significant change in blood pressure. Orthostatic increases in plasma noradrenaline were first observed using the ethylenediamine method (HICKLER et al. 1959). The values obtained were about twice those subsequently reported by VENDSALU (1960), who used the trihydroxyindole method. VENDSALU's results were confirmed by the use of newer radioenzymatic methods (e.g. CHRISTENSEN and BRANDSBORG 1973; CRYER et al. 1974; LAKE et al. 1976). The rise in noradrenaline levels occurs during a 5-minute interval, the delay being consistent with the time required to initiate the reflex and the time necessary to establish a new steady state. The levels attained dur-

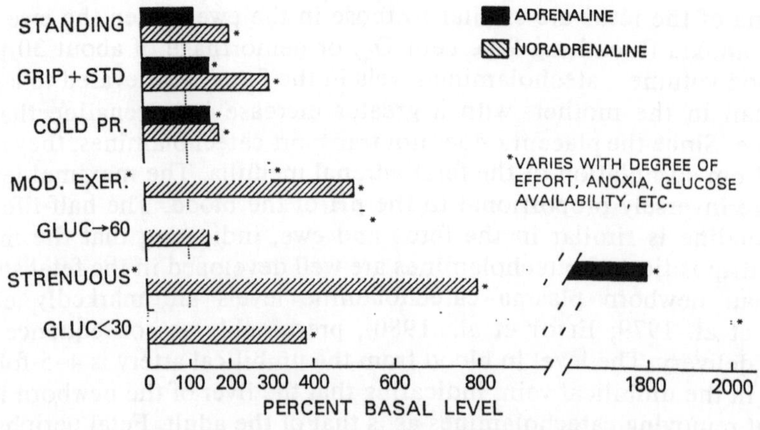

Fig. 7. Evoked increases in plasma catecholamines. The basal levels of catecholamines are taken as 100 percent and the increments during standing, standing with hand grip (see text), a cold pressor test (cold pr.), during moderate exercise (mod. exer.), slight decrease in blood glucose (Gluc→60), strenuous exercise or under severe hypoglycemia (Gluc < 30) are shown. These are typical values. Those which are markedly elevated vary widely at different times as well as among patients

ing quiet standing are about double those found in the supine position (Fig. 6, 7) and are higher in elderly subjects (YOUNG et al. 1980). ROSENTHAL et al. (1978) found that plasma noradrenaline levels gradually increased in normal subjects progressively tilted at 10, 20, and 45 degrees. The maximal levels, usually attained at 45 degrees, were maintained for up to 30 minutes and the increments appeared to be related to mean arterial pessures. There are small relatively insignificant increases in plasma adrenaline and dopamine levels with standing (FRANCO-MORSELLI et al. 1977), although the percent changes may be similar to that of noradrenaline (VAN LOON et al. 1979a).

4. Muscular Work and Exercise Training

For many years it has been known that muscular work and various sorts of exercise evoke increases in urinary and plasma catecholamines (see VENDSALU 1960; CALLINGHAM 1975). CHRISTENSEN et al. (1979b) have recently reviewed the relationships of increases in plasma catecholamines during exercise to hemodynamic parameters, oxygen requirements, and glucose utilization.

The increase in heart rate attending mild exercise is due to a decrease in vagal tone whereas cardiac acceleration evoked by baroreceptor stimulation is a consequence of sympathetic activity (ROBINSON et al. 1966). The relationship of heart rate to plasma catecholamines during exercise is therefore different from that attending postural changes (CHRISTENSEN and BRANDSBORG 1973). During exercise, the initial increase in heart rate is not attended by a rise in plasma catecholamines. Only after the heart rate has increased by about 25 beats/minute do the levels of noradrenaline increase. The slopes of the lines relating increment in heart rate to plasma noradrenaline levels are similar with postural change and exercise, but their intercepts differ because of the early cardiac acceleration due to vagal withdrawal in exercise.

Although sympathetic nerve stimulation results in net release of noradrenaline into blood perfusing the heart (SIEGEL et al. 1961; BRAUNWALD et al. 1964) and the response of the heart parallels the release of catecholamines (YAMAGUCHI et al. 1975), the frequently observed close correlations of increments in heart rate and plasma catecholamines cannot be interpreted as indicating that plasma noradrenaline is derived from the heart. Studies in which catecholamines have been measured in plasma from coronary sinus and from arterial blood (HANEDA et al. 1978; HANSEN et al. 1978a; SCHWART et al. 1979) indicate that there is loss of adrenaline and net gain in noradrenaline during passage of the blood through the heart. The venous-arterial differences in noradrenaline during rest (0.30-0.48 pmol/ml) are increased (0.80-1.18 pmol/ml) during exercise. Even if the plasma flow through the heart reached 1 l/min, the maximal net calculated release of norepinephrine, 1.2 pmol/min (or 0.2 ng/min), would be only a fraction of the total noradrenaline production rate (see above).

After vagal withdrawal in exercise, increased sympathetic outflow to the heart is accompanied by a proportional increase in sympathetic vasomotor activity in the visera (ROWELL 1974). The increase in sympathetic activity throughout the body probably accounts for parallel increases in plasma noradrenaline and heart rate. The plasma noradrenaline elevations parallel the intensity of exercise (WATSON et al. 1980).

Heavy short-term efforts and prolonged periods of work at moderate intensities evoke increases in both noradrenaline and adrenaline levels (Fig. 7). In rats, strenuous exercise evokes striking increases in catecholamines, derived mostly from the adrenal medulla (GALBO et al. 1978; RICHTER et al. 1980). The rise in plasma noradrenaline levels appears to be independent of plasma glucose levels. However a fall in plasma glucose enhances exercise induced secretion of adrenaline as well as other hormones involved in glucohomeostatis (GALBO et al. 1975, 1979); trained rats have lesser responses (GALBO et al. 1977). Similarly in humans, training reduces the noradrenaline and adrenaline responses to both high intensity short-term exertion and prolonged efforts (HICKSON et al. 1979; WINDER et al. 1979) without changing the apparent rate of clearance of the catecholamines (HAGBERG et al. 1979).

A 3-month program of exercise training of patients with ischemic heart disease reduced mean supine noradrenaline levels (from 1.89 ± 0.13 to 1.13 ± 0.12 pmol/ml) but not adrenaline levels (COOKSEY et al. 1978). After training, the maximal capacity for work increased and, at maximal effort, greater plasma catecholamine levels were attained. Recruitment of adrenal medullary secretion is a mechanism for the defense of the internal environment; training results in metabolic changes which enable the organism to endure more strenuous or prolonged exertion before enlisting these mechanisms.

The plasma catecholamine responses of humans to a variety of stimuli, including exercise, have been used as indicators of sympathoadrenal medullary activity, particularly in relation to treatment of hypertension with β-adrenoceptor blocking agents (see below) or for comparison with other stimuli (ROBERTSON et al. 1979b). Isometric hand grip for 5 minutes at 30 percent maximal effort increases plasma noradrenaline levels by about 50 percent above those attained during standing. The sequential elevation in noradrenaline levels with standing and isometric hand grip has been suggested as a standard procedure to evaluate sympathetic responsivity in patients (LAKE et al. 1976).

5. Hypoglycemia

Hypoglycemia is one of the most potent stimuli for release of adrenaline (Fig. 7). CHRISTENSEN (1974a) demonstrated a close correlation between the degree of hypoglycemia and the rise in plasma adrenaline. Plasma levels of noradrenaline are also increased after insulin-induced hypoglycemia, but insulin may increase noradrenaline levels even if the fall in blood glucose is prevented (GUNDERSEN and CHRISTENSEN 1977; LILAVIVATHANA et al. 1979). Even in the absence of changes in insulin infusion rates, rapid decrements of glucose within the physiologic range appear to stimulate adrenal medullary secretion (DEFRONZO et al. 1980; SANTIAGO et al. 1980).

After insulin-induced hypoglycemia, increments in plasma catecholamines precede or coincide with increases in plasma glucose as well as preceding the increments in other hormones (glucagon, cortisol, and growth hormone) involved in the recovery from hypoglycemia (GARBER et al. 1976). There appear to be two phases in the normal sequence of restoration of blood glucose levels, the first of which is rapid, of relatively short duration, and dependent on release of adrenaline (POLINSKY et al. 1980). The second, slower phase of the response appears to be predominantly due to glucagon release. Adrenergic mechanisms are therefore not essential to the restoration of normal glucose, unless glucagon secretion is impaired (RIZZA et al. 1979). The non-metabolizable glucose analogue, 2-deoxyglucose, also evokes striking increases in plasma catecholamine (predominantly adrenaline) levels in humans (THOMPSON et al. 1980) as well as in rats (SUN et al. 1979; DIROCCO and GRILL 1979).

6. Responses to Extremes of Temperature

Maintenance of body temperature in mammals requires efficient mechanisms for generation and conservation of heat during cold exposure and dissipation of heat when the body temperature is elevated. Metabolic and circulatory regulation by the adrenomedullary sympathetic nervous system plays an important role in these processes. When the body temperature is lowered, metabolic processes are accelerated to enhance thermogenesis, and blood flow is diverted from the skin to conserve heat. The sympathoadrenal medullary responses are essential for survival of unacclimated rats exposed to cold (MAICKEL et al. 1967). During six hours of exposure to cold, rat plasma noradrenaline levels reach a peak at about 4 hours (BENEDICT et al. 1979) and then decline, presumably as a result of depletion of catecholamine stores or adaptation. Repeated cold exposure on subsequent days, however, does not alter the noradrenaline response. After rats are adapted to cold, noradrenaline discharged from the sympathetic nerves is mainly responsible for thermogenesis (CARLSON 1966). Adaptation to extremes of temperature may alter the disposition of circulating catecholamines, since DEPOCAS et al. (1978) found that during its infusion, the plateau levels of noradrenaline are much lower in cold-adapted than in warm-adapted rats. The basis of this difference in apparent clearance rates is unknown, but enhanced splanchnic blood flow could be a factor.

In unadapted humans, immersion of a hand in ice-cold water rapidly elicits a complex response involving reflex vasoconstriction and pain which results in a rise in blood pressure without a rise in cardiac output or heart rate. The response to the "cold pressor test," which is excessive in hypertensive patients and possibly those at risk to develop hypertension (see below), is mediated by sympathetic vasoconstriction and attended by increases in plasma noradrenaline (LAKE et al. 1976; WINER and CARTER 1977; JOHNSON et al. 1977) and adrenaline (ROBERTSON et al. 1979a; LEBLANC et al. 1979). The increases in catecholamine levels (Fig. 6), however, are smaller than those obtained with exercise or caffeine (ROBERTSON et al. 1979a).

Increases in body temperature evoke sympathetic-mediated alterations of blood distribution to enhance heat loss. Blood is diverted from the viscera and from muscle to the skin (ROWELL 1974). In anesthetized rabbits, heating increases cutaneous and simultaneously decreases visceral sympathetic activity (RIEDEL et al. 1972). In anesthetized cancer patients treated with hyperthermia, regional changes in noradrenaline levels indicate that the sympathetic nerves in the arm are inactive, whereas the acitivty in visceral sympathetic nerves is markedly increased (KIM et al. 1979).

7. Hypoxia, Hypercapnea and Acidosis

The increase in plasma catecholamines evoked by these potent stimuli of adrenal medullary secretion have long been recognized (CALLINGHAM 1975).

8. Hypotension, Hypovolemia, Hemorrhage and Shock

Decreases in blood pressure evoke compensatory increases in vasomotor tone and heart rate which are mediated by the baroreceptor reflexes. Acute volume depletion induced by administration of furosemide in dogs (CARRIERE et al. 1979) or humans (LAKE and ZIEGLER 1978; HENRY et al. 1980; IBSEN et al. 1980) evokes increases in plasma noradrenaline presumably by increasing baroreceptor reflex activity. Similar effects on plasma noradrenaline have been noted in patients with chronic renal failure who develop hypovolemia during hemodialysis (BRECHT et al. 1975; HENRICH et al. 1977; LAKE et al. 1979c).

Hemorrhage provides a strong stimulus to the sympathoadrenal medullary system (Fig. 8). Awake rats rapidly respond to bleeding with large increases in plasma adrenaline levels which are completely eliminated by adrenalmedullectomy (FREDHOLM et al. 1979). After adrenalmedullectomy, hemorrhage-induced increases in plasma noradrenaline are markedly attenuated, indicating that the adrenal medulla is the predominant source of plasma catecholamines in hemorrhagic shock. The responses are blunted, but not eliminated, in pentobarbital-anesthetized animals (FARNEBO et al. 1979). In anesthetized dogs, controlled hemorrhage causes progressive increases in plasma catecholamines and in plasma DBH concentrations as a result of their release, primarily from the adrenal medulla (PINARDI et al. 1979).

In humans, hemorrhagic shock evokes marked increases in both adrenaline and noradrenaline (BENEDICT and GRAHAME-SMITH 1978). Unlike traumatic or septicemic shock, the levels of adrenaline often exceed those of noradrenaline, particularly in patients who do not survive. In survivors, the plasma catecholamine levels return to the normal range rapidly, whereas, in non-survivors they remain elevated until just before death. Traumatic shock is attended by pain, hypovolemia, hypoxia, and acidosis, all of which contribute to the elevation in plasma adrenaline and noradrenaline. In septicemic shock, hypotension appears to be the major stimulus for sympathoadrenal medullary activity (BENEDICT and GRAHAME-SMITH 1978).

9. Mental Activity, Emotional Reactions and Stress

Increases in alertness and reactions to perceived threats whether in preparation for "fight or flight", evoke sympatho-adrenal medullary responses. Plasma catecholamine levels have been examined in experimental animals to develop models for the study of a variety of stresses and in humans to define the role of sympathoadrenal reactions in psychiatric illness and in the pathogenesis of psychosomatic disorders.

Plasma noradrenaline levels are lowest during sleep (POPPER et al. 1977; DETURK and VOGEL 1980). In rats, alerting responses to approach of humans, handling, restraint, etc. (POPPER et al. 1977; CHIUEH and KOPIN 1978a; KVETNANSKY et al. 1978) make necessary the placement of indwelling cannulae to obtain blood samples by remote syringes without disturbing the animals.

Forced immobilization of rats evokes striking increases in plasma adrenaline and noradrenaline (40-fold or more and about 10-fold, respectively) and about a tripling of dopamine (KVETNANSKY et al. 1978). The initial increase in adrenaline level reaches an early peak (5–20 min) and declines to lower, stable high levels for the remainder of the immobilization period (KVETNANSKY et al. 1978; DETURK and VOGEL 1980). Noradrenaline levels increase less, but are maintained at a high level. The peak levels of plasma catecholamines attained in young male rats are lower during second and third restraints (DETURK and VOGEL 1980).

Spontaneously hypertensive (SHR) and normotensive Wistar-Kyoto (WKY) rats derived from the same strain have similar plasma catecholamine levels when undisturbed, but SHR rats have greater sympathoadrenal medullary responses when exposed to a variety of stressful situations such as confinement to small cages during blood pressure measurement (CHIUEH and KOPIN 1978a), forced immobilization (KVETNANSKY et al. 1979), foot shock (MCCARTY and KOPIN 1978a) or anticipation of foot shock (MCCARTY and KOPIN 1978b) or recovery from stress. In both SHR and WKY rats, foot shock evokes greater behavorial and plasma catecholamine responses than in other strains of rats (McCarty and Kopin 1978c).

When humans perform a visual monitoring task, plasma levels of adrenaline, but not noradrenaline, initially increase, but later decline to normal (O'Hanlon and Horvath 1973). Contrary to expectations, exposure of 18 healthy normal male volunteers to noise failed to alter plasma catecholamine levels, although diastolic and mean blood pressures and peripheral resistance were increased (Andren et al. 1979). During competitive hand-eye coordination games, harassment evokes greater elevations of plasma adrenaline and noradrenaline in Type A (coronary-prone behavior pattern) than in Type B (non-coronary-prone behavior) persons (Glasset al. 1980). In a separate study, young healthy physicians had 2-fold increases in adrenaline (Fig. 6) during the first 3 minutes of an oral presentation before an audience (Dimsdale and Moss 1980). Exercise, however, induced increases primarily in noradrenaline. Mental arithmetic causes greater increases in plasma adrenaline levels (Fig. 6) than does a cold pressor test (Fig. 7), but the cold stimulus elevates noradrenaline more than does mental activity (Leblanc et al. 1979).

In clinically encountered anxiety, plasma catecholamine levels are increased. The plasma catecholamine (adrenaline plus noradrenaline) levels in depressed patients (3.2 ± 1.4 pmol/ml), which were greater than those in normal subjects (1.6 ± 0.5 pmol/ml), are directly related to the degree of anxiety (Wyatt et al. 1971; Louis et al. 1975). Both adrenaline and noradrenaline are increased. The anxiety attending transfer of myocradial infarct patients from intensive care units to the wards may account for increases in plasma noradrenaline (Siggers et al. 1971).

Adrenaline, which is secreted in larger amounts than noradrenaline during anxiety, elicits sympathomimetic effects characteristic of anxiety and, when administered, can evoke anxiety. A self-generating spiraling anxiety reaction may occur if adrenaline released in response to anxiety evokes additional anxiety which evokes further adrenaline release, etc. (Breggin 1964). Such speculations have some support from the results obtained in experimental animals. In rats, blockade of sympathetic responses by administration of either bretylium or chlorisondamine, which do not enter brain, attenuates the behavioral activation during application of intermittent foot shock (McCarty and Kopin 1979). Propranolol, a β-adrenoceptor blocking agent, has been found useful for anxious patients with somatic symptoms (tachycardia, tremor) or acute panic states (e.g. stage fright). This drug presumably interrupts the hypothetical spiral, but appears to be of little use in other forms of anxiety (see Tyrer and Leder 1974; Heiser and Defrancisco 1976). For reviews of central nervous action of β-adrenoceptor blocking agents see the monograph edited by Kielholz (1977).

V. Disease States

1. Phaeochromocytoma

Phaeochromocytoma is a tumor which is derived from sympatho-adrenal medullary tissues and produces catecholamines. The clinical features, which almost invariably include hypertension and tachycardia, are a consequence of the catecholamines secreted into the circulation, and the diagnosis usually rests on the demonstration of increased plasma levels or urinary excretion of the catecholamines and their metabolites (see reviews by ENGELMAN 1977). These patients have markedly increased plasma levels of catecholamines. Although it has been claimed that measurement of plasma levels of noradrenaline and adrenaline presents some advantages (BRAVO et al. 1979), the differentiation of patients with phaeochromocytoma from those with other forms of hypertension can usually be achieved easily by examination of the urinary catecholamines or their metabolites.

Measurement of plasma catecholamines can be useful, however, in localizing the catecholamine-secreting tumor. EULER et al. (1955) first determined the site of a phaeochromocytoma by assaying plasma catecholamines in blood from different levels in the vena cava. This method has been used repeatedly over the years (see SMITH and WINKLER 1972) but newer methods, such as computerized axial tomography or other radiological procedures, are usually sufficient to locate the tumor (GANGULY et al. 1979; DAVIES et al. 1979). When these methods fail, selective venous catheterization for analysis of plasma catecholamines may be useful (DAVIES et al. 1979; JONES et al. 1979).

2. Catecholamines in Essential Hypertension

The importance of the sympathetic nervous system in mediating control of blood pressure has made attractive the hypothesis that abnormalities in sympathetic neuronal function may be important in the development or maintenance of some forms of essential hypertension. The clinical efficacy of some drugs which influence adrenergic function has lent further credence to this hypothesis. With the advent of the new sensitive and specific assays, there have been more clinical studies of plasma catecholamines in hypertension than in any other group of disorders, but the role of catecholamines in hypertension remains an open question.

Of the many studies in which plasma levels of noradrenaline were compared in normotensive and hypertensive subjects, about half reported that the catecholamine levels were significantly higher in the hypertensive patients (see reviews by: KOPIN 1979; KOPIN et al. 1980a,b; GOLDSTEIN et al. 1981). In the positive studies, the ages of the patients tended to be younger, there was generally no correlation of plasma noradrenaline levels with age in either the normal or hypertensive subjects, and medical personnel were included among the control subjects. In the studies in which the normotensive and hypertensive subjects had similar levels of plasma noradrenaline, the mean level for the normotensives was generally higher than the level in the normotensive subjects of the positive studies, older populations were studied, and, among the normotensives, but not in the hypertensives, there was a positive correlation of plasma noradrenaline levels with age. The inclusion of young hypertensives with excessively high plasma noradrenaline levels was generally responsible for the lack of correlation of plasma noradrenaline levels with age among the hypertensive subjects in both positive (hypertensive levels higher than normotensive) and negative (hypertensive levels equal to normotensive) studies. Inclusion of medical personnel among the older controls appears to have been responsible for the lack of relationship of plasma noradrenaline levels to age in the normotensives of the positive studies since older subjects with high levels were rarely encountered. It may be concluded that at rest, except in a few young hypertensives, plasma noradrenaline levels are normal in patients with hypertension.

Studies in SHR rats suggest that adrenergic hyper-responsivity is an early feature of the genetic abnormality. Increases in plasma catecholamine levels in SHR rats subjected to a variety of stresses are higher than in WKY rats (see discussion above). FOLKOW et al. (1970) and FOLKOW (1975) have stressed the importance of this hyper-reactivity in the pathogenesis of established hypertension. In addition to this, however, the pressor response to stimulation of the sympathetic outflow is greater in pithed SHR rats than in the pithed normotensive WKY rats, although there is no difference in the plasma catecholamine responses (YAMAGUCHI and KOPIN 1980). Thus, two distinct abnormalities appear to elevate blood pressure in SHR rats; there is both increased adrenergic responsivity to stress and enhanced reactivity to the released catecholamine (KOPIN et al. 1980a).

Remarkably similar observations have been reported on the responses to mental stress in normal adolescents with hypertensive parents (FALKNER et al. 1979). After a 10-minute difficult mathematics test, 20 control subjects (with no family history of hypertension for at least three generations), had smaller increases in total plasma catecholamines than did 41 children of parents with one hypertensive parent. Of the latter group, 14 adolescents had labile diastolic blood pressures. The pattern of responses among the adolescents with one hypertensive parent with and without labile diastolic hypertension resembles that of SHR rats and WKY rats, respectively, while those of the control group of adolescents is similar to normotensive rats of other strains (MCCARTY and KOPIN 1978a).

It remains to be determined whether increased plasma noradrenaline and elevated blood pressure in a young person are indicators of a high risk for development of hypertension, are part of the pathogenesis of established hypertension, or are indicative of excessive "alerting" with a pressor response appropriate to the level of sympathoadrenal activity.

Although basal catecholamine levels in hypertensive subjects may not be elevated significantly above normal, they are inappropriately high for the blood pressure, indicating some alteration in baroreceptor function. Attempts to determine whether sympathetic responsivity is enhanced in hypertensive patients has led to conflicting reports, depending on the population studied and the method used to evoke a sympathetic response. In some studies (e.g. SEVER et al. 1977) several hypertensives had greater increases than did any of the normotensive patients, suggesting hyperreactivity. In other studies increased sympathetic activity has been stimulated by volume depletion induced by furosemide (LAKE and ZIEGLER 1978; IBSEN et al. 1980), or by a hypotensive agent, nifedipine (PEDERSEN et al. 1980). In these studies the increases in plasma noradrenaline of hypertensives where not significantly greater than normal. The response to hand grip has been reported to be greater in hypertensives (VLACHAKIS and ALEDORT 1980) and results from several laboratories suggest that the elevation of noradrenaline during work on a bicycle (PLANZ et al. 1975; PHILIPP et al. 1978; WATSON et al. 1980) or a treadmill (ROBERTSON et al. 1979b) is greater in hypertensives. BERTEL et al. (1980), however, found similar increases in normotensives and hypertensives during work on a bicycle. Differences in ages, physical condition, hemodynamic changes, etc., may have contributed to the differences between hypertensives and normotensives. The primary role of the sympathetic nervous system in hypertension remains an open question.

3. Orthostatic Hypotension

As indicated earlier, a change of posture from supine to standing evokes baroreflexes to increase sympathetic activity. If the blood pressure is not maintained because receptors are blocked, blood vessels fail to constrict, or blood volume is deficient, the sympathetic response is exaggerated with excessively high plasma noradrenaline (and adrenaline) levels and marked tachycardia. This has been called "hyperadrenergic orthostatic hypotension" (CRYER et al. 1978). Failure of the baro-reflex to evoke a sympathetic response is reflected by absence of the rise in plasma noradrenaline and no increase in heart rate during the fall in blood pressure.

Reflex arc failure may occur at the level of the baroreceptor, from central nervous system damage, with sympathetic ganglion blockade, or it may be due to inability of sympathetic nerve endings to form, store, or release noradrenaline. Receptor-effector failure might be due to blockade of receptors, inability of the blood vessels to constrict or depletion of intravascular volume. As early as 1960, VENDSALU described a patient with orthostatic hypotension and failure to increase plasma noradrenaline levels with standing. In primary autonomic dysfunction with postural hypotension, CRYER and WEISS (1976) reported normal basal noradrenaline levels but deficient responses to standing. ZIEGLER et al. (1977) distinguished between idiopathic orthostatic hypotension (or primary autonomic neuropathy) and the Shy-Drager Syndrome (or multiple system atrophy) in which postural hypotension is associated with a central nervous system disorder characterized by varying degrees of incoordination and Parkinsonian symptoms. In idiopathic orthostatic hypotension peripheral sympathetic nerves appear to be affected and basal noradrenaline levels are low, whereas in the Shy-Drager Syndrome the sympathetic nervous system appears to be intact (basal noradrenaline levels are normal) but not appropriately activated during standing. LEVESTON et al. (1979) confirmed these observations and showed that a patient with the peripheral type of sympathetic failure and four patients with diabetic neuropathy failed to have normal increases in plasma noradrenaline after injection of edrophonium, an acetylcholine esterase inhibitor which evokes stimulation of sympathetic ganglia. Caviezel et al. (1982) found low plasma noradrenaline levels and blunted responses to standing in diabetic patients with autonomic neuropathy, consistent with the changes found in idiopathic orthostatic hypotension. Esler et al. (1980) reported that patients with idiopathic peripheral autonomic insufficiency may have deceptively normal noradrenaline levels; they have diminished plasma clearance rates for the catecholamines, but "spillover" of noradrenaline is low.

The peripheral sympathetic nerves appear to be intact in Familial Dysautonomia, a rare genetic disorder which affects Ashkenazai Jews. In this disorder basal noradrenaline levels are normal, but do not increase with standing (Ziegler et al. 1976b). The reflex arc defect in Familial Dysautonomia may be at the levels of the baroreceptor (Dancis and Smith 1966), since the disease is characterized by sensory deficits and the adrenal medullary response to hypoglycemia is not impaired (Smith and Dancis 1967). Plasma adrenaline responses to insulin-induced hypoglycemia are often absent or blunted in other forms of orthostatic hypotension (Polinsky et al. 1980).

4. Myocardial Infarction

Marked increases in plasma catecholamines have been found in patients with myocardial infarction (Griffiths and Leung 1971; Siggers et al. 1971; Christensen and Videback 1974; Onnesen et al. 1974; Corr and Gillis 1978; Nadeau and de Champlain 1979; Mueller and Ayres 1980). Noradrenaline levels rise, sometimes tenfold or more (Fig. 8), and remain elevated for about two days, whereas adrenaline levels may be increased 20-fold but decrease more rapidly. There appear to be relationships between the severity of the infarct, the occurrence of pulmonary edema, and the rise in plasma catecholamines. In dogs, coronary artery occlusion rapidly (beginning within one minute) evokes increases in plasma adrenaline and noradrenaline which peak (at 10-fold and 3-fold, respectively) within one hour (Karlsberg et al. 1981). There is a direct relationship between the size of the infarct, the lowering of cardiac output and mean blood pressure, and the increases in catecholamines, suggesting that the hemodynamic deterioration triggers reflex activation of sympathoadrenal medullary secretion. A relationship between adrenaline release and arrhythmias has been suggested (Bertel et al. 1982).

5. Congestive Heart Failure

In congestive heart failure plasma noradrenaline levels are elevated in relation to the degree of cardiac decompensation (Thomas and Marks 1978). Furthermore, exercise induces greater increases of plasma noradrenaline levels in patients with congestive heart failure than in normal subjects (Chidsey et al. 1962).

The degree of elevation of plasma levels of noradrenaline is inversely correlated with the cardiac index; levels of the catecholamine decrease and respond less to the postural stress after therapy of congestive heart failure (Cody et al. 1982).

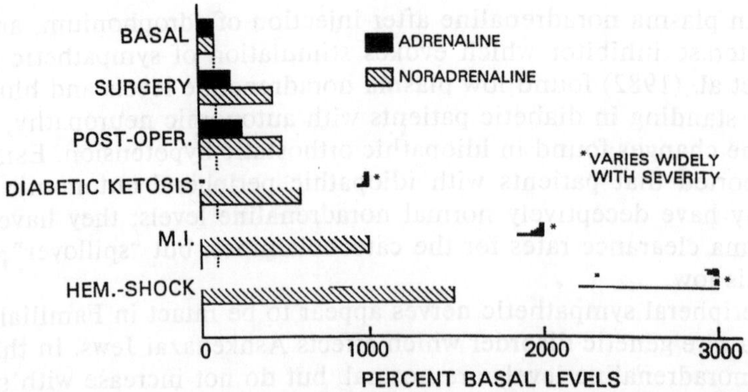

Fig. 8. Plasma catecholamine levels in catastrophic illness. Levels of catecholamines relative to basal (100 percent) are shown for surgery (during incision), postoperatively, during diabetic ketosis, myocardial infarction and hemorrhagic shock. The higher values vary with time in a single individual and among patients

6. Diabetes

The interrelationships of plasma catecholamines and diabetes have been reviewed (Christensen 1979). In untreated ketotic diabetics, there are marked increases in plasma catecholamines (Fig. 8), but these are probably due to compensatory responses to volume depletion (Christensen 1974b). Poorly controlled diabetics without ketosis may have increased catecholamine levels, but the levels return to normal when treatment is adequate (Bolli et al. 1979). Intravenous injection of insulin evokes increases in plasma noradrenaline and augments the postural increase in its levels, presumably as a result of decreased intravascular volume (Gundersen and Christensen 1977; Mogensen et al. 1980). Rats with streptozotocin-induced diabetes have normal noradrenaline levels (Berkowitz et al. 1980). Diabetic patients with autonomic neuropathy may have low supine and standing plasma noradrenaline levels and orthostatic hypotension (hypoadrenergic), whereas others, with reduced red cell mass, may have exaggerated (hyperadrenergic) noradrenaline responses to standing (Christensen 1972b; Cryer et al. 1978 Tohmeh et al. 1979; Caviezel et al. 1982). Six young diabetic subjects with early autonomic neuropathy had deficient catecholamine responses to exercise compared to 7 diabetics of similar age without evidence of neuropathy (Hilsted et al. 1980). Babies born to diabetic mothers are reported to have elevated levels of plasma noradrenaline (Young et al. 1979), but the cause is unknown.

7. Thyroid Disorders

The relationship between thyroid function and catecholamines was unclear until the development of sensitive methods for assay of plasma adrenaline and noradrenaline. There is general agreement that in hypothyroidism plasma noradrenline levels are increased, whether measured by radioenzymatic (Chris-

TENSEN 1972a, 1973) or fluorimetric (STOFFER et al. 1973; COULOMBE et al. 1976; and WERNER et al. 1976) methods. Hyperthyroidism has been associated with decreases in plasma noradrenaline levels (CHRISTENSEN 1973; STOFFER et al. 1973; WERNER et al. 1976), with an inverse relationship between levels of thyroid hormone and noradrenaline, although this was not found in the patients studied by COULOMBE et al. (1976). The similarity in net metabolic and cardiovascular effects of thyroid hormones and of catecholamines makes attractive the hypothesis that reciprocal changes in plasma noradrenaline to alterations in thyroid functions are compensatory. This view is further supported by the effects of thyroid hormone on β-adrenoreceptors (WILLIAMS et al. 1977).

8. Central Nervous System Disorders

Central nervous system disorders which result in rapid increases in intracranial pressure or otherwise interfere with the cerebral circulation have been known to evoke striking increases in blood pressure which were presumed, on the basis of effects of α-adrenoceptor blockade or cervical spinal cord transection, to be mediated by enhanced sympathetic activity. GRAF and ROSSI (1978) showed that, in dogs lightly anesthetized with pentobarbital, acute increases in intracranial pressure produced by inflation of an epidural balloon evoked, within seconds, massive increases in plasma adrenaline (15-215 pmol/ml), noradrenaline (6.5-60 pmol/ml), and dopamine (1-14 pmol/ml) from baseline levels of the operated animals (2.3, 1.4 and 0.2 pmol/ml, respectively). Similarly, in anesthetized rats and cats, acute elevation of central nervous system pressure evoked an immediate pressor response attended by increases in plasma catecholamines (ROOZEKRANS et al. 1979). The increases in adrenaline are somewhat delayed relative to the pressor response, and noradrenaline levels rise even more slowly during the first few minutes. The time course of increase in catecholamine levels is similar to that seen in stimulated pithed rats, as discussed above. Adrenalectomy or reserpine treatment diminishes the pressor and catecholamine responses in rats, but in cats, reserpine pretreatment does not diminish the adrenaline response. Thus, when intracranial pressure increases, adrenal medullary secretion is mainly responsible for the increase in plasma catecholamines.

Acute transection of the spinal cord, whether by decapitation of rats (POPPER et al. 1977; BERKOWITZ and HEAD 1978; TIBBS et al. 1979), in anesthetized dogs, or by section after laminectomy, evokes a sharp increase in both adrenaline and noradrenaline in blood. The adrenaline is derived primarily from the adrenal medulla whereas the noradrenaline is derived largely from sympathetic nerves (KVETNANSKY et al., 1979b; HEAD and BERKOWITZ 1979). In rats, the increase in catecholamines evoked by decapitation is attenuated by anesthetics (ROIZEN et al. 1978) so that the increases in noradreanline with laminectomy and spinal cord section in anesthetized dogs (TIBBS et al. 1979) are probably only a fraction of those which would be attained without anesthesia. In these dogs the catecholamine response was largely dissipated in 30 minutes.

LEVIN and HUBSCHMANN (1980) showed that in three patients with multiple sclerosis, direct stimulation of the dorsal spinal cord via an implanted electrode evoked increases in dopamine, noradrenaline, and adrenaline in plasma. During sustained stimulation for 20 minutes, noradrenaline levels were maintained at about double baseline values, whereas adrenaline and dopamine levels were highest at 3 minutes (50 percent and 150 percent above baseline) and gradually returned to prestimulation levels. Termination of the stimulation also evoked short-lived increases in the catecholamine levels. In patients with spinal cord transection and quadriplegia of less than two weeks duration, plasma levels of noradrenaline and adrenaline are markedly lowered (MATHIAS et al. 1979). Bladder stimulation of these patients does not evoke the pressor responses and corresponding rise in plasma noradrenaline levels which are seen in patients with chronic spinal cord injury and quadriplegia (MATHIAS et al. 1976).

9. Other Disorders

Elevated levels of plasma catecholamines have been reported in diseases in which compensatory sympathetic activity appears to be required to correct abnormalities in fluid volume. In some diseases, however, the mechanisms are unknown. The elevated levels of plasma noradrenaline, but not adrenaline, in patients with duodenal ulcer (BRANDSBORG et al. 1978) do not appear to be related to pain, anemia, blood pressure changes, or vagotomy (CHRISTENSEN et al. 1979a). In patients with anorexia nervosa, plasma noradrenaline levels are lowered (by about one third) in parallel with their low blood pressure (GROSS et al. 1979). The observation that weight loss in obese women is attended by a decrease in blood pressure, urinary VMA, and plasma noradrenaline levels (JUNG et al. 1979) and that oral glucose administration increases plasma noradrenaline levels (YOUNG et al. 1980) may be related to the reversible changes in plasma noradrenaline found in anorexia nervosa.

VI. Effects of Drugs on Plasma Catecholamines

Drugs which inhibit formation, storage, or release of catecholamines may be expected to diminish plasma catecholamine levels, whereas those which evoke release or block removal would be expected to elevate the plasma catecholamine levels. In addition to direct effects, however, these drugs and other agents which alter the effects of the catecholamines may evoke reflex compensatory changes in sympathetic impulse flow. Changes in plasma noradrenaline levels reflect the net effect of drugs on impulse flow, noradrenaline release and overflow into the circulation, and rate of noradrenaline removal from the circulation.

1. Drugs Influencing Synthesis, Storage, Release, Disposition or Action of Catecholamines

Although many drugs are known which influence synthesis, storage, release, uptake and metabolism of catecholamines, few studies have been done in which drug-induced changes in plasma catecholamine levels have been examined.

a. Inhibition of Catecholamine Synthesis

Seven hours after TH is inhibited by α-methyltyrosine, plasma levels of noradrenaline are halved, and dopamine levels are reduced to about one-third of normal, but adrenaline levels are tripled (AVAKIAN and HORVATH 1980). In α-methyltyrosine-treated animals, plasma noradrenaline levels are no longer increased by exposure to cold, and the increase in adrenaline is diminished if the adrenal catecholamines are depleted. Before depletion, adrenal medullary release of catecholamines in response to stress generally is enhanced to compensate for deficient release of noradrenaline from sympathetic nerve endings. During short intervals, inhibition of PNMT, the adrenal medullary enzyme responsible for conversion of noradrenaline to adrenaline, does not alter significantly plasma levels of adrenaline in humans (DUBB et al. 1979) presumably because of the large reserve and slow turnover of adrenal medullary adrenaline.

b. Interference with Storage or Release

Depletion of noradrenaline stores after administration of reserpine is attended by markedly decreased noradrenaline levels in plasma of lightly anesthetized rats (REID and KOPIN 1975). Blockage by bretylium of noradrenaline release from sympathetic nerve endings prevents stress-induced increase in plasma noradrenaline, but not adrenaline (MCCARTY and KOPIN 1979). Bretylium has a similar effect on stimulation-induced increases in plasma catecholamines of pithed rats (YAMAGUCHI and KOPIN 1979). Tyramine releases noradrenaline from tissues and evokes marked increases in plasma noradrenaline levels in rats, probably by inhibiting its uptake as well as causing its release (GROBECKER et al. 1977). In humans, smaller doses of tyramine evoke smaller, but significant, increases in plasma noradrenaline levels (POLINSKY et al. 1980; GOLDSTEIN et al. 1982; SCRIVEN et al. 1983).

c. Chemical Sympathectomy

Chronic treatment of young rats with guanethidine destroys sympathetic nerves so that plasma noradrenaline levels are lowered in undisturbed rats and the plasma catecholamine response to stress is markedly attenuated (KVETNANSKY et al. 1979a). Plasma noradrenaline levels, however, are not decreased 24 hours after administration of 6-hydroxydopamine (MICALIZZI and PALS 1979). In this study, plasma adrenaline levels were increased markedly, presumably because of reflex activation of adrenal medullary secretion, and noradrenaline released from the adrenal medulla may have sustained the plasma levels of this catecholamine. Anesthetized rats previously injected (two and three days before) with 6-hydroxydopamine have moderately (20 percent) decreased plasma noradrenaline without increased adrenaline levels, but both catecholamines increase normally in response to hemorrhage, except when the adrenals are removed, indicating their origin in the adrenal medulla (FREDHOLM et al. 1979).

d. Blockade of Uptake or Metabolism

Noradrenaline released at sympathetic neuro-effector junctions is inactivated in large part by reuptake in the neurone. In rats, interference with reuptake by administration of desipramine increases plasma levels of noradrenaline about two-fold (CHIUEH and KOPIN 1978b). Adrenaline levels are also increased. In enuretic children, imipramine increases supine and standing plasma noradrenaline levels and diastolic blood pressure (LAKE et al. 1979a). Similar effects on plasma noradrenaline levels have been reported in normal adults (ROSS et al. 1983) and in depressed patients treated with tricyclic antidepressants (VEITH et al. 1983), although in adults blood pressure levels decline.

Inhibition of COMT by administration of tropolone to rats has only minor effects on plasma catecholamine levels, presumably because other routes are available for disposing of the catecholamine (CHIUEH and KOPIN 1978b). Cocaine administration markedly increases adrenaline levels in plasma; the noradrenaline levels increase, but the increment is only a fraction of that of the adrenaline (CHIUEH and KOPIN 1978b). The increase is a consequence of the central stimulatory effect of cocaine on adrenal medullary secretion as well as its blocking effect on catecholamine uptake. When tropolone is administered with cocaine, the increase in both adrenaline and noradrenaline is markedly potentiated, indicating that the noradrenaline released in the presence of cocaine is metabolized by O-methylation.

e. α-Adrenoceptor-Mediated Effect

Blockade of α-adrenoceptors with phenoxybenzamine elevates rat plasma noradrenaline levels; this is reversed by ganglionic blockade (REID and KOPIN 1975). The phenoxybenzamine-induced increase in noradrenaline is probably the combined result of several actions. Inhibition of presynaptic α-adrenoceptors prevents feedback inhibition by noradrenaline of its own release and thereby enhances release. Phenoxybenzamine also inhibits both neuronal and extraneuronal uptake. By blockade of post-synaptic α-adrenoceptors, vasoconstriction is decreased and the fall in blood pressure evokes a baroreflex-mediated increase in sympathetic nerve activity. Since increased nerve impulses are essential for increased release of noradrenaline by phenoxybenzamine, ganglionic blockade prevents elevation of plasma noradrenaline. Similar effects, particularly with standing, have been described in hypertensive patients treated with prazosin or phenoxybenzamine (MULVIHILL-WILSON et al. 1979).

Yohimbine, which blocks presynaptic α_2-adrenoceptors, increases blood pressure and plasma noradrenaline in humans (GOLDBERG et al. 1983; GOLDSTEIN et al. 1986). Since yohimbine has little effect on blood pressure or plasma noradrenaline after ganglionic blockade (GOLDSTEIN et al. 1986) it is likely that this effect is dependent upon enhanced sympathetic outflow from the central nervous system. This is supported by the observation that rauwolscine, another α_2-adrenoceptor antagonist, enhances the rate of sympathetic neuronal activity (McCall et al. 1983).

Drugs which stimulate presynaptic α_2-adrenoceptors diminish release of noradrenaline, but many of these also have central actions which decrease sympathetic activity. Bromocriptine, a dopamine agonist, lowers plasma catecholamine levels (van Loon et al. 1979b; Ziegler et al. 1979), possibly by inhibiting release of noradrenaline from nerve endings as well as by its central effects. Clonidine and α-methyldopa have actions at peripheral sympathetic nerve endings (presynaptic α_2-receptor stimulation and formation of α-methylnoradrenaline, respectively) which lower plasma noradrenaline levels; but at doses used clinically their effects are mediated mainly by actions in the central nervous system.

f) β-Adrenoceptor-Mediated Effects

Infusion of isoprenaline, a β-adrenoceptor agonist, elevates plasma noradrenaline levels in both normotensive and hypertensive subjects (Vincent et al. 1982). This effect was attributed to both reflexively induced increase in central sympathetic outflow and activation of presynaptic β-adrenoceptors to facilitate noradrenaline release. Because of their importance in treatment of hypertension and other cardiovascular disorders, the β-adrenoceptor blocking agents have received considerable attention. The mechanism by which these agents lower blood pressure has been the subject of considerable controversy. Stimulation of presynaptic β-receptors enhances noradrenaline release (Adler-Graschinsky and Langer 1975). Lewis and Haeusler (1975) showed that in conscious rabbits administration of ($\pm$)-propranolol, but not (+)-propranolol, lowered the rate of sympathetic nerve impulse flow. But studies of plasma noradrenaline levels do not provide support for the view that there is a decrease in release of noradrenaline. Some studies report increases in plasma noradrenaline within minutes after intravenous administration of propranolol or other β-adrenoceptor blocking agents to normotensive subjects (Benedict et al. 1977), hypertensive patients (Christensen et al. 1978; Pickering et al. 1979; Morganti et al. 1979), or patients with ischemic heart disease (Hansen et al, 1979). In other studies, no significant changes in plasma catecholamine levels were found after acute oral or intravenous administration of β-adrenoceptor blocking agents (Irving et al. 1974; Anavekar et al. 1975; Lutold et al. 1976; Bonelli et al. 1979). In all studies, however, after acute β-adrenoceptor blockade, exercise evoked greater increases in plasma noradrenaline levels. In patients with acute myocradial infarction the elevated levels of catecholamines were lowered within 10 minutes after intravenous administration of propranolol (Mueller and Ayres 1980). In this study, the degree of lowering of the catecholamine levels was related to the inital high levels, suggesting that sympathetic hyperactivity during the early phases of myocardial infarction is acutely reduced by propranolol.

Chronic treatment of hypertensive patients with propranolol produces no change (Pedersen and Christensen 1975; Benedict et al. 1977; Pickering et al. 1979; Franz et al. 1980; de Leeuw et al. 1977; Nilsson et al. 1980; Dean et al. 1980) or small (50 percent) increases (Distler et al. 1978; Rahn et al.

1978; Jones et al. 1980; Vlachakis and Aledort 1980) in plasma noradrenaline levels.

Although in sheep propranolol appears to reduce plasma adrenaline levels during infusion of the catecholamine (Britton et al. 1977; Wood et al. 1979), as indicated earlier, in humans, propranolol appears to reduce plasma clearance of noradrenaline (Cryer et al. 1980). Variable noradrenaline clearance rates and inconsistently reduced sympathetic nerve impulse flow (Lewis and Haeusler 1975) as well as differences in dosages and times of observation might account for the conflicting reports of the effects on plasma catecholamines of β-adrenoceptor blocking agents. Alterations in distribution of cardiac output, as indicated earlier, may influence clearance rates and levels of plasma catecholamines.

Decreased plasma catecholamine levels after treatment of hypertensive patients with β-adrenoceptor blocking agents have been found only in two studies. In these, the pretreatment levels of the catecholamines noradrenaline (Brecht et al. 1976) or adrenaline (Franco-Morselli et al. 1978) were higher than normal and the effects of a placebo were not examined. Cousineau et al. (1978) found that propranolol treatment increased slightly the plasma noradrenaline levels of 16 hypertensive patients with normal basal levels, but decreased to similar levels the elevated noradrenaline levels of 9 "hyperadrenergic" patients with sustained hypertension. As after acute blockade, chronic β-adrenoceptor blockade increases the plasma catecholamine response to exercise, whether or not basal levels are increased.

The effects of β-adrenoceptor blocking agents on plasma catecholamines appear to vary with the state of the patient. Under basal conditions the levels may or may not be increased, but with exercise, particularly if strenuous, the noradrenaline levels attained are higher than in the absence of β-adrenoceptor blockade. By blocking β-adrenoceptors the enhanced clearance produced by increased catecholamines (Cryer et al. 1980) could be prevented; this could elevate catecholamine levels even without enhanced release. The baroreceptor reflex-mediated increases in plasma noradrenaline are unaltered by chronic β-adrenoceptor blockade, whether elicited by standing (Brecht et al. 1976; Hansson and Hökfelt 1976) or nitroprusside-induced hypotension (Dean et al. 1980). This is consistent with the suggestion that the early increase in sympathetic tone elicited by propranolol returns to normal with chronic β-adrenoceptor blockade (Pickering et al. 1979). In situations in which alerting responses may play a role (myocardial infarction, "hyperadrenergic states") increased plasma catecholamines may be lowered by a central action of the adrenoceptor blocking drugs. It is beyond the scope of this review to consider the effects of these drugs on the central nervous system (see the mongraph by Kielholz 1977).

2. Drugs Which Affect Impulse Flow in Sympathetic Nerves

a. Drugs Acting at Sympathetic Ganglia

Transmission of nerve impulses from pre-ganglionic to post-ganglionic neurones in both sympathetic and parasympathetic ganglia is cholinergic. Drugs which block nicotinic cholinoceptors interfere with activation of sympathetic neurones. Chlorisondamine, a long-acting ganglionic blocking agent, decreases plasma catecholamine levels and prevents their increase during stress or by procedures which normally activate baroreflex-mediated catecholamine responses (REID and KOPIN 1975; MCCARTY and KOPIN 1979; MICALIZZI and PALS 1979).

Cholinergic stimulation by administration of edrophonium, a short-acting cholinesterase inhibitor, increases plasma levels of noradrenaline and has been used to demonstrate peripheral sympathetic neural deficiency in patients with diabetic neuropathy (LEVESTON et al. 1979).

b. Drugs Acting in the Central Nervous System

Several drugs which are useful in the treatment of hypertension are believed to act in the central nervous system, particularly at the nucleus tractus solitarii, the central terminus of many of the barorceptors, where α-adrenoceptor stimulation lowers sympathetic outflow. Drugs which diminish blood pressure by actions in the central nervous system would be expected to lower plasma noradrenaline levels.

Clonidine lowers plasma noradrenaline levels in normotensive and hypertensive subjects (HÖKFELT et al. 1975; WING et al. 1977; METZ et al. 1978; MALING et al. 1979) and diminishes reflex-mediated increases in noradrenaline levels (MITCHELL and PETTINGER 1980). This drug stimulates α_2-adrenoceptors and could decrease noradrenaline release by an action on presynaptic adrenoceptors (DOXEY and EVERITT 1977) but its main actions are believed to be in the central nervous system (see reviews by KOBINGER 1975; LOWENSTEIN 1980). Other drugs which have α-adrenoceptor stimulating effects also lower plasma adrenaline levels. Treatment with α-methyldopa results in formation of α-methylnoradrenaline, which displaces noradrenaline from its storage sites in presynaptic vesicles. When α-methylnoradrenaline is released at adrenergic terminals in brain, its effects diminish sympathetic activity and probably account for the lowered levels of plasma noradrenaline in patients with α-methyldopa (THOMAS and MARKS 1978). Release of α-methylnoradrenaline from peripheral sympathetic nerves could not account for the hypotensive effects of α-methyldopa since α-methylnoradrenaline is as potent a pressor agent as is noradrenaline.

Fenfluramine, which predominantly affects serotonergic neurones, lowers plasma noradrenaline levels in normotensive psychiatric patients (DE LA VEGA et al. 1977) and diminishes the elevation of plasma catecholamines which occurs during treatment of hypertensive patients with diuretics (LAKE et al. 1979b). The mechanism of this effect is unknown, but central inhibition of sympathetic activity appears to be involved.

Central cholinergic stimulation appears to evoke increases in sympathetic activity. Oxotremorine, a muscarinic agonist, has a pressor effect and elevates plasma catecholamine levels in conscious rats. Atropine methyl nitrate (which does not enter the brain) reduces the oxotremorine-induced rise in noradrenaline, but not adrenline, whereas atropine abolishes both. Thus central muscarinic receptor stimulation appears to evoke catecholamine release (WEINSTOCK et al. 1979). Carbachol administered intracerebroventricularly also elevates plasma catecholamines (FISHER and BROWN 1980) and it would be expected that the centrally mediated pressor effect of physostigmine (VARAGIC 1955), a cholinesterase inhibitor, is also attended by an increase in plasma catecholamines. Bombesin, a peptide from frog skin which may occur in mammalian brain, when administered intracerebroventricularly evokes increases in plasma adrenaline (BROWN et al. 1979). The effect of bombesin is blocked by somatostatin analogues, suggesting involvement of peptide substances in the modulation by brain of sympatho-adrenalmedullary activity (FISHER and BROWN 1980).

c. Drugs Acting Indirectly

Any drug which, by an action at adrenoceptors or on effector systems, lowers blood pressure will evoke reflex increases in sympathetic activity. Thus, α-adrenoceptor blocking agents, including psychoactive drugs such as clozapine and other neuroleptic drugs (SARAFOFF et al. 1979; NABER et al. 1980) cause increases in plasma noradrenaline. Nifedipine (COREA et al. 1979; PEDERSEN et al. 1980), a calcium antagonist, hydrochlorothiazide (LAKE et al. 1979b), a diuretic, and hydralazine (PEDERSEN and CHRISTENSEN 1975) which acts on arteriolar smooth muscle, by lowering blood pressure, activate baroreflexes and thereby increase plasma catecholamine levels in hypertensive patients. The increment in plasma noradrenaline levels produced when hypotension is increased by infusion of sodium nitroprusside and the decrease in catecholamine levels found when the blood pressure is elevated by infusing phenylephrine has been proposed as a means for measuring baroreceptor-mediated changes in sympathetic vasomotor tone (GROSSMAN et al. 1982).

D. Dopamine-β-Hydroxylase in Plasma

Dopamine-β-hydroxylase (DBH), the enzyme responsible for conversion of dopamine to noradrenaline, is a component of the storage vesicles in the adrenal medulla and sympathetic nerve endings. A portion of the enzyme in the soluble content of the vesicles is released along with catecholamines (see review by AXELROD 1972). The observation that DBH is released into perfusates *in vitro*, led WEINSHILBOUM and AXELROD (1971a) and GOLDSTEIN et al. (1971) to the discovery in plasma of DBH activity; this was followed by a host of other studies of the plasma enzyme. The results of these studies have been extensively reviewed by WEINSHILBOUM (1978b). The enzyme is present in plasma of all species examined (rat, guinea pig, rabbit, dog, cat, etc.) as well as in humans. Although the DBH activity in plasma is derived from the sympatho-adrenal medullary system, in man as well as in animals, plasma levels have been disappointing as an index of acute changes in sympathetic activity in humans and rats. The enzyme levels appear to be more labile in other species. It is the purpose of this section to review briefly methods used for assay of the enzyme in plasma and to summarize current concepts regarding the origin, turnover rate, variation in levels and physiologic significance of DBH activity in plasma.

I. Assay of Plasma DBH

DBH activity in plasma was first detected and assayed by WEINSHILBOUM and AXELROD (1971 a) and GOLDSTEIN et al. (1971) using a twostep radioenzymatic assay based on the method described by MOLINOFF et al. (1971). Cofactors (Cu^{2+}, fumarate, ascorbic acid) required for optimal DBH activity are added to plasma along with a substance—tyramine or phenylethylamine—which is converted to its β-hydroxylated derivative. In the second part of the assay the β-hydroxylated product is enzymatically N-methylated by the action of PNMT and the ^{14}C-N-methylated product obtained is used to determine the amount of β-hydroxylated compound formed. Human plasma was found to have sufficient activity to form about 5.8 pmoles octopamine/min/l, whereas rat plasma had only about 2 percent of this activity.

In humans, plasma DBH activity is several orders of magnitude greater than in experimental animals. The amounts of octopamine formed by human serum DBH from high concentrations of tyramine are readily detectable by spectrophotometric assay of p-hydroxybenzaldehyde, the periodate oxidation product of octopamine (NAGATSU and UDENFRIEND 1972). One unit of enzyme is defined as the amount of activity yeilding 1 mole of octopamine per min at 37 °C. The activity found in human plasma by NAGATSU and UDENFRIEND (1972) was 42.6 ± 27.0 mol/min/l. The higher levels of activity found in the spectrophotometric assay than in the radioenzymatic assay of WEINSHILBOUM and AXELROD (1971 a) may be attributable to the lower concentrations of tyramine or phenylethylamine used as substrate in the latter assay.

By the use of radiolabelled 2-^{14}C-tyramine, the convenience of the p-hydroxybenzaldehyde method is retained, but it is made more sensitive. The radiolabelled octopamine, formed by the action of the enzyme, is oxidized with periodate to the aldehyde; this is extracted into an organic solvent for assay of the product (NAGATSU et al. 1973; WISE and KOPIN 1976). Other methods for assay of DBH, based on release of tritiated water (WILCOX and BEAVEN 1976) or formation of $^{14}CO_2$ (JOH et al. 1974), have not been adopted widely.

The levels of DBH activity appear to reflect adequately the levels of the enzyme molecules present since there is excellent agreement between the enzyme activity and immunoreactive enzyme levels measured by use of specific antibodies to DBH (WOOTON and CIARANELLO 1974; DUNNETTE and WEINSHILBOUM 1976, 1977).

II. Origin of Plasma DBH and Relationship to Sympatho-Adrenal Medullary Activity

Shortly after demonstration of its presence in rat plasma, WEINSHILBOUM and AXELROD (1971 b) showed that partial "chemical sympathectomy" with 6-hydroxydopamine results in a 30 percent decrease in rat plasma levels of the enzyme activity, whereas adrenal medullectomy has no effect. Chronic guanethidine treatment, which also selectively destroys symathetic neurones, also diminishes the enzyme activity in rat plasma (GROBECKER et al. 1977; GRZANNA and COYLE 1978). These results suggest that, in the rat, sympathetic nerve endings rather than the adrenals are the major source of the enzyme. Consistent with this view are the observations that the increase in plasma DBH levels in forcibly immobilized rats is found even after adrenalmedullectomy (WEINSHILBOUM et al. 1971) and that DBH levels increase in blood perfusates from the heart, spleen, and mesentery of pithed rats following selective stimulation of the sympathetic outflow (ALGATE and LEACH 1978). Similar increases occur with swim stress (ROFFMAN et al. 1973). These increases in enzyme levels are small (about 25 percent) and do not attain statistical significance after exposure to other forms of stress, such as cold (BEHRENS and DEPOCAS 1975), hypercapnea induction by inhalation of a mixture of 20 percent CO_2, 25 percent O_2, 45 percent N_2 for 5 hours (ARNAIZ et al. 1978), or in response to hemorrhage or hypoglycemia (CUBEDDU et al. 1977).

After chronic administration to rats of drugs which diminish release of noradrenaline (e.g. bretylium or chlorisondamine) or after drugs, the administration of which is attended by noradrenaline release (e.g. phenoxybenzamine), there are no significant changes in plasma DBH (REID and KOPIN 1975). Furthermore, these changes in sympathetic nerve activity do not alter the turnover rate of the enzyme (GRZANNA and COYLE 1978). Since the enzyme is derived from sympathetic nerves, but its release is relatively independent of sympathetic activity, a mechanism other than stimulation-induced exocytosis must account for release of DBH from the nerves into the plasma.

There are differences among species, however, in the intravesicular distribution, ease of release, and constancy of plasma levels of DBH. ARNAIZ et al. (1978) found that guinea pigs made hypercapneic for 5 hours had 10-fold increases in plasma DBH and noradrenaline. Furthermore, the increase in enzyme levels was diminished by about 2/3 in adrenalectomized guinea pigs and not altered significantly in 6-hydroxydopamine-treated aminals (ARNAIZ et al. 1980). These results and the effects of guanethidine, mecamylamine, and phenoxybenzamine provided strong evidence that in CO_2-stressed guinea pigs, the increase in plasma DBH results from adrenal medullary discharge. In guinea pigs, plasma DBH is increased 4-fold after insulin-induced hypoglycemia (CUBEDDU et al. 1979). ARNAIZ et al. (1978) and CUBEDDU et al. (1979) explained the species differences on the basis of the higher levels of the enzyme in the guinea pig adrenal medulla and the much greater proportion of the enzyme which is released by osmotic shock. The proportion of the total enzyme released by osmotic shock is taken as an index of the "releasable" form of the enzyme. Expressed in terms of plasma volumes, the amount of DBH in the guinea pig adrenals is 102 units "releasable" enzyme/ml plasma, whereas in rats there is only 2.8 units/ml plasma (ARNAIZ et al. 1978). The amount of "releasable" enzyme in rabbit adrenals is similar to that of guinea pigs, whereas the cat adrenal resembles that of the rat. In dogs, CUBEDDU et al. (1977) showed that controlled hemorrhage evokes increases in plasma DBH which are abolished almost completely in adrenalectomized animals. The amount of soluble adrenal DBH relative to the total circulating enzyme in dogs is intermediate between that of rats and guinea pigs (CUBEDDU et al. 1979). In anesthetized dogs, the DBH activity in thoracic lymph increased markedly after baroreceptor denervation and during and after stellate ganglion stimulation (NGAI et al. 1974), suggesting that a considerable proportion of the enzyme enters the circulation via the lymphatics. During controlled hemorrhage in dogs, however, although concentrations of plasma DBH increase, the diminution in lymph flow results in no net change in the output of DBH via the lymph (PINARDI et al. 1979). Restoration of blood volume and augmented lymph flow appear to redistribute DBH from interstitial tissue fluid and thereby increase plasma levels of the enzyme.

DBH levels in humans vary widely, largely due to genetic factors (see below), but they are usually orders of magnitude higher than in experimental animals. Numerous studies have examined, with conflicting conclusions, the effects on human plasma DBH levels of procedures which are known to enhance sympathetic nerve activity (Table 2). The discrepancies in these reports, however, are not as great as they may seem since the increases in enzyme levels, when statistically significant, are only 20-40 percent of the basal levels. These procedures evoke 2-to 10-fold increases in catecholamine levels which are usually unrelated to the magnitude of the increases in DBH. The relative stability of the enzyme levels indicates that DBH is not a statisfactory index of acute changes in sympathetic nerve activity in humans. This might be expected from the relatively slow turnover rate of the enzyme (see below). CUBEDDU et al. (1979) point out that the ratio of soluble adrenal to total circulating DBH is smaller in man than in other species, including rats, so that stress-induced increases in the enzyme level would not be expected. Human adrenal venous blood contains 17 times as much adrenaline and 10 times as much noradrenaline as peripheral venous blood, but equal amounts of DBH (PLANZ and PLANZ 1979). The small, acute changes in enzyme levels reported

Table 2. Reports of effects of various stimuli on human plasma DBH levels

Stimulus	Increase	No Significant Change
Standing	LAKE et al. (1977 b)	WOOTEN and CARDON (1973)
	TAKASHITA et al. (1977)	MUELLER et al. (1974)
	LEVY et al. (1979)	NOTH and MULROW (1976)
Exertion	WOOTEN and CARDON (1973)	FREWIN et al. (1973)
	PLANZ and PALM (1973)	OGIHARA and NUGENT (1974)
	PLANZ et al. (1975)	LAKE et al. (1977b)
Cold Pressor	WOOTEN and CARDON (1973)	OGIHARA and NUGENT (1974)
	STONE et al. (1974 a)	
Hypoglycemia	OKADA et al. (1975)	OGIHARA and NUGENT (1974)
		NISULA and STOLK (1978)

might be related to altered fluid volumes or lymphatic drainage rather than differences in rates of release (see below). Adrenalectomized humans have normal levels of the enzyme; with sodium restriction and enhanced sympathetic activity the levels increase as in normal subjects (NOTH and MULROW 1976). ÅBERG et al. (1974a) found that the level of the enzyme in lymph from the thoracic duct of lightly anesthetized patients was lower than that of plasma; the enzyme in lymph could be derived by slow filtration from plasma into tissue fluids or from sympathetic neurones. Thus, in humans, as in rats and cats, plasma levels of DBH are relatively resistant to change and the enzyme is probably derived mainly from sympathetic nerves.

III. Variations in Plasma Levels of DBH

1. Turnover Rate and Volume of Distribution

As for any plasma constituent—levels of DBH reflect both rates of entry and loss which, at steady-state, are equal. Another factor, which is frequently neglected, is the apparent volume of distribution. This factor may be important in changes in concentrations of plasma constituents which may be present in variable amounts in tissues, or which are subject to alterations with changes in volume of the fluid compartments in which they are distributed. The initial phase of decline in concentration of a tracer molecule or of elevated levels of an endogenous compound may reflect redistribution. For large molecules distribution may be relatively slow and appear to represent turnover.

Various means have been used to estimate the half-life of plasma DBH in different species. In sheep, RUSH and GEFFEN (1972) found that ^{125}I-labelled sheep enzyme disappeared with a half-life of 2.7 hours. This may, however, have represented distribution of the protein into the tissues or rapid removal of damaged enzyme. After termination of a stress the initial decline in elevated levels of the enzyme suggests a half-life of about 3 hours in rats (ROFFMAN et al. 1973) and in guinea pigs (ARNAIZ et al. 1980). It is uncertain as to whether or not a portion of the initial decline of redistribution of the enzyme, and a second, much slower, component (ROFFMAN et al. 1973) reflects true turnover. After intravenous injection of homologous antiserum to inactivate circulating DBH in rats, the rate of reappearance of the enzyme indicates that it has a half-life of about 4.2 days (GRZANNA and COYLE 1977). In humans, plasma levels of DBH decline with a half-life of 8-12 hours after removal of phaeochromocytoma (LOVENBERG et al. 1975). It is possible that the more rapid changes in enzyme levels are related to volume or distribution changes.

Although there is poor agreement and uncertainty regarding the true half-life of DBH in plasma, it is evident that its turnover rate is very slow compared to that of catecholamines. Because of its slow rate of turnover and apparent independence from sympathetic nerve and adrenal medullary activity, it is likely that acute changes in DBH in human or rat plasma are due to alterations in its distribution in tissues or the volumes of the fluid compartments in which it is distributed. During the cold pressor test STONE et al. (1974b) showed that concentrations of other plasma proteins change in parallel with DBH. Complete evaluation of the significance of changes in enzyme levels should include considerations of the compartments or fluid volumes in which the enzyme is distributed, as well as possible alterations in rates of release or removal of the enzyme. Unfortunately this is often difficult and the mechanisms of the changes in concentrations of DBH in plasma are often not well understood.

2. Genetic Control of Plasma DBH Levels

As indicated above, human plasma DBH levels vary widely among individuals in the general population. Genetic factors are important determinants of the levels; the evidence that the very low levels of DBH activity in 3-4 percent of the population is inherited as an autosomal recessive trait has been reviewed (Weinshilboum et al. 1975; Weinshilboum 1978a).

3. Developmental Factors

In rats, during the last few days in utero and the first two days after birth, plasma levels of DBH triple and remain elevated at 5-7 times adult levels until about 2 weeks of age and then decline rapidly during the next 3 weeks (Behrens and Depocas 1975; Behrens et al. 1979). Chemical sympathectomy with 6-hydroxydopamine lowers the enzyme levels to about one-half the two-day old level by 8 days of age (Behrens et al. 1979). The early post-natal elevated level of the enzyme is due almost wholly to a greater rate of entry into the circulation since the properties of the enzyme (thermal stability, K_m for tyramine, pH optimum, and physical properties) and rates of degradation are similar in 15-day and 60-day old rats (Grzanna and Coyle 1977, 1978).

This pattern of development differs from that in primates. In the Japanese monkey, DBH increases 10-fold between the ages of 3 months and 10 years (Kato et al. 1976). Although the absolute levels of activity are 100-fold greater in humans, the pattern of development is similar. In human newborns and during the first year of life plasma DBH levels are low (Wetterberg et al. 1972; Heiss et al. 1980). There are extremely wide ranges of activity, however, in children as well as adults (Weinshilboum et al. 1975; Heiss et al. 1980). There do not appear to be differences between sexes. Comparison studies with plasma frozen for years have been possible because the enzyme is stable during storage and remains remarkably constant in individual adults over periods of up to 15 years (Lamprecht et al. 1975; Heiss et al. 1980). Diurnal variations are minor (Okada et al. 1973; Leon et al. 1974; Markianos and Ickmann 1976) and those that do occur may be related to changes in distribution or fluid volumes.

Since DBH in plasma is derived from the sympathetic nerves, even if genetic and developmental factors influence the plasma levels of the enzyme and the turnover rate of the enzyme in plasma is slow, prolonged increases in sympathetic nerve activity which accelerate the rate of synthesis of the enzyme might be expected to cause some increase in the plasma levels of the enzyme. Treatment of rats with reserpine increases levels of DBH, first in sympathetic ganglia and then, one or two days later, in plasma. This suggested that enhanced synthesis and release of the enzyme might increase its plasma levels (REID and KOPIN 1975). There is however, no definitive evidence to support this hypothesis.

4. DBH Levels in Disease States

Because of the relative ease in measuring plasma levels of DBH and the early hopes that alterations in levels of this enzyme might reflect the level of sympathetic nerve activity, many studies were done which attempted to relate plasma levels of DBH to disease states. The interpretations of many of these studies are questionable because it cannot be assumed that altered levels of DBH always reflect changes in sympathetic nerve activity. Until information on turnover rates and volumes of distribution becomes available and there is a means of evaluating the genetic contribution to differing levels of the enzyme, particularly in heritable diseases, often it will not be possible to establish definitely the bases of abnormal mean levels of the enzyme in a particular disorder.

a. Sympatho-Adrenal Medullary Tumors

Phaeochromocytomas, and frequently neuroblastomas, synthesize catecholamines; in patients with these tumors, levels of DBH in plasma may sometimes be elevated. In neuroblastoma only about half of the patients appear to have elevated levels of DBH (GOLDSTEIN et al. 1972). Although these patients tend to excrete increased levels of noradrenaline metabolites (3-methoxy-4-hydroxymandelic acid) there is not a good quantitative relationship between enzyme levels in plasma and metabolite excretion (BREWSTER and BERRY 1979). The wide breadth of normal DBH levels may obscure increases in the enzyme level in patients with phaeochromocytoma, but there is usually some decrease in the levels after removal of the tumor, sometimes to as low as 10 percent of preoperative levels (LOVENBERG et al. 1975; AUNIS et al. 1976; STONE et al. 1976; KOBAYASHI et al. 1978).

b. Hypertension

Although initial studies in relatively small numbers of subjects suggested that plasma DBH levels were significantly correlated with blood pressure or plasma noradrenaline levels, and that levels of the enzyme might be increased in hypertension, subsequent studies in large populations have not supported these conclusions. In a total community-based sample of 300 subjects, there was no significant relationship between blood pressure and plasma DBH (HEISS et al. 1980). Similarly, in a study of 1194 middle-aged men with diastolic blood pressures ranging from 75–125 mm Hg, there was no relation of pressure to enzyme levels (HUTTUNEN et al. 1979). In five other studies in which plasma DBH was determined in a total of 425 normotensive subjects and 355 hypertensive patients, there was no significant difference in mean enzyme levels (HORWITZ et al. 1973; ÅBERG et al. 1974b; GRANT et al. 1976; LAKE et al. 1977b; LEVY et al. 1979). In one study (LAWTON et al. 1979), hypertensive patients had a mean plasma enzyme level lower than the normotensive subjects. In those studies in which plasma noradrenaline levels were measured (GRANT et al. 1976; LAKE et al. 1977a), there was no relationship between the catecholamine and DBH levels.

Since it is likely that essential hypertension is not a homogenous disorder, attempts have been made to examine subclasses of hypertensives, but in none of the types is established hypertension consistently attended by altered DBH levels. There is some evidence of an association between high DBH levels and labile hypertension (STONE et al. 1974a; LEVY et al. 1979; ISEKI et al. 1979).

c. Other Disease States

As indicated earlier, the association of a disease state with altered plasma levels of DBH may reflect a genetic factor associated with inherited susceptibility to the disease, or may be a consequence of the disease state. Decreased levels of plasma DBH in patients with Familial Dysautonomia and their families (WEINSHILBOUM and ALEXROD 1971c; FREEDMAN et al. 1972) were later found (FREEDMAN et al. 1975) to appear only after age 5. Since some patients with Familial Dysautonomia have high levels of the enzyme, diminished levels of DBH are not a characteristic feature of the disease (FREEDMAN et al. 1975).

Patients with Down's Syndrome have lower levels of plasma DBH than their parents or siblings, and it appears that this is a feature of the disorder (WETTERBERG et al. 1972; COLEMAN et al. 1974), although there is no evidence of disturbed sympathetic function.

d. Experimental Alterations in Plasma DBH

The marked increase in plasma DBH levels in rats made diabetic by treatment with streptozotocin or alloxan (BERKOWITZ et al. 1980) and the glycerol-induced decrease in human serum DBH (KARAHASANOGLU et al. 1978) are two interesting observations which may provide a means for understanding further the mechanisms involved in the release and disposition of the enzyme.

E. Conclusion

Plasma levels of catecholamines usually reflect acute changes in sympathetic nerve ending and adrenal medullary release of noradrenaline and adrenaline, but changes in their rate of removal from blood can produce altered catecholamine levels. DBH levels are a poor index of sympathetic activity. Further understanding of the genetic factors and mechanisms in regulating levels of the enzyme are required before any significance can be attributed to alterations in its levels.

F. References

Åberg HE, Hansson HE, Wetterberg L, Ross SB, Forgen O (1974a) Dopamine-hydroxylase in human lymph. Life Sci 14:65–71

Åberg H, Wetterberg L, Ross SB, et al (1974b) Dopamine-β-hydroxylase in hypertension. Acta Med Scand 196:17–20

Adler-Graschinsky E, Langer SZ (1975) Possible role of a β-adrenoceptor in the regulation of noradrenaline release by nerve stimulation through a positive feedback mechanism. Brit J Pharmacol 53:43–50

Algate DR, Leach GDH (1978) Dopamine-beta-hydroxylase release following acute selective sympathetic nerve stimulation of the heart, spleen and mesentery. J Pharm Pharmacol 30:162–6

Anavekar SN, Louis WJ, Morgan TO, Doyle AE and Johnston CI (1975) The relationship of plasma levels of pindolol in hypertensive patients to effects on blood pressure, plasma renin and plasma noradrenaline levels. Clin Exp Pharmacol Physiol 2:203–212

Andren L, Hansson L, Björkman M, Jonsson A, Borg KO (1979) Hemodynamic and hormonal changes induced by noise. Acta Med Scand (Suppl) 625:13–8

Anton AH, Sayre DF (1962) A study of the factors affecting the aluminum oxide-trihydroxyindole procedure for the analysis of catecholamines. J Pharmacol Exp Ther 138:360–374

Arnaiz JM, Garcia AG, Horga JF, and Kirpekar SM (1978) Tissue and plasma catecholamines and dopamine-β-hydroxylase activity of various animal species after neurogenic sympathetic stimulation differences. J Physiol (London) 285: 515–529

Arnaiz JM, Garcia AG, Horga JF, Pascual R, Sanchez-Garcia P (1980) Origin of guinea-pig plasma dopamine-β-hydroxylase. Br J Pharmacol 69:41–48

Artal R, Glatz TH, Lam R, Nathanielsz PW, Hobel CJ (1979) The effect of acute maternal hemorrhage on the release of catecholamines in the pregnant ewe and the fetus. Am J Obstet Gynecol 135:818–822

Aunis D, Miras-Portugal MT, Coquillat G, Warter JM, Mandel P (1976) Plasma dopamine-β-hydroxylase in a noradrenaline secreting pheochromocytoma. Clin Chim Acta 70:455–458

Avakian EV, Horvath SM (1980) Plasma catecholamine responses to tyrosine hydroxylase inhibition and cold exposure. Life Sci 26:1691–1696

Axelrod J (1972) Dopamine-β-hydroxylase: regulation of its synthesis and release from nerve terminals. Pharmacol Rev 23:233–244

Barnes P, Fitzgerald G, Brown M, Dollery C (1980) Nocturnal asthma and changes in circulating epinephrine, histamine, and cortisol. N. Engl J Med 303:263–267

Barnes RF, Raskind M, Gumbrecht G, Halter JB (1982) The effects of age on the plasma catecholamine response to mental stress in man. J Clin Endocrinol Metab 54:64–69

Behrens WA, Depocas F (1975) Dopamine-β-hydroxylase in rat serum and lymph: changes with age and effect of cold exposure. Can J Physiol Pharmacol 53:1080–1088

Behrens WA, Lacelle S, Depocas F (1979) Blood dopamine-β-hydroxylase activity and noradrenaline levels in the developing white rat. Can J Physiol Pharmacol 57:380–384

Benedict CR, Grahame-Smith DG (1978) Plasma noradrenaline and adrenaline concentrations and dopamine-β-hydroxylase activity in patients with shock due to septicaemia, trauma and hemorrhage. QJ Med N Ser 185:1–20

Benedict CR, Pickering TG, Raine AEG (1977) Acute and long-term effects of α-adrenergic blockade on blood pressure and sympathetic activity in man. J Physiol (Lond) 271:35-36

Benedict CR, Fillenz M, Stanford C (1978) Changes in plasma noradrenaline concentration as a measure of release rate. Br J Pharmacol 64:305-309

Benedict CR, Fillenz M, Stanford C (1979) Noradrenaline release in rats during prolonged cold-stress and repeated swim-stress. Br J Pharmacol 66:521-524

Ben-Jonathan N, Porter JC (1976) A sensitive radioenzymatic assay for dopamine, norepinephrine and epinephrine in plasma and tissue. Endocrinology 98:1497-1507

Berkowitz BA, Head RJ (1978) Adrenal origin of plasma catecholamines after decapitation: a study in normal and diabetic rats. Br J Pharmacol 64:3-5

Berkowitz BA, Head R, John, T, Hempstead J (1980) Experimental diabetes: alterations in circulating dopamine-β-hydroxylase and norepinephrine. J Pharmacol Exp Ther 213:18-23

Bertel O, Buhler ER, Kiowski W, Lutold BE, Fritz R (1980) Decreased beta-adrenoceptor responsiveness as related to age, blood pressure, and plasma catecholamines in patients with essential hypertension. Hypertension 2:130-138

Bertel O, Buhler FR, Baitsch G, Ritz R, Burkart F (1982) Plasma adrenaline and noradrenaline in patients with acute myocardial infarction. Relationship to ventricular arrythmias of varying severity. Chest 82:64-68

Bevan JA (1977) Some functional consequences of variation in adrenergic synaptic cleft width and in nerve density and distribution. Fed Proc 36:2439-2443

Bolli G, Cartechini MG, Compagnucci P, Malvicini S, De Feo P, Santeusanio F, Angeletti G, Brunetti P (1979) Effect of metabolic control on urinary excretion and plasma levels of catecholamines in diabetics. Horm Metab Res 11:493-497

Bonelli J, Hornagel H, Burcke T, Magometschnigg D, Lochs H, Kaik G (1979) Effect of calculation stress on hemodynamics and plasma catecholamines before and after β-blockage with propranolol (inderal) and mepindolol sulfate (Corindolan). Eur J Clin Pharmacol 15:1-8

Brandsborg OK, Brandsborg M, Lövgren NA, Christensen NJ (1978) Increased plasma noradrenaline and serum gastrin in patients with duodenal ulcer. Eur J Clin Invest 8:11-14

Braunwald E, Harrison DC, Chidsey CA (1964) The heart as an endocrine organ. Am J Med 36:1-4

Bravo EL, Tarazi RC, Gifford RW, Stewart BH (1979) Circulating and urinary catecholamines in pheochromocytoma: diagnostic and pathophysiologic implications. N Engl J Med 301:682-686

Brecht HM, Ernst W, Koch K (1975) Plasma norepinephrine concentrations in regular hemodialysis patients. Kidney Int. 8:124

Brecht HM, Banthien F, Schoeppe W (1976) Decrease in plasma noradrenaline levels following long-term treatment with pindolol in patients with essential hypertension. Klin Wschr 54:1095-1105

Breggin PR (1964) The psychophysiology of anxiety. J Nerv Ment Dis 139:558-568

Brewster MA, Berry DH (1979) Serial studies of serum dopamine-β-hydroxylase and urinary vanillylmandelic and homovanillic acids in neuroblastoma. Med Pediatr Oncol 6:93-99

Brick I, Hutchison KJ, Roddie IC (1967) Inactivation of circulating catecholamines in the human forearm. Proc Physiol Soc 190:25-26

Britton BJ, Irving MH, Wood WG (1977) Effects of adrenoceptor blockade on plasma catecholamine levels during adrenaline infusion in sheep. Br J Pharmac 59:3-10

Brown M, Tache Y, Fisher D (1979) Central nervous system action of bombesin: mechanism to induce hyperglycemia. Endocrinology 105:660-665

Bühler HU, Da Prada M, Haefely W, Picotti GB (1978) Plasma adrenaline, noradrenaline and dopamine in man and different animal species. J Physiol (Lond) 276:311–320
Buu NT, Kuchel O (1977) A new method for the hydrolysis of conjugated catecholamines. J Lab Clin Med 90:680–685
Callingham BA (1975) Catecholamines in blood. In: Blaschko H, Sayers G, Smith AE (eds): Adrenal Gland. Am Physiol Soc Washington DC pp 427–445 (Handbook of Physiology, Section 7, Endocrinology. Vol 6)
Callingham BA, Barrand MA (1976) Catecholamines in blood. J Pharm Pharmacol 28:356–360
Cannon WB (1929) Organization for physiological homeostasis. Physiol Rev 9:399–431
Cannon WB, Rosenbluth A (1937) Autonomic Neuroeffector Systems. Macmillan, NY
Carlson LD (1966) The role of catecholamines in cold adaptation. Pharmacol Rev 18:291–301
Carriere S, Cardinal J, Legrimellec C (1979) Carotid sinus reflex and norepinephrine release following acute volume depletion in dogs. Can J Physiol Pharmacol 57:681–687
Catravas JD, Gillis CN (1980) Pulmonary Clearance of (^{14}C)-5-hydroxytryptamine and (^{3}H)norepinephrine in vivo: effects of pretreatment with imipramine or cocaine. J Pharmacol Exp Ther 312:120–127
Caviezel F, Picotti GB, Margonato A, Slaviero G, Galva MD, Camagna P, Bondiolotti GP, Carruba MO, Pozza G (1982) Plasma adrenaline and noradrenaline concentrations in diabetic patients with and without autonomic neuropathy at rest. Diabetologia 23:19–23
Chang PC, van Brummelen P, Vermeij P, and van der Krogt JA (1985) Neuronal and extraneuronal uptake of norepinephrine in the human forearm. J Hypertens 3:S141–S143
Chidsey CA, Harrison DC, and Braunwald E (1962) Augmentation of the plasma norepinephrine response to exercise in patients with congestive heart failure. N Engl J Med 267:650–654
Chiueh CC, Kopin IJ (1978a) Hyper-responsivity of spontaneously hypertensive rat to indirect measurement of blood pressure. Am J Physiol 3:H690–H695
Chiueh CC, Kopin IJ (1978b) Centrally mediated release by cocaine of endogenous epinephrine and norepinephrine from the sympatho-adrenal medullary system of unanesthesized rats. J Pharmacol Exp Ther 205:148–154
Christensen NJ (1972a) Increased levels of plasma noradrenaline in hypothyroidism. J Clin Endocrinol Metab 35:359–363
Christensen NJ (1972b) Plasma catecholamine in long-term diabetes with and without neuropathy in hyophysectomized subjects. J Clin Invest 51:779–787
Christensen NJ (1973) Plasma noradrenaline and adrenaline in patients with thyrotoxicosis and myxoedma. Clin Sci Mol Med 42:163–171
Christensen NJ (1974a) Hypoadrenalinemia during insulin hypoglycemia in children with ketotic hypoglycemia. J Clin Endocrinol Metab 38:107–112
Christensen NJ (1974b) Plasma norepinephrine and epinephrine in untreated diabetes, during fasting and after insulin administration. Diabetes 23:1–8
Christensen NJ (1979) Catecholamines and diabetes mellitus. Diabetologia 16:211–224
Christensen NJ (1982) Sympathetic nervous activity and age. Eur J Clin Invest 12:91–92
Christensen NJ, Brandsborg O (1973) The relationship between plasma catecholamine concentration and pulse rate during exercise and standing. Eur J Clin Invest 3:299–306

Christensen NJ, Videback J (1974) Plasma catecholamines and carbohydrate metabolism in patients with acute myocardial infarction. J Clin Invest 54: 278-286

Christensen NJ, Mathias CJ, Frankel HL (1976) Plasma and urinary dopamine: studies during fasting and exercise and in tetraplegic man. Eur J Clin Invest 6:403-409

Christensen NJ, Trap-Jensen J, Svendsen TL, Rasmussen S, and Neilsen PE (1978) Effect of labetalol on plasma noradrenaline and adrenaline in hypertensive man. Eur J Clin Pharmacol 14:227-230

Christensen NJ, Brandsborg O, Lovgreen NA, Brandsborg M (1979a) Elevated plasma noradrenaline concentrations in duodenal ulcer patients are not normalized by vagotomy. J Clin Endocrinol Metab 49:331-334

Christensen NJ, Galbo H, Hansen JF, Hesse B, Richter EA, Trap-Jensen J (1979b) Catecholamines and exercise. Diabetes 28, Suppl No 1, 58-62

Clutter WE, Bier DM, Shah SD, Cryer PE (1980) Epinephrine plasma metabolic clearance rates and physiologic thresholds for metabolic and hemodynamic actions in man. J Clin Invest 66:94-101

Cody RJ, Franklin KW, Kluger J, Laragh JH (1982) Sympathetic responsiveness and plasma norepinephrine during therapy of chronic congestive heart failure with captopril. Am J Med 72 (5):791-797

Cohen G, Goldenberg M (1957) The simultaneous fluorimetric determintion of adrenaline and noradrenaline in plasma. II Peripheral venous plasma concentrations in normal subjects and in patients with pheochromocytoma. J Neurochem 2:71-80

Cohen G, Holland B, Sha J, and Goldenberg M (1959) Plasma concentrations of epinephrine and norepinephrine during intravenous infusions in man. J Clin Invest 38:1935-1941

Coleman M, Campbell M, Freedman LS, et al (1974) Serum dopamine-β-hydroxylase levels in Down's syndrome. Clin Genet 5:312-315

Cooksey JD, Reilly P, Brown S, Bomze H, and Cryer PE (1978) Exercise training and plasma catecholamines in patients with ischemic heart disease. Am J Cardiol 42:372-376

Corea L, Miele N, Bentivoglio M, Boschetti E, Agabitirosei E, and Muiesan G (1979) Acute and chronic effects of nifedipine on plasma renin activity and plasma adrenaline and noradrenaline in controls and hypertensive patients. Clin Sci 57:115s-117s

Corr PB, Gillis RA (1978) Autonomic neural influences on the dysrhythmias resulting from myocardial infarction. Cir Res 43:1-9

Coulombe, Dussault JH, Walker P (1976) Plasma catecholamine concentrations in hyperthyroidism and hypothyroidism. Metabolism 25:973-979

Cousineau D, Dechamplain J, and Lapointe L (1978) Circulating catecholamines and systolic time intervals in labile and sustained hypertension. Clin Sci Mol Med 55:65s-68s

Cryer PE, Weiss S (1976) Reduced plasma norepinephrine response to standing in autonomic dysfunction. Arch Neurol 33:275-277

Cryer PE, Santiago JV, Shah SD (1974) Measurement of norepinephrine in small volumes of human plasma by a single isotope derivative method: response to the upright posture. J Clin Endocrinol Metab 39:1025-1029

Cryer PE, Haymond MW, Santiago JV, Shah SD (1976) Norepinephrine and epinephrine release and adrenergic mediation of smoking associated hemodynamic and metabolic events. N Engl J Med 295-573

Cryer PE, Silverberg AB, Santiago JV, Shah SD (1978) Plasma catecholamines in diabetes. The syndromes of hypoadrenergic and hyperadrenergic postural hypotension. Am J Med 64:407-416

Cryer PE, Rizza RA, Hammond MW, and Gerich JE (1980) Epinephrine and norepinephrine are cleared through β-adrenergic but not α-adrenergic mechanisms in man. Metabolism 30:1114–1118
Cubeddu LX, Santiago E, Talmaciu R, Pina Di G (1977) Adrenal origin of the increase in plasma dopamine β-hydroxylase and catecholamines induced by hemorrhagic hypotension in dogs. J Pharmacol Exp Ther 203:587–597
Cubeddu LX, Barbella YR, Marrero A, Trifaro J, Israel AS (1979) Circulating pool and adrenal soluble content of dopamine beta-hydroxylase (DBH), in rats, guinea pigs, dogs and humans: their role in determining acute stress-induced changes in plasma enzyme levels. J Pharmacol Exp Ther 211:271–279
Dancis J, Smith AA (1966) Familial dysautonomia. N Engl J Med 274:207–209
Da Prada M, Zürcher G (1976) Simultaneous radio-enzymatic determination of plasma and tissue adrenaline, noradrenaline and dopamine within the femtomole range. Life Sci 19:1161–1174
Davies RA, Patt NL, Sole MJ (1979) Localization of pheochromocytoma by selective venous catheterization and assay of plasma catecholamines. J Canadian Med Assoc J 120:539–542
Dean CR, Maling T, Dargie HJ, Reid JL, and Dollery CT (1980) Effect of propranolol on plasma norepinephrine during sodium nitroprusside-induced hypotension. Clin Pharmacol Ther 27:156–164
Defronzo RA, Hendler R, Christensen N (1980) Stimulation of counterregulatory hormonal responses in diabetic man by a fall in glucose concentration. Diabetes 29:125–131
De La Vega CE, Slater S, Ziegler MG, Lake CR, and Murphy DL (1977) Reduction in plasma norepinephrine during fenfluramine treatment. Clin Pharmacol Ther 21:216–221
De Leeuw PW, Falke HE, Kho TL, Vandongen R, Wester A, Birkenhager WH (1977) Effects of beta-adrenergic blockade on diurnal variability of blood pressure and plasma noradrenaline levels. Act Med Scand 202:389–392
Depocas F, Behrens WA, Foster DO (1978) Noradrenaline-induced calorigenesis in warm- and in cold-acclimated rats: the interrelation of dose of noradrenaline, its concentration in arterial plasma, and calorigenic response. Can J Physiol Pharmacol 56:168–174
Deturk KH, Vogel WH (1980) Factors influencing plasma catecholamine levels in rats during immobilization. Pharmacol Biochem Behav 13:129–131
Dimsdale JE, Moss J (1980) Plasma catecholamines in stress and exercise. JAMA 243:340–342
Dirocco RJ, Grill HJ (1979) The forebrain is not essential for sympathoadrenal hyperglycemic response to glucoprivation. Science 204:1112–1113
Distler A, Keim HJ, Cordes U, Philipp T, Wolff HP (1978) Sympathetic responsiveness and antihypertensive effect of beta-receptor blockade in essential hypertension. Am J Med 64:446–451
Doxey JC, Everitt J (1977) Inhibitory effects of clonidine on responses to sympathetic nerve stimulation in the pithed rat. Br J Pharmacol 61:559–566
Dubb JW, Stote RM, Alexander F, Intoccia AP, Geczy M, Pendleton RG (1979) Studies with a PNMT inhibitor. Clin Pharmacol Ther 25:837–843
Dunnette J, Weinshilboum R (1976) Human serum dopamine-β-hydroxylase: correlation of enzymatic activity with immunoreactive protein in genetically defined samples. Am J Hum Genet 28:155–166
Dunnette J, Weinshilboum R (1977) Inheritance of low immunoreactive human plasma dopamine-β-hydroxylase: radioimmunoassay studies. J Clin Invest 60:1080–1087

Ehrhardt JD, Schwartz J (1978) A gas chromatography-mass spectrometry assay of human plasma catecholamines. Clin Chim Act 88:71–79

Eliot RJ, Lam R, Leake RD, Hobel CJ, Fisher DA (1980) Plasma catecholamine concentrations in infants at birth and during the first 48 hours of life. J Pediatr 96:311–315

Endert E (1979) Determination of noradrenaline and adrenaline in plasma by a radioenzymatic assay using high pressure liquid chromatography for the separation of the radiochemical products. Clin Chem Acta 96:233–239

Engelman K (1977) Phaeochromocytoma. Clin Endocrinol Metab 6:769–797

Engelman K, Portnoy B (1970) Sensitive double-isotope derivative assay for norepinephrine and epinephrine. Circ Res 26:53–57

Esler M, Jackman G, Bobik A, Kelleher D, Jennings G, Leonard P, Skews H, Korner P (1979) Determination of norepinephrine apparent release rate and clearance in humans. Life Sci 25:1461–1470

Esler M, Jackman G, Kellerher D, Skews H, Jennings G, Bobik A, Korner P (1980) Norepinephrine kinetics in patients with idiopathic autonomic insufficiency. Circ Res 46:I-47–I-48

Esler M, Jackman G, Leonard P, Skews H, Bobik A, Korner P (1981) Effect of norepinephrine uptake blockers on norepinephrine kinetics. Clin Pharmacol Ther 29:No 1, 12–10

Esler M, Leonard P, O'Dea K, Jackman G, Jennings G, and Korner P (1982) Biochemical quantification of sympathetic nervous activity in humans using radiotracer methodology: fallibility of plasma noradrenaline measurements. J Cardiovascular Pharmacol 4:S152–S157

Esler M, Jennings G, Leonard P, Sacharias N, Burke F, Johns J, Blombery P (1984) Contribution of individual organs to total noradrenaline release in humans. Acta Physiol Scand, Suppl 527:11–16

Euler, US von (1948) Identification of the sympathomimetic ergone in adrenergic nerves of cattle (sympathin N) with laevo-noradrenaline. Acta Physiol Scan 16:63–74

Euler US, von (1959) The development and applications of the trihydroxyindole method for catecholamine. Pharmacol Rev 11:262–268

Euler US von, Gemzell CA, Strom G, Westman A (1955) Report of a case of phaeochromocytoma, with special regard to preoperative diagnostic problems. Acta Med Scand 153:127–136

Evans MI, Halter JB, Porte D (1978) Comparison of double- and single-isotope enzymatic derivative methods for measuring catecholamines in human plasma. Clin Chem 24:657–670

Falkner B, Onseti G, Angelakos ET, Fernandes M, Langman C (1979) Cardiovascular response to mental stress in normal adolescents with hypertensive parents. Hypertension 1:23–30

Farnebo LO, Hallman H, Hamberger B, Jonsson G (1979) Catecholamines and hemorrhagic shock in awake and anesthetized rats. Cir Shock 6:109–118

Fisher DA, Brown MR (1980) Somatostatin analog: plasma catecholamine suppression mediated by the central nervous system. Endocrinol 107:714–718

Fitzgerald GA, Hossmann V, Hamilton CA, Reid JL, Davies DS, Dollery CT (1979) Interindividual variation in kinetics of infused epinephrine. Clin Pharmacol Therap 26:669–674

Folkow B (1975) Relationship between physical vascular properties and smooth muscle function; its importance for vascular control and reactivity. Clin Exp Pharmacol Physiol 2:55–61

Folkow B, Hallback M, Lundgren Y, Weiss L (1970) Background of increased flow resistance and vascular reactivity in spontaneously hypertensive rats. Acta Physiol Scand 80:93-106

Franco-Morselli R, Elghozi JL, Joly E, Di Giuilio S, Meyer P (1977) Increased plasma adrenaline concentrations in benign essential hypertension. Brit Med J 2:1251-1254

Franco-Morselli R, Baudouin-Legros M, Meyer P (1978) Plasma adrenaline and noradrenaline in essential hypertension and after longterm treatment with α-adrenoreceptor-blocking agents. Clin Sci Mol Med 55:97s-100s

Franz IW, Lohmann FW, Koch G (1980) Differential effects of longterm cardioselective and nonselective beta-receptor blockade on plasma catecholamines during and after physical exercise in hypertensive patients. J. Cardiovasc Pharmacol 2:35-44

Fredholm BB, Farnebo LO, Hamberger B (1979) Plasma catecholamines, cyclic AMP and metabolic substrates in hemorrhagic shock of the rat. The effect of adrenal demedullation and 6-OH-dopamine treatment. Acta Physiol Scand 105:481-495

Freedman LS, Ohuchi T, Goldstein M, Axelrod F, Fish I, Dancis J (1972) Changes in serum dopamine-β-hydroxylase with age. Nature (Lond.) 236:310-311

Freedman LS, Ebstein RP, Goldstein M, Axelrod FB, Dancis J (1975) Serum dopamine-beta-hydroxylase in familial dysautonomia. J Lab Clin Med 85:1008-1010

Frewin DB, Downey JA, Levitt M (1973) The effect of heat, cold, and exercise on plasma dopamine-β-hydroxylase activity in man. Can J Physiol Pharmacol 51:986-989

Gaddum JH (1959) Bioassay procedures. Pharmacol Rev 11:241-249

Galbo H, Holst JJ, Christensen NJ (1975) Glucagon and plasma catecholamine responses to graded and prolonged exercise in man. J Appl Physiol 38:70-75

Galbo H, Richter EA, Holst JJ, Christensen NJ (1977) Diminished hormonal responses to exercise in trained rats. J Appl Physiol 43:953-958

Galbo H, Richter EA, Christensen NJ, Holst JJ (1978) Sympathetic control of metabolic and hormonal responses to exercise in rats. Acta Physiol Scand 102:441-449

Galbo H, Holst JJ, Christensen NJ (1979) The effect of different diets and of insulin on the hormonal response to prolonged exercise. Acta Physiol Scand 107:19-32

Ganguly A, Henry DP, Yune HY, Pratt IH, Grim CE, Donohue JP, Weinberger MH (1979) Diagnosis and localization of phaeochromocytoma: detection by measurement of urinary norepinephrine excretion during sleep, plasma norepinephrine concentration and computerized axial tomography (CT-Scan). Am J Med 67:21-26

Garber AJ, Cryer PE, Santiago JV, Haymond MW, Pagliara AS, Kipnis DM (1976) The role of adrenergic mechanisms in the substrate and hormonal response to insulin-induced hypoglycemia in man. J Clin Invest 58:7-15

Ghione S, Palombo C, Pellegrini M, Fommei E, Pilo A, Donato L (1978) The kinetics of plasma noradrenaline in normal and hypertensive subjects. Clin Sci Mol Med 55:89s-92s

Gillespie JS, Muir TC (1967) A method of stimulating the complete sympathetic outflow from the spinal cord to blood vessels. Br J Pharmacol 30:78-87

Gillis CN, Green NM, Cronau LH, Hammond GL (1972) Pulmonary extraction of 5-hydroxytryptamine and norepinephrine before and after cardiopulmonary bypass in man. Cir Res 30:666-674

Gillis CN, Huxtable RJ, Roth RA (1978) Effects of monocrotaline pretreatment of rats on removal of 5-hydroxytryptamine and noradrenaline by perfused lung. Br J Pharmacol 63:435-443

Gitlow SE, Mendlowitz M, Wilk EK, Wilk S, Wolf RL, Naftchi NE (1964) Plasma clearance of dl-7-H^3-norepinephrine in normal human subjects and patients with essential hypertension. J Clin Invest 43:2009-2015

Glass DC, Karakoff LR, Contrada R, Hilton WF, Kehoe K, Mannucci EG, Collins C, Snow B, Elting E (1980) Effect of harassment and competition upon cardiovascular and plasma catecholamine responses in Type A and Type B individuals Psychophysiology 17:453-462

Goldberg MR, Hollister AS, Robertson D (1983) Influence of yohimbine on blood pressure, autonomic reflexes, and plasma catecholamines in humans. Hypertension 5:772-778

Goldstein DS, Feuerstein G, Izzo JL, Kopin IJ, and Keiser HR (1981) Validity and reliability of high pressure liquid chromotography with electrochemical detection for measuring plasma levels of norepinephrine and epinephrine in man. Life Sci 28:467-475

Goldstein DS, Nurnberger J JR, Simmons S, Gershon ES, Polinsky R, and Keiser HR (1982) Effects of injected sympathomimetic amines on plasma catecholamines and circulatory variables in man. Life Sci.

Goldstein DS, Horwitz D, Keiser HR, Polinsky RJ, Kopin IJ (1983) Plasma l-(^{3}H)-norepinephrine, d-(^{14}C)-norepinephrine, and d,l-(^{3}H)-isoproterenol kinetics in essential hypertension. J Clin Invest 72:1748-1758

Goldstein DS, Zimlichman R, Stull R, Folio J, Levinson PD, Keiser HR, Kopin IJ (1985) Measurement of regional neuronal removal of norepinephrine in man. J Clin Invest 76:15-21

Goldstein DS, Zimlichman R, Stull R, Keiser HR, Kopin IJ (1986) Estimation of intrasynaptic norepinephrine. Hypertension 8:471-475

Goldstein DS, Eisenhofer G, Sax F, Keiser HR, Kopin IJ (1987) Plasma norepinephrine pharmacokinetics during mental challenge. Psychosomatic Medicine in press.

Goldstein M, Freedman LS, Bonnay M (1971) An assay for dopamine-β-hydroxylase activity in tissues and serum. Experientia 27:632-633

Goldstein M, Freedman L, Bohuon AC, Guerinot (1972) Serum dopamine-β-hydroxylase activity in neuroblastoma. N Engl J Med 286:1123-1125

Graf CJ, Rossi NP (1978) Catecholamine response to intracranial hypertension. J. Neurosurg. 49:862-868

Grant C, Routh JI, Lawten W, and Witte D (1976) The effects of therapy for mild hypertension on circulating levels of dopamine-beta-hydroxylase. Chem. Acta 69:333-340

Griffiths J, Leung F (1971) The sequential estimation of plasma catecholamines and whole blood histamine in myocardial infarction. Am Heart J 82:171-179

Grobecker H, Roizen MF, Kopin IJ (1977) Effect of tyramine and guanethidine on dopamine-β-hydroxylase activity and norepinephrine concentrations, in vesicular fraction of the heart and plasma of rats. Life Sci 20:1099-1016

Gross HA, Lake CR, Ebert MH, Ziegler MG, Kopin IJ (1979) Catecholamine metabolism in primary anorexia nervosa. J Clin Endocrinol Metab 49:805-809

Grossman SH, Davis D, Cunnells JC, Shand DG (1982) Plasma norepinephrine in the evaluation of baroreceptor function in humans. Hypertension 4:566-571

Grzanna R, Coyle JT (1977) Immunochemical studies on the turnover of rat dopamine-β-hydroxylase. Mol Pharmacol 13:956-964

Grzanna R, Coyle JT (1978) Absence of a relationship between sympathetic neuronal activity and turnover of serum dopamine-β-hydroxylase. Naunyn-Schmiedeberg's Arch Pharmacol 304:231-236

Gundersen HJG, Christensen NJ (1977) Intravenous insulin causing loss of intravascular water and albumin and increased adrenergic nervous activity in diabetics. Diabetes 26:551-557

Hagberg JM, Hickson RC, McLane JA, Ehsani AA, Widner WW (1979) Disappearance of norepinephrine from the circulation following strenuous exercise. J Appl Physiol 47(6):1311–1314

Häggendal J (1963) An improved method for fluorimetric determination of small amounts of adrenaline and noradrenaline in plasma and tissues. Acta Physiol Scand 59:242–254

Hallman H, Farnebo LO, Hamberger B, Jonsson G (1978) A sensitive method for the determination of plasma catecholamines using liquid chromatography with electrochemical detection. Life Sci 23:1049–1052

Halter JB, Pflug AE, Tolas AG (1980) Arterial-venous differences of plasma catecholamines in man. Metab 29:9–12

Haneda T, Miura Y, Miyazawa K, Honna T, Arai T, Nakajima T, Miura T, Yoshinaga K, Takishima T (1978) Plasma norepinephrine concentration in the coronary sinus in cardiomyopathies. Cathet Cardiovasc Diagn 4:399–405

Hansen JF, Christensen NJ, Hesse B (1978) Determinants of coronary sinus noradrenaline in patients with ischaemic heart disease: coronary sinus catecholamine concentration in relation to arterial catecholamine concentration, pulmonary artery oxyen saturation and left ventricular end-diastolic pressure. Cardiovasc Res 12:415–421

Hansen JF, Hesse B, and Christensen NJ (1979) Enhanced sympathetic nervous activity after intravenous propranolol in ischaemic heart disease: plasma noradrenaline, splanchinic blood flow and mixed venous oxygen saturation at rest and during exercise. Eur J Clin Invest 8:31–36

Hansson BG, Hökfelt B (1976) Long term treatment of moderate hypertension with penbutolol (Hoe 893d). II. Effect on the response of plasma catecholamines and plasma renin activity to insulin-induced hypoglycemia. Eur J Clin Pharmacol 9(4):245–251

Head RJ, Berkowitz BA (1979) Vascular release of catecholamines. J Pharm Pharmacol 31(4):266–277

Heiser JF, Defrancisco D (1976) The treatment of pathological panic states with propranolol. Am J Psychiatry 133:1389–1394

Heiss G, Tyroler HA, Gunnells JC, McGuffin WL, Hames CG (1980) Dopamine-beta-hydroxylase in a biracial community: demographic, cardiovascular and familiar factors. J Chronic Dis 33:301–310

Henrich W, Katz F, Molinoff P, Schrier R (1977) Competitive effect of hypokalemia and volume depletion on plasma renin activity, aldosterone and catecholamine concentrations in hemodialysis patients. Kidney Int 12:279–284

Henry DP, Starman BJ, Johnson DG, Williams RH (1975) A sensitive radioenzymatic assay for norepinephrine in tissue and plasma. Life Sci 16: 375–384

Henry DP, Luft FC, Weinberger MH, Fineberg NS, Grim CE (1980) Norepinephrine in urine and plasma following provocative maneuvers in normal and hypertensive subjects. Hypertension 2:20–28

Hickler RB, Hamlin JT, Wells RE (1959) Plasma norepinephrine response to tilting in essential hypertension. Circulation 20:422–426

Hickson RC, Hagberg JM, Conlee RK, Jones DA, Ehsani AA, Winder WW (1979) Effect of training on hormonal responses to exercise in competitive swimmers. Eur J Appl Physiol 41:211–219

Hilsted J, Galbo H, Christensen NJ (1980) Impaired responses of catecholamines, growth hormone, and cortisol to graded exercise in diabetic autonomic neuropathy. Diabetes 29:257–262

Hjemdahl P, Belfrage E, Daleskog M (1979a) Vascular and metabolic effects of circulating epinephrine and norepinephrine concentration: concentration-effect study in dogs. J Clin Invest 64:1221–1228

Hjemdahl P, Daleskog M, Kahan T (1979b) Determination of plasma catecholamines by high performance liquid chromatography with electrochemical detection: comparison with a radioenzymatic method. Life Sci 25:131-138

Hjemdahl R, Freyschuss U, Juhlin-Danfelt A, Linde B (1984) Differentiated sympathetic activation during mental stress evoked by the Stroop test. Acta Physiol Scand, Suppl 572:25-29

Hökfelt B, Hedeland H, Hansson BG (1975) The effect of clonidine and penbutolol, respectively, on catecholamines in blood and urine, plasma renin activity and urinary aldosterone in hypertensive patients. Arch. Int. Pharmacodyn 213:307-321

Horwitz D, Alexander RW, Lovenberg W, Keiser HR (1973) Human serum dopamine-β-hydroxylase. Circ Res 32:594-599

Huttunen JK, Pispa J, Kumlin T, Mattila S, Naukkarinen V, Miettinen T (1979) Lack of correlation between serum dopamine-beta-hydroxylase activity and blood pressure in middle-aged men. Hypertension 1:47-52

Ibsen H, Christensen NJ, Hollangel H, Leth A, Kappelgaard AM, Giese J (1980) Plasma noradrenaline concentration in hypertensive and normotensive forty-year-old individuals: relationship to plasma renin concentration. Scand J Clin Lab Invest 40:333-339

Irving MH, Britton BJ, Wood WG, Padgham C, Carruthers M (1974) Effects of adrenergic blockade on plasma catecholamines in exercise. Nature 248:531-533

Iseki F, Kuchii M, Nishio I, Masuyama Y (1979) The evaluation of plasma dopamine-beta-hydroxylase activity in essential and secondary hypertension. Jpn Heart J 20:307-320

Iversen LL, Jarrot B (1970) Modification of an enzyme radiochemical assay procedure for noradrenaline. Biochem Pharmacol 19:1841-1844

Iwasawa Y, Gillis CN (1974) Pharmacological analysis of norepinephrine and 5-hydroxy-tryptamine removal from the pulmonary circulation: differentiation of uptake sites. J Pharmacol Exp Ther 188:386-393

Joh TH, Ross RA, Reis DJ (1974) A simple sensitive assay for dopamine-β-hydroxylase. Anal Chem 62:248-254

Johnson DG, Hayward JS, Jacobs TP, Collis ML, Eckerson JD, Williams RH (1977) Plasma norepinephrine responses in man in cold water. J Appl Physiol Respirat Environ Exercise Physiol 43:216-221

Jones CT, Robinson RO (1975) Plasma catecholamines in foetal and adult sheep. J Physiol (Lond) 248:15-25

Jones DH, Hamilton CA, Reid JL (1978) Plasma noradrenaline, age and blood pressure: a population study. Clin Sci Mol Med 55:73s-75s

Jones DH, Allison DJ, Hamilton CA, Reid JL (1979) Selective venous sampling in the diagnosis and localization of phaeochromocytoma. Clin Endocrinol (Oxf) 10:179-186

Jones DH, Daniel J, Hamilton CA, Reid JL (1980) Plasma noradrenaline concentration in essential hypertension during long-term adrenoceptor blockade with oxprenolol. Br J Clin Pharmacol 9:27-31

Jung RT, Shetty PS, Barrand M, Callingham BA, James WPT (1979) Role of catecholamines in hypotensive response to dieting. Br Med J 1:12-13

Kadowitz PJ, Hyman AL (1973) Effect of sympathetic nerve stimulation on pulmonary vascular resistance in the dog. Cir Res 32:221-227

Karahasanoglu AM, Tildon JJ, Ozand PT, Maclaren NK (1978) Glycerol-induced changes in human serum dopamine-beta-hydoxylase activity. Biochem Pharmacol 27(19):2369-2371

Karlsberg RP, Cryer PE, Roberts R (1981) Serial plasma catecholamine response early in the course of clinical acute myocardial infarction: relationship to infarct extent and mortality. Am Heart J 102(1):27-31

Kato T, Ikuta K, Nagatso T, Takahashi H (1976) Changes in dopamine-β-hydroxylase activity of monkey plasma with age. Experientia (Basel) 32:834-835

Kato T, Hashimoto Y, Nagatsu T, Shinoda T, Okada T, Takeuchi T, Umezawa H (1980) 24-hour rhythm of human plasma noradrenaline and the effect of fusaric acid, a dopamine-β-hydroxylase inhibitor. Neuropsychobiology 6:61-65

Kielholz P (Ed) (1977) Beta-blockers and the central nervous system. University Park Press Baltimore-London-Tokyo

Kim Young D, Lake CR, Lees DE, Schuette WH, Bull JM, Weise V, Kopin IJ (1979) Hemodynamic and plasma catecholamine responses to hyperthermic cancer therapy in humans. Am J Physiol 237:H570-H574

Kissinger PT, Refshauge C, Dreiling R, Adams RN (1973) An electrochemical detector for liquid chromatography with picogram sensitivity. Anal Ltrs 6:465-477

Klaniecki TS, Corder CN, McDonald RH, Feldman JA (1977) High-performance liquid chromatographic radioenzymatic assay for plasma catecholamines. J Lab Clin Med 90:604-612

Kobayashi K, Miura Y, Tomioka H, Sakuma H, Adachi M, Sato T, Yoshinaga K (1978) Exocytosis plays an important role in catecholamine secretion from human pheochromocytoma. Clin Chim Acta 85:159-165

Kobinger W (1975) Central cardiovascular actions of clonidine. In: Central Action of Drugs in Blood Pressure Regulation. Turnbridge Well, Kent, Pitman Medical Publishing Company Ltd

Kopin IJ (1979) Plasma catecholamines in human and experimental hypertension. In Meyer P, Schmitt H (eds) Nervous System and Hypertension. John Wiley and Sons New York, pp 267-276

Kopin IJ, McCarty R, Yamaguchi I (1980a) Plasma catecholamines in human and experimental hypertension. Clin Exp Hypertens 2:379-394

Kopin IJ, Goldstein DS, Feuerstein GZ (1980b) The sympathetic nervous system and hypertension. In Laraugh JH, Bühler FH, Selden EW (eds) Frontiers in Hypertension Research. Springer New York

Kopin IJ, Zukowska-Grojec Z, Bayorh MA, Goldstein DS (1984) Estimation of intrasynaptic norepinephrine concentrations at vascular neuroeffector junctions in vivo. Naunyn-Schmiedeberg's Arch Pharmacol 325:298-305

Kvetnansky R, Sun CL, Lake CR, Thoa NB, Torda T, Kopin IJ (1978) Effect of handling and forced immobilization on rat plasma levels of epinephrine, norepinephrine, and dopamine-β-hydroxylase. Endocrinology 103:1868-1874

Kvetnansky R, McCarty R, Thoa NB, Lake CR, Kopin IJ (1979a) Sympatho-adrenal responses of spontaneously hyertensive rats to immobilization stress. Am J Physiol 236:H457-H462

Kvetnansky R, Weise VK, Thoa NB, Kopin IJ (1979b) Effects of chronic guanethidine treatment and adrenal medullectomy on plasma levels of catecholamines and corticosterone in forcibly immobilized rats. J Pharmacol Exp Ther 209:287-291

Labrosse EH, Axelrod J, Kopin IJ, Kety SS (1961) Metabolism of 7-H^3-epinephrine-D-bitartrate in normal young men. J Clin Invest 40:253-260

Lake CR, Ziegler MG (1978) Effect of acute volume alterations on norepinephrine and dopamine-β-hydroxyplase in normotensive and hypertensive subjects. Circulation 57:774-778

Lake CR, Ziegler MG, Kopin IJ (1976) Use of plasma norepinephrine for evaluation of sympathetic neuronal function in man. Life Sci 18:1315-1325

Lake CR, Ziegler MG, Coleman MD, Kopin IJ (1977a) Age-adjusted plasma norepinephrine levels are similar in normotensive and hypertensive subjects. N. Engl J Med 296:208-209

Lake CR, Ziegler MG, Coleman MD, Kopin IJ (1977b) Lack of correlation of plasma norepinephrine and dopamine-β-hydroxylase in hypertensive and normotensive subjects. Circ Res 41:865-869

Lake CR, Mikkelsen EJ, Rapoport JL, Zavadil AP, Kopin IJ (1979a) Effect of imipramine on norepinephrine and blood pressure in enuretic boys. Clin Pharmacol Ther 26:647-653

Lake CR, Ziegler MG, Coleman MD, Kopin IJ (1979b) Hydrochlorothiazide-induced sympathetic hyperactivity in hypertensive patients. Clin Pharmacol Ther 26:428-432

Lake CR, Ziegler M, Coleman MD, Kopin IJ (1979c) Plasma levels of norepinephrine and dopamine-β-hydroxylase in CRF patients treated with dialysis. Cardiovasc Med 4:1099-1111

Lawton WJ, Fitz A, Grant C, Witte DL (1979) Dopamine-betahydroxylase and plasma renin activity in patients with low-, normal-, and high-renin essential hypertension. Circulation 59:1063-1069

Leblanc J, Cote J, Jobin M, Labrie A (1979) Plasma catecholamines and cardiovascular responses to cold and mental activity. J Appl Physiol: Respirat Environ Exercise Physiol 47:1207-1211

Leon AS, Thomas PE, Sernatinger E, Canlas A (1974) Serum dopamine-beta-hydroxylase activity as an index of sympathetic activity. J Clin Pharmacol 14:354-362

Leveston SA, Shah SD, Cryer PE (1979) Cholinergic stimulation of norepinephrine release in man: evidence of a sympathetic postganglionic axonal lesion in diabetic adrenergic neuropathy. J Clin Invest 64:374-380

Levin BE, Hubschmann OR (1980) Dorsal column stimulation: effect on human cerebrospinal fluid and plasma catecholamines. Neurology (NY) 30:65-71

Levy SB, Frigon RF, Stone RA (1979) Plasma dopamine-β-hydroxylase activity and blood pressure variability in hypertensive man. Clin Endocrinol 11:187-199

Lewis PJ, Haeusler G (1975) Reduction on sympathetic nervous activity as a mechanism for the hypotensive effect of propranolol. Nature 256:440-441

Lilavivathana U, Brodows RG, Woolf PD, Campbell RG (1979) Counterregulatory hormonal responses to rapid glucose lowering in diabetic man. Diabetes 28:873-877

Louis WJ, Doyle AE, Anavekar SN (1975) Plasma noradrenaline concentration and blood pressure in essential hypertension, phaeochromocytoma and depression. Clin Sci Mol Med 48:239s-242s

Lovenberg W, Goodwin JR, Wallace EF (1975) Molecular properties and regulation of dopamine-β-hydroxylase. In: Neurobiological Mechanisms of Adaptation and Behavior. New York, Raven Press

Lowenstein J (1980) Drugs five years later: clonidine. Ann Intern Med 92:74-77

Lund A (1950) Simultaneous fluorimetric determinations of adrenaline and noradrenaline in blood. Acta Pharmacol Toxicol 6:137-146

McCall RB, Schuette MR, Humphrey SJ, Lahti RA, Barsuhn C (1983) Evidence for a central sympathoexcitatory action of alpha-2 adrenergic agonists. J Pharmacol Exp Therap 224:501-507

McCarty R, Kopin IJ (1978b) Changes in plasma catecholamines and behavior of rats during the anticipation footshock. Hormones Behav 11:248-257

McCarty R, Kopin IJ (1978c) Sympatho-adrenal medullary activity and behaviour during exposure to footshock stress: a comparison of seven rats strains. Physiol Behav 21:567-572

McCarty R, Kopin IJ (1979) Stress-induced alterations in plasma catecholamines and behavior of rats: effects of chlorisondamine and bretylium. Behav Neurl Biol 27:249-265

Maickel RP, Matussek N, Stern DN, and Brodie BB (1967) The sympathetic nervous system as a homeostatic mechanism. I. Absolute need for sympathetic nervous function in body temperature maintenance of cold-exposed rats. J Pharmacol Exp Ther 157:103-110

Maling TJB, Dollery CT, and Hamilton CA (1979) Clonidine and sympathetic activities during sleep. Clin Sci 57:509-514

Markianos E, Beckmann H (1976) Diurnal changes in dopamine-β-hydroxylase, homovanillic acid and 3-methoxy-4-hydroxyphenylglycol in serum of man. J Neural Transm 39:79-93

Mathias CJ, Christensen NJ, Corbett JL, Phil D, Frankel HL, and Spalding JMK (1976) Plasma catecholamines during paroxysmal neurogenic hypertension in quadriplegic man. Circ Res 39:204-208

Mathias CJ, Christensen NJ, Frankel HL, and Spalding JM (1979) Cardiovascular control in recently injured tetraplegics in spinal shock. Q J Med 48(190):273-287

Metz SA, Halter JB, Porte D, and Robertson RP (1978) Suppression of plasma catecholamines and flushing by clonidine in man. J Clin Endocrinol Metab 46:83

Micalizzi ER, Pals DT (1979) Evaluation of plasma norepinephrine as an index of sympathetic neuron function in the conscious, unrestrained rat. Life Sci 24:2071-2076

Miller JZ, Luft FC, Grim CE, Henry DP, Christian JC, and Weinberger MH (1980) Genetic influences on plasma and urinary norepinephrine after volume expansion and contraction in normal men. J Clin Endorcrinol Metab 50:219-222

Mitchell HC, Pettinger WA (1980) The hypernoradrenergic state in vasodilator drug-treated hypertension patients: effect of clonidine. J Cardiovasc Pharmacol 2:1-7

Miura Y, Haneda T, Sato T, Miyazawa K, Sakuma H, Kobayashi K, Minai K, Shirato K, Honna T, Takishima T, and Toshinoga K (1976) Plasma catecholamine levels in the coronary sinus, aorta, and femoral vein of subjects undergoing cardiac catheterization at rest and during exercise. Jpn Circ J 40:929-934

Mogensen CE, Christensen NJ, and Gunderson HJG (1980) The acute effect of insulin on heart rate, blood pressure, plasma noradrenaline and urinary albumin excretion. Diabetologia 18:453-457

Molinoff PB, Weinshilboum R, and Axelrod J (1971) A sensitive enzymatic assay for dopamine-β-hydroxylase. J Pharmacol Exp Ther 178:425-431

Morganti A, Pickering TG, Lopez-Ovejero JA, and Laragh JH (1979) Contrasting effects of acute beta blockade with propranolol on plasma catecholamines and renin in essential hypertension: a possible basis for the delayed antihypertensive response. Am Heart J 98:490-494

Mueller HS, Ayres SM (1980) Propranolol decreases sympathetic nervous activity reflected by plasma catecholamines during evolution of myocardial infarction in man. J Clin Invest 65:338-346

Mullen, PE, Lightman S, Linsell C, McKeon P, Sever PS, and Todd K (1981) Rhythms of plasma noradrenaline in man. Psychoneuroendocrinology. 6(3):213-222

Muller T, Hofschuster E, Kuss H, and Welter D (1979) A highly sensitive and precise radioenzymatic assay for plasma epinephrine and norepinephrine. J Neural Transm 45:219-225

Mulvihill-Wilson J, Graham RM, Pettinger W, Muckleroy C, Anderson S, Gaffney FA, and Blomquvist CG (1979) Comparative effects of prazosin and phenoxybenzamine on arterial blood pressure, heart rate, and plasma catecholamines in essential hypertension. J Cardiovas Pharmacol (suppl) 1:S1-S7

Naber D, Finkbeiner C, Fischer B, Zander KJ, and Ackenheil M (1980) Effect of long-term neuroleptic treatment of prolactin and norepinephrine levels in serum of chronic schizophrenics: relations to psychopathology and extrapyramidal symptoms. Neuropsychobiology 6:181-189

Nadeau RA, De Champlain J (1979) Plasma catecholamines in acute myocardial infarction. Am Heart J 98:548-554

Nagatsu T, Udenfriend S (1972) Photometric assay of dopamine-β-hydroxylase activity in human blood. Clin Chem 18:980-983

Nagatsu T, Thomas P, Rush R, and Udenfriend S (1973) A radioassay for dopamine-β-hydroxylase activity in rat serum. Anal Biochem 55:615-619

Ngai SH, Dairman W, Marchelle M, and Spector S (1974) Dopamine-β-hydroxylase in dog lymph—effect of sympathetic activation. Life Sci 14:2431-2439

Nicholas TE, Strum JM, Angelo LS, and Junod AF (1974) Site and mechanism of uptake of ^{3}H-l-norepinephrine by ioslated perfused rat lungs. Circ Res 35:670-680

Nilsson OR, Karlberg BE, and Soderberg A (1980) Plasma catecholamines and cardiovascular responses to hypoglycemia in hyperthyroidism before and during treatment with metoprolol or propranolol. J Clin Endocrinol Metab 50:906-911

Nisula BC, Stolk JM (1978) Serum dopamine-β-hydroxylase activity does not reflect sympathetic activation during hypoglycemia. J Clin Endocrinol Metab 47:902-905

Noth RH, Mulrow PJ (1976) Serum dopamine-beta-hydroxylase as an index of sympathetic nervous system activity in man. Cir Res 38(1):1-5

Ogihara T, Nugent CA (1974) Serum dopamine-beta-hydroxylase in three forms of acute stress. Life Sci 15:923-930

O'Hanlon JF, Horvath SM (1973) Interrelationships among performance, circulating concentrations of adrenaline, noradrenaline, glucose, and the free fatty acids in men performing in monitoring task. Psychophysiology 10(3):251-259

Okada T, Fujita T, Ohta T, *et al* (1973) A 24-hour rhythm in human serum dopamine-β-hydroxylase activity (letter). Experientia 30:605-607

Okada F, Yamashita I, Suwa N, Kunita H, and Hata S (1975) Elevation of plasma dopamine-β-hydroxylase activity during insulin induced hypoglycemia in man. Experientia 31:70-71

O'Neill MJ, Pennock JL, Seaton JF, Dortimer AC, Waldhausen JA, and Harrison TS (1978) Regional endogenous plasma catecholamine concentrations in pulmonary hypertension. Surgery 84:140-146

Onnesen KH, Kindskov J, and Amtorp O (1974) Hand blood flow, plasma norepinephrine, and central hemodynamics in myocardial infarction. Scand J Clin Lab Invest 34(2):133-139

Palmer GJ, Ziegler MG, and Lake CR (1978) Response of norepinephrine and blood pressure to stress increases with age. J Gerontol 33(4):482-487

Passon PG, Peuler JD (1973) A simplified radiometric assay for plasma norepinephrine and epinephrine. Anal. Biochem 51:518-631

Peart WS (1949) The nature of splenic sympathin. J Physiol (Lond) 108:491-501

Pedersen EB, Christensen NJ (1975) Catecholamines in plasma and urine in patients with essential hypertension determined by double-isotope derivative techniques. Act Med Scand 198:373-377

Pedersen OL, Christensen NJ, and Ramsch KD (1980) Comparison of acute effects of nifedipine in normotensive and hypertensive man. J Cardiovasc Pharmacol 2:357-366

Pettersson J, Hussi E, and Janne J (1980) Stability of human plasma catecholamines. Scand J Clin Lab Invest 40:297-303

Peuler JD, Johnson GA (1977) Simultaneous single isotope radioenzymatic assay of plasma norepinephrine, epinephrine and dopamine. Life Sci 21:625-635

Philipp T, Distler A, and Cordes U (1978) Sympathetic nervous system and blood pressure control in essential hypertension. Lancet 959-963

Pickering TG, Raine AE, Levitt M, Morganti A, Niarchos AP, and Larash JH (1979) Immediate and delayed hypotensive effects of propranolol at rest and during exercise. Trans Assoc Am Physicians 92:277-285

Pinardi G, Talmaciu RK, Santiago E, and Gubeddu LX (1979) Contribution of adrenal medulla, spleen and lymph, to the plasma levels of dopamine-β-hydroxylase and catecholamines induced by hemorrhagic hypotension in dogs. J Pharmacol Exp Ther 209:176-184

Planz G, Palm D (1973) Acute enhancement of dopamine-β-hydroxylase activity in human plasma after maximum work load. Eur. J Clin Pharmacol 5:255-258

Planz G, Planz R (1979) Dopamine-β-hydroxylase, adrenaline, noradrenaline and dopamine in the venous blood of adrenal gland of man: a comparison with levels in periphery of the circulation. Experientia 35:207-208

Planz G, Wiethold G, Appel E, Bohmer D, Palm D, and Grobecker H (1975) Correlation between increased dopamine-β-hydroxylase activity and catecholamine concentrations in plasma: determination acute changes in sympathetic activity in man. Eur J Clin Pharmacol 8:181-188

Polinsky RJ, Kopin IJ, Ebert MH, and Weise V (1980) The adrenal medullary response to hypoglycemia in patients with orthostatic hypotension. J Clin Endocrinol Metab 51:401-406

Popper CW, Chiueh CC, and Kopin IJ (1977) Plasma catecholamine concentrations in unanesthetized rats during sleep, wakefulness, immobilization and after decapitation. J Pharmacol Exp Ther 202:144-148

Price HL, Price ML (1957) The chemical estimation of epinephrine and norepinephrine in human and canine plasma. II. A critique of the trihydroxyindole method. J Lab Clin Med 50:769-777

Prinz PN, Halter J, Benedetti C, and Raskind M (1979) Circadian variation of plasma catecholamines in young and old men: relation to rapid eye movement and slow wave sleep. J Clin Endocrinol Metab 49:300-304

Quigley ME, Sheehan KL, Wilkes MM, and Yen SSC (1979) Effects of maternal smoking on circulating catecholamine levels and fetal heart rates. Am J Obstet Gynecol 133:685-690

Rahn KH, Gierlichs HW, Planz G, Planz R, Schols M, and Stephany W (1978) Studies on the effects of propranolol on plasma catecholamine levels in patients with essential hypertension. Eur J Clin Invest 8:143-148

Refshauge, C, Kissinger PT, Drieling R, Blank L, Freeman R, and Adams RN (1974) New high performance liquid chromatographic analysis of brain catecholamines. Life Sci 14:311-322

Reid JL, Kopin IJ (1975) The effects of ganglionic blockade, reserpine and vinblastine on plasma catecholamines and dopamine-β-hydroxylase in the rat. J Pharmacol Exp Ther 193:748-756

Reid JL, Fones DH, Fitzgerald G, Davies D, and Boobis A (1980) Catecholamine turnover in essential hypertension. Clin Exper Hyper 2:395-408

Renzini V, Brunori CA, and Valori C (1970) A sensitive and specific method for determination of noradrenaline and adrenaline in human plasma. Clinica Chimica Acta 30:587-594

Richter EA, Galbo H, Sonne B, Holst JJ, and Christensen NJ (1980) Adrenal medullary control of muscular and hepatic glycogenolysis and of pancreatic hormonal secretion in exercising rats. Acta Physiol Scand 108:235-242

Riedel W, Iriks S, and Simon E (1972) Regional differentiation of sympathetic activity

during peripheral heating and cooling in anesthetized rabbits. Pflügers Arch 332:239-247
Ring-Larsen H, Hesse B, Henriksen JH and Christensen NJ (1982) Sympathetic nervous activity and renal and systemic hemodynamics in cirrhosis: plasma norepinephrine concentration, hepatic extraction, and renal release. Hepatology 2:304-310
Rizza RA, Cryer PE, and Gerich JE (1979) Role of glucagon, catecholamines, and growth hormone in human glucose counterregulation. J Clin Invest 64:62-71
Robertson D, Johnson GA, Robertson RM, Nies AS, Shand DG, and Oates JA (1979a) Comparative assessment of stimuli that release neuronal and adrenomedullary catecholamines in man. Circulation 59:637-743
Robertson D, Shand DG, Hollifield JW, Nies AS, Jurgen C, Frolich JS, and Oates JA (1979b) Alterations in responses of the sympathetic nervous system and renin in borderline hypertension. Hypertension 1:118-124
Robinson BF, Epstein SE, Beiser GD, and Braunwald E (1966) Control of heart rate by the autonomic nervous system. Circ Res 19:400-411
Roffman M, Freedman LS, and Goldstein M (1973) The effects of acute and chronic swim stress on dopamine-beta-hydroxylase activity. Life Sci Oxford 12:369-376
Roizen MF, Moss J, Henry DP, Weise V, and Kopin IJ (1978) Effect of general anesthetics on handling- and decapitation-induced increases in sympathoadrenal discharge. J Pharmacol Exp Ther 204:11-18
Roozekrans NT, Porsius AJ, and Van Zwieten PA (1979) Comparison between the pressor response and the rise in plasma catecholamines induced by acutely elevated intracranial pressure. Arch Int Pharmacodyn Ther 240:143-157
Rosenthal T, Birch M, Osikowska B, and Sever PS (1978) Changes in plasma noradrenaline concentration following sympathetic stimulation by gradual tilting. Cardiovasc Res 12:144-147
Ross RJ, Zavadil AP, Calil HM, Linnoila M, Kitanaka I, Blombery P, Kopin IJ and Potter WZ (1983) Effects of desmethylimipramine on plasma norepinephrine, pulse, and blood pressure. Clin Pharmacol Ther 33:429-437
Rowell LB (1974) Human cardiovascular adjustments to exercise and thermal stress. Physiol Rev 54:75-159
Rubin PC, Scott PJW, McLean K, Reid JL (1982) Noradrenaline release and clearance in relation to age and blood pressure in man. Eur J Clin Invest 12:121-125
Rush RA, Geffen LB (1972) Radioimmunoassay and clearance of circulating dopamine β-hydroxylase. Circ Res 31:444-452
Saelens JK, Schoen MS and Kovacsics GB (1967) An enzyme assay for norepinephrine in brain tissue. Biochem Pharmacol 16:1403-1409
Santiago JV, Clarke WL, Suresh DS, and Cryer PE (1980) Epinephrine, norepinephrine, glucagon, and growth hormone release in association with physiological decrements in the plasma glucose concentration in normal and diabetic man. J Clin Endocrinol Metab 51:877-883
Sarafoff M, Davis L, and Ruther E (1979) Clozapine induced increase of human plasma norepinephrine. J Neural Transm 46(2):175-180
Schwartz L, Sole MJ, Vaughan-Neil EF, and Hussain NM (1979) Catecholamines in coronary sinus and peripheral plasma during pacing induced angina in man. Circulation 59:37-43
Scriven AJI, Dollery CT, Murphy MB, MacQuin I and Brown MJ (1983) Blood pressure and plasma norepinephrine concentrations after endogenous norepinephrine release by tyramine. Clin Pharmacol Ther 33:710-716
Sever PS, Osikowska B, Birch M, and Turnridge RDG (1977) Plasma noradrenaline in essential hypertension. Lancet, 1078-1081

Sever PS, Peart WS, Meade TW, Davies IB, Gordon D, and Turnbridge RDG (1978) Are racial differences in essential hypertension due to different pathogenetic mechanisms? Clin Sci Mol Med 55:383s-386s

Sever PS, Peart WS, Davies IB, Turnbridge RDG, and Gordon D (1979) Ethnic differences in blood pressure with observations on noradrenaline and renin. Clin Exper Hypertens. 1:745-760

Siegel JH, Gilmore JP, and Sarnoff SJ (1961) Myocardial extraction and production of catecholamines. Circ Res 9:1336-1350

Siggers DC, Salter C, and Fluck DC (1971) Serial plasma adrenaline and noradrenaline levels in myocardial infarction using a new double isotope technique. Br Heart J 33:878-883

Silverberg AB, Shah SD, Haymond MW, and Cryer PE (1978) Norepinephrine: hormone and neurotransmitter in man. Am J Physiol 234(3):E252-E256

Smith AA, Dancis J (1967) Catecholamine release in familial dysautonomia. N Engl J Med 277:61-64

Smith AD, Winkler H (1972) Fundamental mechanisms in the release of catecholamines. In: Blaschko H, Muscholl E (Eds) Catecholamines. Berlin Heidelberg New York Tokyo (Handbook of Experimental Pharmacology, vol 33)

Smith RT (1980) Effect of *in vitro* hemolysis on assay of plasma catecholamines and DOPA. Clin Chem 26(9):1354-1356

Sole MJ, Hussain MN (1977) A simple specific radioenzymatic assay for the simultaneous measurement of picogram quantities of norepinephrine, epinephrine, and dopamine in plasma and tissue. Biochem Med 18:301-307

Sole MJ, Drobac M, Schwartz L, Hussain MN, and Vaughan-Neil EF (1979) The extraction of circulating catecholamines by the lungs in normal man and in patients with pulmonary hypertension. Circulation 60:160-163

Stjärne L, Kaijser L, Math EA, and Birke G (1975) Specific and unspecific removal of circulating noradrenaline in pulmonary and systemic vascular beds in man. Act Physiol Scand 95:46-53

Stoffer SS, Juang NS, Geomar CA, and Pikler GM (1973) Plasma catecholamines in hypothyroidism and hyperthyroidism. J Clin Endocrinol Metab 36:587-589

Stone RA, Gunnells JC, Robinson RR, Schanberg SM, and Kirshner N (1974a) Dopamine-beta-hydroxylase in primary and secondary hypertension. Circ Res Suppl I 34-35:147-156

Stone RA, Kirshner N, Gunnels JC, and Robinson J (1974b) Changes of plasma dopamine-β-hydroxylase activity and other plasma constituents during the cold pressor test. Life Sci 14:1797-1805

Stone RA, Lilley JJ, and Golden J (1976) Plasma dopamine-beta-hydroxylase activity in phaeochromocytoma. Clin Endocrinol 5:181-185

Sun CL, Thoa NB, and Kopin IJ (1979) Comparison of the effects of 2-deoxy-glucose and immobilization on plasma levels of catecholamines and corticosterone in awake rats. Endocrinology 105:306-311

Sundlöf G, Wallin BG (1978) Human muscle nerve sympathetic activity at rest. Relationship to blood pressure and age. J Physiol (Lond) 274:621-637

Takishita S, Fukiyama K, Kumamoto K, Noda Y, Kawaskai T, and Omae T (1977) Plasma dopamine-β-hydroxylase activity in normal young men: its responsiveness to manipulation of sodium balance and upright posture. Jpn Circ J 41:895-901

Thomas JA, Marks BH (1978) Plasma norepinephrine in congestive heart failure. Am J Cardiol 41(2):233-243

Thompson DA, Lilavivathana U, Campbell RG, Welle SW, and Craig AB (1980) Thermoregulatory and related response to 2-deoxy-glucose administration in humans. Am J Physiol 239:R291-R295

Tibbs PA, Young B, Ziegler MG, and McAllister RG (1979) Studies of experimental cervical spinal cord transection. Part II. Plasma norepinephrine levels after acute cervical spinal cord transection. J Neurosurg 50(5):629-632

Tohmeh JF, Shah SD, and Cryer PE (1979) The pathogenesis of hyperadrenergic postural hypotension in diabetic patients. Am J Med 67:772-778

Turton MB, Deegan T (1974) Circadian variations of plasma catecholamine, cortisol and immunoreactive insulin concentrations in supine subjects. Clin Chem Acta 55:389-397

Tyrer J, Leder MH (1974) Physiological response to propranolol and diazepam in chronic anxiety. Br J Clin Pharmacol 1:387-390

Van Loon GR, Schwartz L, and Sole MJ (1979a) Plasma dopamine responses to standing and exercise in man. Life Sci 11:24:2273-2277

Van Loon GR, Sole MJ, Bain J, and Ruse JL (1979b) Effecs of bromocriptine on plasma catecholamines in normal men. Neuroendocrinology 28:425-434

Varagic V (1955) Action of eserine on blood pressure of rat. Br J Pharmacol Chemother 10:349-353

Vecht RJ, Graham GW, and Sever PS (1978) Plasma noradrenaline concentrations during isometric exercise. Br Heart J 40:1216-1220

Veith RC, Raskind MA, Barnes RF, Gumbrecht G, Ritchie JL and Halter JB (1983) Tricyclic antidepressants and supine, standing, and exercise plasma norepinephrine levels. Clin Pharmacol Ther 33:763-769

Velasquez MT, Alexander N (1979) Plasma and lymph dopamine-β-hydroxylase and norepinephrine during carotid occlusion. Am J Physiol 236:H96-H100

Vendasalu A (1960) Studies on adrenaline and noradrenaline in human plasma. Acta Physiol Scand (Suppl) 49:39-41

Vincent HH, Manin'tVeld AJ, Boomsma F, Wenting GJ and Schalekamp MADH (1982) Elevated plasma noradrenaline in response to beta-adrenoceptor stimulation in man. Br J Clin Pharmacol 13:717-721

Vlachakis ND, Aledort L (1980) Hypertension and propranolol theraphy: effect on blood pressure, plasma catecholamines and platelet aggregation. Am J Cardiol 45:321-325

Wallin BG, Sundlöf G (1979) A quantitative study of muscle nerve sympathetic activity in resting normotensive and hypertensive subjects. Hypertension 1:67-77

Wallin BG, Sundlöf G, Eriksson BM, Dominiak P, Grobecker H and Lindblad LE (1981) Plasma noradrenaline correlates to sympathetic muscle nerve activity in normotensive man. Acta Physiol Scand 111:69-73

Watson RD, Page AJ, Littler WA, Jones DH, and Reid JL (1979) Plasma noradrenaline concentrations at different vascular sites during rest and isometric and dynamic exercise. Clin Sci 57:545-547

Watson RD, Hamilton CA, Jones DH, Reid JL, Stallard TJ, and Littler WA (1980) Sequential changes in plasma noradrenaline during bicycle exercise. Clin Sci 58:37-43

Weil-Malherbe H, Bone AD (1953) Adrenergic amines of human blood. Lancet 1:974-977

Weinshilboum RM (1978a) Human biochemical genetics of plasma dopamine-beta-hydroxylase and erythrocyte catechol-O-methyltransferase. Hum Genet, Suppl 1:101-112

Weinshilboum RM (1978b) Serum dopamine-beta-hydroxylase. Pharmacol Rev 30(2):133-166

Weinshilboum RM, Axelrod J (1971a) Serum dopamine-beta-hydroxylase activity. Cir Res 28:307-315

Weinshilboum RM, Axelrod J (1971b) Serum dopamine-β-hydroxylase: decrease of chemical sympathectomy. Science 173:931-934

Weinshilboum RM, Axelrod J (1971c) Reduced plasma dopamine-beta-hydroxylase activity in familial dysautonomia. N Eng J Med 285:938-942

Weinshilboum RM, Kvetnansky R, Axelrod J, and Kopin IJ (1971) Elevation of serum dopamine-β-hydroxylase activity with forced immobilization. Nature New Biol 230:287-288

Weinshilboum RM, Schrott HG, Raymond RA, Weidman WH, and Elveback LR (1975) Inheritance of very low serum dopamine-β-hydroxylase activity. Am J Hum Genetics 27:573-585

Weinstock M, Zavadil AP, Chiueh CC, and Kopin IJ (1979) The effect of oxotremorine on blood pressure and plasma catecholamines in conscious and anesthetized rats. Life Sci 24:301-310

Werner U, Hackenberg K, Schley G, and Reinwein D (1976) Plasma catecholamines in patients with thyroid gland function disorders. Med Welt 27(24):1187-1188

Wetterberg L, Gustavson KH, Backstrom M, Ross SB, and Froden O (1972) Low dopamine-β-hydroxylase in Down's syndrome. Clin Genet 3:152-153

Wilcox G, Beaven MA (1976) A sensitive and specific tritium release assay for dopamine-β-hydroxylase in serum. Anal. Biochem 75:484-497

Williams LT, Lefkowitz RJ, Watanabe AM, Hathaway DR, and Besch HR (1977) Thyroid hormone regulation of β-adrenergic receptor number. J Biol Chem 252:2787-2789

Winder WW, Hickson RC, Hagberg JM, Ehsani AA, and McLane JA (1979) Training-induced changes in hormonal and metabolic responses to submaximal exercise. J Appl Physiol Respirat Environ Exercise Physiol 46:766-771

Winer N, Carter C (1977) Effect of cold pressor stimulation on plasma norepinephrine, dopamine-β-hydroxylase, and renin activity. Life Sci 20:887-894

Wing LMH, Reid JK, Hamilton CA, Sever P, Davies DS, and Dollery CT (1977) Effect of clonidine on biochemical indices of sympathetic function and plasma renin activity in normotensive man. Clin Sci Mol Med 53:33-45

Wise VK, Kopin IJ (1976) Assay of catecholamines in human plasma: studies of a single isotope radioenzymatic procedure. Life Sci 19:1673-1686

Wood WG, Britton BJ, Irving MH (1979) Effects of adrenergic blockade on plasma catecholamine levels during adrenaline infusion. Horm Metab Res 11:52-57

Wooten GF, Cardon PV (1973) Plasma dopamine-β-hydroxylase activity: elevation in man during cold pressor test and exercise. Arch Neurol 28:103-106

Wooten GF, Ciaranello RD (1974) Proportionality between dopamine-beta-hydroxylase activity and immunoreactive protein concentration in human serum. Pharmacology 12(4-5):272-282

Wyatt RJ, Portnoy B, Kupfer DJ, Snyder F, and Engelman K (1971) Resting plasma concentrations in patients with depression and anxiety. Arch Gen Psychiatry 24:65-70

Yamaguchi I, Kopin IJ (1979) Plasma catecholamine and blood pressure responses to sympathetic stimulation in pithed rats. Am J Physiol 237:H305-H310

Yamaguchi I, Kopin IJ (1980) Blood pressure, plasma catecholamines, and sympathetic outflow in pithed SHR and WKY rats. Am J Physiol 238:H365-H372

Yamaguchi N, De Champlain J, and Nadeau R (1975) Correlation between the response of the heart to sympathetic stimulation and the release of endogenous catecholamines into the coronary sinus of the dog. Circ Res 36:662-668

Young JB, Cohen WR, Rappaport EB, and Landsberg L (1979) High plasma norepinephrine concentrations at birth in infants of diabetic mothers. Diabetes 28:697-699

Young JB, Rowe JW, Pallotta JA, Sparrow D, and Landsberg L (1980) Enhanced plasma norepinephrine response to upright posture and oral glucose administration in elderly human subjects. Metabolism 29:532-539

Ziegler MG, Lake CR, and Kopin IJ (1976a) Plasma noradrenaline increases with age. Nature 261:5558-333-335

Ziegler MG, Lake CR, and Kopin IJ (1976b) Deficient sympathetic nervous response in familial dysautonomia. N Engl J Med 294:630-633

Ziegler MG, Lake CR, and Kopin IJ (1977) The sympathetic nervous system defect in primary orthostatic hypotension. N Engl J Med 296:293-297

Ziegler MG, Lake CR, Williams AC, Teychenne PF, Shoulson I, and Steinsland O (1979) Bromocriptine inhibits norepinephrine release. Clin Pharmacol Therp 25:137-142

CHAPTER 16

Dopaminergic Neurons: Basic Aspects

G. Bartholini, B. Zivkovic, and B. Scatton

A. Introduction

In 1958, Carlsson and collaborators (Carlsson et al. 1958) discovered that dopamine is present in high concentrations in the rat corpus striatum and is depleted by reserpine. Since the hypokinesia associated with reserpine administration was reversed by L-dopa, Carlsson (1959) postulated that dopamine is a neurotransmitter involved in extrapyramidal motor function. In 1960, Ehringer and Hornykiewicz found that the Parkinsonian brain contains reduced amounts of dopamine due to degeneration of dopaminergic neurons. These findings have led to the replacement therapy of Parkinson's disease with L-dopa (Birkmayer and Hornykiewicz 1961; Barbeau et al. 1962; Cotzias et al. 1967) and by L-dopa plus peripheral decarboxylase inhibitors (Bartholini et al. 1967). Since then, investigations on the physiology and pharmacology of the dopaminergic systems have resulted in major progress in the understanding of the pathogenesis of neuropsychiatric disorders involving dopaminergic mechanisms. These include schizophrenia and extrapyramidal disorders.

In the following, some basic aspects of the function of the dopaminergic systems will be discussed. In particular, the most recent developments in dopamine neuron regulation, receptors and interactions of dopaminergic cells with other neurons will be reviewed. Some aspects, such as synthesis, storage, release, uptake and metabolism, the mechanisms of which are common to other catecholamine neurons, will be briefly discussed; for further details the reader is referred to other chapters of this volume.

B. Synthesis, Storage, Release, Uptake and Metabolism

I. Synthesis

As is the case for the other catecholamines, the biosynthesis of dopamine is initiated by the hydroxylation of the essential amino acid tyrosine in position 3 of the phenyl ring to form dopa. This reaction is mediated by the enzyme tyrosine hydroxylase (TH) (Nagatsu et al. 1964).

The kinetic properties and substrate requirements of TH extracted from brain regions receiving a rich dopaminergic innervation do not differ substantially from those of the enzyme isolated from regions containing predominantly noradrenaline (ZIVKOVIC et al. 1974). However, the molecular weight of TH isolated from dopaminergic nerve terminals appears to be lower than that of the enzyme localized in noradrenergic neurons (JOH and REIS 1975). The significance of this difference is not known, but it may account for differences in the regulation of TH activity in dopaminergic and noradrenergic neurons (REIS et al. 1975). Thus, shortly after treatment with reserpine, which depletes both dopamine and noradrenaline, an allosteric activation of TH takes place in dopaminergic neurons, while the activity of the enzyme is unchanged in noradrenergic neurons (ZIVKOVIC et al. 1974). However, 72 h later, TH activity in dopaminergic neurons has returned to control values, while the enzyme activity from noradrenergic neurons is increased (ZIVKOVIC et al. 1974; SORIMOCHI 1975; REIS et al. 1975). This delayed increase of TH activity in noradrenergic neurons is not due to an allosteric activation, but to an increased number of enzyme molecules (REIS et al. 1975). Thus, TH from dopaminergic, in contrast to that from noradrenergic, neurons does not appear to be inducible.

The rate of tyrosine hydroxylation *in vivo* depends on the concentration and availability of specific substrates and endogenous inhibitors. Under physiological conditions, the striatal concentration of tyrosine appears to fully saturate the enzyme and therefore is not considered to be a rate-limiting factor in dopamine synthesis (MURRIN et al. 1976.) Indeed, striatal tyrosine can be depleted by 70% to 80% before a decrease in dopamine synthesis is observed (BIGGIO et al. 1976). However, it has been reported that large doses of tyrosine slightly increase striatal dopamine synthesis (WURTMAN et al. 1974), suggesting that an increase in substrate concentration enhances the rate of the enzyme activity. This effect of elevated plasma and brain tyrosine on striatal dopamine synthesis has not been confirmed (CARLSSON and LINDQVIST 1978). However, when dopamine synthesis is increased due to allosteric activation of TH, the availability of tyrosine may become limiting. Thus, after treatment with neuroleptics, which increase the firing rate of dopaminergic neurons (BUNNEY et al. 1973b), increase of brain tyrosine produces a further elevation of dopamine synthesis (SCALLY et al. 1977; CARLSSON and LINDQVIST 1978; SVED et al. 1979).

The pteridine cofactor is assumed to play a regulatory role in tyrosine hydroxylation. Although the concentration of the reduced pteridine cofactor in dopaminergic neurons is not known (it is proposed to be 10^{-5} *M;* BULLARD et al. 1978), this assumption is founded on the observation that intracerebroventricular administration of tetrahydrobiopterin (the proposed endogenous cofactor) increases the rate of striatal dopamine synthesis in a dose-dependent manner (KETTLER et al. 1974). Similarly, synthetic pteridines accelerate dopamine synthesis when added *in vitro* to striatal synaptosomal preparations (PATRICK and BARCHAS 1976). Moreover, drugs may change the concentration of the pteridine cofactor. Thus, amphetamine has been shown to decrease the concentration of striatal pteridine cofactor when given in doses which also decrease dopamine synthesis (BULLARD et al. 1978). These findings strongly suggest that a pteridine cofactor is the rate-limiting factor in the tyrosine hydroxylation reaction.

Like other regulatory enzymes, TH is also susceptible to end-product inhibition which is one of the major mechanisms of the regulation of enzyme activity (IKEDA et al. 1966; COSTA and NEFF 1966; SPECTOR et al. 1967). Among endogenous inhibitors which may play a regulatory role, both dopamine and its precursor dopa (which are competitive inhibitors of tyrosine hydroxylase with respect to the cofactor) are of importance (see MASSERANO et al., this volume).

Decarboxylation of dopa by L-aromatic amino acid decarboxylase is the second and the final step in dopamine biosynthesis. The enzyme is non-specific with respect to both localization and substrates (LOVENBERG et al. 1962). It is localized not only in catecholaminergic but also in serotoninergic neurons where it transforms 5-hydroxytryptophan into 5-HT (GOLDSTEIN et al. 1973). The enzyme is also present in endothelial cells of brain capillaries where it constitutes an enzymatic blood-brain barrier for aromatic amino acids (BERTLER et al. 1966); therefore inhibitors of this enzyme largely facilitate the penetration of exogenous L-dopa into the brain (BARTHOLINI and PLETSCHER 1975). The fact that the decarboxylation reaction is not rate-limiting makes possible the use of L-dopa as a treatment for Parkinson's disease.

II. Storage

Newly synthesized dopamine is protected from immediate intraneuronal degradation by uptake and storage into synaptic vesicles (GLOWINSKI and IVERSEN 1966) from where it is released upon arrival of a depolarizing stimulus (FARNEBO and HAMBERGER 1971). Though various techniques have been used to study the intraneuronal distribution of dopamine, our knowledge of this subject remains scanty. The notion that multiple pools of dopamine are present in dopamine nerve terminals comes from the observation that, after inhibition of TH by α-methyl-p-tyrosine (α-MT), the concentration of dopamine decreases in a biphasic manner: a short-lasting fast phase and a long-lasting slow phase (JAVOY and GLOWINSKI 1971). Thus, releasable and storage pools of dopamine have been postulated. The releasable pool is proposed to contain newly synthesized dopamine which is released in preference to the older dopamine contained in the storage pool. Moreover, it is supposed that the releasable pool, which is also called "functional", is relatively small, containing only 20% of the total dopamine (JAVOY and GLOWINSKI 1971). Since certain dopaminergic functions can be inhibited by α-MT at the time when 80% of tissue dopamine levels remain, it was postulated that the equilibrium between the storage and functional pools is very slow (SHORE and DORRIS 1975; MCMILLEN et al. 1980). However, a number of studies are not consistent with the idea that newly synthesized dopamine is released preferentially (COSTA et al. 1975; PAPESCHI 1977) and alternative explanations for the biphasic dopamine decline after α-MT are offered. Thus, it is proposed that the first rapid phase of dopamine decline is due to the release induced by p-hydroxyamphetamine and p-hydroxynorephedrine, metabolites of α-MT (DOTEUCHI et al. 1974). Alternatively, the fact that newly synthesized and newly taken up (see below) dopamine are released together (CERRITO et al. 1980) indicates that the releasable pool of dopamine represents the fraction of the transmitter which is formed and/or taken up by vesicles which are close to the membrane and therefore have a higher chance to be discharged into the synaptic cleft (LANGEN et al. 1979).

Although the question of dopamine compartmentation remains open, the multiple pool idea appears to be a useful concept in explaining various effects of centrally acting drugs.

III. Release

The release of dopamine has been studied by various methods, both *in vivo* and *in vitro*. However, the basic mechanisms have not yet been clearly elucidated. By analogy to the calcium-dependent exocytotic release of catecholamines from sympathetic nerves and adrenal medulla (see WINKLER, this volume), it is assumed that the central dopaminergic neurons also release dopamine by exocytosis. In support of this assumption are the findings that the dopamine release from striatal slices or synaptosomes evoked by potassium or electrical stimulation is calcium-dependent (FARNEBO and HAMBERGER 1971; MULDER et al. 1975), and that depolarization-induced release of dopamine is not associated with the liberation of its metabolites, which are not stored in synaptic vesicles (MULDER et al. 1975).

Besides the release from vesicles, dopamine may also be released from extravesicular pools. Thus phenylethylamines (e.g. amphetamine) release dopamine when vesicular stores of the transmitter are completely depleted by reserpine (CHIUEH and MOORE 1975; NIDDAM et al. 1985; PARKER and CUBEDDU 1986). Whether this extravesicular dopamine requires nerve impulses for release, and whether it is released under physiological conditions remains speculative. In fact, PARKER and CUBEDDU (1986) found that the depletion of endogenous dopamine by reserpine abolishes the release of endogenous dopamine induced by electrical stimulation but fails to affect the release of the transmitter by amphetamine, suggesting that release of cytoplasmic dopamine is not driven by nerve impulses. However, this finding contradicts the results obtained *in vivo* by CHIUEH and MOORE (1975) showing that in reserpine-pretreated cats, amphetamine increases the release of newly synthetized dopamine from the caudate nucleus more effectively in the presence of an increased impulse flow. Along the same line is the observation that amphetamine-induced increase in striatal 3-methoxytyramine is strikingly decreased by the acute transaction of nigro-striatal neurons at a time when dopamine synthesis is increased (SPECKENBACH and KEHR 1976).

After being released into the synaptic cleft, dopamine elicits biological responses by interacting with its specific receptors (see Sect. D). As the magnitude of the response depends on the amount of dopamine released *in situ*, various methods have been developed to study quantitatively dopamine liberation *in vivo*. One reliable method is the measurement of endogenous dopamine itself which is washed out from a tissue by means of a push-pull cannula (for review, see BARTHOLINI et al. 1976a). More recently, intracerebral dialysis has been introduced (ZETTERSTRÖM et al. 1983; IMPERATO and DI CHIARA 1984; ZETTERSTRÖM and UNGERSTEDT 1984). The latter technique involves collection of perfusates from a loop of dialysis tubing implanted in the brain tissue and subsequent analysis of the perfusates using high-performance liquid chromatography with electrochemical detection. Another recent methodological approach is the use of *in vivo* voltammetry (ADAMS 1978; GONON et al. 1980, 1983; STAMFORD 1985; SCATTON et al. 1986b). This technique involves application of a potential to the surface of a microelectrode (generally made of a carbon fiber) implanted in the brain. The flow of electrons resulting from the *in situ* oxidation of monoamines and/or their metabolites at the surface of the working electrode (as the applied potential attains their redox potential) is measured as current, which is directly proportional to the concentration of oxidizable species present in the extracellular fluid.

The major factor determining the release of dopamine is the activity of dopaminergic cells which depends on the balance between the excitatory and inhibitory inputs to these neurons. Many studies indicate a correlation between neuronal activity and dopamine release. Thus, electrical stimulation of dopaminergic cell bodies or axons produces a frequency-dependent release of endogenous dopamine from the striatum perfused by means of a push-pull cannula (McLENNAN 1964) or with dialysis tubing (IMPERATO and DI CHIARA 1984) and an increase in the HVA concentration in the ventricular perfusate (PORTIG and VOGT 1968). Similarly, electrical stimulation of dopaminergic neurons releases into the striatal push-pull cannula perfusate ^{3}H-dopamine newly formed from locally perfused ^{3}H-tyrosine (NIEOULLON et al. 1977a). *In vivo* voltammetric studies have also shown that electrical bursted stimulation of the medial forebrain bundle elicits a frequency-dependent release of dopamine from the striatum of pargyline—pretreated rats (GONON 1986). Changes in endogenous dopamine release in the striatum are also observed after treatment with drugs which alter the activity of dopaminergic neurons. Thus, neuroleptic drugs, which increase the firing rate of dopaminergic neurons, enhance the release of dopamine (STADLER et al. 1973; IMPERATO and DI CHIARA 1985; ZETTERSTRÖM et al. 1985), while drugs which decrease the activity of these neurons (apomorphine, progabide) have an opposite effect (STADLER et al. 1973; BARTHOLINI et al. 1976b, 1979; ZETTERSTRÖM and UNGERSTEDT 1984).

Dopamine is also released from dendrites and/or cell bodies in substantia nigra. Thus, potassium-induced depolarization of substantia nigra *in vitro* (GEFFEN et al. 1976) or *in vivo* (by means of push-pull cannula) (NIEOULLON et al. 1977b) enhances the local release of dopamine. Moreover, electrical stimulation of the nigrostriatal dopaminergic neurons increases the dopamine metabolites HVA and DOPAC in the substantia nigra (KORF et al. 1976b; ZIVKOVIC 1979). The fact that the release of dendritic dopamine from the substantia nigra is calcium-dependent (GEFFEN et al. 1976) suggests that an exocytotic mechanism is involved. However, in the substantia nigra, dendritic dopamine appears to be released from the smooth endoplasmic reticulum rather than from vesicles, which are the classical storage sites for dopamine at nerve terminals (MERCER et al. 1979). Dopamine released from dendrites in the substantia nigra may play an important role in the regulation of dopaminergic neuron activity. By acting on somatodendritic autoreceptors located on dopaminergic neurons or on presynaptic dopamine receptors located on the terminals of neurons afferent to nigral dopaminergic cells, dopamine would inhibit the firing of dopaminergic neurons.

One would assume that the dopamine release is solely dependent on the firing rate of dopaminergic neurons. However, this is not likely as dopamine autoreceptors as well as receptors for other neurotransmitters have been found at dopamine nerve terminals: these receptors probably modulate the release of dopamine independently of the firing rate of the neuron (see Sect. D and F and LANGER and LEHMANN, this volume).

IV. Uptake and Metabolism

The interaction of dopamine with its receptor (see Sect. D) is terminated by diffusion out of the synaptic cleft, by extraneuronal metabolism and by re-uptake into the presynaptic dopaminergic terminals; the latter mechanism appears to be the most efficient. Thus, by analogy to the noradrenergic system, it is presumed that about 70% to 80% of the released dopamine is removed from the synaptic cleft by re-uptake into dopaminergic nerve terminals (HORN 1979).

Dopamine uptake has been extensively studied (for review see HORN 1979). Most studies of dopamine uptake have monitored the accumulation of radiolabelled dopamine into nerve terminals in brain slices or synaptosomal preparations. Many features of the dopamine uptake process are similar to those of noradrenaline. The major difference is that those tricyclic antidepressants that are potent inhibitors of noradrenaline uptake, are rather poor inhibitors of dopamine uptake (GLOWINSKI and IVERSEN 1966; HORN et al. 1971). A number of drugs are potent in inhibiting dopamine uptake. Among these drugs are benztropine (COYLE and SNYDER 1969), nomifensine (HUNT et al. 1974), bupropion (COOPER et al. 1980), methylphenidate (ROSS 1977), amphetamine, amfonelic acid (FULLER et al. 1978) and cocaine (HEIKKILA et al. 1975). Many of these drugs, however also enhance dopamine release; it is therefore difficult to estimate precisely the contribution of the dopamine uptake blockade and dopamine release enhancement induced by these drugs (BAUMANN and MAITRE 1976). Cocaine appears most specific in this regard causing little or no dopamine liberation (HEIKKILA et al. 1975).

High affinity binding sites for radiolabelled dopamine uptake inhibitors (e.g. ^{3}H-nomifensine, ^{3}H-cocaine, ^{3}H-methylphenidate, ^{3}H-mazindol) have been identified in membrane preparations from several mammalian species (DUBOCOVICH and ZAHNISER 1985; JANOWSKY et al. 1985; SCHOEMAKER et al. 1985; JANOWSKY et al. 1986). These binding sites appear to be primarily associated with the neuronal uptake of dopamine in dopamine-rich brain areas. Like the carrier-mediated transport of dopamine, the binding of these radioligands is sodium-dependent and the relative potencies of various drugs in competing for this binding correlates well with their potencies at inhibiting dopamine uptake in striatal synaptosomes. Chemical lesion of the ascending dopaminergic pathways causes a large decrease in the binding of these ^{3}H-ligands in the rat striatum; this confirms the association of this binding with the presynaptic uptake site. It is as yet unclear, however, whether the dopamine transporter and the dopamine uptake inhibitor binding sites are the same entity.

The precise localization of dopamine uptake inhibitor binding sites has been studied in both rats and humans by using the technique of receptor autoradiography (SCATTON et al. 1985; JAVITCH et al. 1985; SCATTON et al. 1986a). The regional autoradiographic distribution of desipramine-insensitive ^{3}H-nomifensine and ^{3}H-mazindol binding in rat brain sections coincides to a large extent with that of dopamine terminals as determined by histofluorescence studies: the highest density of sites is found in the striatum, nucleus accumbens, olfactory tubercle, lateral septum and amygdala. ^{3}H-Nomifensine also labels the dopamine uptake recognition site in postmortem human brain (SCATTON et al. 1986a). While dense ^{3}H-nomifensine binding is observed in the caudate nucleus of normal subjects, relatively little ^{3}H-ligand binding is found in parkinsonian patients suggesting that in the human brain ^{3}H-nomifensine labels dopaminergic terminals as well. ^{3}H-Dopamine uptake inhibitor autoradiography thus offers the possibility of providing detailed maps of dopaminergic nerve densities in the human CNS and of studying their alterations in neuropsychiatric disorders.

Since, under physiological conditions, released dopamine is rapidly metabolized, changes in the concentration of its metabolites are often used as indicators for dopamine release. It is generally believed that released dopamine undergoes O-methylation by COMT, an extraneuronal enzyme leading to the formation of 3-methoxytyramine. Kinetic studies indicate, however, that 3-methoxytyramine represents only a minor product of dopamine metabolism (DEDEK et al. 1979).

The major route for dopamine metabolism is oxidative deamination by MAO (see YOUDIM et al. this volume). Since MAO is a mitochondrial enzyme, deamination of dopamine occurs predominantly intraneuronally (ROFFLER-TARLOV et al. 1971). Therefore, in order to be metabolized, released dopamine must be taken up by presynaptic dopamine nerve terminals.

In experimental situations where the release of the amine is changed DOPAC is the first metabolite to be altered. Thus, after treatment with neuroleptics an increase in DOPAC concentration precedes that of HVA (WESTERINK 1979). DOPAC may be O-methylated to give HVA or excreted in unchanged or conjugated form (DEDEK et al. 1979).

The rate of HVA formation closely follows that of DOPAC (WESTERINK 1979). Since HVA is mainly formed from DOPAC it also is believed to be a good index of dopamine release. Like DOPAC, HVA is removed from brain in unchanged or conjugated form. About one third of HVA and DOPAC are sulfate conjugated (DEDEK et al. 1979) by a rather nonspecific phenolsulfotransferase (FOLDES and MEEK 1973). Sulfate conjugation of dopamine does not seem to play a significant role in terminating the action of dopamine as the rate of dopamine sulfation at physiological pH is very slow (MEEK and NEFF 1973).

DOPAC and HVA are eliminated from the brain into the blood by an active transport mechanism which is shared by the 5-HT metabolite 5-hydroxyindoleacetic acid. This transport is inhibited by probenecid and, as a consequence, the acidic metabolites accumulate linearly with time in the brain (WERDINIUS 1966; NEFF et al. 1967).

A portion of HVA diffuses into cerebrospinal fluid from where it is actively removed into the blood by the choroid plexus or eliminated by bulk absorption of fluid (ASHCROFT et al. 1968).

Very little DOPAC has been found in cerebrospinal fluid (BERGER et al. 1980). Thus, the acid is probably O-methylated into HVA by ependymal COMT before reaching the ventricles (KAPLAN et al. 1979). Since the cerebrospinal fluid concentration of HVA closely mirrors that in the brain, its analysis in cerebrospinal fluid is often employed for studying brain dopamine metabolism in man (WOOD 1980).

C. Regulation of Dopamine Synthesis and Release

It is generally accepted that the regulation of dopamine synthesis and release is mediated by dopamine receptors through a homeostatic feedback mechanism. This basic concept originates from early observations of CARLSSON and LINDQVIST (1963) that the neuroleptics haloperidol and chlorpromazine increase dopamine metabolism. These authors postulated that neuroleptics block dopamine receptors in the striatum: the blockade of dopaminergic transmission activates dopamine neuron firing and dopamine turnover via feedback neuronal circuits which link dopamine target cells with cell bodies of dopaminergic neurons. Subsequently, it was demonstrated that the neuroleptic-induced increase in dopamine turnover does depend on the firing of dopaminergic neurons. Thus, the blockade of nerve impulse flow by an acute transection of dopaminergic neurons decreases the rate of, and antagonizes the neuroleptic effect on, dopamine turnover in striatum (ANDÉN et al. 1971; NYBÄCK and SEDVALL 1971). Also, reversible inhibition of the dopamine neuron activity by γ-hydroxybutyrate curtails the neuroleptic-induced increase in striatal HVA (ROTH 1973). These biochemical data indicate that neuroleptics increase the activity of dopaminergic neurons. This has been directly demonstrated electrophysiologically by an enhancement of their firing rate (BUNNEY et al. 1973b; IWATSUBO and CLOUET 1977). The direct evidence that these changes induced by neuroleptics are triggered by the blockade of dopamine receptors has been provided by the finding that these drugs bind to striatal membranes with high affinity and with a relative potency which corresponds to their pharmacological activity (CREESE et al. 1976; SEEMAN 1977, 1980).

Direct evidence exists that an increase in the firing rate of dopaminergic neurons is associated with an acceleration of dopamine turnover. Thus, electrical stimulation of the nigrostriatal dopaminergic neurons results in an increased conversion of tritiated tyrosine into dopamine as well as in a frequency-dependent acceleration of striatal dopa accumulation (MURRIN and ROTH 1976a). The increased dopamine synthesis induced by electrical stimulation is accompanied by an enhancement in striatal HVA and DOPAC concentration (KORF et al. 1976a; ZIVKOVIC 1979). As discussed above (see Sect. B), electrical stimulation of dopaminergic neurons also enhances dopamine liberation from axonal and dendritic terminals.

The molecular events leading to increased dopamine turnover during elevated firing of dopaminergic neurons may involve two seperate mechanisms which are not mutually exclusive. The first mechanism involves an allosteric activation of TH. It has been observed that TH isolated from dopamine-rich brain regions (striatum, olfactory tubercle, nucleus accumbens) of rats treated with neuroleptics displays a higher affinity for the pteridine cofactor than the enzyme from saline-treated animals (ZIVKOVIC et al. 1974, 1975c; LERNER et al. 1977; TISSARI et al. 1978; GALE et al. 1978). Several pieces of evidence indicate that the neuroleptic-induced allosteric activation of TH is related to the firing rate of dopaminergic neurons. Thus, cessation of the activity of dopaminergic neurons induced by γ-hydroxybutyrate or acute cerebral hemitransection abolishes the neuroleptic-induced activation of striatal TH (ZIVKOVIC et al. 1975a,b). Similarly, apomorphine antagonizes the neuroleptic-induced increase in the firing of nigrostriatal dopaminergic cells (IWATSUBO and CLOUET 1977) and the activation of striatal TH (ZIVKOVIC and GUIDOTTI 1974). Also, electrical stimulation of the nigrostriatal neurons produces changes in the kinetic properties of striatal TH similar to those observed after treatment with neuroleptics (MURRIN et al. 1976; ZIVKOVIC 1981). This is the most direct evidence that the changes in the enzyme activity are related to the firing of dopaminergic neurons.

Although there is not yet direct evidence that allosteric activation of TH constitutes the molecular basis for the dopamine turnover increase during enhanced firing of dopaminergic neurons, this is suggested by the excellent correlation between the potencies of neuroleptics in increasing dopamine turnover and in activating tyrosine hydroxylase (ZIVKOVIC 1979; ZIVKOVIC et al. 1980, 1983).

The second mechanism is the product inhibition of TH. As TH is inhibited *in vitro* by its immediate product, dopa, as well as by the end product, dopamine (IKEDA et al. 1966), it was hypothesized that the concentration of dopamine in the vicinity of the enzyme regulates the rate of the hydroxylation reaction (COSTA and NEFF 1966). According to this hypothesis, lowering of the axoplasmic concentration of dopamine (due to release) decreases the enzyme inhibition, resulting in an increased rate of hydroxylation. Experimental data are consistent with the above hypothesis. Thus, MAO inhibitors, e.g. pargyline, which increase the axoplasmic concentration of dopamine, decrease the turnover rate of the transmitter (COSTA and NEFF 1966; JAVOY et al. 1972). Similarly, addition of dopamine to striatal synaptosomes *in vitro* inhibits dopamine synthesis, this effect being blocked by the dopamine uptake inhibitor benztropine (IVERSEN et al. 1976). On the other hand, agents which release dopamine in the absence of increased (and rather decreased) firing of dopaminergic neurons, e.g. amphetamine (BUNNEY et al. 1973a) enhance the synthesis of striatal dopamine (COSTA et al. 1972).

The allosteric activation of TH and the release of the enzyme from product inhibition during increased firing of dopaminergic neurons are not mutually exclusive as control mechanisms for dopamine synthesis. After activation by neuroleptics (Costa et al. 1975) or following electrical stimulation of dopaminergic neurons (Murrin et al. 1976) TH still retains its property of being susceptible to dopamine inhibition, indicating that the two mechanisms are complementary.

Increase in dopamine synthesis is not only associated with increased firing rate of dopaminergic neurons and increased release of the transmitter. An increased rate of dopamine synthesis is also observed during reduced firing of these neurons. Compounds like γ-hydroxybutyric acid (Walters and Roth 1972), baclofen (Andén and Wachtel 1977) or 1-hydroxy-3-amino-pyrrolidinone (HA-966) (van Zwieten-Boot and Noach 1975), which diminish neuronal activity of dopaminergic neurons, as well as acute transection of these neurons (Kehr et al. 1972), produce an increase in the rate of synthesis of dopamine. As a consequence of decreased exocytotic release, dopamine accumulates. Indeed, electrical stimulation of nigrostriatal dopaminergic neurons at physiological frequencies reverses the γ-hydroxybutyric acid-induced increase in striatal dopamine levels (Murrin and Roth 1976b). Also, an increase in both dopamine levels and dopamine synthesis can be antagonized by small doses of dopamine receptor agonists such as apomorphine, piribedil or ergot alkaloids (Kehr et al. 1972; Walters and Roth 1974; Marek and Roth 1980), and this effect is reversed by some neuroleptics (Kehr et al. 1972; Walters and Roth 1976). The antagonism between dopamine receptor stimulants and blockers in modifying dopamine synthesis and metabolism in the absence of nerve impulse flow suggests the existence of dopamine inhibitory receptors on axon terminals of dopaminergic neurons which regulate synthesis and release of the transmitter (Kehr et al. 1972; Carlsson 1975; see Sect. D).

D. Dopamine Receptors

Electrophysiological, biochemical and behavioural data indicate that the action of dopamine is exerted on specific dopamine receptors. These receptors, however, do not appear to constitute a uniform population as they can be differentiated according to anatomical localization, pharmacological properties and possibly functional significance (Kebabian and Calne 1979; Stoof and Kebabian 1984).

Most of the knowledge concerning the pharmacological properties of dopamine receptors comes from biochemical experiments. In the early seventies it was found that dopamine stimulates adenylate cyclase in homogenates of tissues which receive a dopaminergic innervation (Kebabian et al. 1972). Since dopamine-induced stimulation of adenylate cyclase was mimicked by apomorphine and antagonized by a variety of neuroleptic drugs (but not by α- and β-adrenoceptor antagonists), it was postulated that adenylate cyclase is a part of the dopamine receptor complex. Using this *in vitro* model of the dopamine receptor, structural requirements for its interaction with agonists have been largely elucidated (MILLER et al. 1974; IVERSEN 1975). The catechol group of the dopamine molecule appears important while minor changes in the structure of the ethylamine chain may render the molecule more or less active. Moreover, by studying the semi-rigid analogue of dopamine 2-amino-6,7-dihydroxy-1,2,3,4 tetrahydronaphthalene (ADTN), it was found that an extended form of the β rotamer is the most active conformation of the otherwise flexible dopamine molecule (IVERSEN 1975). Apomorphine and its analogues appear to be partial agonists of dopamine-sensitive adenylate cyclase: in low concentrations they stimulate the enzyme, while in higher concentrations they antagonize the dopamine-induced stimulation (KEBABIAN et al. 1972). With the exception of pergolide and LSD (DA PRADA et al. 1975; GOLDSTEIN et al. 1980), dopamine receptor agonists of the ergot alkaloid class (bromocriptine, lisuride, lergotrile) do not stimulate adenylate cyclase (KEBABIAN et al. 1977; SCHMIDT and HILL 1977; GOLDSTEIN et al. 1978); rather they antagonize the dopamine-induced stimulation of the enzyme. Thus, it was suggested that a population of dopamine receptors is not coupled to adenylate cyclase (KEBABIAN et al. 1977). This suggestion has also been supported by experiments in which the effects of dopamine antagonists on the dopamine-induced stimulation of adenylate cyclase were studied. While phenothiazine neuroleptics are particularly potent in blocking dopamine-stimulated adenylate cyclase (CLEMENT-CORMIER et al. 1974), the neuroleptics of the benzamide class are virtually devoid of this activity (TRABUCCHI et al. 1975; SCATTON et al. 1977a). Neuroleptics of the butyrophenone and of the diphenylbutylamine series have relatively low potency in antagonising dopamine-induced stimulation of the enzyme, despite being potent neuroleptics (CLEMENT-CORMIER et al. 1974). These discrepancies between the abilities of drugs to induce or to block dopamine-like effects and to affect adenylate cyclase has led to the hypothesis of the existence of two dopamine receptor subtypes; the D-1 dopamine receptor which is coupled to adenylate cyclase and the D-2 dopamine receptor which is not, but is recognized by butyrophenones and benzamides (SPANO et al. 1979; for review see KEBABIAN and CALNE 1979). Subsequent studies have shown that, at least in some tissues, the D-2 dopamine receptor is coupled in a negative manner to adenylate cyclase (STOOF and KEBABIAN 1984; ONALI et al. 1985).

Another widely used biochemical method to study dopamine receptor properties is the radioligand binding. A large number of tritiated ligands of the agonist type (dopamine, apomorphine, N-propylnorapomorphine, ADTN) or with antagonist properties (haloperidol, spiroperidol, (cis)-flupenthixol, domperidone, thioproperazine, sulpiride) have been used to characterize dopamine recognition site(s) (SEEMAN et al. 1975; HYTTEL 1978; BAUDRY et al. 1979; CREESE et al. 1979a; THEODOROU et al. 1979; TITELER and SEEMAN 1979; BLANCHARD et al. 1980; DAVIS et al. 1980). In general, dopamine receptor agonists and antagonists do not bind to the same recognition site (TITLERER et al. 1978; SOKOLOFF et al. 1980a); agonists are poor displacers of antagonist ligands and vice versa.

In the presence of guanyl nucleotides, dopamine receptor agonists bind less to their recognition sites and are less potent in displacing antagonists

(CREESE et al. 1979b). Since guanyl nucleotides increase the stimulation by dopamine of adenylate cyclase (CLEMENT-CORMIER et al. 1978; CHEN et al. 1980), it is suggested that at least a fraction of the recognition sites for dopamine agonists is coupled to this enzyme.

Potencies of neuroleptics in displacing antagonist ligands (e.g. ^{3}H-haloperidol or ^{3}H-spiroperidol) correlate well with their potencies in inducing pharmacological and biochemical effects (except inhibition of dopamine-sensitive adenylate cyclase) in experimental animals (CREESE at al. 1976; SEEMAN 1977, 1980). Their displacing potencies also correlate with average clinical doses for the treatment of psychoses (SEEMAN 1977, 1980).

Radioligand binding studies have generated much controversy concerning the existence of multiple categories of dopamine receptors. As many as four distinct receptors for dopamine have been proposed (SEEMAN 1980; SKOLOFF et al. 1980a). The current general consensus is that there exist only two distinct categories of dopamine receptors: the D-1 dopamine receptor that mediates the stimulation of adenylate cyclase and the D-2 dopamine receptor which is either unassociated with this enzyme or inhibits it (STOOF and KEBABIAN 1984). Both D-1 and D-2 receptors possess high and low interconvertible affinity states for agonists (CREESE 1985; SEEMAN et al. 1986). The characterization of these dopamine receptor populations has been greatly aided by the identification of tissues possessing only one dopamine receptor subtype (parathyroid gland contains D-1, whereas pituitary gland contains D-2 receptors) and, above all, by the discovery of drugs that discriminate with high selectivity between the two receptor categories. At present the most selective D-1 agonist and antagonist are the benzazepines, SKF 38393 and SCH 23390, respectively (WEINSTOCK et al. 1980; IORIO et al. 1983), whereas the most selective D-2 agonist and antagonist are the ergoline, LY 171555, and the benzamide, sulpiride, respectively (STOOF and KEBABIAN 1984). These compounds have recently been radiolabelled and used as ligands to label the D-1 and D-2 receptors in the mammalian CNS (BILLARD et al. 1984; MARTRES et al. 1985; DUBOIS et al. 1986; SAVASTA et al. 1986).

Several recent studies have suggested that a functional interaction exists between striatal D-1 and D-2 receptors: for instance, D-1 receptor stimulation may modulate biochemical and behavioral effects of D-2 receptor stimulation and blockade, and D-1 receptor blockade can attenuate the effects of D-2 receptor blockade (SALLER and SALAMA 1986). Conversely, D-2 receptor stimulation antagonizes the effect of D-1 receptor stimulation on adenylate cyclase (STOOF and KEBABIAN 1981; ONALI et al. 1985).

Thus, from the above studies it appears that dopamine receptors exist in multiple forms. Their role regarding the specific physiological response to dopamine has not yet been completely elucidated, but is beginning to be unravelled. For instance, the dopamine receptors which modulate acetylcholine release in the striatum are of the D-2 type (SCATTON 1982). Moreover, although neuroleptics are specific antagonists of dopamine action, the crucial proof that the pharmacological effects of these drugs are elicited by interaction with physiologically active dopamine receptors is still missing.

The localization of dopamine receptors has been extensively studied in the mammalian CNS. In general, the regional distribution of D-1 and D-2 receptors in the brain correlates closely with the distribution of dopaminergic terminals. Autoradiographic studies have shown that the distribution of D-1 receptors is more widespread than that of D-2 receptors in the rat brain and that there are some brain areas which contain D-1 but not D-2 receptors (DUBOIS et al. 1986; SAVASTA et al. 1986). This adds further support to the view that D-1 and D-2 receptors are separate molecular entities.

With respect to dopamine receptor localization, the nigrostriatal dopaminergic system has been studied in detail. D-2 dopamine receptors are found both in axon terminal (striatum) and perikarya (substantia nigra) regions. Most of the striatal D-1 and D-2 dopamine receptors are localized on dopamine target cells. Thus, after chemical destruction of striatal neurons by kainic or ibotenic acid, dopamine-sensitive adenylate cyclase, ^{3}H-SKF 38393 and ^{3}H-SCH 23390 binding virtually disappear and ^{3}H-spiroperidol binding sites decrease by 40 to 50% (SCHWARCZ et al. 1978; QUICK et al. 1979; DUBOIS et al. 1986; PORCEDDU et al. 1986). Mechanical lesion of cortical afferents produces a further 20 to 30% decrease in the ^{3}H-spiroperidol binding sites, suggesting D-2 dopamine receptor localization at the axon terminals of corticostriatal (possibly glutamatergic) projections (SCHWARCZ et al. 1978). As the latter procedure fails to affect striatal dopamine-sensitive adenylate cyclase, it is suggested that the dopamine receptor at cortico-striatal afferents is of the D-2 type. However, the interpretation of these results should be made with caution, since HATTORI and FIBIGER (1982) demonstrated that similar lesions also produce a transsynaptic degeneration of dendritic spines of striatal target neurons.

A population of D-1 dopamine receptors has also been found in substantia nigra (PHILLIPSON and HORN 1976; GALE et al. 1976; SPANO et al. 1977; PORCEDDU et al. 1986). These D-1 receptors seem to be localized on terminals of striatal projections, since the adenylate cyclase response of substantia nigra to dopamine disappears after lesion of striato-nigral neurons but not after 6-hydroxydopamine-induced degeneration of dopaminergic cells (GALE et al. 1976; PHILLIPSON et al. 1977). These receptors are probably stimulated by dopamine released from dendrites of dopaminergic neurons (see Sect. B).

Evidence has accumulated indicating that dopamine receptors are not only localized on dopamine target cells but also on dopaminergic neurons. These receptors are called autoreceptors as they are activated by dopamine itself during release (CARLSSON 1975). Thus, iontophoretically applied dopamine receptor agonists diminish the activity of dopaminergic cells in the substantia nigra and ventral tegmental area, this effect being antagonized by systemic administration of a neuroleptic (BUNNEY et al. 1973b; GROVES et al. 1975; WHITE and WANG 1984). The existence of autoreceptors at dopamine cell bodies and/or dendrites is supported by the finding that a substantial portion of ^{3}H-spiroperidol binding sites in substantia nigra is lost after 6-hydroxydopamine lesion of nigrostriatal neurons (MURRIN et al. 1979). As these lesions do not affect dopamine-sensitive adenylate cyclase (GALE et al. 1976), dopamine autoreceptors appear to be of the D-2 type. Microiontophoretic studies with various dopamine agonists also showed that somatodendritic dopamine autoreceptors exhibit the pharmacological characteristics of the D-2 receptor (White and Wang 1984).

Dopamine autoreceptors seem to be present also on axon terminals (presynaptic or preterminal autoreceptors). The major support for their existence comes from pharmacological findings showing that dopamine receptor agonists and antagonists affect dopamine synthesis and release in the absence of nerve impulse flow (see Sect. C). Kinetic studies of apomorphine binding to striatal membranes (Sokoloff et al. 1980b) as well as results demonstrating that lesion by 6-hydroxydopamine decreases striatal apomorphine binding (Nagy et al. 1978) would support the existence of autoreceptors on dopamine axon terminals. The detection of the dopamine autoreceptor on axon terminals with receptor binding techniques remains, however, controversial.

The function of dopamine autoreceptors is not completely elucidated, although it has been postulated to involve an autoregulatory mechanism. At the cell bodies, the dopamine released from dendrites seems to inhibit the firing of dopaminergic neurons (Aghajanian and Bunney 1973; Groves et al. 1975). In the axon terminal region, the released dopamine inhibits its own liberation independently of the nerve impulse flow (Carlsson 1975). The functional aspects of dopamine autoreceptors are discussed in detail by Langer and Lehman (this volume).

E. Adaptative Changes

When the activity of pre- and/or postsynaptic dopamine receptors is chronically altered, both the dopamine neuron and its target cells undergo adaptive changes, which, teleologically, can be considered homeostatic, inasmuch as they tend to normalize dopaminergic transmission. Thus, repeated treatment with dopamine receptor blockers leads to supersensitivity of the dopamine target cells to dopamine-agonists and tolerance to the cataleptogenic action of neuroleptics (Asper et al. 1973; Tarsy and Baldessarini 1974; Ezrin-Waters and Seeman 1977). Supersensitivity of dopamine target cells is also shown by electrophysiological studies: chronic haloperidol treatment enhances the response of striatal cells to iontophoretically applied dopamine (Skirboll and Bunney 1979).

There is evidence that supersensitivity is associated with an increased number of receptors rather than with changes in their affinity for the agonist. Thus, Scatchard analysis of binding of labelled ligands to membranes prepared from dopamine-rich brain areas of rats repeatedly treated with neuroleptics possessing a preferential affinity for the D-2 dopamine receptors, reveals an increase in maximal number of D-2 sites without changes in the binding affinity constant (Burt et al. 1977; Allen et al. 1980; MacKenzie and Zigmond 1985). Similarly, chronic treatment with the selective D-1 receptor antagonist SCH 23390 causes an increase in striatal D-1 receptor density (Creese and Chen 1985; Porceddu et al. 1985). Therefore, it can be assumed that the increased number of dopamine receptors enhances the action of dopamine agonists on the target cell and reduces the blockade by dopamine antagonists. This hypothesis, however, can only partly explain the increased response of the dopamine target system to dopamine agonists, since the relative increase in the number of binding sites for ligands is small as compared to the enhanced behavioural response. Thus, an amplifying mechanism should be postulated, such as orther changes in the sensitivity of dopamine target cells for other neurotransmitters which may contribute to the overall response, or alterations of events beyond the receptor, which determine physiological responses. For instance, it has been shown that repeated treatment with neuroleptics enhances the response of striatal adenylate cyclase to dopamine (GNEGY et al. 1977a, b). In addition, sustained blockade of dopamine receptors increases the concentration of calmodulin (GNEGY et al. 1977a, b), a protein which has been proposed to regulate, by a calcium-dependent mechanism, the activities of cyclic nucleotide phosphodiesterase (CHEUNG, 1971) and adenylate cyclase (BROSTROM et al. 1975; CHEUNG et al. 1975). The mechanism(s) which may amplify the response to dopamine agonists in supersensitive conditions remain(s), however, speculative.

Supersensitivity of dopamine target cells also occurs after lesion of the dopaminergic systems by 6-hydroxydopamine (UNGERSTEDT et al. 1975; CREESE et al. 1977) as well as after depletion of dopamine storage sites by reserpine (TARSY and BALDESSARINI 1974) or α-MT (DOMINIC and MOORE 1969). Thus, it can be concluded that all manipulations leading to a decrease in dopaminergic transmission induce supersensitivity of dopamine target systems. This holds true not only in experimental animals but also in man, since an increase in D-2 and possibly D-1 dopamine receptor density has been observed in parkinsonian patients (SEEMAN 1980; PIMOULE et al. 1985; RAISMAN et al. 1985; RINNE et al. 1985).

The opposite effect, namely subsensitivity, is observed when dopaminergic transmission is increased for prolonged periods of time, e.g. by repeated administration of dopamine receptor agonists such as apomorphine. Under such conditions, the sterotyped behaviour induced by this compound is attenuated and the action of dopamine receptor antagonists is enhanced (SCATTON and WORMS 1978). These behavioural changes, reflecting subsensitivity of dopamine target system, are accompanied by a decrease in the number of binding sites for ^{3}H-spiroperidol (SCATTON et al. 1979) or haloperidol (MISHRA et al. 1978). However, as in the case of supersensitivity, changes in the receptor number are not sufficient to explain the behavioural alterations. Obviously, it is likely that (unknown) changes opposite to those occurring in the supersensitive state take place. Indeed, administration of appropriate doses of dopamine receptor agonists to rats which have been made supersensitive by repeated treatment with neuroleptics shortens the duration of the supersensitive state (LIST and SEEMAN 1979; ALLEN et al. 1980), suggesting that the same homeostatic mechanism is affected by the two types of drugs. In man, too, chronic dopa therapy of parkinsonian patients may lead to a normalization of the sensitivity of at least D-1 receptors (RAISMAN et al. 1985).

The changes in the sensitivity of the dopamine target system are accompanied by modification of dopamine neuron activity. Thus, repeated treatment with neuroleptics reduces the degree of activation of dopamine neurons that is seen after single administration of a neuroleptic (BUNNEY and GRACE 1978). Also, the number of dopaminergic cells spontaneously firing in the substantia nigra and ventral tegmental area is diminished after repeated neuroleptic treatment (BUNNEY and GRACE 1978; CHIODO and BUNNEY 1983; WHITE and WANG 1983), and those neurons which are spontaneously firing are more sensitive to inhibition by dopamine agonists (NOWYCKY and ROTH 1977; GALLAGER et al. 1978).

The decreased activity of dopaminergic neurons is reflected by the alterations in dopamine metabolism and release. As compared to the enhancement induced by acute administration, repeated treatment with neuroleptics attenuates i) the release of dopamine (as measured by means of the push-pull cannula (LLOYD et al. 1980)), ii) the dopamine metabolite levels (ASPER et al. 1973; SCATTON 1977), iii) the TH activity as measured both *in vitro* (LERNER et al. 1977; GUIDOTTI et al. 1978; RELJA and ZIVKOVIC 1978) and *in vivo* (NOWYCKY and ROTH 1977). Opposite changes occur following repeated treatment with dopamine agonists (SCATTON and WORMS 1978).

The mechanisms whereby changes in dopaminergic transmission alter dopamine synthesis, release and metabolism probably involve i) feedback neuronal circuits which connect dopamine target cells with dopaminergic neurons and/or ii) dopamine autoreceptors (see Sect. D).

Alterations of the activity of these feedback neurons may be shown by changes in the release of their transmitters. It has been suggested that striatal neurons utilizing acetylcholine, GABA, enkephalin or substance P are part of the dopamine target system and influence in turn the activity of the nigrostriatal dopaminergic neurons (see Sect. F). For instance, the enhancement in acetylcholine turnover after acute treatment with neuroleptics is attenuated after repeated administration of these drugs (MAO et al. 1977). Similar results have been observed after degeneration of dopaminergic neurons by 6-hydroxydopamine (AGID et al. 1975). In contrast, chronic treatment with dopamine receptor agonists diminishes the reduction in acetylcholine turnover observed after acute administration (SCATTON and WORMS 1979). Single injection of certain neuroleptics increases GABA turnover (MAO et al. 1977) and decreases the level of the neurotransmitter in the substantia nigra (LLOYD and HORNYKIEWICZ 1977). However, when neuroleptics are given repeatedly, GABA turnover is decreased and GABA levels are unchanged (MAO et al. 1977; LLOYD and HORNYKIEWICZ 1977). While acute treatment with neuroleptics does not affect striatal levels of met-enkephalin, chronic treatment induces an increase in the striatal concentration of this polypeptide (HONG et al. 1978b). Substance P levels in substantia nigra are not changed after a single injection of a neuroleptic but are decreased after repeated treatment (HONG et al. 1978a; HANSON et al. 1981; OBLIN et al. 1984).

There is also evidence that changes in dopamine turnover during sustained alterations in dopaminergic transmission involve dopamine autoreceptors. In animals chronically treated with neuroleptics, apomorphine antagnizes the γ-hydroxybutyric acid-induced increase in dopa synthesis more potently than after single neuroleptic administration (NOWYCKY and ROTH 1977). Since γ-hydroxybutyric acid suppresses the firing of dopaminergic neurons, apomorphine is assumed to act via stimulation of dopamine autoreceptors. Also, electrical stimulation of the nigrostriatal dopaminergic pathway causes an activation of dopa synthesis which, in animals pretreated with a long-acting neuroleptic, is smaller than in control rats (NOWYCKY and ROTH 1978). A supersensitivity of the dopamine autoreceptor involved in the control of dopamine release has also been observed in rabbit striatal slices after repeated neuroleptic treatment (NOWAK et al. 1983). The above results suggest that dopamine autoreceptors develop supersensitivity to dopamine receptor agonists as a result of their sustained blockade by neuroleptics.

Dopamine neurons in other brain areas differ from the nigrostriatal pathway inasmuch as they are less susceptible to development of the adaptive changes described above. Mesolimbic and mesocortical dopaminergic neurons show supersensitivity to dopamine agonists only after repeated doses of neuroleptics which are much higher than those which provoke changes in the nigrostriatal dopaminergic system (SCATTON 1977; SCATTON and ZIVKOVIC 1984). The reasons for these regional differences are unknown but they may involve interactions with different neuronal networks. The lack of tolerance to the increase in dopamine metabolism in the frontal cortex after repeated neuroleptic treatment has been attributed to the absence of dopamine autoreceptors in this brain region (BANNON and ROTH 1983).

The tuberoinfundibular and retinal dopaminergic systems respond in a different way to acute and chronic treatment with neuroleptics. Thus, in the hypothalamus, acute neuroleptic treatment produces a delayed prolactin-mediated activation of dopaminergic neurons (MOORE et al. 1978) which does not seem to be attenuated after repeated treatment (ANNUNZIATO et al. 1980). However, pituitary prolactin-secreting cells on which dopamine receptors are localized develop supersensitivity to the transmitter during repeated neuroleptic or α-MT administration. This is indicated by a decrease of the apomorphine threshold dose producing prolactin release inhibition (LAL et al. 1977; ANNUNZIATO and MOORE 1977). The retinal dopaminergic system is activated by the acute administration of neuroleptics; however, no tolerance has been observed after repeated treatment with these drugs (SCATTON et al. 1977b). The reasons for this different responsiveness of the tuberoinfundibular and retinal dopaminergic systems as compared to nigrostriatal and mesolimbic dopaminergic systems are unknown.

F. Interaction of Dopamine and Other Neurons

Knowledge of the (functional) links between dopamine, acetylcholine and GABA neurons has led to a greater understanding of the physiological mechanisms involved in striatal and limbic function as well as of some aspects of the pathogenesis of neuropsychiatric disorders. Although experimental and clinical data are consistent, these links obviously represent only a partial picture. Indeed, a great deal of data point to interactions of dopaminergic pathways with many neuronal systems other than the cholinergic and GABAergic. However, it is impossible as yet to integrate the available data into a complete picture so far as physiological functions and, particularly, pathological mechanisms are concerned. In the following, therefore, evidence for the existence of these additional interconnections will be given with only limited discussion of the possible functional implications.

I. Extrapyramidal System

1. Interaction of Dopamine and Acetylcholine Neurons

In experiments of *in vivo* perfusion of the cat caudate nucleus by means of the push-pull cannula it was first demonstrated that an intrastriatal functional link exists between dopamine and acetylcholine neurons (STADLER et al. 1974; BARTHOLINI et al. 1976a). Thus, impairment of dopaminergic transmission (e.g. by neuroleptics) leads to enhanced release of striatal acetylcholine in the absence of changed acetylcholine levels and acetylcholinesterase activity. These data indicate that the turnover of striatal acetylcholine is enhanced, possibly as a result of an enhanced firing rate of striatal cholinergic interneurons (STADLER et al. 1973; BARTHOLINI et al. 1976b). As some neuroleptic drugs possess cholinoceptor blocking properties it could be argued that the enhanced acetylcholine turnover is the result of a feedback activation of cholinergic neurons triggered by the blockade of cholinoceptors. However, one of the most potent neuroleptics in enhancing acetylcholine turnover is haloperidol, a compound virtually devoid of cholinoceptor antagonism; also, atropine or scopolamine are only weakly effective in increasing acetylcholine liberation (STADLER et al. 1973; BARTHOLINI et al. 1976b).

Involvement of dopamine receptors in the neuroleptic-induced acetylcholine turnover activation is indicated by the fact that compounds which diminish the firing of dopamine neurons (e.g. γ-butyrolactone) potentiate (STADLER et al. 1974), and dopamine receptor agonists (e.g. apomorphine) or L-dopa reverse or prevent the neuroleptic effect. Also, apomorphine or L-dopa, as well as electrical stimulation of dopamine cell bodies (which liberate dopamine in the striatum) cause a decreased release of striatal acetylcholine (BARTHOLINI et al. 1976b).

These data indicate that dopamine neurons inhibit the activity of cholinergic cells in the striatum (STADLER et al. 1974; BARTHOLINI et al. 1976b). Some histological data indicate a direct link (MCGEER et al. 1976). The influence of dopamine on cholinergic neurons is probably tonic in nature as suggested by the increased acetylcholine utilization following transection of the nigrostriatal dopaminergic pathway (GUYENET et al. 1975). The dopamine receptors which modulate cholinergic neuron activity exhibit the pharmacological characteristics of D-2 receptors (EUVRARD et al. 1979; SCATTON 1982; STOOF et al. 1982).

The opposite regulation, namely, of dopamine neurons by cholinergic cells, appears to exist: perfusion of the cat caudate nucleus (by means of the push-pull cannula) with physiological media containing acetylcholine, physostigmine or oxotremorine increases the striatal dopamine liberation, this effect being antagonized by cholinoceptor antagonists (BARTHOLINI and STADLER

1975a). These initial results have been confirmed in *in vitro* studies (see CHESSELET 1984, for review). Moreover, it has been demonstrated that both muscarinic and nicotinic receptors control the acetylcholine-mediated release of dopamine. More recently, the M-1 subtype of the muscarinic receptor has been suggested to play a crucial role in the facilitation of the release of dopamine by acetylcholine (MARCHI and RAITERI 1985). The existence of an acetylcholine-dopamine link is also supported by the fact that acetylcholine applied iontophoretically enhances the firing rate of dopaminergic cells (BUNNEY and AGHAJANIAN 1976). Finally, the dopamine turnover and its enhancement caused by neuroleptic drugs are attenuated by cholinoceptor antagonists (O'KEEFE et al. 1970; BARTHOLINI and PLETSCHER 1971; BARTHOLINI et al. 1975), opposite effects are observed with cholinomimetic agents (ANDÉN 1974; BARTHOLINI et al. 1975).

According to these results it appears that the activity of dopamine neurons is modulated—inter alia—by their own target cholinergic cells (BUNNEY and AGHAJANIAN 1976) via an excitatory input. It is possible that the regulation of dopamine neurons by cholinergic cells subserves a feedback function. Thus, the activation of dopamine turnover due to blockade of dopamine receptors by neuroleptics might occur via the enhanced liberation of acetylcholine which excites the dopamine neurons. This mechanism—in teleological terms—may be initiated in order to overcome the impairment of dopaminergic transmission. Other feedback circuits (utilizing GABA or substance P neurons) are involved in dopamine neuron regulation (see below).

2. Interaction of Dopamine and GABA Neurons

Electrical stimulation of the cat striatum decreases the firing rate of putative dopamine cells in the substantia nigra (YOSHIDA and PRECHT 1971) probably due to release of GABA (VAN DEN HEYDEN et al. 1979). A similar effect is observed after microiontophoretic application of GABA into the rat substantia nigra (AGHAJANIAN and BUNNEY 1973). Also, intranigral GABA injection lowers rat striatal dopamine turnover ipsilaterally (ANDÉN and STOCK 1973) while bicuculline increases the release of dopamine in the cat caudate nucleus (BARTHOLINI and STADLER 1975b). It appears, therefore, that GABA inhibits the nigrostriatal dopamine pathway by an input on dopamine cell bodies or dendrites in the pars reticulata. The inhibitory GABA influence is also exerted on dopaminergic terminals, as *in vivo* perfusion of the cat caudate nucleus with bicuculline or picrotoxin enhances striatal dopamine liberation, this effect being blocked by the addition of GABA to the perfusing medium. The amino acid alone decreases the spontaneous liberation of dopamine (BARTHOLINI and STADLER 1977). The GABAergic presynaptic input on dopamine neurons is probably mediated by striatal GABA interneurons or collaterals of the strio-nigral GABAergic pathway. The existence of GABA synapses is, indeed, well documented by the presence in the striatum of i) GABA uptake in nerve terminals (IVERSEN and SCHON 1973), ii) glutamic acid decarboxylase acitivity in synaptosomes (STORM-MATHISEN, 1975) and iii) GABA binding sites in membrane preparations (ZUKIN et al. 1974).

The GABA-mediated inhibition of dopamine neurons is also shown by the effect of drugs. Thus, GABA receptor agonists such as progabide (BARTHOLINI et al. 1979; LLOYD et al. 1979; KAPLAN et al. 1980; SCATTON et al. 1982; WOOD 1982) and muscimol (BARTHOLINI et al. 1979; SCATTON et al. 1980) reduce the dopamine synthesis as indicated by the reduction of i) striatal dopa accumulation after inhibition of aromatic L-amino acid decarboxylase; ii) *in vitro* formation of $^{14}CO_2$ from L-^{14}C-tyrosine in striatal slices, this effect being blocked by picrotoxin. GABA receptor agonists also reduce the utilization of striatal dopamine (reduction of the αMT-induced dopamine disappearance and diminished liberation of dopamine in the cat caudate nucleus perfused by means of the push-pull cannula (BARTHOLINI et al. 1979).

The inhibitory action of GABA receptor agonists on nigrostriatal dopamine neurons is more marked after activation of dopamine turnover (e.g. by neuroleptics) (BARTHOLINI et al. 1979). This suggests that under resting conditions, dopamine neurons are in a state of tonic (possibly GABAergic) inhibition; accordingly, the efficacy of GABAergic agents may depend on the basal firing rate of dopaminergic neurons.

Although a great deal of evidence indicates that GABA inhibits dopamine neurons, some reports point to an activation by the amino acid of the nigrostriatal dopaminergic neurons. *In vitro* studies suggest that depending on the concentration, GABA can both inhibit and facilitate ^{3}H-dopamine release in the striatum (REIMANN et al. 1982). Moreover, nigral application (pars reticulata) of compounds supposed to increase GABAergic transmission (muscimol, aminooxyacetic acid, benzodiazepines) has led to results compatible with an activation of dopamine neurons (DRAY and STRAUGHAN 1976; CHERAMY et al. 1977a, 1978a; DI CHIARA et al. 1979; SCHEEL-KRUEGER et al. 1979). This activation might occur via a GABA-mediated inhibition of inhibitory (possibly glycinergic) interneurons impinging on dopamine cell bodies (GRACE and BUNNEY 1979). However, several studies suggest that the excitatory action of GABA on dopamine neurons is independent of primary changes in dopaminergic transmission (SCATTON and BARTHOLINI 1979, 1980) being rather related to modifications of striatal output circuits. These, probably, involve cholinergic neurons (see above).

The opposite influence, namely that of nigrostriatal dopamine neurons on GABA cells, is suggested by some results. Thus, a decrease in dopaminergic transmission by neuroleptics has been reported to enhance GABA turnover in extrapyramidal centers (MAO et al. 1977). Also, chronic L-dopa treatment increases, in the rat, glutamic acid decarboxylase activity and in parkinsonian patients results in a less marked diminution of the enzyme activity in extrapyramidal centers (LLOYD and HORNYKIEWICZ 1973). This has been interpreted as suggesting that reduction of dopaminergic transmission results in a "biochemical atrophy" of GABA neurons which is prevented by L-dopa therapy. More recent studies have demonstrated that, in the rat striatum, dopamine exerts an inhibitory influence on GABA release (VAN DER HEYDEN et al. 1980; TOSSMAN and UNGERSTEDT 1986).

3. Interaction of Dopamine and Substance P Neurons

Substance P neurons originate in the head of the striatum and project to the substantia nigra (BROWNSTEIN et al. 1977; GALE et al. 1977). These neurons have an excitatory influence on dopamine cells as indicated by the fact that microiontophoretic application of Substance P to dopamine cell bodies enhances their firing rate (DAVIES and DRAY 1976). Also, nigral application of Substance P enhances the release of newly synthesized ^{3}H-dopamine in the cat caudate nucleus (CHERAMY et al. 1977b) and increases the HVA levels in the ipsilateral striatum and elicits contralateral turning (JAMES and STARR 1977; WALDMEIER et al. 1978). Intranigral infusion of substance P antagonists or antibodies induces opposite changes (CHERAMY et al. 1978b; MELIS and GALE 1984).

Substance P neurons may participate in the regulation of dopaminergic neurons, being together with the GABA projections one of the long feedback circuits which project from the striatum to the substantia nigra (GALE et al. 1977; MELIS and GALE 1984). This type of regulation also takes place in the meso-limbic and meso-cortical dopaminergic systems (STINUS et al. 1978; BANNON et al. 1983). A possible balance of the influence of these two pathways on dopamine neuron activity may exist: thus, electrical stimulation of the striatum results in a decrease in, and, in the presence of bicuculline, in an increase of the dopamine cell firing rate (KANAZAWA and YOSHIDA 1980). These effects are interpreted as being due to the liberation of both substance P and GABA, the excitatory action of the former being masked by the prevalent inhibitory action of GABA.

Substantial evidence has also been provided for an influence of dopamine neurons on Substance P cells. Nigral Substance P concentrations are decreased following chronic treatment with neuroleptic drugs or chemical lesion of the nigrostriatal dopaminergic pathway (HONG et al. 1978a; HANSON et al. 1981; OBLIN et al. 1984). The dopamine receptor involved in this modulation appears to be akin to the D-2 subtype (OBLIN et al. 1984).

4. Interaction of Dopamine and Enkephalin Neurons

The presence of opiate receptors has been demonstrated in the substantia nigra (LLORENS-CORTES et al. 1979) as well as on dopamine nerve terminals, where naloxone binding sites are decreased following 6-hydroxydopamine-induced degeneration of dopaminergic neurons (POLLARD et al. 1978). Intranigral injection of morphine decreases the neuroleptic-induced activation of tyrosine hydroxylase (GUIDOTTI et al. 1979). These data suggest a possible inhibitory influence of enkephalin neurons on dopamine cells. However, it has also been reported that morphine or enkephalins increase both dopamine turnover (KUSCHINSKY and HORNYKIEWICZ 1972; BIGGIO et al. 1978) and dopamine neuron firing rate (IWATSUBO and CLOUET 1977), these actions being antagonized by naloxone. Iontophoretic studies failed to reveal any effect of morphine on dopamine neurons in substantia nigra (PERT et al. 1979), indicating an indirect enkephalin influence on dopamine neuron activity. A presynaptic localization for opiate receptors modulating striatal dopamine release has been suggested in a number of studies reviewed by LANGER and LEHMANN (this volume). In addition, the striatal function may also be influenced by indirect mechanisms. Thus, catalepsy induced by opiates appears to be related to an extrastriatal mechanism, as it occurs after injection of morphine in the reticular formation (DUNSTAN et al. 1981) and in animals with lesioned striata (KOFFER et al. 1978).

The opposite influence on enkephalin neurons by dopamine cells is suggested by the increased met-enkephalin concentration in striatum after chronic neuroleptic treatment (HONG et al. 1978b).

5. Clinical Implications

a. Dopamine-Acetylcholine Neuron Relation

Several neurological (primary or iatrogenic) disorders of the extrapyramidal system involve primary or secondary changes in dopaminergic transmission. Due to the dopamine-acetylcholine neuron link, alterations of cholinergic activity invariably occur. Indeed, striatal cholinergic neurons appear to exert a key role in extrapyramidal function by translating changes in dopaminergic transmission into motor patterns.

α. Enhanced Dopaminergic Transmission—Cholinergic Hypoactivity

L-dopa-induced involuntary movements in parkinsonian patients, neuroleptic-induced tardive dyskinesia and Huntington's chorea appear to be connected to an exaggerated dopaminergic transmission and a decreased activity of cholinergic neurons.

L-dopa-induced abnormal movements in parkinsonian patients have been proposed to be connected to supersensitivity of dopamine receptors (GERLACH et al. 1974). However, while some indirect evidence suggests supersensitivity in non-L-dopa-treated patients (SHIBUYA 1979), chronic administration of the amino acid should lead to subsensitivity of dopamine target cells. It is, therefore, more likely that the syndrome involves an exaggerated dopamine formation in, and liberation from the remaining, hyperactive (LLOYD 1977) dopaminergic neurons. As a consequence and as suggested by the experimental evidence discussed above, a reduction of striatal cholinergic activity should result (STADLER et al. 1973; BARTHOLINI et al. 1974). This is supported by the fact that cholinoceptor antagonists aggravate the L-dopa-induced abnormal movements (BIRKET-SMITH 1974).

The tardive dyskinesias which appear in schizophrenic patients during, or after withdrawal from, chronic neuroleptic medication are probably the result of the development of supersensitivity to dopamine of dopamine target cells (KLAWANS and RUBOVITS 1972) (see Sect. E). The exaggeration of dopaminergic transmission consequent to supersensitivity probably leads to a decreased striatal cholinergic activity. Evidence for this view is provided by the fact that during repeated treatment with neuroleptics, choline acetyltransferase activity is reduced (LLOYD et al. 1977) and acetylcholine release is diminished within the cat caudate nucleus (BARTHOLINI et al. 1976b); also, in the rat, tolerance to the cataleptogenic action of neuroleptics develops (see Sect. E).

The association of increased dopaminergic transmission and reduction of cholinergic activity in tardive dyskinesia is also supported by the effect of drugs in patients. Thus, the dyskinetic symptoms are ameliorated by a more potent neuroleptic or reserpine (GERLACH et al. 1974) as well as by physostigmine or possibly deanol (DAVIS et al. 1975; DE SILVA and HUANG 1975); conversely, cholinoceptor antagonists aggravate the syndrome (ZIRKLE and KAISER 1974; DAVIS et al. 1975).

Some dyskinetic symptoms of Huntington's chorea may result from degeneration of strio-nigral GABAergic neurons which, under physiological conditions, inhibit dopamine cells (see above). This degeneration leads to an exaggerated dopaminergic transmission and a consequent decreased cholinergic function: thus, reserpine or neuroleptics, on one hand, and cholinomimetics, on the other, ameliorate the choreiform movements; in contrast, the symptoms are aggravated by cholinoceptor antagonists or by L-dopa (KLAWANS 1970; AQUILONIUS and SJOESTROEM 1971; KLAWANS and RUBOVITS 1972).

β. Decreased Dopaminergic Transmission—Cholinergic Hyperactivity

Striatal cholinergic neuron hyperactivity occurs as a consequence of reduced dopaminergic transmission. Thus, Parkinson's disease (primary degeneration of dopaminergic neurons) and neuroleptic-induced parkinsonism are ameliorated by cholinoceptor antagonists (GREENBLATT and SHADER 1973; MUNKVAD et al. 1976); similarly, in the rat, the cataleptogenic action of neuroleptics (ASPER et al. 1973) is antagonized by cholinoceptor antagonists. These drugs mainly ameliorate rigidity and tremor but do not affect hypokinesia which is improved only by L-dopa; this feature of Parkinsonism, therefore, does not appear to have a cholinergic component (BARBEAU 1974; BARTHOLINI and STADLER 1975b).

b. Dopamine-GABA Neuron Relation

The inhibition of dopamine neurons by GABA possibly intervenes in the regulation of striatal function. Thus, an increased GABAergic transmission as that induced by GABA-mimetics (e.g. muscimol, progabide, BARTHOLINI et al. 1979) leads to behavioural effects such as potentiation of neuroleptic-induced catalepsy (WORMS et al. 1978; LLOYD et al. 1979). This is explained by the decrease in dopamine release induced by GABA-mimetics which enhances the effect of dopamine receptor blockade. Conversely, GABA receptor antagonists diminish the cataleptogenic action of neuroleptic drugs (WORMS and LLOYD 1978).

Alterations of the inhibitory GABAergic input on dopamine neurons might be involved in the genesis of symptoms of Huntington's chorea and of neuroleptic-induced tardive dyskinesias.

As stated above, hyperactivity of dopaminergic neurons in Huntington's chorea appears to result from suppression of an inhibitory GABAergic input due to degeneration of striatal and strio-nigral GABAergic cells. Thus, GABA levels and glutamic acid decarboxylase activity are decreased in both striatum and substantia nigra of choreic patients (DE SILVA 1977; ENNA et al. 1977). Also, GABAergic agents ameliorate dyskinetic symptoms of Huntington's chorea in patients at an early stage of the disease (MORSELLI et al. 1980).

The neuroleptic-induced tardive dyskinesias are related to a relative enhancement of dopaminergic transmission resulting from supersentivity of dopamine target cells to dopamine (see Sect. E). A decreased GABAergic transmission might be involved in these events via a dual mechism: i) reduction of the inhibitory GABAergic input on dopamine neurons (see above) and ii) alterations of events postsynaptically to dopamine neurons, leading to the development of dopamine target cell supersensitivity. This latter mechanism, still obscure, is suggested by the fact that administration of the GABA receptor agonist progabide together with haloperidol prevents the neuroleptic-induced supersensitivity to dopamine receptor agonists (LLOYD and WORMS 1980). Also, GABAergic agents (valproate, progabide) ameliorate neuroleptic-induced tardive dyskinesia (LINNOILA et al. 1976; MORSELLI et al. 1980).

II. Limbic System

Mesolimbic dopamine neurons which originate in the ventral tegmental area of the mesencephalon (DAHLSTROEM and FUXE 1964) and project to limbic areas including the frontal cortex (THIERRY et al. 1973) form a (functional) network with other neurons which is different from that of the nigrostriatal pathway. Thus, it appears that mesolimbic dopamine neurons do not influence limbic cholinergic activity: blockade of dopaminergic transmission by a neuroleptic does not result in changes in acetylcholine release in cat septal areas or the ventral and dorsal hippocampal formation perfused by means of the push-pull cannula (LLOYD et al. 1973; BARTHOLINI et al. 1976b). Although these results have been confirmed by various investigators (see CONSOLO et al. 1977), a report points to the enhancement of acetylcholine turnover in some limbic areas by neuroleptic drugs (MAO et al. 1977).

The inverse influence of cholinergic neurons on dopamine cells appears to exist. Thus, limbic dopamine turnover is accelerated by cholinomimetic agents (ANDÉN 1974; BARTHOLINI et al. 1975), whereas the opposite effect is observed with cholinoceptor antagonists; the latter also counteract the neuroleptic-induced activation of dopamine neurons (O'KEEFE et al. 1970). As a consequence, it has to be assumed that limbic dopaminergic transmission is under an excitatory influence of cholinergic neurons.

Blockade of dopaminergic transmission in limbic areas appears to enhance GABA turnover (MAO et al. 1977). Conversely, GABAergic agents reduce both synthesis and utilization of dopamine in limbic regions (BARTHOLINI et al. 1979) indicating that, as in the extrapyramidal system, limbic dopamine neurons are under a GABAergic inhibitory input. However, in this respect, a marked difference exists between nigrostriatal and mesolimbic dopamine neurons, as the latter, both under resting conditions and in an activated state (e.g. after neuroleptics), are far less sensitive to GABAergic inhibition (BARTHOLINI et al. 1979). The reasons for this difference are unknown.

1. Clinical Implications

It is thought today that some schizophrenic symptoms are connected with an absolute or relative exaggeration of dopaminergic transmission in the limbic system (see BARTHOLINI et al. 1976b). If mesolimbic dopamine neurons were to influence cholinergic cells in the limbic system as in the striatum, it could be assumed that i) limbic dopaminergic transmission is mediated by (inter alia) cholinergic neurons and ii) the latter are involved in the antipsychotic action of neuroleptic drugs. However, the finding that, in animals, alterations of dopaminergic transmission are without effect on cholinergic activity makes these hypotheses unlikely (if the extrapolation to humans is allowed).

In addition, although many contradictory papers have been published on this matter, administration to schizophrenic patients of drugs affecting cholinergic transmission (agonists or antagonists) does not dramatically change either the basic syndrome or the antipsychotic action of neuroleptics (SINGH and LAL 1981). These findings also indicate that the excitatory input of acetylcholine cells on limbic dopamine neurons does not play a major role in modifying dopaminergic transmission in man and, therefore, in affecting schizophrenia. It is, however, beyond any doubt that changes in cholinergic transmission affect animal operant behaviour (see SINGH and LAL 1981) but this may be unrelated to dopaminergic transmission and/or have no relevance for schizophrenia.

Similarly, clinical results do not support the involvement of GABA in schizophrenia. Thus, GABAergic agents (progabide, muscimol) administered alone or together with neuroleptic drugs do not change schizophrenic symptoms (TAMMINGA et al. 1978; MORSELLI et al. 1980) suggesting that GABAergic transmission does not play a fundamental role in the regulation of limbic dopaminergic function in humans. This is supported by the experimental finding that limbic dopamine neurons show a low sensitivity to GABAergic inhibition (see above). The clinical data with GABAergic agents indicate also that GABA-mediated events are not involved in the antipsychotic action of neuroleptics.

G. References

Adams RN (1978) In vivo electrochemical recording: a new neurophysiological approach. Trends Neurosci 1:160-163

Aghajanian GK, Bunney BS (1973) Central dopaminergic neurons: neurophysiological identification and responses to drugs. In: Usdin E, Snyder SH (Eds) Frontiers in Catecholamine Research, Pergamon Press, New York, pp 643-648

Agid Y, Guyenet P, Glowinski J, Beaujouan JC, Javoy F (1975) Inhibitory influence of the nigrostriatal dopamine system on the striatal cholinergic neurons in the rat. Brain Res 86:488-492

Allen RM, Lane JD, Branchi JT (1980) Amantadine reduces haloperidol-induced dopamine receptor hypersensitivity in the striatum. Eur J Pharmacol 65:313-315

Andén NE (1974) Effects of oxotremorine and physostigmine on the turnover of dopamine in the corpus striatum and the limbic system. J Pharm Pharmacol 26:738-740

Andén NE, Stock G (1973) Inhibitory effect of gamma-hydroxybutyric acid and gamma-aminobutyric acid on the dopamine cells in the substantia nigra. Naunyn-Schmiedeberg's Arch Pharmacol 279:89-92

Andén NE, Wachtel H (1977) Biochemical effects of baclofen (β-parachlorophenyl-GABA) on the dopamine and the noradrenaline in the rat brain. Acta Pharmacol Toxicol 40:310-320

Andén NE, Corrodi H, Fuxe K, Ungerstedt U (1971) Importance of nervous impulse flow for the neuroleptic induced increase in amine turnover in central dopamine neurons. Eur J Pharmacol 15:193-199

Annunziato L, Moore KE (1977) Increased ability of apomorphine to reduce serum concentrations of prolactin in rats treated chronically with α-methyltyrosine. Life Sci 21:1845-1850

Annunziato L, Quattrone A, Schettini G, Di Renzo G (1980) Supersensitivity of pituitary dopamine receptors involved in the inhibition of prolactin secretion. In: Cattabeni F, Racagni G, Spano PF, Costa E (Eds) Long-Term Effects of Neuroleptics. Raven Press, New York, pp 379-385

Aquilonius SM, Sjöström R (1971) Cholinergic and dopaminergic mechanisms in Huntington's Chorea. Life Sci 10:405-408

Ashcroft GW, Dow RC, Moir ATB (1968) The active transport of 5-hydroxyindol-3-yl-acetic acid and 3-methoxy-4-hydroxyphenylacetic acid from a recirculatory perfusion system of the cerebral ventricles of the unanaesthetized dog. J Physiol (Lond) 199:397-425

Asper H, Baggiolini M, Burki HR, Lauener H, Ruch W, Stille G (1973) Tolerance phenomena with neuroleptics: catalepsy, apomorphine stereotypies and striatal dopamine metabolism in the rat after single and repeated administration of loxapine and haloperidol. Eur J Pharmacol 22:287-294

Bannon MJ, Roth RH (1983) Pharmacology of mesocortical dopamine neurons. Pharmacol Rev 35:53-68

Bannon MJ, Elliott PJ, Alpert JE, Goedert M, Iversen SD, Iversen LL (1983) Role of endogenous substance P in stress-induced activation of mesocortical dopamine neurones. Nature 306:791-792

Barbeau A (1974) The clinical physiology of side effects in long-term L-DOPA therapy. In: McDowel F, Barbeau A (Eds) Advances Neurol, vol 5. Raven Press, New York, pp 347-353

Barbeau A, Sourkes TL, Murphy GF (1962) Les catécholamines de la maladie de Parkinson. In: de Ajuriagueraa J (Ed) Monoamines et Système Nerveux Central. Masson, Paris, pp 247-262

Bartholini G, Pletscher A (1971) Atropine-induced changes of cerebral dopamine turnover. Experientia 27:1302-1303

Bartholini G, Pletscher A (1975) Decarboxylase inhibitors. Pharmacol Ther (B) 1:407-421

Bartholini G, Stadler H (1975a) Neurotransmitter interaction in the basal ganglia: relation to extrapyramidal disorders. In: Birkmayer W, Hornykiewicz O (Eds) Advances in Parkinsonism. Editions Roche, Basel, pp 115-123

Bartholini G, Stadler H (1975b) Cholinergic and GABAergic influence on the dopamine release in extrapyramidal centers. In: Almgren O, Carlsson A, Engel J (Eds) Chemical Tools in Catecholamine Research II. American Elsevier Amsterdam, New York, North-Holland, pp 235-241

Bartholini G, Stadler H (1977) Evidence for an intrastriatal GABAergic influence on dopamine neurons of the cat. Neuropharmacology 16:343-347

Bartholini G, Burkard WP, Pletscher A, Bates HM (1967) Increase of cerebral catecholamines caused by 3,4-dihydroxyphenylalanine after inhibition of peripheral decarboxylase. Nature 215:852-853

Bartholini G, Lloyd KG, Stadler H (1974) Dopaminergic regulation of striatal cholinergic neurons: relation to parkinsonism. In: McDowel F, Barbeau A (Eds) Advances in Neurology, vol 5. Raven Press, New York, pp 11-14

Bartholini G, Keller H, Pletscher A (1975) Drug induced changes of dopamine turnover in striatum and limbic system of the rat. J Pharm Pharmacol 27:439-441

Bartholini G, Stadler H, Gadea-Ciria M, Lloyd KG (1976a) The use of the push-pull cannula to estimate the dynamics of acetylcholine and catecholamines within various brain areas. Neuropharmacology 15:515-519

Bartholini G, Stadler H, Gadea-Ciria M, Lloyd KG (1976b) The effect of antipsychotic drugs on the release of neurotransmitters in various brain areas. In: Sedvall G, Uvnäs B, Zotterman Y (Eds) Antipsychotic Drugs: Pharmacodynamics and Pharmacokinetics. Pergamon Press, Oxford, pp 105-116

Bartholini G, Scatton B, Zivkovic B, Lloyd KG (1979) On the mode of action of SL 76.002, a new GABA receptor agonist. In: Kofod H, Scheel-Krüger J, Krogsgaard-Larsen P (Eds) GABA-Neurotransmitters. Munksgaard, Copenhagen, pp 326-339

Baudry M, Martres MP, Schwartz JC (1979) ^{3}H-domperidone: a selective ligand for dopamine receptors. Naunyn-Schmiedeberg's Arch Pharmacol 308:231-237

Baumann PA, Maitre L (1976) Is drug inhibition of dopamine uptake a misinterpretation of *in vivo* experiments? Nature 264:789-790

Berger PA, Faull KF, Kilkowski J, Anderson PJ, Kraemer H, Davis KL, Barchas JD (1980) CSF monoamine metabolites in depression and schizophrenia. Am J Psychiatry 137:174-180

Bertler A, Falck B, Owman C, Rosengren E (1966) The localization of monoaminergic blood brain barrier mechanism. Pharmacol Rev 18:369-385

Biggio G, Porceddu ML, Gessa GL (1976) Decrease of homovanillic dihydroxyphenylacetic acid and cyclic-adenosine-3′,5′-monophosphatase content in the rat caudate nucleus induced by the acute administration of an aminoacid mixture lacking tyrosine and phenylalanine. J Neurochem 26:1253-1255

Biggio G, Casu M, Corda MG, Di Bello C, Gessa GL (1978) Stimulation of dopamine synthesis in caudate nucleus by intrastriatal enkephalins and antagonism by naloxone. Science 200:552-554

Billard W, Ruperto V, Crosby G, Iorio LC, Barnett A (1984) Characterization of the binding of ^{3}H-SCH 23390, a selective D_1 receptor antagonist ligand, in rat striatum. Life Sci 35:1885-1893

Birket-Smith E (1974) Abnormal involuntary movements induced by anticholinergic therapy. Acta Neurol Scandinav 50:801-811

Birkmayer W, Hornykiewicz O (1961) Der L-Dioxyphenylalanin-(= L-DOPA)-Effekt bei Parkinson-Akinese. Wien. Klin Wochenschr 73:787-788

Blanchard JC, Boireau A, Garret C, Julou L (1980) The use of thioproperazine, a phenothiazine derivative, as a ligand for neuroleptic receptor. Biochem Pharmacol 29:2933-2938

Brostrom CO, Huang Y-C, Breckenridge B, Wolff DJ (1975) Identification of a calcium-binding protein as a calcium-dependent regulator of brain adenylate cyclase. Proc Nat Acad Sci USA 72:64-68

Brownstein MJ, Mroz EA, Tappaz ML, Leeman SE (1977) On the origin of substance P and glutamic acid decarboxylase (GAD) in the substantia nigra. Brain Res 135:315-323

Bullard WP, Guthric PB, Russo PV, Mandell AJ (1978) Regional and subcellular dis-

tribution and some factors in the regulation of reduced pterins in rat brain. J Pharmacol Exp Ther 206:4–20
Bunney BS, Aghajanian GK (1976) The effects of antipsychotic drugs on the firing of dopaminergic neurons: a reappraisal. In: Sedvall G, Uvnäs B, Zotterman Y (Eds) Antipsychotic Drugs: Pharmacodynamics and Pharmacokinetics. Pergamon Press, Oxford, pp 305–318
Bunney BS, Grace AA (1978) Acute and chronic haloperidol treatment: comparison of effects on nigral dopaminergic cell activity. Life Sci 23:1715–1728
Bunney BS, Aghajanian GK, Roth RH (1973a) Comparison of effects of L-DOPA, amphetamine and apomorphine on firing rate of rat dopaminergic neurons. Nature New Biol 245:123–125
Bunney BS, Walters JR, Roth RH, Aghajanian GK (1973b) Dopaminergic neurons: effect of antipsychotic drugs and amphetamine on single cell activity. J Pharmacol Exp Ther 185:560–571
Burt DR, Creese I, Snyder SH (1977) Antischizophrenic drugs: chronic treatment elevates dopamine receptor binding in brain. Science 196:326–328
Carlsson A (1959) The occurence, distribution and physiological role of catecholamines in the nervous system. Pharmacol Rev 11:490–493
Carlsson A (1975) Receptor-mediated control of dopamine metabolism. In: Usdin E, Bunney WE (Eds) Pre- and Postsynaptic Receptors. Dekker I, New York, pp 49–63
Carlsson A, Lindqvist M (1963) Effect of chlorpromazine or haloperidol on formation of 3-methoxytyramine and normetanephrine in mouse brain. Acta Pharmacol Toxicol 20:140–144
Carlsson A, Lindqvist M (1978) Dependence of 5-HT and catecholamine synthesis on concentrations of precursor amino-acids in rat brain. Naunyn-Schmiedeberg's Arch Pharmacol 303:157–164
Carlsson A, Lindqvist M, Magnusson T, Waldeck B (1958) On the presence of 3-hydroxytyramine in brain. Science 127:471–472
Cerrito F, Casazza G, Levi G, Raiteri M (1980) Evidence for a similar compartmentation of recaptured and endogenously synthesized dopamine in striatal synaptosomes. Neurochem Res 5:115–121
Chen TC, Cote TE, Kebabian JW (1980) Endogenous components of the striatum confer dopamine-sensitivity upon adenylate cyclase activity: the role of endogenous guanyl nucleotides. Brain Res 181:139–149
Cheramy A, Nieoullon A, Glowinski J (1977a) Effects of peripheral and local administration of picrotoxin on the release of newly synthesized ^{3}H-dopamine in the caudate nucleus of the cat. Naunyn-Schmiedeberg's Arch Pharmacol 297:31–37
Cheramy A, Nieoullon A, Michelot R, Glowinski J (1977b) Effects of intranigral application of dopamine and substance P on in vivo release of newly synthesized [^{3}H]-dopamine in the ipsilateral caudate nucleus of the cat. Neurosci Lett 4:105–109
Cheramy A, Nieoullon A, Glowinski J (1978a) GABAergic processes involved in the control of dopamine release from nigrostriatal dopaminergic neurons in the cat. Eur J Pharmacol 48:281–295
Cheramy A, Michelot R, Leviel V, Nieoullon A, Glowinski J, Kerdelhue B (1978b) Effect of the immunoneutralization of substance P in the cat substantia nigra on the release of dopamine from dendrites and terminals of dopaminergic neurons. Brain Res 155:404–408
Chesselet M-F (1984) Presynaptic regulation of neurotransmitter release in the brain: facts and hypothesis. Neuroscience 12:347–375

Cheung WY (1971) Cyclic 3',5'-nucleotide phosphodiesterase. Evidence for and properties of a protein activator. J Biol Chem 246:2859-2869

Cheung WY, Bradham LS, Lynch TJ, Lin YM, Tallant EA (1975) Protein activator of cyclic 3',5'-nucleotide phosphodiesterase of bovine or rat brain also activates its adenylate cyclase. Biochem Biophys Res Comm 66:1055-1062

Chiodo LA, Bunney BS (1983) Typical and atypical neuroleptics: differential effects of chronic administration on the activity of A9 and A10 midbrain dopaminergic neurons. J Neurosci 3:1607-1619

Chiueh CC, Moore KE (1975) d-Amphetamine-induced release of "newly synthesized" and "stored" dopamine from the caudate nucleus *in vivo.* J Pharmacol Exp Ther 192:642-653

Clement-Cormier YC, Kebabian JW, Petzold GL, Greengard P (1974) Dopamine sensitive adenylate cyclase in mammalian brain: a possible site of action of antipsychotic drugs. Proc Nat Acad Sci USA 71:1113-1117

Clement-Cormier YC, Rudolph FB, Robison GA (1978) Dopamine-sensitive adenylate cyclase from the rat caudate nucleus: regulation by guanyl nucleotides and the interaction of magnesium and magnesium ATP. J Neurochem 30:1163-1172

Consolo S, Ladinsky H, Bianchi S, Ghezzi D (1977) Apparent lack of a dopaminergic-cholinergic link in the rat nucleus accumbens septi-tuberculum olfactorium. Brain Res 135:255-263

Cooper BR, Hester TJ, Maxwell A (1980) Behavioral and biochemical effects of antidepressant bupropion (Wellbutrin): evidence for selective blockade of dopamine uptake *in vivo.* J Pharmacol Exp Ther 215:127-134

Costa E, Neff NH (1966) Isotopic and non-isotopic measurements of catecholamine synthesis. In: Costa E, Cot' L, Yahr MD (Eds) Biochemistry and Pharmacology of the Basal Ganglia. Raven Press, New York, pp 141-156

Costa E, Groppetti A, Naimzada MK (1972) Effect of amphetamine on the turnover of brain catecholamines and motor activity. Br J Pharmacol 44:742-751

Costa E, Carenzi A, Cheny D, Guidotti A, Racagni G, Zivkovic B (1975) Compartmentation of striatal dopamine: problems in assessing the dynamics of functional and storage pools of transmitters. In: Berl S, Clarke DD, Schneider D (Eds) Metabolic Compartmentation and Neurotransmission. Plenum Press, New York, pp 167-186

Cotzias GC, Van Woert MH, Schiffer LM (1967) Aromatic amino acids and modification of parkinsonim. New Engl J Med 276:374-379

Coyle JT, Snyder SH (1969) Antiparkinsonian drugs: inhibition of dopamine uptake in the corpus striatum as a possible mechanism of action. Science 166:899-901

Creese I (1985) Binding interactions of neuroleptic drugs with dopamine receptors and their implications. In: Seiden LS, Balster RL (Eds) Behavioral Pharmacology: The Current Status Alan Liss, New York, pp 221-241

Creese I, Chen A (1985) Selective D-1 dopamine receptor increase following chronic treatment with SCH 23 390. Eur J Pharmacol 109:127-130

Creese I, Burt DR, Snyder SN (1976) Dopamine receptor binding predicts clinical and pharmacological potencies of antischizophrenic drugs. Science 192:481-483

Creese I, Burt DR, Snyder SH (1977) Dopamine receptor binding enhancement accompanies lesion-induced behavioural supersensitivity. Science 197:596-598

Creese I, Padgett L, Fazzini E, Lopez F (1979a) ^{3}H-N-n-propylnorapomorphine: a novel agonist ligand for central dopamine receptors. Eur J Pharmacol 56:411-412

Creese I, Usdin T, Snyder SH (1979b) Guanine nucleotides distinguish two dopamine receptors. Nature 278:577-578

Dahlström A, Fuxe K (1964) Evidence for the existence of monoamine-containing neurons in the central nervous system. I. Demonstration of monoamines in the cell bodies of brain stem neurons. Acta Physiol Scand 62, Suppl 232

Da Prada M, Saner A, Burkard WP, Bartholini G, Pletscher A (1975) Lysergic acid diethylamide: evidence for stimulation of cerebral dopamine receptors. Brain Res 94:67-73

Davies J, Dray A (1976) Substance P in the substantia nigra. Brain Res 107:623-627

Davis A, Poat JA, Woodruff GN (1980) The use of R(+)-ADTN in dopamine receptor binding assays. Eur J Pharmacol 63:237-238

Davis KL, Hollister LE, Berger PA, Barchas JD (1975) Cholinergic imbalance hypotheses of psychoses and movement disorders: strategies for evaluation. Psychopharmacol Comm 1:533-543

Dedek J, Baumes R, Tien-Duc N, Gomeni R, Korf J (1979) Turnover of free and conjugated (sulphonyloxy) dihydroxyphenylacetic acid and homovanillic acid in rat striatum. J Neurochem 33:687-695

De Silva L (1977) Biochemical mechanisms and management of choreiform movement disorders. Drugs 14:300-307

De Silva L, Huang CY (1975) Deanol in tardive dyskinesia. B Med J 3:466-468

Di Chiara G, Porceddu ML, Morelli M, Mulas ML, Gessa GL (1979) Strio-nigral and nigro-thalamic GABAergic neurons as output pathways for striatal responses. In: Krogsgaard-Larsen P, Scheel-Krüger J, Kofod H (Eds) GABA-Neurotransmitters, Munksgaard, Copenhagen, pp 465-481

Dominic JA, Moore KE (1969) Supersensitivity to the central stimulant actions of adrenergic drugs following discontinuation of a chronic diet of α-methyl tyrosine. Psychopharmacologia (Berl) 15:96-101

Doteuchi M, Wang C, Costa E (1974) Compartmentation of dopamine in rat striatum. Mol Pharmacol 10:225-234

Dray A, Straughan DW (1976) Synaptic mechanism in the substantia nigra. J Pharm Pharmacol 28:400-405

Dubocovich ML, Zahniser NR (1985) Binding characteristics of the dopamine uptake inhibitor [^{3}H]nomifensine to striatal membranes. Biochem Pharmacol 34:1137-1144

Dubois A, Savasta M, Curet O, Scatton B (1986) Autoradiographic distribution of the D_1 agonist [^{3}H] SKF 38 393 in the rat brain and spinal cord. Comparison with the distribution of D_2 dopamine receptors. Neuroscience 19:125-137

Dunstan R, Broekkamp CLK, Lloyd KG (1981) Involvement of caudate nucleus, amygdala or reticular formation in neuroleptic and narcotic catalepsy. Pharmacol Biochem Behav 14:169-174

Ehringer H, Hornykiewicz O (1960) Verteilung von Noradrenalin und Dopamin (3-Hydroxytyramin) im Gehirn des Menschen und ihr Verhalten bei Erkrankungen des extrapyramidalen Systems. Klin Wochenschr 38:1236-1239

Enna SJ, Stern LZ, Wastek GJ, Yamamura HI (1977) Neurobiology and pharmacology of Huntington's disease. Life Sci 20:205-212

Euvrard C, Premont J, Oberlander C, Boissier JR, Bockaert J (1979) Is dopamine-sensitive adenylate cyclase involved in regulating the activity of striatal cholinergic neurons? Naunyn Schmiedeberg's Arch Pharmacol 309:241-245

Ezrin-Waters C, Seeman P (1977) Tolerance to haloperidol catelepsy. Eur J Pharmacol 41:321-327

Farnebo LO, Hamberger B (1971) Drug-induced changes in the release of ^{3}H-monoamines from field stimulated rat brain slices. Acta Physiol Scand 35, Suppl 371:35-44

Foldes A, Meek JL (1973) Rat brain phenolsulfotransferase—partial purification and some properties. Biochem Biophys Acta 327:365-374

Fuller RW, Perry KW, Bymaster FP, Wong DT (1978) Comparative effects of pemoline, amfonelic acid and amphetamine on dopamine uptake and release *in vitro* and on brain 3,4-dihydroxyphenylacetic acid concentration in spiperone-treated rats. J Pharm Pharmacol 30:197-198

Gale K, Guidotti A, Costa E (1976) Dopamine sensitive andenylate cyclase: location in substantia nigra. Science 195:503-505

Gale K, Hong J-S, Guidotti A (1977) Presence of substance P and GABA in separate striatonigral neurons. Brain Res 136:371-375

Gale K, Costa E, Toffano G, Hong J-S, Guidotti A (1978) Evidence for a role of nigral γ-aminobutyric acid and substance P in the haloperidol-induced activation of striatal tyrosine hydroxylase. J Pharmacol Exp Ther 306:29-37

Gallager DW, Pert A, Bunney WE, Jr (1978) Haloperidol-induced presynaptic dopamine supersentivity is blocked by chronic lithium. Nature 273:309-312

Geffen LB, Jessell TM, Cuello AC, Iversen LL (1976) Release of dopamine from dendrites in rat substantia nigra. Nature 260:258-260

Gerlach J, Reisby N, Randrup A (1974) Dopaminergic hypersensitivity and cholinergic hypofunction in the pahtophysiology of tardive dyskinesia. Psychopharmacologia (Berl) 34:21-35

Glowinski J, Iversen LL (1966) Regional studies of catecholamines in the rat brain. III. Subcellular distribution of endogenous and exogenous catecholamines in various brain regions. Biochem Pharmacol 15:977-987

Gnegy ME, Lucchelli A, Costa E (1977a) Correlation between drug-induced supersensitivity of dopamine dependent striatal mechanisms and the increase in striatal content of the Ca^{2+} regulated protein activator of cAMP phosphodiesterase. Naunyn-Schmiedeberg's Arch Pharmacol 301:121-127

Gnegy M, Uzunov P, Costa E (1977b) Participation of an endogenous Ca^{2+}-binding protein activator in the development of drug-induced supersensitivity of striatal dopamine receptors. J Pharmacol Exp Ther 202:558-564

Goldstein M, Anagnoste B, Freedman LS, Roffman M, Ebstein RP, Park DH, Fuxe K, Hökfelt T (1973) Characterization, localization and regulation of catecholamine synthesizing enzymes. In: Usdin E, Snyder SH (Eds) Frontiers in Catecholamine Research. Pergamon Press, New York, pp 69-78

Goldstein M, Lew JY, Nakamura S, Battista AF, Leberman A, Fuxe K (1978) Dopamine-philic properties of ergot alkaloids. Fed Proc 37:2202-2206

Goldstein M, Liberman A, Lew JY, Asano T, Rosenfeld MR, Makman M (1980) Interaction of pergolide with central dopaminergic receptors. Proc Nat Acad Sci USA 77:3725-3728

Gonon F (1986) Control of dopamine release by dopamine receptors and by impulse flow a studied by in vivo voltammetry. Ann New York Acad Sci, 473:160-169

Gonon F, Buda M, Cespuglio R, Jouvet M, Pujol J-F (1980) In vivo electrochemical detection of catechols in the neostriatum of anaesthetized rats: dopamine or DOPAC? Nature 286:902-904

Gonon F, Cespuglio R, Buda M, Pujol JF (1983) In vivo electrochemical detection of monoamine derivatives. In Parvez S, Nagatsu T, Nagatsu I, Parvez H (Eds) Methods in Biogenic Amine Research. Elsevier, Amsterdam, pp 165-188

Grace AA, Bunney BS (1979) Paradoxical GABA excitation of nigral dopaminergic cells: indirect mediation through reticulata inhibitory neurons. Eur J Pharmacol 59:211-218

Greenblatt DJ, Shader RJ (1973) Drug therapy—anticholinergics. New Engl J Med 288:1215-1219

Groves PM, Wilson CJ, Young SJ, Rebec CV (1975) Self inhibition by dopaminergic neurons. Science 190:522–529

Guidotti A, Gale K, Toffano G, Vargas FM (1978) Tolerance to tyrosine hydroxylase activation in n. accumbens and c. striatum after repeated injections of "classical" and "atypical" antischizophrenic drugs. Life Sci 23:501–506

Guidotti A, Moroni F, Gale K, Kumakura K (1979) Opiate receptor stimulation blocks the activation of striatal tyrosine hydroxylase (TH) induced by haloperidol. In: Usdin E, Kopin IJ, Barchas J (Eds) Catecholamines: Basic and Clinical Frontiers. Pergamon Press, New York, pp 1035–1037

Guyenet PG, Agid Y, Javoy F, Beaujouan JC, Rossier J, Glowinski J (1975) Effects of dopaminergic receptor agonists and antagonists on the activity of the neo-striatal cholinergic system. Brain Res 84:227–244

Hanson JS, Alphs L, Wolf W, Levine R, Lovenberg W (1981) Haloperidol-induced reduction of nigral substance P-like immunoreactivity: a probe for the interactions between dopamine on substance P neuronal systems. J Pharmacol Exp Ther 218:568–574

Hattori T, Fibiger HC (1982) On the use of lesions of afferents to localize neurotransmitter receptor site in the striatum. Brain Res 238:245–250

Heikkila R, Orlansky H, Cohen G (1975) Studies on the distinction between uptake inhibition and release of ^{3}H-dopamine in rat brain tissue slices. Biochem Pharmacol 24:847–852

Hong JS, Yang, H-YT, Costa E (1978a) Substance P content of substantia nigra after chronic treatment with antischizophrenic drugs. Neuropharmacology 17:83–85

Hong JS, Yang H-YT, Fratta W, Costa E (1978b) Rat striatal methionine-enkephalin content after chronic treatment with cataleptogenic and non-cataleptogenic antischizophrenic drugs. J Pharmacol Exp Ther 205:141–147

Horn AS (1979) Characteristics of dopamine uptake. In: Horn AS, Korf J, Westerink BHC (Eds) The Neurobiology of Dopamine. Academic Press, London, pp 217–235

Horn AS, Coyle JT, Snyder SH (1971) Catecholamine uptake by synaptosomes from rat brain. Mol Pharmacol 7:66–80

Hunt P, Kannengiesser M-H, Raynaud J-P (1974) Nomifensine: a new potent inhibitor of dopamine uptake into synaptosomes from rat brain corpus striatum. J Pharm Pharmacol 26:370–371

Hyttel J (1978) Effects of neuroleptics on ^{3}H-haloperidol and ^{3}H-Cis(Z)-flupenthixol binding and on adenylate cyclase activity in vitro. Life Sci 23:551–556

Ikeda M, Fahien LA, Udenfriend S (1966) A kinetic study of bovine adrenal tyrosine hydroxylase. J Biol Chem 241:4452–4456

Imperato A, Di Chiara G (1984) Trans-striatal dialysis coupled to reverse phase high performance liquid chromatography with electrochemical detection: a new method for the study of the in vivo release of endogenous dopamine and metabolites. J Neurosci 4:966–977

Imperato A, Di Chiara G (1985) Dopamine release and metabolism in awake rats after systemic neuroleptics as studied by trans-striatal dialysis. J Neurosci 5:297–306

Iorio LC, Barnett A, Leitz FH, Houser VP, Korduba CA (1983) SCH 23 390, a potential benzazepine antipsychotic with unique interactions on dopaminergic systems. J Pharmacol Exp Ther 226:462–468

Iversen LL (1975) Dopamine receptors in the brain. A. Dopamine-sensitive adenylate cyclase models. Synaptic receptors, illuminating antipsychotic drug action. Science 188:1084–1089

Iversen LL, Schon FE (1973) The use of autoradiographic techniques for the identification and mapping of transmitter-specific neurones in CNS. In: Mandel A (Ed)

New Concepts in Neurotransmitter Regulation. Plenum Press, New York, pp 153-159

Iversen LL, Ragowski MA, Miller RJ (1976) Comparison of the effects of neuroleptic drugs on pre- and postsynaptic dopaminergic mechanism in the rat striatum. Mol Pharmacol 12:251-262

Iwatsubo K, Clouet DH (1977) Effect of morphine and haloperidol on the electrical activity of rat nigrostriatal neurons. J Pharmacol Exp Ther 202:429-436

James TA, Starr MS (1977) Behavioural and biochemical effects of substance P injected into the substantia nigra. J Pharm Pharmacol 29:181-182

Janowsky A, Schweri M, Berger P, Long R, Skolnick P, Paul S (1985) The effects of surgical and chemical lesions on striatal [^{3}H] threo-(±)-methylphenidate binding: correlation with [^{3}H] dopamine uptake. Eur J Pharmacol 108:187-191

Janowsky A, Berger P, Vocci F, Labarca R, Skolnick P, Paul SM (1986) Characterization of sodium-dependent [^{3}H] GBR-12935 binding in brain: a radioligand for selective labelling of the dopamine transporter complex. J Neurochem 46:1272-1275

Javitch JA, Strittmatter SM, Snyder SH (1985) Differential visualization of dopamine and norepinephrine uptake sites in rat brain using [^{3}H]-mazindol autoradiography. J Neurosci 5:1513-1521

Javoy F, Glowinski J (1971) Dynamic characteristics of the "functional compartment" of dopamine in dopaminergic terminals of the rat striatum. J Neurochem 18: 1305-1311

Javoy F, Agid Y, Bouvet D, Glowinski J (1972) Feedback control of dopamine synthesis in dopaminergic terminals of the rat striatum. J Pharmacol Exp Ther 182:454-463

Joh TH, Reis DJ (1975) Different forms of tyrosine hydroxylase in central dopaminergic and noradrenergic neurons, sympathetic ganglia and adrenal medulla. Brain Res 85:146-151

Kanazawa I, Yoshida M (1980) Electrophysiological evidence for the existence of excitatory fibers in the caudato-nigral pathway in the cat. Neurosci Lett 20:301-306

Kaplan GP, Hartman BK, Creveling CR (1979) Non-neuronal localization of catechol-O-methyltransferase in brain. In: Usdin E, Kopin IJ, Barchas J (Eds) Catecholamines: Basic and Clinical Frontiers. Pergamon Press, New York, pp 1354-1356

Kaplan JP, Raison B, Desarmenien M, Feltz P, Haedley PM, Worms P, Lloyd KG, Bartholini G (1980) New anticonvulsants: Schiff bases of γ-aminobutyric acid and γ-aminobutyramide. J Med Chem 23:702-704

Kebabian JW, Calne DB (1979) Multiple receptors for dopamine. Nature 277:93-96

Kebabian JW, Petzold GL, Greengard P (1972) Dopamine-sensitive adenylate cyclase in caudate nucleus of the rat brain and its similarity to the "dopamine receptor". Proc Nat Acad Sci USA 69:2145-2149

Kebabian JW, Calne DB, Kebabian PR (1977) Lergotrile mesylate: an *in vivo* dopamine agonist which blocks dopamine receptors *in vitro*. Comm Psychopharmacol 1:311-318

Kehr W, Carlsson A, Lindqvist M, Magnusson T, Atack C (1972) Evidence for a receptor-mediated feedback control of striatal tyrosine hydroxylase activity. J Pharm Pharmacol 24:744-747

Kettler R, Bartholini G, Pletscher A (1974) In vivo enhancement of tyrosine hydroxylation in rat striatum by tetrahydrobiopterin. Nature 249:476-478

Klawans HL (1970) A pharmacologic analysis of Huntington's chorea. Eur Neurol 4:148-163

Klawans HL, Rubovits R (1972) Central cholinergic-anticholinergic antagonism in Huntington's chorea. Neurology 22:107-113

Koffer KB, Berney S, Hornykiewicz O (1978) The role of the corpus striatum in neuroleptic- and narcotic-induced catalepsy. Eur J Pharmacol 47:81–86

Korf J, Grasdijk L, Westerink BHC (1976a) Effects of electrical stimulation of the nigrostriatal pathway of the rat on dopamine metabolism. J Neurochem 26:579–584

Korf J, Zieleman M, Westerink BHC (1976b) Dopamine release in substantia nigra? Nature 260:257–258

Kuschinsky K, Hornykiewicz O (1972) Morphine catalepsy in the rat: relation to striatal dopamine metabolism. Eur J Pharmacol 19:119–122

Lal H, Brown W, Drawbaugh R, Hynes M, Brown G (1977) Enhanced prolactin inhibition following chronic treatment with haloperidol and morphine. Life Sci 20:101–106

Langen CDJ, De Stoff JC, Mulder AH (1979) Studies on the nature of the releasable pool of dopamine from rat corpus striatum: depolarization-induced release of ^{3}H-dopamine from superfused synaptosomes labelled under various conditions. Naunyn-Schmiedeberg's Arch Pharmacol 308:41–49

Lerner P, Nose P, Gordon EK, Lovenberg W (1977) Haloperidol: effect of long-term treatment on rat striatal dopamine synthesis and turnover. Science 197:181–183

Linnoila M, Viukari M, Hietala O (1976) Effect of sodium valproate on tardive dyskinesia. Br J Psychiat 129:114–119

List SJ, Seeman P (1979) Dopamine agonists reverse the elevated ^{3}H-neuroleptic binding in neuroleptic-pretreated rats. Life Sci 24:1447–1452

Llorens-Cortes C, Pollard H, Schwartz JC (1979) Localization of opiate receptors in substantia nigra: evidence by lesion studies. Neurosci Lett 12:165–170

Lloyd KG (1977) Neurochemical compensation in Parkinson's disease. In: Lakke JPWF, Korf J, Weselling H (Eds) Parkinson's disease, Concepts and Prospects. Excerpta Medica, Amsterdam, pp 61–72

Lloyd KG, Hornykiewicz O (1973) L-glutamic acid decarboxylase in Parkinson's disease: effect of L-Dopa therapy. Nature 243:521–523

Lloyd KG, Hornykiewicz O (1977) Effect of chronic neuroleptic or L-DOPA administration on GABA levels in the rat substantia nigra. Life Sci 21:1489–1496

Lloyd KG, Worms P (1980) Sustained gamma-aminobutyric acid receptor stimulation and chronic neuroleptic effects. Cattabeni F, Racagni G, Spano PF, Costa E (Eds) Advances in Biochemical Psychopharmacology, vol 24. Raven Press, New York, pp 253–258

Lloyd KG, Stadler H, Bartholini G (1973) Dopamine and acetylcholine neurons in striatal and limbic structures: Effect of neuroleptic drugs. In: Usdin E, Snyder SH (Eds) Frontiers in Catecholamine Research. Pergamon Press, New York, pp 777–779

Lloyd KG, Shibuya M, Davidson L, Hornykiewicz O (1977) Chronic neuroleptic therapy: tolerance and GABA system. In: Costa E, Gessa GL (Eds) Advances in Biochemical Psychopharmacology, vol. 16. Raven Press, New York, pp 409–415.

Lloyd KG, Worms P, Depoortere H, Bartholini G (1979) Pharmacological profile of SL 76.002, a new GABA-mimetic drug. In: Krogsgaard-Larsen P, Scheel-Krüger J, Kofod H (Eds) GABA-Neurotransmitters. Munksgaard, Copenhagen, pp 308–325

Lloyd KG, Le Roux B, Bartholini G (1980) The dynamics of endogenous dopamine (DA) release from the cat caudate nucleus *in vivo*. Society for Neuroscience, 10th Annual Meeting, Abstracts, P 142

Lovenberg W, Weissbach H, Udenfriend S (1962) Aromatic L-amino acid decarboxylase. J Biol Chem 237:89–93

MacKenzie RG, Zigmond MJ (1985) Chronic neuroleptic treatment increases D-2 but not D-1 receptors in rat striatum. Eur J Pharmacol 113:159–165

Mao CC, Cheney DL, Marco E, Revuelta A, Costa E (1977) Turnover times of gamma-aminobutyric acid and acetylcholine in nucleus caudatus, nucleus accumbens, globus pallidus and substantia nigra: effects of repeated administration of haloperidol. Brain Res 132:375-379

Marchi M, Raiteri M (1985) Differential antagonism by dicyclomine, pirenzepine and secoverine at muscarinic receptor subtypes in the rat frontal cortex. Eur J Pharmacol 107:287-288

Marek KL, Roth RH (1980) Ergot alkaloids: interaction with presynaptic dopamine receptors in the neostriatum and olfactory tubercles. Eur J Pharmacol 62:137-146

Martres MP, Bouthenet ML, Sales N, Sokoloff P, Schwartz JC (1985) Widespread distribution of brain dopamine receptors evidenced with [^{125}I] iodosulpiride, a highly selective ligand. Science 228:752-755

McGeer PL, Grewaal DS, McGeer EG (1976) Effect on extrapyramidal GABA levels of drugs which influence dopamine and acetylcholine metabolism. In: Birkmayer W, Hornykiewicz O (Eds) Advances in Parkinsonism. Editions Roches, Basel, pp 132-140

McLennan H (1964) The release of acetylcholine and 3-hydroxytyramine from the caudate nucleus. J Physiol (Lond) 174:152-161

McMillen BA, German DC, Shore PA (1980) Functional and pharmacological significance of brain dopamine and norepinephrine storage pools. Biochem Pharmacol 29:3045-3050

Meek JL, Neff NH (1973) Biogenic amines and their metabolites as substrates for phenol sulphotransferase (EC 2.8.2.1) of brain and liver. J Neurochem 21:1-9

Melis MR, Gale K (1984) Intranigral application of substance P antagonists prevents the haloperidol-induced activation of striatal tyrosine hydroxylase. Naunyn Schmiedeberg's Arch Pharmacol 326:83-86

Mercer L, Del Fiacco M, Cuello AC (1979) The smooth endoplasmic reticulum as a possible storage site for dendritic dopamine in substantia nigra neurones. Experientia 35:101-103

Miller R, Horn A, Iversen LL, Pinder R (1974) Effects of dopamine-like drugs on rat striatal adenyl cyclase have implications for CNS dopamine receptor topography. Nature 250:238-241

Mishra RK, Wong YW, Varmuza SL, Tuff L (1978) Chemical lesion and drug induced supersensitivity and subsensitivity of caudate dopamine receptors. Life Sci 23:443-446

Moore KE, Annunziato L, Gudelsky GA (1978) Studies on tuberoinfundibular dopamine neurons. In: Roberts PJ, Woodruff GN, Iversen LL (Eds) Advances in Biochemical Psychopharmacology, vol 19. Raven Press, New York, pp 193-204

Morselli PL, Bossi L, Henry JF, Zarifian E, Bartholini G (1980) On the therapeutic action of SL 76.002, a new GABA mimetic agent: preliminary observations in neuropsychiatric disorders. Brain Res Bull 5 (Suppl 2):411-414

Mulder AH, Van den Berg WB, Stoof JC (1975) Calcium-dependent release of radiolabelled catecholamines and serotonin from rat brain synaptosomes in a superfusion system. Brain Res 99:419-424

Munkvad I, Fog R, Kristijansen P (1976) The drug approach to therapy. Long-term treatment of schizophrenia. In: Kemali D, Bartholini G, Richter D (Eds) Schizophrenia Today. Pergamon Press, Oxford, pp 173-182

Murrin LC, Roth RH (1976a) Dopaminergic neurons: effect of electrical stimulation on dopamine biosynthesis. Mol Pharmacol 12:463-475

Murrin LC, Roth RH (1976b) Dopaminergic neurons: reversal of effects elicited by gamma-butyrolactone by stimulation of the nigro-striatal pathway. Naunyn-Schmiedeberg's Arch. Pharmacol 295:15-20

Murrin LC, Morgenroth VH 3rd, Roth RH (1976) Dopaminergic neurons: effects of electrical stimulation on tyrosine hydroxylase. Mol Pharmacol 12:1070–1081

Murrin LC, Gale K, Kuhar MJ (1979) Autoradiographic localization of neuroleptic and dopamine receptors in the caudate-putamen and substantia nigra: effects of lesions. Eur J Pharmacol 60:229–235

Nagatsu T, Levitt M, Udenfriend S (1964) Tyrosine hydroxylase. The initial step in norepinephrine biosynthesis. J Biol Chem 239:2910–2917

Nagy JY, Lee T, Seeman P, Fibiger HC (1978) Direct evidence for presynaptic dopamine receptors in brain. Nature 274:278–281

Neff NH, Tozer TN, Brodie BB (1967) Application of steady-state kinetics to studies of the transfer of 5-hydroxyindoleacetic acid from brain to plasma. J Pharmacol Exp Ther 158:214–218

Niddam R, Arbilla S, Scatton B, Dennis T, Langer SZ (1985) Amphetamine induced release of endogenous dopamine in vitro is not reduced following pretreatment with reserpine. Naunyn Schmiedeberg's Arch Pharmacol 329:123–127

Nieoullon A, Cheramy A, Glowinski J (1977a) An adaptation of the push-pull cannula method to study *in vivo* release of ^{3}H-dopamine synthesized from ^{3}H-tyrosine in the cat caudate nucleus: Effects of various physical and pharmacological treatments. J Neurochem 28:819–828

Nieoullon A, Cheramy A, Glowinski J (1977b) Release of dopamine *in vivo* from cat substantia nigra. Nature 266:375–377

Nowak JZ, Arbilla S, Galzin AM, Langer SZ (1983) Changes in sensitivity of release modulating dopamine autoreceptors following chronic treatment with haloperidol. J Pharmacol Exp Ther 226:558–564

Nowycky MC, Roth RH (1977) Presynaptic dopamine receptors. Development of supersensitivity following treatment with fluphenazine decanoate. Naunyn-Schmiedeberg's Arch Pharmacol 300:247–254

Nowycky MC, Roth RH (1978) Dopaminergic neurons: role of presynaptic receptors in the regulation of transmitter synthesis. Prog Neuro-Psychopharmacol 2:139–158

Nybäck H, Sedvall G (1971) Effect of nigral lesion on chlorpromazine-induced acceleration of dopamine synthesis from ^{14}C-tyrosine. J Pharm Pharmacol 23:322–326

Oblin A, Zivkovic B, Bartholini G (1984) Involvement of the D-2 dopamine receptor in the neuroleptic-induced decrease in nigral substance P. Eur J Pharmacol 105:175–177

Onali P, Olianas MC, Gessa GL (1985) Characterization of dopamine receptors mediating inhibition of adenylate cyclase activity in rat striatum. Mol Pharmacol 28:138–145

O'Keefe R, Sharman DF, Vogt M (1970) Effect of drugs used in psychoses on cerebral dopamine metabolism. Br J Pharmacol 38:287–295

Papeschi R (1977) The functional pool of brain catecholamines: its size and turnover rate. Psychopharmacology 55:1–7

Parker EM, Cubeddu LX (1986) Effects of d-amphetamine and dopamine synthesis inhibitors on dopamine and acetylcholine neurotransmission in the striatum. I. Release in the absence of vesicular transmitter stores. J Pharmacol Exp Ther 237:179–192

Patrick RL, Barchas JD (1976) Dopamine synthesis in striatal synaptosomes. II. Dibutyryl cyclic adenosine 3′5′-monophosphoric acid and 6-methyltetrahydropterin-induced synthesis increases without an increase in endogenous dopamine release. J Pharmacol Exp Ther 197:97–104

Pert A, De Wald LA, Gallager DW (1979) Effects of opiates on nigrostriatal dopaminergic activity: electrophysiological and behavioral analyses. In: Usdin E, Kopin IJ,

Barchas J (Eds) Catecholamines: Basic and Clinical Frontiers. New York pp 1041–1043

Phillipson OT, Horn AS (1976) Substantia nigra of the rat contains a dopamine sensitive adenylate cyclase. Nature 261:418–420

Phillipson OT, Emson PC, Horn AS, Jessel T (1977) Evidence concerning the anatomical location of the dopamine-stimulated adenylate cyclase in the substantia nigra. Brain Res 136:45–58

Pimoule C, Schoemaker H, Reynolds GP, Langer SZ (1985) [^{3}H]-SCH 23 390 labeled D_1 dopamine receptors are unchanged in schizophrenia and Parkinson's disease. Eur J Pharmacol 114:235–237

Pollard H, Llorens C, Schwartz JC, Gross C, Dray F (1978) Localization of opiate receptors and enkephalins in the rat striatum in relationship with the nigro-striatal dopaminergic system. Brain Res 151:392–398

Porceddu ML, Ongini E, Biggio G (1985) [^{3}H]-SCH 23 390 binding sites increase after chronic blockade of D-1 dopamine receptors. Eur J Pharmacol 118:367–370

Porceddu ML, Giorgi O, Ongini E, Mele S, Biggio G (1986) ^{3}H-SCH 23 390 binding sites in the rat substantia nigra: evidence for a presynaptic localization and innervation by dopamine. Life Sci 39:321–328

Portig PJ, Vogt M (1968) Release into the cerebral ventricles of substances with possible transmitter function in the caudate nucleus. J Physiol (Lond) 204:687–715

Quick M, Emson PC, Joyce E (1979) Dissociation between the presynaptic dopamine-sensitive adenylate cyclase and ^{3}H-spiroperidol binding sites in rat substantia nigra. Brain Res 355–365

Raisman R, Cash R, Ruberg M, Javoy-Agid F, Agid Y (1985) Binding of [^{3}H]-SCH 23 390 to D-1 receptors in the putamen of control and Parkinsonian subjects. Eur J Pharmacol 113:467–468

Reimann W, Zumstein A, Starke K (1982) γ-Aminobutyric acid can both inhibit and facilitate dopamine release in the caudate nucleus of the rabbit. J Neurochem 39:961–969

Reis DJ, Ross RA, Pickel VM, Lewander T, Joh TH (1975) Some differences in tyrosine hydroxylase in central noradrenergic and dopaminergic neurons. In: Almgren O, Carlsson A, Engel J (Eds) Chemical Tools in Catecholamine Research. American Elsevier, New York, pp 53–60

Relja M, Zivkovic B (1978) Repeated treatment with haloperidol and the function of dopaminergic neurons. Iugoslav Physiol Pharmacol Acta 14:233–235

Rinne JO, Rinne JK, Laakso K, Lonnberg P, Rinne UK (1985) Dopamine D-1 receptors in the Parkinsonian brain. Brain Res 359:306–310

Roffler-Tarlov S, Sharman DF, Tegerdine P (1971) 3,4-Dihydroxyphenylacetic acid and 4-hydroxy-3-methoxyphenylacetic acid in the mouse striatum: a reflection of intra- and extra-neuronal metabolism of dopamine. Br J Pharmacol 42:343–351.

Ross SB (1977) On the mode of action of central stimulatory agents. Acta Pharmacol Toxicol 41:392–396

Roth RH (1973) Inhibition by gamma-hydroxybutyrate of chlorpromazine-induced increase in homovanillic acid. Br J Pharmacol 47:408–414

Saller CF, Salama AI (1986) D-1 and D-2 dopamine receptor blockade: interactive effects in vitro and in vivo. J Pharmacol Exp Ther 236:714–720

Savasta M, Dubois A, Scatton B (1986) Autoradiographic localization of D_1 dopamine receptors in the rat brain with [^{3}H]-SCH 23 390. Brain Res 375:291–301

Scally MC, Ulus I, Wurtman RJ (1977) Brain tyrosine level controls striatal dopamine synthesis in haloperidol treated rats. J Neural Transm 41:1–6

Scatton B (1977) Differential regional development of tolerance to increase in dopamine turnover upon repeated neuroleptic administration. Eur J Pharmacol 46:363–369

Scatton B (1982) Further evidence for the involvement of D_2, but not D_1 dopamine receptors in dopaminergic control of striatal cholinergic transmission. Life Sci 31:2883–2890

Scatton B, Bartholini G (1979) Increase in striatal acetylcholine levels by GABA mimetic drugs: lack of involvement of the nigrostriatal dopaminergic neurons. Eur J Pharmacol 56:181–182

Scatton B, Bartholini G (1980) Modulation by GABA of cholinergic transmission in striatum. Brain Res 183:211–216

Scatton B, Worms P (1978) Subsensitivity of striatal and mesolimbic dopamine target cells after repeated treatment with apomorphine dipivaloyl ester. Naunyn-Schmiedeberg's Arch Pharmacol 303:271–278

Scatton B, Worms P (1979) Tolerance to increases in striatal acetylcholine concentrations after repeated administration of apomorphine dipivaloyl ester. J Pharm Pharmacol 31:861–863

Scatton B, Zivkovic B (1984) Neuroleptics and the limbic system. In: Trimble MR, Zarifian E (Eds) Psychopharmacology of the Limbic System. Oxford University Press, Oxford, pp 174–197

Scatton B, Bischoff S, Dedek J, Korf J (1977a) Regional effects of neuroleptics on dopamine metabolism and dopamine-sensitive adenylate cyclase activity. Eur J Pharmacol 44:287–292

Scatton B, Dedek J, Korf J (1977b) Effect of single and repeated administration of haloperidol and sulpiride on striatal and retinal dopamine turnover in the rat. Brain Res 135:374–377

Scatton B, Briley M, Worms P (1979) Subsensitivity of striatal dopamine target cells after repeated treatment with apomorphine dipivaloyl ester. In: Usdin E, Kopin IJ, Barchas J (Eds) Catecholamines: Basic and Clinical Frontiers. Pergamon Press, New York, pp 595–597

Scatton B, Zivkovic B, Bartholini G (1980) Differential influence of GABAergic agents on dopamine metabolism in extrapyramidal and limbic systems of the rat. Brain Res Bull 5, Suppl 2:421–425

Scatton B, Zivkovic B, Dedek J, Lloyd KG, Constantinidis J, Tissot R, Bartholini G (1982) γ-Aminobutyric acid (GABA) receptor stimulation. III. Effect of progabide (SL 76002) on norepinephrine, dopamine and 5-hydroxytryptamine turnover in rat brain areas. J Pharmacol Exp Ther 220:678–688

Scatton B, Dubois A, Dubocovich ML, Zahniser NR, Fage D (1985) Quantitative autoradiography of ^{3}H-nomifensine binding sites in rat brain. Life Sci 36:815–822

Scatton B, Dubois A, Camus A, Zahniser NR, Dubocovich ML, Agid Y, Cudennec A (1986a) Autoradiographic visualization and quantification of dopamine receptors and uptake sites in the rat and human CNS. In: Woodruff GN, Poat JA, Roberts PJ (Eds) Dopaminergic Systems and their Regulation. MacMillan Press, London, pp 111–130

Scatton B, Serrano A, Degueurce A (1986b) The use of in vivo voltammetry to investigate functional recovery with transplants and neurotransmitter interactions in the rat brain. Ann New York Acad Sci, 473:284–301

Scheel-Krüger J, Arnt A, Braestrup C, Christensen AV, Magelund G (1979) Development of new animal models for GABAergic actions using muscimol as a tool. In: Krogsgaard-Larsen P, Scheel-Krüger J, Kofod H (Eds) GABA-Neurotransmitters. Munksgaard, Copenhagen, pp 447–464

Schmidt MJ, Hill LE (1977) Effects of ergots on adenylate cyclase activity in the corpus striatum. Life Sci 20:789–798

Schoemaker H, Pimoule C, Arbilla S, Scatton B, Javoy-Acid F, Langer SZ (1985) Sodium-dependent [^{3}H] cocaine binding associated with dopamine uptake sites in the rat striatum and human putamen decrease after dopaminergic denervation and in Parkinson's disease. Naunyn Schmiedeberg's Arch Pharmacol 329:227-235

Schwarcz R, Creese I, Coyle JT, Snyder SH. (1978) Dopamine receptors localized on cerebral cortical afferents to rat corpus striatum. Nature 271:766-768

Seeman P (1977) Anti-schizophrenic drugs. Membrane receptor sites of action. Biochem Pharmacol 26:1741-1748

Seeman P (1980) Brain dopamine receptors. Pharmacol Rev 32:229-313

Seeman P, Chan-Wong M, Tedesco J, Wong K (1975) Brain receptors for antipsychotic drugs and dopamine: direct binding assays. Proc Nat Acad Sci USA 72:4376-4380

Seeman P, Grigoriadis D, George SR, Watanabe M, Ulpian C (1986) Functional states of dopamine receptors. In: Woodruff GN, Poat JA, Roberts P (Eds) Dopaminergic Systems and their Regulation. MacMillan Press, London, pp 97-110

Shibuya A (1979) Dopamine-sensitive adenylate cyclase activity in the striatum in Parkinson's disease. J Neural Transm 44:287-295

Shore PA, Dorris RL (1975) On a prime role for newly synthesized dopamine in striatal function. Eur J Pharmacol 30:315-318

Singh MM, Lal H (1982) Central cholinergic mechanism, neuroleptic action and schizophrenia. In: Essman WB, Valzelli L (Eds) Neuropharmacology: Clinical Applications. Kluwer Publishers Group, Burdrecht pp 337-389

Skirboll LR, Bunney BS (1979) The effects of acute and chronic haloperidol treatment on spontaneously firing neurons in the caudate nucleus of the rat. Life Sci 25:1419-1434

Sokoloff P, Martres M-P, Schwartz JC (1980a) Three classes of dopamine receptor (D-2, D-3, D-4) identified by binding studies with ^{3}H-apomorphine and ^{3}H-domperidone. Naunyn-Schmiedeberg's Arch Pharmacol 315:89-102

Sokoloff P, Martres M-P, Schwartz JC (1980b) ^{3}H-Apomorphine labels both dopamine postsynaptic receptors and autoreceptors. Nature 288:283-286

Sorimochi M (1975) Increase of tyrosine hydroxylase activity after reserpine: evidence for the selective response of noradrenergic neurons. Brain Res 99:400-404

Spano PF, Trabucchi M, Di Chiara G (1977) Localization of nigral dopamine-sensitive adenylate cyclase on neurons originating from corpus striatum. Science 196:1343-1345

Spano PF, Stefanini E, Trabucchi M, Fresia P (1979) Stereospecific interaction of sulpiride with striatal and neostriatal dopamine receptors. In: Spano PF, Trabucchi M, Corsini GU, Gessa GL (Eds) Sulpiride and Other Benzamides. Raven Press, New York pp 11-31

Speckenbach W, Kehr W (1976) Effect of (+) amphetamine on monoamine synthesis and metabolism after axotomy in rat forebrain. Naunyn-Schmiedeberg's Arch Pharmacol 296:25-30

Spector S, Gordon R, Sjoerdsma A, Udenfriend S (1967) End-product inhibition of tyrosine hydroxylase as a possible mechanism for regulation of norepinephrine synthesis. Mol Pharmacol 3:549-555

Stadler H, Lloyd KG, Gadea-Ciria M, Bartholini G (1973) Enhanced striatal acetylcholine release by chlorpromazine and its reversal by apomorphine. Brain Res 55:476-480

Stadler H, Lloyd KG, Bartholini G (1974) Dopaminergic inhibition of striatal cholinergic neurons. Synergistic blocking action of γ-butyrolactone and neuroleptic drugs. Naunyn-Schmiedeberg's Arch Pharmacol 283:129-134

Stamford JA (1985) In vivo voltammetry: promise and perspective. Brain Res Rev 10:119-135

Stinus L, Kelley AE, Iversen SD (1978) Increased spontaneous activity following substance P infusion into A10 dopaminergic area. Nature 276:616-618
Stoof JC, Kebabian JW (1981) Opposing roles for D-1 and D-2 dopamine receptors in efflux of cyclic AMP from rat striatum. Nature (Lond) 294:366-368
Stoof JC, Kebabian JW (1984) Two dopamine receptors: biochemistry, physiology and pharmacology. Life Sci 35:2281-2296
Stoof JC, de Boer T, Sminia P, Mulder AH (1982) Stimulation of D_2-dopamine receptors in rat neostriatum inhibits the release of acetylcholine and dopamine but does not affect the release of γ-aminobutyric acid, glutamate or serotonin. Eur J Pharmacol 84:211-214
Storm-Mathisen J (1975) Accumulation of glutamic acid decarboxylase in the proximal parts of presumed GABAergic neurons after axotomy. Brain Res 87:107-109
Sved AF, Fernstrom JD, Wurtman RJ (1979) Tyrosine administration decreases serum prolactin levels in chronically reserpinized rats. Life Sci 25:1293-1299
Tamminga CA, Crayton JW, Chase TN (1978) Muscimol: GABA agonist therapy in schizophrenia. Am J Psychiat 135:746-747
Tarsy D, Baldessarini RJ (1974) Behavioural supersensitivity to apomorphine following chronic treatment with drugs which interfere with the synaptic function of catecholamines. Neuropharmacology 13:927-940
Theodorou A, Crockett M, Jenner P, Marsden CD (1979) Specific binding of ^{3}H sulpiride to rat striatal preparations. J Pharm Pharmacol 31:424-426
Thierry AM, Blanc G, Sobel A, Stinus L, Glowinski J (1973) Dopaminergic terminals in the rat cortex. Science 182:499-501
Tissari AH, Casu M, Gessa GL (1978) Chronic haloperidol: tolerance to the stimulating effect on striatal tyrosine hydroxylase. Life Sci 24:411-416
Titeler M, Seeman P (1979) Selective labeling of different dopamine receptors by a new agonist ^{3}H-ligand: ^{3}H-N-n-propylnorapomorphine. Eur J Pharmacol 56:291-292
Titeler M, Weinreich P, Sinclair D, Seeman P (1978) Multiple receptors for brain dopamine. Proc Nat Acad Sci USA 75:1153-1156
Tossman U, Ungerstedt U (1986) The effect of apomorphine and pergolide on the potassium-evoked overflow of GABA in rat striatum studied by microdialysis. Eur J Pharmacol 123:295-298
Trabucchi M, Longoni R, Fresia P, Spano PF (1975) Sulpiride: a study of the effects on dopamine receptors in rat neostriatum and limbic forebrain. Life Sci 17:1551-1556
Ungerstedt U, Ljungberg T, Hoffer B, Siggins G (1975) Dopaminergic supersensitivity in the striatum. In: Calne DB, Chase TN, Barbeau A (Eds) Advances in Neurology vol 9. Raven Press, New York, pp 57-65
Van der Heyden JAM, Venema K, Korf J (1979) In vivo release of endogenous GABA from rat substantia nigra measured by a novel method. J Neurochem 32:469-476
Van der Heyden JA, Venema K, Korf J (1980) Biphasic effects of dopamine and apomorphine on endogenous GABA release in rat substantia nigra. J Neurochem 34:119-124
Van Zwieten-Boot BJ, Noach EL (1975) Effect of blocking dopamine release on synthesis rate of dopamine in the striatum of the rat. Eur J Pharmacol 33:247-254
Waldmeier PC, Kam R, Stöcklin K (1978) Increased dopamine metabolism in rat striatum after infusions of substance P into the substantia nigra. Brain Res 159:223-227
Walters JR, Roth RH (1972) Effects of gamma-hydroxybutyrate on dopamine and dopamine metabolites in the rat striatum. Biochem Pharmacol 21:2111-2121

Walters JR, Roth RH (1974) Dopaminergic neurons: drug-induced antagonism of the increase in tyrosine hydroxylase activity produced by cessation of impulse flow. J Pharmacol Exp Ther 191:82-91

Walters JR, Roth RH (1976) Dopaminergic neurons: an in vivo system for measuring drug interactions with presynaptic receptors. Naunyn-Schmiedeberg's Arch Pharmacol 296:5-14

Weinstock J, Wilson JM, Ladd DL, Brush CK, Pfeiffer FR, Kuo GY, Holden KG, Yin NCF (1980) Separation of potent central and renal dopamine agonist activity in substituted 6-chloro-2,3,4,5-tetrahydro-7,8-dihydroxyl-1-phenyl 1-H-3-benzazepines. J Med Chem 23:973-975

Werdinius B (1966) Effect of probenecid on the level of homovanillic acid in the corpus striatum. J Pharm Pharmacol 18:546-547

Westerink BHC (1979) Further studies on the sequence of dopamine metabolism in the rat brain. Eur J Pharmacol 56:313-322

White FJ, Wang RY (1983) Comparison ot the effects of chronic haloperidol treatment on A9 and A10 dopamine neurons in the rat. Life Sci 32:983-993

White FJ, Wang RY (1984) Pharmacological characterization of dopamine autoreceptors in the rat ventral tegmental area. J Pharmacol Exp Ther 231:275-280

Wood JH (1980) Neurochemical analysis of cerebrospinal fluid. Neurology 30:645-651

Wood PL (1982) Actions of GABAergic agents on dopamine metabolism in the nigrostriatal pathway of the rat. J Pharmacol Exp Ther 222:674-679

Worms P, Lloyd KG (1978) Influence of GABA-agonists and antagonists on neuroleptic-induced catalepsy in rats. Life Sci 23:475-478

Worms P, Willigens MT, Lloyd KG (1978) GABA involvement in neuroleptic-induced catalepsy. J Pharm Pharmacol 30:716-719

Wurtman RJ, Larin F, Mostafapour S, Fernstrom JD (1974) Brain catechol synthesis: control by brain tyrosine concentration. Science 185:183-184

Yoshida M, Precht W (1971) Monosynaptic inhibition of neurons of the substantia nigra by caudato-nigral fibers. Brain Res 32:225-228

Zetterström T, Ungerstedt U (1984) Effects of apomorphine on the in vivo release of dopamine and its metabolites studied by brain dialysis. Eur J Pharmacol 97:29-34

Zetterström T, Sharp T, Marsden CA, Ungerstedt U (1983) In vivo measurement of dopamine and its metabolites by intracerebral dialysis: changes after d-amphetamine. J Neurochem 41:1769-1773

Zetterström T, Sharp T, Ungerstedt U (1985) Effect of neuroleptic drugs on striatal dopamine release and metabolism in the awake rat studied by intracerebral dialysis. Eur J Pharmacol 106:27-37

Zirkle CL, Kaiser C (1974) Antipsychotic agents (tricyclic). In: Gordon M (Ed) Psychopharmacological agents. Academic Press, New York, pp 39-128

Zivkovic B (1979) Significance of tyrosine hydroxylase activation in the regulation of dopamine synthesis. In: Usdin E, Kopin IJ, Barchas JD (Eds) Catecholamines: Basic and Clinical Frontiers. Pergamon Press, New York, pp 109-111

Zivkovic B (1981) Role of tyrosine hydroxylase activation in the regulation of dopamine synthesis. In: Usdin E, Weiner N, Youdim MBH (Eds) Function and Regulation of Monoamine Enzymes: Basic and Clinical Aspects. MacMillan, London, pp 35-43

Zivkovic B, Guidotti A (1974) Changes of kinetic constant of striatal tyrosine hydroxylase elicited by neuroleptics that impair the function of dopamine receptors. Brain Res 79:505-509

Zivkovic B, Guidotti A, Costa E (1974) Effects of neuroleptics on striatal tyrosine hydroxylase: changes in affinity for the pteridine cofactor. Mol Pharmacol 10:727-735

Zivkovic B, Guidotti A, Costa E (1975a) The regulation of striatal tyrosine hydroxylase: effects of gamma-hydroxybutyric acid and haloperidol. Naunyn-Schmiedeberg's Arch Pharmacol 291:193-200

Zivkovic B, Guidotti A, Costa E (1975b) The regulation of the kinetic state of striatal tyrosine hydroxylase and the role of postsynaptic dopamine receptors. Brain Res 92:516-521

Zivkovic B, Guidotti A, Revuelta A, Costa E (1975c) Effect of thioridazine, clozapine and other antipsychotics on the kinetic state of tyrosine hydroxylase and on the turnover rate of dopamine in striatum and nucleus accumbens. J Pharmacol Exp Ther 194:37-46

Zivkovic B, Scatton B, Dedek J, Worms P, Lloyd KG, Bartholini G (1980) Involvement of different types of dopamine receptors in the neuroleptic action. 12th CINP Congress, Abstract 729, Göteborg

Zivkovic B, Worms P, Scatton B, Dedek J, Oblin A, Lloyd KG, Bartholini G (1983) Functional similarities between benzamides and other neuroleptics. In Biggi o Biggio G, Costa E, Gessa GL, Spano PF (Eds) Receptors as Supramolecular Entities. Pergamon Press, Oxford, pp 155-170

Zukin SR, Young AB, Snyder SH (1974) Gamma-aminobutyric acid binding to receptor sites in the rat central nervous system. Proc Nat Acad Sci USA 71:4802-4807

CHAPTER 17

Catecholamines and Blood Pressure

J. L. Reid and P. C. Rubin

A. Introduction

Towards the end of the nineteenth century two observations established a relationship between catecholamines and blood pressure. The first of these was that injection of adrenal medulla extract to experimental animals produced a striking elevation of blood pressure and heart rate (Oliver and Schafer 1895). Only a short period elapsed before the active principle responsible for these cardiovascular changes was identified as a catecholamine: adrenaline (Abel 1897). Nearly half a century later another catecholamine was isolated from peripheral sympathetic nerves (von Euler 1946) and this substance was also found to raise blood pressure (Goldenberg et al. 1948). The past few years have witnessed an enormous increase in research on the relationship between catecholamines and blood pressure control—both normal and deranged. That relationship in its diverse forms constitutes the subject of this chapter.

B. Morphological Aspects of the Central and Peripheral Nervous System Relevant to Blood Pressure

Blood pressure homeostasis is maintained by a negative feedback mechanism which is largely under the control of the autonomic nervous system. The physiological factors which combine to determine blood pressure are cardiac output (heart rate x stroke volume) and peripheral resistance, particularly that in the arterioles. Changes in blood pressure are detected by baroreceptors in the heart, carotid sinus, aortic arch and other large vessels. Afferent impulses are transmitted from these structures via the carotid sinus nerve and the glossopharyngeal and vagus nerves to the brain stem (Korner 1971; Scher 1977). The carotid sinus nerve terminates in two areas which play a major role in blood pressure control: the nuclei of the tractus solitarius (NTS) (Crill and Reis 1968) and a part of the reticular formation known as the paramedian nucleus (Miura and Reis 1969a, 1972).

The major role played by the NTS in blood pressure control has been clarified by stereotaxically placed electrolytic lesions. In the rat, when the NTS are ablated bilaterally, baroreflexes are abolished and arterial pressure rises substantially as the result of increased peripheral resistance (Doba and

Reis 1973). Since the fulminating rise in blood pressure was prevented by administering α-adrenoceptor antagonists, the inference to be drawn is that destruction of the NTS results in an increased sympathetic outflow with α-adrenoceptor mediated increases in arterial resistance and pressure. Electrical stimulation of the NTS produces the opposite response with a fall in blood pressure (Seller and Illert 1969). It can reasonably be concluded from these studies that the NTS are intimately involved in blood pressure control, particularly with respect to integrating information from the baroreceptors.

The paramedian nucleus receives fibres not only from the carotid sinus but also from the fastigial nucleus of the cerebellum (Thomas et al. 1956). Electrical stimulation of either the paramedian nucleus or the fastigial nucleus results in increased blood pressure and tachycardia (Miura and Reis 1969b, 1970; Achari and Downman 1970). Ablation of the rostral fastigial nucleus in the anaesthetized and paralysed cat results in impairment of the reflex changes in response to a head-up tilt with resulting postural hypotension (Doba and Reis 1974a). This suggests that the fastigial nucleus is involved in the control of blood pressure during postural changes.

From the NTS, neuronal connections are made both with efferent pathways (the secondary neurones of the baroreceptor reflex arc) and with ascending neurones which carry information to higher structures in the brain. A small lesion in the medial part of the NTS produces diffuse degeneration of fibres in the dorsal vagal nucleus, the nucleus reticularis lateralis, the nucleus reticularis medullae oblongatae and the nucleus reticularis gigantocellularis (Palkovits and Zaborsky 1977). Fibres from these various nuclei project directly to the intermedio-lateral nucleus of the spinal cord, and it seems possible that baroreflex information reaches the preganglionic sympathetic cells of the spinal cord by this multi-synaptic pathway. The NTS also communicate with the nucleus ambiguus; although baroreceptor neurones might be located in this nucleus (McAllen and Spyer 1976) definitive evidence has not so far been forthcoming.

The precise neuroanatomical pathways connecting the NTS to higher cardiovascular control centres have not yet been clearly delineated. However, considerable evidence accumulated over many years points to higher brain centres being actively involved in the modulation of the lower cardiovascular reflex control areas.

The hypothalamus has been implicated as a site of higher cardiovascular control for over half a century (Karplus and Kreidl 1927). Electrical stimulation of the anterior hypothalamus or of the pre-optic area produces bradycardia, fall in blood pressure and inhibition of the baroreceptor reflexes (Hilton 1963; McAllen 1976). Conversely, bilateral destruction of these areas decreases the baroreflex response to afferent stimulation (Hilton and Spyer 1971). Also the highest brain centres can influence blood pressure control. For example, mental arithmetic has been shown to elevate blood pressure (Brod et al. 1959), while various relaxation techniques such as biofeedback and transcendental meditation have been shown to lower blood pressure. However, the detailed anatomical relationship between these highest centres and the lower cardiovascular control areas remain to be defined.

C. Central and Peripheral Catecholamines in the Control of Blood Pressure in Animals and Man

A detailed study of the distribution of catecholamines has been made possible by histofluorescence techniques (FALCK et al. 1962). A high concentration of noradrenergic nerve fibres has been demonstrated in the baroreceptors of the carotid sinus (REIS and FUXE 1968) and it would certainly seem possible that these nerves modulate the sensitivity of the pressure receptor.

At the central level those anatomical areas which appear to be involved in blood pressure control have been shown to contain a large number of catecholaminergic fibres. Higher concentrations of catecholamines and PNMT activity have been detected in the NTS and dorsal vagal nucleus than in other nuclei of the medulla oblongata (Van der GUGTEN et al. 1976; VERSTEEG et al. 1976a, b; SAAVEDRA et al. 1979; CHALMERS et al. 1984). Within the NTS the highest catecholamine concentrations are found in the regions which are presumed to be involved in baroreflex control (Van der GUGTEN et al. 1976; VERSTEEG et al. 1976a, b). The nucleus reticularis lateralis in the medulla oblongata also contains high numbers of both noradrenergic (DAHLSTRÖM and FUXE 1964) and adrenergic neurones (HÖKFELT et al. 1973).

In the spinal cord high concentrations of catecholamine-containing nerve fibres have been located in the intermediolateral nucleus which synapses with sympathetic preganglionic neurones (CARLSSON et al. 1964; DAHLSTRÖM and FUXE 1965).

In the periphery most postganglionic sympathetic fibres have been shown to be noradrenergic. These fibres have been observed in the majority of vascular beds (NORBERG 1967; BURNSTOCK et al. 1970) with particularly dense innervation being found in the small arterioles which play such an important role in determining peripheral resistance (MELLANDER and JOHANSSON 1968).

There is some dispute as to whether descending catecholaminergic tracts play a role in the baroreflex arc and if they do, whether this role is excitatory or inhibitory. One approach to the question of whether catecholaminergic fibres are involved in this reflex has been to destroy the central adrenergic neurones by injecting 6-hydroxydopamine into one of the cerebral ventricles. When this drug is injected into the lateral brain ventricle of the rat there is no detectable alteration in the fall in blood pressure which results from bilateral electrical stimulation of the sinus nerves (HAEUSLER and LEWIS 1975). Injection of 6-hydroxydopamine into the ventricles of rabbits had no effect on the relationship between drug-induced rise in blood pressure and bradycardia (HAEUSLER and LEWIS 1975). Although these results do not support involvement of the central adrenergic pathways in the baroreflex, it must be remembered that 6-hydroxydopamine produces, at best, only a partial destruction of these neurones and sufficient nerve structure could have presisted to subserve the reflex. The same authors have presented other data to support their contention that central noradrenergic fibres are not involved in the baroreflex arc. For instance, pretreatment of cats with reserpine to deplete noradrenaline stores and with α-methyl-p-tyrosine to prevent noradrenaline synthesis did not influence the cardiovascular response to bilateral electrical stimulation of the carotid sinus nerve (HAEUSLER 1974; HAEUSLER and LEWIS 1975).

Other workers, however, have produced evidence to implicate catecholamine-containing neurones in the baroreflex arc. For instance, using the disappearance of intracisternally administered tritiated noradrenaline as an index of noradrenaline turnover, CHALMERS and WURTMAN (1971) showed that sinoaortic denervation in rabbits produces a selective increase in turnover in the thoracolumbar cord. No change in noradrenaline turnover was found in the cervical or lumbar cord, from which these workers infer that the increased activity was located in the bulbospinal noradrenergic nerves which terminate in the lateral sympathetic horn. The activity of tyrosine hydroxylase is also selectively increased in the thoracolumbar cord. In contrast to the findings of other groups, these workers found that intracisternal administration of 6-hydroxydopamine prevented or reversed the hypertension resulting from sinoaortic denervation in the rabbit (CHALMERS and REID 1972). Central deafferentation of baroreflexes by means of stereotaxic destruction of the NTS in the rat ordinarily produces neurogenic hypertension. However, prior treatment with 6-hydroxydopamine has been shown to prevent the increase in pressure following destruction of the NTS (DOBA and REIS 1973, 1974b).

The nature of the control exerted by descending catecholaminergic nerve fibres is a subject of some dispute. On the basis of the evidence presented above, CHALMERS and his colleagues have proposed that the descending bulbospinal neurones exert a positive influence on sympathetic outflow. This has been supported by the electrophysiological and pharmacological experiments of NEUMAYR et al. (1974). Conversely, other electrophysiological studies imply that the descending noradrenergic pathways are inhibitory to sympathetic outflow (COOTE and MACLEOD 1974, 1975; HENRY and CALARESU 1974). There is some evidence that inhibitory fibres could synapse with the descending tracts. For instance, destruction of catecholaminergic nerves in the NTS, by local stereotaxic injection of 6-hydroxydopamine, produces hypertension (DOBA and REIS 1974b). In addition, injection of noradrenaline into the NTS causes a decrease in arterial pressure in rats (DE JONG 1974). It would, therefore, seem possible that the descending catecholaminergic bulbospinal tracts are excitatory but that there is an inhibitory link interposed between them and the afferent neurones. There is now persuasive evidence that adrenaline-containing cells are also involved in central cardiovascular control (FUXE et al. 1979, 1980).

Adrenaline-containing neurones have been identified in the medualla oblongata, particularly in the NTS (BOLME et al. 1974; FUXE et al. 1979). These adrenaline-containing fibres have been found to project to the hypothalamic areas associated with blood pressure control (HÖKFELT et al. 1974). High PNMT activity (SAAVEDRA et al. 1979) and high adrenaline concentrations (VAN DER GUGTEN et al. 1976) are present in these hypothalamic centres. A significant role for adrenaline in central cardiovascular control is also suggested by pharmacological studies. When adrenaline is injected into the anterior hypothalamus of the rat, it not only produces a fall in blood pressure and heart rate, but it is about ten times more potent in this regard than noradrenaline (STRUYKER-BOUDIER and BEKER 1975).

Physiological studies on the role of the peripheral sympathetic nervous system in blood pressure control have been limited by the great difficulty in achieving complete sympathectomy. Various techniques have been attempted e.g. surgical denervation, immunosympathectomy and adrenergic neurone blockers but none has proven consistently reliable. More recently the use of 6-hydroxydopamine has produced promising results. As mentioned earlier, this drug selectively destroys noradrenergic nerve fibres and has the added advantage of not crossing the blood brain barrier. In addition it seems not to affect the adrenal medulla (KOSTOZEWA and JACOBOWITZ 1974). Within a few hours of injecting 6-hydroxydopamine intravenously to normotensive rats the mean arterial pressure falls from around 100 mm Hg to approximately 70 mm Hg (DE CHAMPLAIN and VAN AMERINGEN 1972). In the absence of sympathetic nervous activity it appears that the adrenal medulla assumes an important role in blood pressure control, since adrenalectomy in previously sympathectomised animals results in a further substantial fall in blood pressure. Similar observations were made by the same group in dogs (GAUTHIER et al. 1972). Two pieces of evidence suggest that the compensatory role of the adrenal is exerted via an increase in catecholamine release from the medulla. First, sympathectomised rats show a virtually identical fall in blood pressure to that which occurs following adrenalectomy when they are given an α-adrenoceptor antagonist (DE CHAMPLAIN and VAN AMERINGEN 1972). Second, chemical sympathectomy has been shown to result in a doubling of the activity of tyrosine hydroxylase in the adrenal medulla (MUELLER et al. 1969).

The integrity of the sympathetic nervous system is essential to the maintenance of blood pressure on standing, and patients with various neurological disorders show substantial postural hypotension which is accompanied by inappropriately low plasma concentrations of catecholamines. The most common disease associated with postural hypotension is diabetes mellitus, since an autonomic neuropathy is a common finding in patients who have been diabetic for many years. Diabetic patients who do not have autonomic neuropathy have normal plasma concentrations of catecholamines in both supine and standing positions and also do not show a fall in blood pressure on standing. However, those patients who do have autonomic neuropathy and who do show a postural fall in blood pressure have plasma catecholamine concentrations which are less than 50% of those of the control population (CHRISTENSEN 1972). The reason for the low concentrations of catecholamines in this type of patient would appear to be decreased synthesis rather than impaired release of catecholamines, since the cardiovascular tissues of longstanding diabetics with neuropathy contain less than 10% of the noradrenaline found in control subjects (NEUBAUER and CHRISTENSEN 1976).

Further evidence for the relationship between sympathetic nervous system, catecholamines and blood pressure control in man comes from studies on patients with various rare dysautonomias. The Shy Drager syndrome is characterised by numerous central nervous system defects giving rise to symptoms and signs similar to those of Parkinsonism, with severe postural hypotension (SHY and DRAGER 1960). These patients have normal plasma noradrenaline concentrations when they are supine. On standing the marked fall in blood pressure is accompanied by no increase in plasma noradrenaline (ZIEGLER et al. 1977). Patients with Shy Drager syndrome do show a pressor response to infused tyramine which acts by releasing noradrenaline from sympathetic nerve endings. These findings suggest that the defect is different from that in diabetic autonomic neuropathy. The Shy Drager patients do not appear to possess a functional baroreceptor reflex arc. This suggests that, amongst the multiple central nervous system defects, there are some which affect the central cardiovascular control centres. This has important implications for treatment, since the Shy Drager patients will respond to drugs which release noradrenaline while patients whose peripheral noradrenaline stores are depleted will not respond to such drugs. MAO inhibitors and tyramine have been used to increase blood pressure in Shy Drager patients (NANDA et al. 1976). Another rare condition associated with postural hypotension is familial dysautonomia (RILEY et al. 1949). In common with the Shy Drager patients, these individuals have normal plasma noradrenaline levels while supine, but on standing they exhibit a fall in blood pressure without a rise in plasma noradrenaline (ZIEGLER et al. 1976). The difference between Shy Drager patients and those with familial dysautonomia is that the former have a virtually normal pressor response to infused noradrenaline (KONTOS et al. 1975), but the patients with familial dysautonomia have a substantially increased response (SMITH and DANCIS 1964). The aetiology of this hypersensitivity is not known but it presumably plays a role in the episodes of hypertension which commonly accompany periods of anxiety in this condition (DANCIS and SMITH 1966). The occurrence of these episodes of hypertension indicates that pathways do exist which can mediate a release of noradrenaline in this condition, but it must be inferred that the neurones involved in cardiovascular control are defective at a critical site. Huntington's chorea is another central nervous system disease accompanied by hypotension and low concentrations of plasma noradrenaline (SHOULSON et al. 1976).

Cardiovascular homeostasis is influenced by many complex and interacting systems. It is not realistic to consider in isolation the role of catecholamines in blood pressure control while ignoring other physiological mechanisms which themselves interact with the catecholaminergic system. This point is illustrated by the octapeptide, angiotensin II, which interacts with catecholamines both peripherally and centrally. In the periphery angiotensin II stimulates the adrenal medulla to release adrenaline (ROBINSON 1967) and acts on sympathetic nerve terminals to increase noradrenaline synthesis (BOADLE et al. 1969), increase transmitter release (MCCUBBIN and PAGE 1963) and decrease re-uptake (PEACH et al. 1969). In the central nervous system, angiotensin II can increase heart rate, blood pressure and cardiac output by a combination of reduced parasympathetic, and increased sympathetic, tone through an action on the area postrema (JOY and LOWE 1970).

Several other peptides have possible roles in blood pressure control (RUBIN 1984) but the exact nature of their action and any involvement with catecholamines remain largely speculative. The NTS has been shown to contain opioid, Substance P, somatostatin and neurotensin immunoreactive nerve terminals (HÖKFELT et al. 1978). Injection of β-endorphins into the cisterna magna of dogs produces a fall in both heart rate and blood pressure (LAUBIE et al. 1977); opioid peptide content is decreased in the sympathetic ganglia of spontaneously hypertensive rats (DI GUILIO et al. 1979) and naloxone, a specific opioid antagonist, prevents the fall in blood pressure which occurs during sleep in man (RUBIN et al. 1981). It is not clear whether opioids interact with catecholamines to produce these effects on blood pressure, but there is evidence that the two systems influence each other (RUBIN et al. 1983a; REID et al. 1984). For instance, met-enkephalin decreases the release of ^{3}H-noradrenaline from rat occipital cortex (GÖTHERT et al. 1979), while morphine decreases noradrenaline release from mouse vas deferens (HENDERSON and HUGHES 1976).

Substance P has been proposed as a neurotransmitter at the first synapse of the baroreflex arc (HAEUSLER and OSTERWALDER 1980).

D. Animal Models of Hypertension and Their Relevance to Man

Animal models of experimental hypertension serve two distinct functions:

1) to investigate pathophysiological mechanisms of hypertension and to test mechanistic hypotheses,
2) to screen for activity and evaluate possible usefulness of antihypertensive drugs developed for ultimate use in man.

A wide range of animal species have been examined, including sheep and turkeys. The most commonly used include dogs, cats, rabbits, rats and mice. There is considerable heterogeneity in aetiological factors in human hypertension (Table 1) which is paralleled by the wide variety of experimental models which have been employed (Table 2).

In most populations, identifiable causes of hypertension are rare but include severe renal impairment, renal vascular obstruction, and overproduction of mineralocorticoids or catecholamines by tumours of the adrenal cortex or medulla, respectively. These secondary forms of hypertension in man provide the rationale for the animal models of renoprival hypertension, renal artery

Table 1. Pathogenesis of human hypertension

Primary		
Essential hypertension	90 % +	Mechanism unknown
Secondary		
Renal disease	5-7 %	Reduced renal function
Renal artery stenosis	3 %	Renin-angiotensin
Hyperaldosteronism	< 1 %	Mineralocorticoid
Phaeochromocytoma	< 1 %	Catecholamines

Table 2. Models of experimental hypertension

Secondary hypertension
Renoprival hypertension
Renal artery constriction
Deoxycorticosterone + Saline
Behavioural modification
Immobilization
Auditory stress (or sound deprivation)
Social isolation etc.
"Neurogenic" hypertension
Hypothalamic stimulation
Baroreceptor deafferentation
Solitary tract nuclei (N.T.S.) lesions
Genetic hypertension
Spontaneous hypertension (several strains)
Salt sensitive rats

constriction with a silver clip, and deoxycorticosterone and saline (DOCA salt) hypertension.

It is easy to appreciate that these models may provide useful information on the pathogenesis of particular forms of human hypertension. However, even when comparing renal artery stenosis in man with experimental renal artery constriction with a silver clip, caution is advisable. The time course of the rise in blood pressure differs, and the factors maintaining hypertension may change during the transition from early to established hypertension (CARAVAGGI et al. 1976). Several authors have shown that other factors, including the presence or absence of the normal contralateral kidney, will affect the development of hypertension (BRUNNER et al. 1974; REID et al. 1976).

However, the vast majority of cases of human hypertension, termed "essential hypertension" is of unknown etiology. Despite extensive investigation of known hypertensive mechanisms, there is still no evidence that qualitative or quantitative disturbances of any single factor underlie the increase in blood pressure. Our lack of understanding of essential hypertension makes it difficult to justify any particular animal model as uniquely relevant. In the absence of measurable changes in renal function, renal vascular obstruction, mineralocorticoid or catecholamine excess, it would seem that the secondary models of hypertension are not directly relevant to essential hypertension.

Human essential hypertension has a strong genetic component with polygenic inheritance (PICKERING 1955). Blood pressure is unimodally distributed in the population and essential hypertensives represent the extreme right side of the distribution. Several groups have sought genetic models of hypertension in small animals by selective cross breeding. These include strains which are sensitive or resistant to dietary sodium chloride (DAHL et al. 1962), behavioural models (HENRY et al 1975) and the spontaneously hypertensive rat (SHR)

(Okamoto and Aoki 1963), where no additional external factors are required. Although the salt sensitive strains of rats have been extensively evaluated and used to provide support for a hypothesis that human essential hypertension is related to dietary sodium intake (Dahl et al. 1962), this view remains controversial, as there are great discrepancies between the sodium intake required by animal models and the intake of hypertensive patients on an average Western diet.

Several behavioural models of hypertension have been studied. Most of these involve profound interference with activity and could be considered, in human terms, very stressful conditions. An example of such a stress is prolonged immobilization. However, others, such as alteration of the social conditions in mice (Henry et al. 1975), while also stressful, have more obvious human parallels. Most behavioural models of hypertension do not result in sustained elevation of blood pressure after removal of the stimulus.

Spontaneously hypertensive rats (SHR), first developed at Kyoto, have been extensively studied over the last 10–12 years (Okamoto and Aoki 1963). Early studies were confounded by the use of inappropriate control groups. More systematic studies on the Japanese SHR with Wistar Kyoto controls together with other strains of SHR developed in New Zealand, Milan, Paris, Lyon and other centres have resulted in a range of contradictory or inconsistent publications examining hypertensive mechanisms in these strains. Marked inter-strain differences have been noted and the lack of consistent relationships between hypertensive strains suggest that many earlier observations of changes of assumed pathogenetic relevance merely reflected genetic overlap with other factors not directly related to blood pressure control (Lovenberg et al. 1973). However, spontaneously hypertensive rats do share some features with human essential hypertension, in particular, the gradual development of hypertension from early life and the cardiac and cerebrovascular complications associated with longstanding hypertension (Okamoto et al. 1974). If SHR are used as a model, the specificity of changes observed should be confirmed by study of the time course and the absence of similar changes in control strains.

Animal models of hypertension are widely used to screen for antihypertensive activity and to evaluate potential usefulness of a new drug for human hypertension. Drugs which are developed specifically to interact with known mechanisms of secondary hypertension can be most appropriately tested in these models. Angiotensin converting enzyme inhibitors or anti-renin drugs could be evaluated in the renovascular model of hypertension, and aldosterone antagonists in DOCA salt hypertension. However, these models will be of less value in routine screening or in evaluating compounds with other mechanisms of action known or unknown.

Spontaneous hypertensive rats are now widely used as a screening test model. While this is justified by the common appearance of vascular complications and the genetic nature of human hypertension, there have been occasions in which useful antihypertensive efficacy was missed on routine screening in animals and only detected by careful clinical observation in man. Propranolol, the β-adrenoceptor antagonist, and the thiazide diuretics, the mainstay of present antihypertensive therapy, were not identified as such in preclinical animal testing. If blood pressure can be measured directly in unrestrained animals, the responses to a wide dose range of a drug in normotensive animals may provide at least as dependable results as studies with indirect pressure measurement in models of hypertension. It is inadvisable to depend on a single secondary model of hypertension or a single species for screening new drugs.

E. Peripheral and Central Catecholamines in Hypertension in Animals and Man

I. Experimental Hypertension

In the 1950s indirect support for an important role of catecholamines in hypertension was derived from the observation of the efficacy of drugs modifying efferent sympathetic outflow. Ganglion blockers, adrenergic neurone blockers and adrenoceptor antagonists lowered blood pressure in experimental hypertensive animals and in man. Conceptual developments in biochemical pharmacology and technical developments in methodology have permitted a more direct assessment of peripheral catecholamine mechanisms in hypertension, while the demonstration of central catecholaminergic pathways in the brain and the establishment of a central site of action of clonidine and α-methyldopa encouraged subsequent studies on brain mechanisms (REIS and DOBA 1974; CHALMERS 1975; DAVIES and REID 1975; DE JONG et al. 1977; MEYER and SCHMITT 1978). The early studies of DE CHAMPLAIN and his colleagues suggested that, in several models of experimental hypertension, peripheral catecholamine mechanisms may contribute directly to the rise in blood pressure (DE CHAMPLAIN et al. 1968, 1969a, 1969b). In this section the role of peripheral and central catecholamines in animal and human hypertension will be reviewed. Observations on plasma catecholamines will be presented briefly only in as much as they reflect peripheral sympathoadrenal activity. Plasma catecholamines, their measurement and relevance as an index of sympathetic activity are discussed in another chapter in this volume (see KOPIN, Chap 15). Other strategies to investigate the role of catecholamines in hypertension will be reviewed, including measurement of tissue amines, synthetic and degradative enzymes, central and peripheral sympathectomy and more recent studies of receptor binding.

1. Deoxycorticosterone Salt (DOCA salt) Hypertension

Hypertension develops over 2–4 weeks when large amounts of mineralocorticoid are administered and 1% sodium chloride is given in place of drinking water to rats or rabbits. This regimen results in changes in catecholamine metabolism with increased noradrenaline turnover in peripheral tissues and an apparent defect in the storage of the neurotransmitter (DE CHAMPLAIN et al.

1968a, 1969a, 1969b). Employing sensitive radioenzymatic assays for catecholamines, several authors described increases in plasma noradrenaline in DOCA salt hypertensive rats (REID et al. 1975; DE CHAMPLAIN et al. 1976). However, the plasma levels of catecholamines observed in restrained or anaesthetized animals are much higher than those in unrestrained, unstressed rats (POPPER et al. 1977; RASCHER et al. 1980). The pathophysiological significance of the elevated levels during stress in not clear but parallels the changes reported during stress in SHR (MCCARTY et al. 1978).

Adrenalectomy does not modify the development of hypertension in DOCA salt rats unless combined with sympathectomy (DE CHAMPLAIN and AMERINGEN 1972). Studies on the role of sympathectomy alone are contradictory, complicated by the use of different techniques (surgical, immunological or chemical) and differences in the completeness and duration of destruction and the time at which sympathectomy was performed. Partial destruction with adaptive changes in adrenoceptor sensitivity and/or regeneration of nerve fibres could explain the variable results, which range from prevention of hypertension (AYITEY-SMITH and VARMA 1970; CLARKE 1971) to no effect (FINCH et al. 1973) or to an accelerated development of high blood pressure (CLARKE et al. 1970; PROVOOST and DE JONG 1978). There is, thus, some evidence to implicate peripheral catecholamine mechanisms in the maintenance of hypertension in this model.

Observations of increased peripheral catecholamine turnover in DOCA salt hypertension contrast with reports of reduced turnover of noradrenaline in brain stem areas (VAN AMERINGEN et al. 1977). Destruction of brain catecholamine pathways with centrally administered 6-hydroxydopamine prevents the development of DOCA salt hypertension (FINCH et al. 1972; REID et al. 1975). This prevention is not secondary to changes in salt intake (REID et al. 1975; LEWIS et al 1975). Increases in adrenaline and in the activity of the adrenaline synthesizing enzyme (PNMT) have been described in specific brain stem nuclei (SAAVEDRA et al. 1976; SAAVEDRA 1979). These increases may indicate increased synthesis of adrenaline which, like noradrenaline, has been implicated in central cardiovascular regulation (HÖKFELT et al. 1974; BOLME et al. 1974).

Other workers have reported that lesions of tissues surrounding the anterior third of the third ventricle of the forebrain may modify the development of experimental hypertension in a manner similar to centrally applied 6-hydroxydopamine (BUGGY et al. 1977). It is possible that this lesion indirectly destroys catecholaminergic pathways or pathways controlling central catecholamine mechanisms. Although there is now considerable evidence of changes in catecholamine metabolism in DOCA salt hypertension, the mechanistic link between sodium retention and raised blood pressure is uncertain.

2. Renovascular Hypertension

The blood pressure rise occurs progressively over a few days after restriction of blood flow in the renal artery. The assessment of mechanisms in renovascular hypertension is complicated by the observation of different factors contributing at different times in the one kidney one clip and two kidney one clip models of Goldblatt hypertension (BRUNNER et al. 1974). After the first few hours the one kidney model shows little evidence of dependence on angiotensin and has haemodynamic similarities to other so-called volume models of hypertension, including DOCA salt hypertension. Earlier reports suggested that the mechanisms maintaining blood pressure in one kidney one clip hypertension might include increases in noradrenaline turnover (VOLICER et al. 1968; HENNING 1969). Plasma noradrenaline levels were elevated in restrained rats with one kidney one clip hypertension compared to uninephrectomised litter mates (DARGIE et al. 1977a). Animals with two kidney one clip hypertension had no significant changes in plasma noradrenaline levels (DARGIE et al. 1977b).

Immunosympathectomy delayed or prevented the development of renovascular hypertension (WILLARD and FULLER 1969; AYITEY SMITH and VARMA 1970). However, although chemical sympathectomy with 6-hydroxydopamine in weanling rats was reported to attenuate the blood pressure rise after bilateral renal artery clipping (GREWAL and KAUL 1971), this was not confirmed in subsequent studies (PROVOOST and DE JONG 1978) in animals treated in the neonatal period. Differences in the renovascular models used may underly some of these apparent contradictions.

Evidence for a permissive role of the sympathetic nervous system in one kidney one clip renovascular hypertension is supported by evidence of a central catecholamine contribution probably mediated by changes in efferent sympathetic outflow. Centrally applied 6-hydroxydopamine prevents the development of one kidney renovascular hypertension (FINCH et al. 1972; DARGIE et al. 1977a) but not two kidney one clip hypertension (DARGIE et al. 1977b). Turnover in whole brain and spinal cord is not altered (HENNING 1969). In studies combining microdissection techniques with sensitive radiometric assays, early (3-7 days) transient falls in noradrenaline and tyrosine hydroxylase in several brain stem and hypothalamic nuclei were associated with a reduced disappearance of noradrenaline from hypothalamic nuclei after inhibition of synthesis with α-methyl-p-tyrosine (PETTY and REID 1977, 1979). However, with the full development of hypertension, more prominent and longlasting changes at 7 and 28 days in PNMT activity in the NTS and the locus coeruleus were observed (PETTY and REID 1979). Thus, in established renovascular and DOCA salt hypertension, the synthesis of adrenaline in localised brain stem areas may be increased. While these changes in synthesis and turnover could be causally related to hypertension, they may merely reflect secondary changes following attempts to buffer the rise in pressure by baroreceptors or other mechanisms. The finding that PNMT activity is reduced in brain regions after baroreceptor deafferentation, a procedure which increases blood pressure, favours a secondary role for the changes in brain adrenaline synthesis (CHALMERS et al. 1979).

3. Spontaneously Hypertensive Rat (SHR)

There is considerable evidence that in several strains of SHR peripheral and central catecholamine mechanisms are either directly involved or play a permissive role in the full development of hypertension. The Japanese Kyoto strain has been most extensively studied. Early biochemical and pharmacological studies which indicated enhanced central efferent sympathetic activity (OKAMOTO et al. 1967) have been supported by subsequent work.

During the development of hypertension in young SHR there are changes both in adrenal catecholamine enzyme activity and amine levels (GROBECKER et al. 1975; NAGATSU et al. 1976) and increases in circulating catecholamines (ROIZEN et al. 1976; NAKAOKA and LOVENBERG 1976). Although plasma catecholamine levels in unrestrained rats were not significantly elevated (MCCARTY et al. 1978), these animals did show greater increases in noradrenaline plasma levels and blood pressure than controls in response to stress (MCCARTY et al. 1978). Similar abnormal haemodynamic responses had been previously reported (HALLBÄCK and FOLKOW 1974). More recently, using high affinity ligands for adrenoceptors, there have been reports of decreased β-adrenoceptor numbers in the hearts of SHR (LIMAS and LIMAS 1978; WOODSTOCK et al. 1979). At present no satisfactory observations on α-receptors are available.

Several studies have reported increased peripheral noradrenaline turnover in the heart and other tissues of SHR, particularly during the early development of hypertension (LOUIS et al. 1970; NAKAMURA et al. 1971). However, there have been several negative and contradictory reports, perhaps reflecting different ages of animals studied and the choice of normotensive control used (LOVENBERG et al. 1973).

Further direct evidence of the importance of peripheral sympathetic mechanisms in the maintenance of blood pressure in SHR comes from studies in which peripheral sympathetic activity was found to be increased (OKAMOTO et al. 1967; IRIUCHIJIMA 1973).

The effect of sympathectomy on the development of hypertension in SHR has been studied by several groups. There is general agreement that extensive sympathectomy performed early in life attenuates the rise in blood pressure in SHR. Similar results have been reported for the New Zealand strain (CLARKE et al. 1971). Neonatal sympathectomy with 6-hydroxydopamine has proved the most effective means of preventing the rise in blood pressure (PROVOOST and DE JONG 1978). However, the blood brain barrier is imcomplete in neonatal animals. Thus some 6-hydroxydopamine may enter the brain and destroy central catecholamine neurones involved in cardiovascular regulation. This may explain the lack of effect of 6-hydroxydopamine in young or adult animals (FINCH et al. 1973).

Central catecholaminergic pathways do appear to be concerned not only in blood pressure regulation in normal animals but also in spontaneous hypertension. Although centrally applied 6-hydroxydopamine given to adult SHR causes only a transient fall in blood pressure (FINCH et al. 1972), if young animals are treated, the rise in blood pressure is abolished or markedly attenuated (HAEUSLER et al. 1972). Selective destruction of spinal cord catecholaminergic pathways, however, does not prevent the development of hypertension (LOEWY et al. 1980).

Transient changes have been reported in young animals of 4 weeks of age in both brain stem and forebrain nuclei (SAAVEDRA et al. 1978; Le QUAN BUI et al. 1980). Although alterations have been reported in noradrenaline tissue levels in established hypertension (VERSTEEG et al. 1976a) these were not seen by others (SAAVEDRA et al. 1978). However, there are difficulties in interpreting tissue levels in isolation and it may be preferable to look at amine turnover by assessing synthetic enzyme activities or the effect of synthesis inhibition (NAKAOKA and LOVENBERG 1977; NAKAMURA and NAKAMURA 1978; WIJEN et al. 1980; FUXE et al. 1980). The last two groups of authors describe a reduction in noradrenaline turnover in the anterior hypothalamus during the development of SHR and they speculate that this may reflect a genetically induced failure of development or expression of a noradrenergic depressor pathway in the anterior hypothalamus as proposed from micro injection studies (STRUYKER-BOUDIER et al. 1975). This might be a trigger factor in initiating hypertension. Although there have been few reports of measurements of dopamine, there is little to support a major contribution of central dopaminergic pathways to the development of genetic or other models of hypertension (SAAVEDRA et al. 1979; FUXE et al. 1980).

Recently there has been considerable interest in the possibility that brain adrenaline pathways may be involved directly or indirectly in hypertension (CHALMERS et al. 1984). Unfortunately, at present there are inconsistent and contradictory reports. Early reductions in brain stem adrenaline together with increases in activity of PNMT have been reported, suggesting that localized increases in adrenaline turnover occur in brain stem during the development of hypertension (SAAVEDRA et al. 1978, 1979). An early increase in PNMT activity has been confirmed in another strain of SHR (DENOROY et al. 1979). However, others have reported increased adrenaline in brain stem nuclei at similar times (WIJNEN et al. 1977). Adrenaline turnover, measured with synthesis inhibition in a brain stem region including the nucleus of the solitary tract and adjacent nuclei, confirmed a reduction in adrenaline turnover both in young rats and in older animals with established blood pressure elevation (FUXE et al. 1980). These authors postulate the disturbance of a vasopressor system in the medulla which is mediated by adrenaline and usually facilitates baroreceptor reflex activity. It is difficult to reconcile the reported reduction in adrenaline turnover (FUXE et al. 1980) with the increased PNMT activity (SAAVEDRA et al. 1978), as the latter has been proposed to indicate increased adrenaline synthesis and turnover. These studies illustrate the problems of interpreting not only tissue levels but also extrapolating indirect or direct information on the rate of transmitter synthesis or utilisation to the functional state of individual neuronal pathways.

4. Salt-Sensitive Hypertension (Dahl Strain)

Central and peripheral catecholamine metabolism in the salt-sensitive model of genetic hypertension developed by DAHL has recently been reviewed (SAAVEDRA et al. 1979). Changes in noradrenaline and adrenaline were noted in several forebrain and medullary areas. However, some changes appeared to be related to high salt intake and not to blood pressure, while others were genetically determined and not related to blood pressure. As in the other models investigated, it remains difficult to distinguish primary mechanisms involved in blood pressure regulation from secondary effects resulting from baroreflex or other adaptive changes consequent to a rise in pressure, electrolyte or humoral changes.

5. Neurogenic Hypertension

Sinoaortic denervation to achieve baroreceptor deafferentation leads to increased lability of blood pressure. Under some conditions and in some species the median blood pressure increases. There is some doubt as to whether, in the dog (COWLEY et al. 1973) or the rat (JONES and HALLBÄCK 1978) in the absence of sustained pressure increases, this can truly be considered a model of hypertension. Bilateral lesion of the nucleus of the solitary tract and the brain stem lead to qualitatively similar cardiovascular changes (DOBA and REIS 1973). Less extensive lesions can lead to chronic labile hypertension in the cat (NATHAN and REIS 1977) and chronically labile blood pressure in the rat (SNYDER et al. 1978). Hypertension may also be induced by localized brain stem lesions lateral to the nucleus of the solitary tract (de JONG and PALKOVITS 1976).

Peripheral sympathetic activity is increased after sinoaortic denervation (De QUATTRO et al. 1969), and destruction of central catecholamine neurons with 6-hydroxydopamine will prevent the blood pressure rise or reverse the increase after denervation in rabbits (CHALMERS and REID 1972). Changes in hypothalamic and spinal cord noradrenaline turnover have been described (CHALMERS and WURTMAN 1971). Recently, local changes in catecholamines and synthetic enzymes after sinoaortic denervation in rats have been reported (CHALMERS et al. 1979; SAAVEDRA et al. 1979). Although sinoaortic denervation may not be an entirely satisfactory model of neurogenic hypertension, it does permit some assessment of the primary and secondary nature of changes in brain catecholamines in other models of hypertension. The fall in PNMT activity noted in anterior and posterior hypothalamic nuclei after sinoaortic denervation (CHALMERS et al. 1979) contrasts with the increases in PNMT activity in brain stem in spontaneous hypertension, DOCA salt hypertension (SAAVEDRA et al. 1978) and renovascular hypertension (PETTY and REID 1979).

II. Human Hypertension

The neurogenic contribution to human hypertension and, in particular, the role of catecholamines has been reassessed since the development and application of more reliable assays of catecholamines and their metabolites in body fluids. Both short term and long term contributions appear likely. However, controversy remains as to whether abnormalities of catecholamine-mediated sympathetic nervous system in the brain or the periphery are directly responsible for elevated blood pressure in some or all human hypertensives (for review see de QUATTRO and MIURA 1973; AXELROD 1975; de CHAMPLAIN et al. 1977). Clinical studies have been directed mainly at peripheral catecholamines because of the lack of suitable approaches to the assessment of central catecholamine function in man. Strategies developed in animals have had limited application to man because of practical and ethical considerations. Approaches are usually indirect and many of the conflicting and controversial reports are a consequence of differences in protocol.

Assays of catecholamines and their principal metabolites in urine have allowed reliable biochemical assessment of sympathetic activity. Although urinary catecholamine excretion would appear an attractive integrated index of sympathetic activity, the multiplicity of routes of metabolism, widespread inter-individual differences in metabolising enzyme activities and common terminal metabolites of adrenaline and noradrenaline present difficulties. Urinary catecholamines, metanephrines, MOPEG and VMA would need to be measured to obtain an overall assessment of the principal metabolites. Plasma catecholamines, which are fully reviewed in another section of this volume, show rapid changes in response to physical activity, stress and anxiety and this may limit their usefulness.

Direct recording of sympathetic nerve activity in man is difficult, particularly recording from splanchnic or visceral sites. Cutaneous and muscle nerves do contain sympathetic efferent fibres, and recordings from these have been related to short term haemodynamic changes (WALLIN et al. 1973; SUNDLÖF and WALLIN 1978).

Assessment of central catecholamine turnover is even less satisfactory. There is no simply measured accessible biochemical marker. In spite of previous claims (SCHANBERG et al. 1968) it is unlikely that urinary excretion of MOPEG or its sulphate conjugate can be used in man as an index of central noradrenaline turnover (BOOBIS et al. 1980). MOPEG levels in lumbar cerebrospinal fluid may provide an index of overall brain noradrenaline turnover but may require repeated lumbar puncture in patients without known neurological or psychiatric disease. Similar limitations apply to the measurement of noradrenaline itself in cerebrospinal fluid (ZIEGLER et al. 1976). More complex strategies to examine catecholamine turnover noninvasively in human brain are being developed using radioactive or stable isotope labelled precursors (SJÖQVIST et al. 1979; BLOMBERY et al. 1979). These approaches, although of great interest, have not yet reached the stage of application to hypertensive patients.

1. Essential Hypertension

Essential hypertension is a diagnosis reached by exclusion of known secondary causes of high blood pressure. It is now accepted that in any population blood pressure is distributed in a unimodal Gaussian manner (PICKER-

ING 1955). Hypertension can only be defined empirically and, in view of the influences of age, sex and variability within individuals, these criteria must be considered in any definition. As discussed earlier, in spite of extensive investigations, no single abnormality or characteristic, apart from raised blood pressure, has been useful in defining hypertensive patients. It increasingly appears that factors maintaining blood pressure in the majority of hypertensive patients are the same as those in so called "normotensives". There may be as yet undefined differences in the regulation of the interrelationships between these mechanisms which were genetically determined and from the earliest years of life determine the actual level of blood pressure achieved (DE SWIET et al. 1976). There is, of course, no doubt that long-standing hypertension leads to secondary humoral, neurogenic and morphological consequences which themselves may magnify the pressure increase (FOLKOW 1971).

Studies on plasma catecholamines in essential hypertension have led to conflicting and confusing results. Soon after the development of sensitive assays there were several reports of increased plasma noradrenaline in up to 50% of essential hypertensives (ENGLEMAN et al. 1970; de QUATTRO and CHAN 1972; LOUIS et al. 1973; de CHAMPLAIN et al. 1977). LOUIS and his colleagues observed a positive direct relationship between resting plasma noradrenaline and blood pressure in their hypertensive subjects (LOUIS et al. 1973). However, other groups were unable to confirm these results of elevated levels (PEDERSEN and CHRISTIANSEN 1975) and others reported a significant association of plasma noradrenaline with age but not with the level of blood pressure (LAKE et al. 1977). In addition of age, the choice of clinical or paraclinical staff as controls may confound results (JONES et al. 1979b). Several other groups have now failed to observe differences between plasma noradrenaline of established hypertensives and age-matched controls (SEVER et al. 1977; FRANCO-MORSELLI et al. 1977; BUHLER et al. 1978), although the latter two groups report increased plasma adrenaline levels in their hypertensives. Some groups have chosen to subdivide populations either on the basis of their catecholamine levels (DE CHAMPLAIN et al. 1977) or plasma renin activity (ESLER et al. 1977; TAYLOR et al. 1978). Although, in a small study, a direct relation between plasma renin activity and plasma noradrenaline was observed (ESLER et al. 1977), in a larger study with groups of low, normal and high renin hypertensives, this relationship was not substantiated (TAYLOR et al. 1978). Recently, in a comparison of adolescent hypertensives with appropriate controls from the same population (HOFMAN et al. 1979), elevated plasma noradrenaline was noted in the hypertensive group, supporting the view that, in younger hypertensives, plasma noradrenaline levels may be inappropriately elevated (SEVER et al. 1977). Within hypertensive individuals, plasma noradrenaline serves as a reliable index of these sympathetically mediated changes in blood pressure and heart rate (WATSON et al. 1979). GOLDSTEIN (1981; 1983) has comprehensively reviewed published studies. There is, overall, a small non-significant rise in plasma noradrenaline. However, younger subjects may show relatively higher levels than older hypertensives. Longitudinal studies of younger subjects may determine whether there is an early contribution of sympathetic hyperactivity to the development of hypertension.

Urinary catecholamine measurements are equally unrevealing. Early studies suffered from lack of specificity of the methodology (de Champlain 1976), but, even more recently, there has been no general agreement that urinary catecholamines or metabolites are increased in essential hypertension. In a population study, which examined the relationship of blood pressure to plasma noradrenaline, urinary catecholamines and urinary metanephrines, only small, insignificant contributions were revealed on multivariate analysis (Jones et al. 1978).

Turnover of noradrenaline in the periphery has been studied with contradictory results (Rubin and Reid 1984). Early studies with ^{3}H-(±)-noradrenaline are difficult to interpret because of contributions from the physiologically inactive (+)-isomer (Gitlow et al. 1971; de Quattro 1971). More recent studies using the (−)-isomer (Esler et al. 1979) suggest differences in uptake or metabolism of the amine in some essential hypertensives (Esler et al. 1980) and between different organs (Esler et al. 1984). Another approach has been to study sensitivity and clearance of infused unlabelled (−)-noradrenaline (Fitz-Gerald et al. 1979) or to label noradrenaline pools with precursors labelled with stable isotopes (Blombery et al. 1979). These approaches may help in defining changes in peripheral catecholamine turnover in essential hypertension.

Although attention initially focused on increased sympathetic activity and increased release of catecholamine transmitters, changes in the responsiveness of the heart and vascular smooth muscle may amplify the influences of the sympathetic nervous system (Philipp et al. 1978). The study of binding of *α*- and *β*-adrenoceptor agonists or antagonists of high specific activity to *in vitro* membrane preparations may provide important information on sensitivity and number of receptors. However, readily accessible tissues which include *β*-receptors located on leucocytes (Williams et al. 1976) and *α*-receptors on platelets (Hoffman and Lefkowitz 1980) may not provide an adequate index of receptors located pre- and postsynaptically in heart and vascular smooth muscle.

Involvement of central catecholaminergic pathways in essential hypertension has been proposed on the basis of results obtained with SHR, reviewed above, and the clinical responses to centrally acting anti-hypertensive drugs. However, the conclusions in man are tentative and preliminary. Concentrations of MOPEG, the principal metabolite of noradrenaline in brain, have been reported to be increased in cerebrospinal fluid of hypertensive patients (Saran et al. 1978). Although these authors noted a direct relationship between MOPEG in cerebrospinal fluid and blood pressure, in another study in a group of unselected patients undergoing diagnostic lumbar puncture, MOPEG, free or total, measured by mass fragmentography, was not related to blood pressure levels (Murray et al. 1977). Increases in noradrenaline in the cerebrospinal fluid of hypertensive, compared to normotensive, neurological patients have been reported (Eide et al. 1979; Ziegler et al. 1980). These observations provide an interesting approach to the investigation of central noradrenergic mechanisms. However, validation of the use of noradrenaline in the cerebrospinal fluid as an index of central amine turnover is needed.

2. Other Forms of Hypertension

With the exception of phaechromocytoma, there are few large scale systematic studies of catecholamines in secondary hypertension. There has been some interest in the possible contribution of the sympathetic nervous system to the genesis or maintenance of hypertension associated with pregnancy (Rubin 1983). While this is a special type of secondary hypertension, it presents the attractive possibility of studying hypertension in evolution and devolution in the human.

There were initial indications that increased activity of the sympathetic nervous system was involved in this condition, since urinary catecholamines were found to be elevated (Cession 1966; Zuspan 1979). However, the introduction of more sensitive and specific assay techniques has generally failed to support this earlier suggestion. Only two studies have provided evidence of increased sympathetic activity in pregnancy hypertension. One involved Cape coloured women in South Africa and found that in the last trimester of pregnancy those patients with pregnancy-associated hypertension had higher plasma adrenaline, noradrenaline and dopamine concentrations than normotensive controls (Davey and McNab 1981). However, this difference was found only in blood samples collected soon after admission to the hospital and subsequent analyses in these women failed to confirm a distinction between groups. A second study found that COMT activity was higher in the erythrocytes of women with pregnancy associated hypertension compared to controls (Bates et al. 1982).

In contrast, the remaining studies in this area suggest that the sympathetic nervous system responds appropriately to the elevated blood pressure in pregnancy associated hypertension. In one investigation plasma noradrenaline and adrenaline concentrations were measured in patients with pregnancy associated hypertension, essential hypertension during pregnancy, normal pregnant women and non-pregnant control subjects. Blood was drawn in the second and third trimesters and again five days and three months following delivery (Pedersen et al. 1982). Noradrenaline and adrenaline concentrations were the same in all groups and at all times. Similar results were found in our own laboratory (Rubin et al. 1983). In addition, one study found that plasma noradrenaline concentrations were significantly lower in hypertensives compared to normotensive women of the same gestation (Tunbridge and Donnai 1981). A similar observation was made in a study which followed women serially during pregnancy with regular collection of blood samples (Natrajan et al. 1982). Sixty-one normotensive women were initially involved and of these, nine developed hypertension. Noradrenaline und adrenaline concentrations fell significantly during the pregnancies and there was a trend towards lower catecholamine concentrations in the hypertensive group. These data suggest that the sympathetic nervous system is not involved in pregnancy hypertension.

Autonomic neuropathy and sympathetic degeneration have been described in patients with chronic renal failure. However, there is no evidence that abnormality of catecholamine metabolism or responsiveness is a common feature (NAIK et al. 1981).

Phaeochromocytoma tumours of neuroectodermal tissue usually occur in the adrenal medulla but may be found in any sympathetic ganglia from the neck to the pelvis. The tumours synthesize catecholamines, primarily noradrenaline and adrenaline and less commonly dopamine, together with certain polypeptides. The clinical aspects of diagnosis, investigation and management of phaeochromocytoma have been extensively reviewed elsewhere (MANGER and GIFFORD 1977; GANGULY et al. 1979; JONES et al. 1980b). Phaeochromocytoma, in which autonomously increased catecholamine production leads to long term increases in circulating catecholamines and increased blood pressure, would appear to be a model of catecholamine-induced hypertension. However, the level of blood pressure in phaeochromocytoma is not always directly related to plasma levels of noradrenaline. Hypertension may be paroxysmal or sustained and occasionally absent. Plasma noradrenaline levels although showing peaks and troughs are generally grossly increased at all times. Surgical removal of the tumour does not invariably reverse hypertension (MODLIN et al. 1979). Phaeochromocytoma illustrates that catecholamine levels and blood pressure will not always be directly related and that adaptive changes in adrenoceptors or activation of other reflexes will tend to buffer blood pressure effect.

Plasma noradrenaline is elevated in the vast majority of phaeochromocytomata (GANGULY et al. 1979; BRAVO et al. 1979; JONES et al. 1979a), although rarely it may be normal when adrenaline levels are raised. Measurement of plasma catecholamines is a valuable adjunct to and may even replace urinary catecholamine or metabolite excretion in diagnosis (BRAVO et al. 1979). Simultaneous measurement of plasma noradrenaline and its metabolite DOPEG may be useful in the diagnosis of these tumors (BROWN 1984). Measurement of catecholamine levels in plasma from multiple venous sites can further assist in the localisation of multiple or extra adrenal tumours (JONES et al. 1979a).

F. Adrenoceptor Agonists and Antagonists in Blood Pressure Regulation

Several drugs currently or previously used in the control of high blood pressure in man act via central or peripheral adrenoceptors and the actions of these various agents serve to illustrate involvement of catecholamines in blood pressure control. It is interesting to note that most of these drugs were discovered by chance to have blood pressure lowering properties: an interesting reflection of current understanding of cardiovascular pharmacology.

I. Drugs Acting on Peripheral α-Adrenoceptors

Several types of peripheral α-adrenoceptor antagonists have been used in man but only one of these, prazosin, is currently used extensively in clinical practice.

The haloalkylamines were first studied towards the end of the 1940s (NICKERSON and GOODMAN 1947; NICKERSON and GUMP 1949). The α-adrenoceptor blockade produced by these compounds is delayed in onset, suggesting that a metabolite is the actual antagonist in each case. However, once blockade is established, it is relatively stable, since the compounds form a covalent bond at the receptor (NICKERSON 1957). In addition to the adrenoceptor antagonism, haloalkylamines also influence noradrenaline release and uptake. Several years ago it was demonstrated that, in the presence of phenoxybenzamine, there was an increase in noradrenaline release from isolated preparations subjected to low stimulation frequencies (POTTER et al. 1971). Presumably, phenoxybenzamine prevents released noradrenaline from exerting its usual negative feedback control on subsequent release. In addition, phenoxybenzamine and many other compounds in this series prevent the uptake of noradrenaline into nerve terminals ($uptake_1$) and into extraneuronal tissue ($uptake_2$) (IVERSEN et al. 1972).

Administration of phenoxybenzamine by slow intravenous infusion to healthy, normovolaemic subjects results in very little change in blood pressure, so long as they remain in the supine position. In contrast, standing or tilting results in a substantial fall in pressure, accompanied by a reflex tachycardia. However, if a subject is volume-depleted, then infusion of phenoxybenzamine even in the supine position produces a dramatic fall in pressure. These interesting observations suggest that, in the supine position, plasma volume is normally of more importance than the adrenergic tone in maintaining blood pressure. However, on standing the α-adrenoceptors play a more important role. When plasma volume is depleted, α-adrenoceptors become important in cardiovascular control in the supine as well as the standing position.

Two substituted imidazolines, phentolamine and tolazoline, also possess α-adrenoceptor blocking activity but this differs from that produced by phenoxybenzamine in being competitive and of short duration. Phentolamine is the only member of this group of drugs which is still in clinical practice, though its use is somewhat limited (see below). Following administration to man, this drug produces changes in blood pressure and heart rate similar to those described for phenoxybenzamine.

The α-adrenoceptor blocking activity of prazosin was discovered by chance and resulted from a study which had begun as a search for drugs with sedative activity. During these investigations, it was observed that compounds with a quinazoline ring lower blood pressure (HAYAO et al. 1965). The earliest observations on the pharmacology of prazosin suggested that this drug might act in a somewhat unusual manner, since it produced an effect considered characteristic of α-adrenoceptor blockade (reversal of the adrenaline pressor effect) but lowered blood pressure without producing the reflex increase in heart rate seen with other α-adrenoceptor blockers (SCRIABINE et al. 1968). It was suggested at that time that prazosin might be exerting an unconventional type of α-adrenoceptor blockade. At concentrations similar to those seen in human plasma during therapy, prazosin has a marked affinity for α_1-adrenoceptors of rabbit pulmonary artery strips (CAMBRIDGE et al. 1977, 1978). On the basis of this observation it was suggested that the absence of reflex tachycardia during the prazosin-induced fall in blood pressure could be explained by the absence of blockade of α_2-adrenoceptors which could then mediate the negative feedback control of noradrenaline release. While this theory is attractive, various observations suggest that it might have limited application to species other than the rabbit. For instance, studies in the rat confirmed the selectivity for α_1-adrenoceptors seen in the rabbit but this selectivity was not seen in the dog (CAVERO et al. 1977). In addition, studies with prazosin, using human digital arteries obtained at autopsy, suggest that the drug does not act as a competitive antagonist while phentolamine used under the same conditions behaved as expected (JAUERNIG et al. 1978).

When prazosin is given to man, its effects on the cardiovascular system are similar to those observed with other α-adrenoceptor blockers. In the supine position, there is a modest fall in blood pressure, but on standing the pressure falls substantially and, contrary to earlier reports, continuous monitoring of heart rate reveals a substantial reflex tachycardia (RUBIN and BLASCHKE 1980). Both in the supine and standing positions, plasma noradrenaline concentration rises significantly following administration of prazosin to normal young subjects (RUBIN and BLASCHKE 1980). This observation is not consistent with the suggestion that the α_1-adrenoceptor selectivity demonstrated by prazosin in certain animals is also applicable to man. It certainly appears that noradrenaline release is increased by prazosin, since, even in the supine position when blood pressure is unchanged, plasma noradrenaline concentration is significantly above control values. It would seem that prazosin is acting as an α-adrenoceptor antagonist in man but that it is probably not demonstrating the same selectivity for α_1-receptors as is seen in certain other species.

Prazosin is now widely used in the treatment of high blood pressure either alone or in combination with a β-receptor antagonist. However, phenoxybenzamine and phentolamine in conjunction with a β-adrenoceptor antagonist, are used only in the short term management of patients with phaeochromocytoma.

II. Drugs Acting on Central α-Adrenoceptors

In the late 1960s several groups independently demonstrated that drugs with α-adrenoceptor agonist properties could lower blood pressure by direct action on the brain stem. Clonidine and other imidazoline derivatives lowered blood pressure after injection into the cerebrospinal fluid (KOBINGER and WALLAND 1967; SCHMITT and SCHMITT 1969) and after infusion into the vertebral artery (SATTLER and van ZWIETEN 1967). In further studies it was proposed that α-methyldopa also had a central hypotensive effect (HENNING and van

ZWIETEN, 1968). This central hypotensive effect of α-methyldopa (HENNING 1969) and levodopa in rats (HENNING and RUBENSON 1970) was dependent on decarboxylation in the brain and subsequent hydroxylation to α-methyl-noradrenaline or noradrenaline respectively. The antagonism of the hypotensive effect of clonidine (SCHMITT et al. 1973) and α-methyldopa (HEISE and KRONEBERG 1972) by some, but not all, centrally applied α-adrenoceptor antagonists led to the hypothesis that these drugs lower blood pressure by acting as agonists at α-adrenoceptor sites in the brain stem which inhibit sympathetic outflow (VAN ZWIETEN 1975). This central mechanism of action is now well accepted (KOBINGER 1975; REID 1979). However, α-methyldopa and clonidine can interact at other more peripheral sites to influence blood pressure. These sites include spinal cord interneurones (SINHA et al. 1973), postganglionic nerve endings (LANGER 1974) and vascular smooth muscle either directly (ZAIMIS and HANNINGTON 1969) or indirectly via α-adrenoceptors. Although these peripheral effects may contribute to the actions of α-methyldopa and clonidine during chronic long term therapy in man, it appears likely that the central hypotensive effect is quantitatively the most important. Patients who have an interruption of neural connections between the brain stem and the intermediolateral horn cells of the spinal cord do not show a fall in blood pressure after oral clonidine (REID et al. 1977a). These results and other clinical and experimental evidence for a central action of α-methyldopa and clonidine have been recently reviewed elsewhere (REID et al. 1977c, 1980).

The fall in plasma noradrenaline and urinary catecholamines after clonidine administration to hypertensive patients (HÖKFELT et al. 1975) and normotensive subjects (WING et al. 1977) is consistent with a sympathoinhibitory effect, although it does not conclusively localise the effect to the brain. This reduction in sympathetic activity persists with chronic drug treatment but is rapidly reversed on drug withdrawal. The latter is often accompanied by an overshoot of catecholamine release which may be associated with a hypertensive reaction (HÖKFELT et al. 1975; REID et al. 1977b). The clonidine-induced reduction in catecholamine release and turnover observed in the periphery has also been described in the central nervous system. Clonidine reduces the formation of MOPEG in animals (STONE 1976) and cerebrospinal fluid MOPEG concentrations in hypertensives (BERTILSSON et al. 1977).

Although clonidine and α-methyldopa, via α-methylnoradrenaline, act as α-agonists in the brain to lower blood pressure, they can also raise blood pressure when given intravenously (MROCZEK et al. 1973). Clonidine in high doses orally may elicit a paradoxical hypertensive effect (HUNYOR et al. 1975; WING et al. 1977), presumably as a result of peripheral α-adrenoceptor stimulation. The paradox regarding why α-adrenoceptor agonists with peripheral pressor effects actually lowered blood pressure was partially resolved by the demonstration that these drugs which contributed to the regulation of noradrenaline release (LANGER 1974, 1977) exhibited a lower affinity for classical postsynaptic α-receptors (STARKE et al. 1975). Furthermore the central α-receptors stimulated by clonidine were found to be similar to presynaptic receptors and unlike the classical postsynaptic α_1-receptors (LANGER 1977). In view of the difficulties in determining anatomical localization of receptors, particularly in the brain, it is preferable that the terminology α_1 and α_2-receptors should be used for responses first described as postsynaptic and presynaptic, respectively, regardless of their anatomical localization (BERTHELSEN and PETTINGER 1977). Developments in the classification and understanding of α-receptors and the localization of α_1 and α_2 receptors are discussed in another chapter in this volume.

Clonidine and α-methylnoradrenaline as α_2 selective agonists modify efferent sympathetic outflow by stimulating central α_2 receptors in the brain stem (ZANDBERG et al. 1979) and hypothalamus (PHILIPPU et al. 1973; STRUYKER-BOUDIER et al. 1975). The precise location of these receptors is uncertain. Direct microinjections of α-methyl-noradrenaline and clonidine into the nucleus of the solitary tract (NIJKAMP and de JONG 1977; ZANDBERG et al. 1979) caused a fall in blood pressue abolished by α_2-adrenoceptor blockers such as yohimbine or the mixed α_1/α_2-adrenoceptor blocker, phentolamine. LAUBIE et al. (1976) present further results favouring the nucleus of the solitary tract as a site of action of clonidine. However, clonidine also lowered blood pressure when applied to the locus coeruleus in the mid brain (ZANDBERG et al. 1979). Iontophoretic application of clonidine to neurones of the locus coeruleus (CEDARBAUM and AGHAJANIAN 1976) have given some support to the hypothesis that α_2 receptors may be located on central noradrenergic neurones in the locus coeruleus and solitary tract nuclei. Although these receptors would be presynaptic with respect to the noradrenergic neurones, they may be postsynaptic to a proposed central adrenaline input (BOLME et al. 1974) which may modulate noradrenergic activity (FUXE et al. 1980).

Thus, the evidence at present favours an action of clonidine and α-methyldopa at mid brain and medullary sites, particularly the nucleus of the solitary tract. Observations on centrally acting drugs with α_2-adrenoceptor agonist activity indicate the important role of brain catecholamine mechanisms in cardiovascular regulation and blood pressure control.

III. β-Adrenoceptor Antagonists, Central and Peripheral Actions

In many countries β-adrenoceptor blocking drugs are the first line drugs in the treatment of essential hypertension (CONOLLY et al. 1976). Competitive antagonists of β-adrenoceptors were not developed as antihypertensive agents, and it was clinical experience in early studies with propranolol which revealed their effects on blood pressure (PRITCHARD and GILLAM 1969). There are now many β-adrenoceptor antagonists available with a range of other pharmacological effects including cardioselectivity (β_1-selectivity), intrinsic sympathomimetic and local anaesthetic activity. However, these properties are of minor clinical importance, as all these drugs lower blood pressure as a consequence of their common β-adrenoceptor blocking action (CONNOLLY et al. 1976). It is not clear which β-receptors are principally involved. Several alternative sites and mechanisms of action have been proposed but none of these is entirely satisfactory (LEWIS 1976). The demonstration of β-receptors in the central nervous system which participated in central blood pressure control presented an attractive alternative explanation (DAY and ROACH 1973; REID et al. 1974). Although considerable brain levels of lipid soluble drugs like propranolol are found in man (MYERS et al. 1975), a central nervous site of action appears unlikely, since more polar agents (practolol or atenolol), which have at least equivalent hypotensive efficacy to propranolol, would not be expected to cross the blood brain barrier as readily. An alternative site of action of β-adrenoceptor blockers, which would also have the result of reducing peripheral sympathetic tone and catecholamine release, is on facilitatory presynaptic β-adrenoceptors at peripheral sympathetic nerve endings (LANGER 1974, 1977). Such a site would presumably be accessible to both polar and non-polar drugs. Evidence for presynaptic β-receptors and their effect on catecholamine overflow *in vitro* (LANGER 1977) and *in vivo* (YAMAGUCHI et al. 1977) has been presented in animals. It would be anticipated that either a central nervous effect of β-adrenoceptor blockers or a peripheral presynaptic action would lead to a reduction in peripheral catecholamine release and overflow into plasma. In some studies, β-adrenoceptor blockers lowered plasma noradrenaline either in the group as a whole (BRECHT et al. 1976) or in sub-groups who responded with a fall in blood pressure (de CHAMPLAIN et al. 1977; BUHLER et al. 1978). In contrast, in other studies in responders and non-responders, hypertensives and normotensives, β-adrenoceptor blockers have been shown to increase or not change plasma noradrenaline (HÖKFELT et al. 1975; MALING et al. 1979; JONES et al. 1980a). In addition, possible effects of β-adrenoceptor blockers on reflexly mediated changes in sympathetic tone were not observed when the changes in blood pressure and plasma noradrenaline following intravenous sodium nitroprusside were examined with and without block of β-adrenoceptors (DEAN et al. 1980).

The site and precise mechanism of action of β-adrenoceptor blockers is still obscure and, although β-receptors in brain do take part in cardiovascular regulation, and some β-adrenoceptor blockers have other central nervous effects, there is no good clinical evidence that a central nervous or peripheral presynaptic mechanism contributes substantially to the antihypertensive action.

IV. Miscellaneous Drugs Which Illustrate the Role of Catecholamines in Blood Pressure Control

A number of drugs which are no longer widely used in the treatment of high blood pressure can still usefully be discussed because they illustrate in different ways the importance of catecholamines in blood pressure control.

Adrenergic neurone blocking drugs were once extensively used in clinical practice, examples being guanethidine, bethanidine and debrisoquin. They all have in common the impairment of post-ganglionic adrenergic neurone conduction by inhibiting the release of catecholamines from nerve terminals in response to neuronal depolarisation. Following long-term treatment there is also a decline in tissue catecholamine stores but this does not seem to be an important factor in the mechanism of action of these drugs. Following the administration of an effective dose of one of these drugs, both blood pressure and heart rate decrease. The fall in blood pressure is considerably greater in the standing position. This postural fall in blood pressure and other adverse effects of diarrhoea and impaired sexual function has led to this group of drugs being used less frequently.

Another hypotensive drug which interacts with the adrenergic nervous system is reserpine. This drug depletes stores of catecholamines and 5-HT in a wide range of tissues including the brain and adrenal medulla. Adverse effects, notably sedation, have led to a substantial decline in the use of this drug in many countries.

Acknowledgements. We are most grateful to Mrs. Mary Wood for her considerable and valuable secretarial assistance in preparing this manuscript. Professor Rubin was supported as a British Heart Foundation Junior Research Fellow during the initial preparation of this Chapter and as a Wellcome Trust Senior Fellow in Clinical Science during subsequent minor updating.

G. References

Abel JJ (1897) On the blood pressure raising constituents of the suprarenal capsule. Johns Hopkins Hosp Bull 8:151–157

Achari NK, Downman CB (1970) Autonomic effector responses to stimulation of nuclear fastigium. J Physiol (Lond) 210:637–650

Axelrod J (1975) Central actions of drugs in blood pressure regulation. In: Davies DS Reid JL (Eds) Wells Pitmans Medical Tunbridge, pp 1–7

Ayitey-Smith E, Varma DR (1970) An assessment of the role of the sympathetic nervous system in experimental hypertension using normal and immunosympathectomized rats. Br J Pharmacol 40:175–185

Bates GW, Whitworth NS, Jackson E (1982) Erythrocyte catechol-o-methyltransferase activity in pregnant women with pregnancy-induced hypertension. Amer J Obstet Gynaecol 142:177–178

Berthelsen S, Pettinger WA (1977) A functional basis for classification of α-adrenergic receptors. Life Sci 21:595–606

Bertilsson L, Haylind K, Ostman J, Rawlins M, Ringberger V, Sjoqvist F (1977) Monoamine metabolites in cerebrospinal fluid during treatment with clonidine or alprenolol. Eur J Clin Pharmacol 11:125–128

Blombery P, Kopin IJ, Gordon EK, Ebert MM (1979) Metabolism and turnover of M.H.P.G. in the monkey. In Usdin E, Kopin IJ, Barchas J (Eds) Catecholamines Pergamon Press Oxford, pp 1875–1877

Boadle MC, Hughes J, Roth RH (1969) Angiotensin accelerates catecholamine biosynthesis in sympathetically innervated tissue. Nature 222:987–988

Bolme P, Corrodi H, Fuxe K, Hökfelt T, Lidlink P, Goldstein M (1974) Possible involvement of central adrenaline neurons in vasomotor and respiratory control. Eur J Pharmacol 28:89–94

Boobis AR, Murray S, Jones DH, Reid JL, Davies DS (1980) Urine conjugates of 4 hydroxy 3 methoxy phenylethylene glycol do not provide an index of brain amine turnover in man. Clin Sci 58:311–316

Bravo EL, Tarazi RC, Gifford RW, Stewart BH (1979) Circulating and urinary catecholamines in phaeochromocytoma. N Engl J Med 301:682–686

Brecht HM, Banthien F, Ernst W, Schoeppe W (1976) Increased plasma noradrenaline concentration in essential hypertension and their decrease after long term treatment with a receptor blocking agent (Pindolol). Clin Sci Mol Med 51:485-488s

Brod J, Fencl V, Hezl Z, Jirka J (1959) Circulatory changes underlying blood pressure elevation during acute emotional stress (mental arithmetic) in normotensive and hypertensive subjects. Clin Sci 18:269-279

Brown MJ (1984) Simultaneous assay of noradrenaline and its deaminated metabolite dihydroxyphenylethyleneglycol in plasma: a simplified approach to the exclusion of phaeochromocytoma. Eur J Clin Invest 14:67-72

Brunner HR, Gavras H, Laragh JH (1974) Specific inhibitions of the renin angiotensin system: a key to the understanding of blood pressure regulation. Proj Cardiovasc Dis 17:87-98

Buggy J, Fink GD, Johnson AK, Brody MJ (1977) Prevention of the development of renal hypertension by anteroventral third ventricular tissue lesions. Circ Res 40 (5 suppl 1):110-117

Buhler F, Bertel O, Kiowski (1978) Plasma noradrenaline and adrenaline and β-adrenoceptor responsiveness in renin subgroups of essential hypertension. Clin Sci Mol Med 55 Suppl 4:57-60s

Burnstock G, Cannon B, Iwayama T (1970) Sympathetic innervation of vascular smooth muscle in normal and hypertensive animals. Circ Res 26, 27 (Suppl 2):5-24

Cambridge D, Davey MJ, Massingham R (1977) Prazosin, a selective antagonist of postsynaptic alpha-adrenoceptors. Br J Pharmacol 59:14P

Cambridge D, Davey MJ, Massingham R (1978) Further evidence for a selective postsynaptic alpha-adrenoceptor blockade with prazosin in vascular smooth muscle. Naunyn-Schmiedeberg's Arch Pharmacol 302:R52

Carlsson A, Falck B, Fuxe K, Hillarp NA (1964) Cellular localisation of monoamines in the spinal cord. Acta Physiol Scand 60:112-119

Cavero I, Lefevre F, Roach AG (1977) Differential effects of prazosin on the pre- and postsynaptic α-adrenoceptors in the rat and dog. Br J Pharmacol 61:469P

Cedarbaum JM, Aghajanian GK (1976) Noradrenergic neurons of the locus coeruleus: inhibition by epinephrine and activation by the antagonist piperoxane. Brain Res 112:413-419

Cession G (1966) Sur l'élimination urinaire des catecholamines et de leur metabolites au cours de la dysgravidie due 3me trimestre de la gestation. Bull Soc Roy Belge Gyn Obstet 36:196-216

Chalmers JP (1975) Brain amines and models of experimental hypertension. Circ Res 36:469-480

Chalmers JP, Reid JL (1972) Participation of central noradrenergic neurones in arterial baroreceptor reflexes in the rabbit: a study with intracisternally administered 6-hydroxydopamine. Circ Res 31:789-804

Chalmers JP, Wurtman RJ (1971) Participation of central noradrenergic neurons in arterial baroreceptor reflexes in the rabbit. Circ Res 28:480-491

Chalmers JP, Petty MA, Reid JL (1979) Participation of adrenergic and noradrenergic connections of arterial baroreceptor reflexes in the rat. Circ Res 45:516-522

Chalmers JP, Minson J, Deneroy L, Stead B, Howe PRC (1984) Brain stem PNMT neurons and experimental hypertension in the rat. Clin Exp Hypertension A6:243-258

Christensen NJ (1972) Plasma catecholamines in long-term diabetes with and without neuropathy. J Clin Invest 51:779-787

Clarke DE, Smookler HH, Barry H (1970) Sympathetic nerve function and DOCA Na-Cl induced hypertension. Life Sci 9:1097-1108

Clarke DWJ (1971) Effects of immunosympathectomy on development of high blood pressure in genetically hypertensive rats. Circ Res 28:330-336

Connolly ME, Kerstung F, Dollery CT (1976) The clinical pharmacology of beta adrenergic drugs. Proj Cardiovasc Dis 19:203-234
Coote JH, Macleod VH (1974) The influence of bulbospinal monoaminergic pathways on sympathetic nerve activity. J Physiol (Lond.) 241:453-475
Coote JH, Macleod VH (1975) The spinal route of sympathoinhibitory pathways descending from the medulla oblongata. Pflügers Arch 359:335-347
Cowley AW, Liard JF, Guyton AC (1973) Role of the baroreceptor reflex in daily control of blood pressure and other variables in the dog. Circ Res 32:564-576
Crill WE, Reis DJ (1968) Distribution of carotid sinus and depressor nerves in cat brain stem. Am J Physiol 214:269-276
Dahl LK, Heine M, Tassinari L (1962) Role of genetic factors in susceptibility to experimental hypertension due to chronic excess salt ingestion. Nature (Lond.) 194:480-482
Dahlström A, Fuxe K (1964) Evidence for the existence of monoamine neurones in the central nervous system. I. Demonstration of monoamines in the cell bodies of brain stem neurones. Acta Physiol Scand 62 Suppl. 232:1-55
Dahlström A, Fuxe K (1965) Evidence for the existence of monoamine neurones in the central nervous system. II. Experimentally induced changes in the intraneuronal amine levels of bulbospinal neuron systems. Acta Physiol Scand 64 Suppl. 247:1-36
Dancis J, Smith AA (1966) Familial dysautonomia. N Engl J Med 274:207-209
Dargie HJ, Franklin SS, Reid JL (1977a) Plasma noradrenaline concentrations in experimental renovascular hypertension in the rat. Clin Sci Mol Med 52:477-483
Dargie HJ, Franklin SS, Reid JL (1977b) Central and peripheral noradrenaline in the two kidney model of renovascular hypertension in the rat. Br J Pharmacol 61:213-215
Davey DA, Macnab MF (1981) Plasma adrenaline, noradrenaline and dopamine in pregnancy hypertension. Br J Obstet Gynaecol 88:611-618
Davies DS, Reid JL (1975) Central action of drugs in the regulation of blood pressure. Wells Pitmans Tunbridge, Medical Press
Day MD, Roach AG (1973) β-Adrenergic receptors in the central nervous system of the cat concerned with control of arterial pressure and heart rate. Nature New Biol 242:30-31
Dean CR, Maling T, Dargie HJ, Reid JL, Dollery CT (1980) Effect of propranolol on plasma norepinephrine during sodium nitroprusside induced hypotension. Clin Pharm Ther 27:156-164
de Champlain J (1976) Evaluation of the neurogenic component in human hypertension. In Julius S, Esler M (Eds) The Nervous System in Arterial Hypertension. Thomas, Springfield, pp 267-300
de Champlain J, van Ameringen MR (1972) Regulation of blood pressure by sympathetic nerve fibres, adrenal medulla in normotensive and hypertensive rats. Circ Res 36:617-628
de Champlain J, Krakoff LR, Axelrod J (1968) Increased monoamine oxidase during the development of cardiac hypertrophy in the rat. Circ Res 23:361-369
de Champlain J, Krakoff LR, Axelrod J (1969a) Interrelationships of sodium intake, hypertension and norepinephrine storage in the rat. Circ Res 24-25 Suppl. 1:75-92
de Champlain J, Mueller RA, Axelrod J (1969b) Turnover and synthesis of norepinephrine in experimental hypertension in rats. Circ Res 25:285-291
de Champlain J, Farley L, Cousineau D, van Ameringen MR (1976) Circulating catecholamine levels in human and experimental hypertension. Circ Res 38:109-114
de Champlain J, Cousineau D, van Ameringen MR, Marc Aurele J, Yamaguchi N

(1977) The role of the sympathetic nervous system in experimental and human hypertension. Posgrad Med J 53 (Suppl. 3):15-30
de Jong W (1974) Noradrenaline: central inhibitory control of blood pressure and heart rate. Europ J Pharmacol 29:179-181
de Jong W, Palkovits M (1976) Hypertension after localized transection of brain stem fibres. Life Sci 18:61-64
de Jong W, Provoost AP, Shapiro AP (1977) Hypertension and brain mechanisms. Prog Brain Res 47 Elsevier Amsterdam
Denoroy L, Fourniere S, Vincent M, Renand B, Pujol JF, Sassard J (1979) Phenylethanolamine-N-methyl transferase and dopamine β-hydroxylase alterations in young spontaneously hypertensive rats. Brain Res 162:184-187
de Quattro V (1971) Evaluation of increased norepinephrine excretion using L-dopa-^{3}H. Circ Res 28:84-97
de Quattro V, Chan S (1972) Raised plasma catecholamines in some patients with primary hypertension. Lancet 1:806-809
de Quattro V, Miura Y (1973) Neurogenic factors in human hypertension: mechanism or myth? Am J Med 55:362-378
de Quattro V, Nagatsu T, Maronde R, Alexander N (1969) Catecholamine synthesis in rabbits with neurogenic hypertension. Circ Res 24:545-555
de Swiet M, Fayers P, Shinebourne EA (1976) Blood pressure survey in a population of newborn infants. Brit Med J 2:9-11
Di Guilio AM, Young HY, Fratta W, Costa E (1979) Decreased content of immunoreactive enkephalin-like peptide in peripheral tissues of spontaneously hypertensive rats. Nature 278:646-647
Doba N, Reis DJ (1973) Acute fulminating neurogenic hypertension produced by brainstem lesions in the rat. Circ Res 32:584-593
Doba N, Reis DJ (1974a) Role of the cerebellum and the vestibular apparatus in regulation of orthostatic reflexes in the cat. Circ Res 34:9-18
Doba N, Reis DJ (1974b) Role of central and peripheral adrenergic mechanisms in neurogenic hypertension produced by brainstem lesion in rats. Circ Res 34:293-301
Eide I, Kolloch R, de Quattro V, Miano L, Dugger R, van der Meulen J (1979) Raised cerebrospinal fluid norepinephrine in some patients with primary hypertension. Hypertension, 1:225-260
Engelman K, Portroy B, Sjoerdsma A (1970) Plasma catecholamine concentrations in patients with hypertension. Circ Res 26 and 27 Suppl 1:141-145
Esler M, Julius S, Zweifler A, Randall O, Harburg E, Gardinger H, de Quattro V (1977) Mild high-renin essential hypertension. N Engl J Med 296:405-411
Esler M, Jackman G, Bobik A, Kellerher D, Jenning G, Leonard P, Skews H, Korner P (1979) Determination of norepinephrine apparent release rate and clearance in humans. Life Sci 25:1461-1470
Esler M, Jackman G, Leonard P, Bobik A, Skews H, Jenning G, Kelleher D, Korner P (1980) Determination of noradrenaline uptake, spillover to plasma and plasma concentration in patients with essential hypertension. Clin Sci 59 Suppl 6:311-314
Esler M, Jennings G, Korner P, Blombery P, Burke F, Willett I, Leonard P (1984) Total and organ specific noradrenaline plasma kinetics in essential hypertension. Clin Exp Hypertension. A6:493-506
Falck B, Hillarp NA, Thieme G, Torp A (1962) Fluorescence of catecholamine and related compounds condensed with formaldehyde. J Histochem Cytochem 10:348-354
Finch L, Haeusler G, Thoenen H (1972) Failure to induce experimental hypertension

in rats after intraventricular injection of 6-hydroxydopamine. Br J Pharmacol 44:356-357

Finch L, Cohen M, Hurst WD (1973) Effects of 6-hydroxydopamine at birth on the development of hypertension in the rat. Life Sci 13:1403-1410

Fitzgerald GA, Hossman V, Hamilton CA, Reid JL, Davies DS, Dollery CT (1979) Interindividual variation in the kinetics of infused noradrenaline in man. Clin Pharmacol Ther 26:669-675

Folkow B (1971) The haemodynamic consequences of adaptive structural changes of the resistance vessel in hypertension. Clin Sci 41:1-12

Franco-Morselli R, Elghozi JL, Joly E, di Guilio S, Meyer P (1977) Increased plasma adrenaline concentration in benign essential hypertension. Br Med J 4:1251-1254

Fuxe K, Bolme P, Jannson G, Agnata L, Goldstein M, Hökfelt T, Schwarz R, Engel J (1979) On the cardiovascular role of noradrenaline, adrenaline and peptide containing neurone systems in the brain. In: Meyer P, Schmitt H (Eds) Nervous System and Hypertension Wiley-Flammarion, Paris, ppl-17

Fuxe K, Ganten D, Bolme P, Agnati LF, Hökfelt T, Andersson K, Goldstein M, Harfstrand A, Unger T, Rascher W (1980) The role of central catecholamine pathways in spontaneous and renal hypertension in rats. In Fuxe K, Goldstein M, Hökfelt B, Hökfelt T (Eds) Central adrenaline neurons: basic aspects and their role in cardiovascular function. Wenner-Gren Symposium 33 Pergamon Press, Oxford

Ganguly A, Henry DP, Yune HY, Pratt JH, Grim CE, Donohue JP Weinburger MH (1979) Diagnosis and localization of phaeochromocytoma. Am J Med 67:21-26

Gauthier P, Nadeau RA, de Champlain J (1972) Acute and chronic cardiovascular effects of 6-hydroxydopamine in dogs. Circ Res 31:207-217

Gitlow SE, Mendleowitz M, Bertani LM (1971) Human norepinephrine metabolism: its evaluation by administration of tritiated norepinephrine. J Clin Invest 50:859-869

Goldenberg M, Pines KL, Baldwin E, Greene DG, Rob CE (1948) The haemodynamic response of man to norepinephrine and epinephrine and its relation to the problem of hypertension. Am J Med 5:792-806

Goldstein DS (1981) Plasma norepinephrine in essential hypertension. A study of the studies. Hypertension 3:48-52

Goldstein DS (1983) Plasma catecholamines and essential hypertension. An analytical review. Hypertension 5:86-99

Göthert M, Pohl IM, Wehking E (1979) Effects of presynaptic modulation on Ca induced noradrenaline release from central noradrenergic neurons. Naunyn-Schmiedeberg's Arch Pharmacol 307:21-27

Grewal RS, Kaul CL (1971) Importance of the sympathetic nervous system in the development of renal hypertension in the rat. Br J Pharmacol 42:497-504

Grobecker H, Roizen MF, Weise V, Saavedra JM, Kopin IJ (1975) Sympathoadrenal medullary activity in young spontaneously hypertensive rats. Nature 258:267-268

Haeusler G (1974) Organisation of central cardiovascular pathways in the cat and the question of an involvement of adrenergic neurones. Naunyn-Schmiederberg's Arch Pharmacol 285:R28

Haeusler G, Lewis P (1975) The baroreceptor reflex and its relation to central adrenergic mechanisms. In Milliez P, Safar M (Eds) Recent Advances in Hypertension, vol 2. Laboratoires Boehringer Ingelheim, Reims, pp 17-26

Haeusler G, Osterwalder R (1980) Is substance P the transmitter at the first synapse of the baroreceptor reflex? Clin Sci 59, Suppl 6:295-298

Haeusler G, Finch L, Thoenen H (1972) Central adrenergic neurons and the initiation and redevelopment of experimental hypertension. Experientia 28:1200-1203

Hallbäck M, Folkow B (1974) Cardiovascular responses to acute stress in spontaneously hypertensive rats. Acta Physiol Scand 90:684–698

Hayao S, Havera HJ, Strycker WG, Leipzig TJ, Kalp RA, Hartzler HE (1965) New sedative and hypotensive 3-substituted 2,4 (1H;3H)-quinazolinediones. J Med Chem 8:807–811

Heise A, Kroneberg G (1972) Alpha sympathetic receptor in the brain and the hypotensive action of methyldopa. Eur J Pharmacol 17:315–317

Henderson G, Hughes J (1976) The effects of morphine on the release of noradrenaline from the mouse vas deferens. Br J Pharmacol 57:551–557

Henning M (1969) Interactions of dopa decarboxylase inhibition with the effect of alpha methyldopa on blood pressure and tissue monoamines of rats. Acta Pharmacol Toxicol 27:135–148

Henning M (1969) Noradrenaline turnover in renal hypertensive rats. J Pharm Pharmacol 21:61–62

Henning M, Rubenson A (1970) Evidence for a centrally mediated hypotension effect of L-dopa in the rat. J Pharm Pharmacol 22:241–243

Henning M, van Zwieten PA (1968) Central hypotensive effect of alpha methyldopa. J Pharm Pharmacol 20:409–417

Henry JL, Calaresu FR (1974) Pathways from medullary nuclei to spinal cardioacceleratory neurons in the cat. Exp Brain Res 20:505–514

Henry JP, Stevens PM, Santisteban GA (1975) A model of psychosocial hypertension showing reversibility and progression of cardiovascular complications. Circ Res 36:156–164

Hilton SM (1963) Inhibition of baroreceptor reflexes on hypothalamic stimulation. J Physiol (Lond) 165:56–57p

Hilton SM, Spyer KM (1971) Participation of the anterior hypothalamus in the baroreceptor reflex. J Physiol (Lond) 218:279–293

Hoffman BB, Lefkowitz RJ (1980) Alpha adrenergic receptor subtypes. N Engl J Med 302:1390–1396

Hofman A, Boomsma F, Schalekamp MAD, Valkenburg HA (1979) Raised blood pressure and plasma noradrenaline concentrations in teenagers and young adults selected from an open population. Br Med J 1:1536–1538

Hökfelt T, Fuxe K, Goldstein M, Johansson O (1973) Evidence for adrenergic neurones in the rat brain. Acta Physiol Scand 89:286–288

Hökfelt T, Fuxe K, Goldstein M, Johannson O (1974) Immunological evidence for the existence of adrenaline neurons in the rat brain. Brain Res 66:235–251

Hökfelt B, Hedeland H, Hanson B-G (1975) The effects of clonidine and penbutolol on catecholamines in blood and urine, plasma renin activity and urinary aldosterone in hypertensive patients. Arch Int Pharmacodyn 213:307–321

Hökfelt T, Elde R, Johannson O (1978) Psychopharmacology: a generation of progress (ed. Lipton, Di Mascio and Killan) 39–66, New York, Raven Press

Hunyor SN, Bradstock K, Sommerville PJ, Lucas N (1975) Clonidine overdose. Br Med J, IV, 23

Iriuchijima J (1973) Role of splanchnic nerves in spontaneously hypertensive rats. Jap Circ J 37:1251–1253

Iversen LL, Salt PJ, Wilson HA (1972) Inhibition of catecholamine uptake in the isolated rat heart by haloalkylamines related to phenoxybenzamine. Br J Pharmacol 46:647–657

Jauernig RA, Moulds RFW, Shaw J (1978) The action of prazosin in human vascular preparations. Arch Int Pharmacodyn 231:81–89

Jones DH, Hamilton CA, Reid JL (1978) Plasma noradrenaline, age and blood pressure: a population study. Clin Sci Mold Med 55:73–75s

Jones DH, Allison DJ, Hamilton CA, Reid JL (1979a) Selective venous sampling in the diagnosis and localization of phaeochromocytoma. Clin Endocrinol 10:179–186
Jones DH, Hamilton CA, Reid JL (1979b) Choice of control groups in the appraisal of sympathetic nervous activity in essential hypertension. Clin Sci 57:339–344
Jones DH, Daniel J, Hamilton CA, Reid JL (1980a) Plasma noradrenaline concentration in essential hypertension during long term β-adrenoceptor blockade with oxprenolol. Br J Clin Pharmacol 9:27–31
Jones DH, Reid JL, Hamilton CA, Allison DJ, Welboum RB, Dollery CT (1980b) Biochemical diagnosis, localization and follow up of phaeochromocytoma: the role of plasma and urinary catecholamine measurements. Quart J Med 49:341–361
Jones JV, Hallbäck M (1978) Cardiovascular reactivity and design in rats with experimental "neurogenic hypertension". Acta Physiol Scand 102:41–49
Joy MD, Lowe RD (1970) Evidence that the area postrema mediates the central cardiovascular response to angiotensin II. Nature, 228:1303–1304
Karplus JP, Kreidl A (1927) Gehirn und Sympathicus. VII. Über Beziehungen der Hypothalamus Zentren zu Blutdruck und innerer Sekretion. Pflügers Arch 215:667–672
Kobinger W (1975) Central cardiovascular actions of clonidine In: Davies DS, Reid JL (Eds) Central Actions of drugs in blood pressure regulation. U.K. Pitmans Medical Tunbridge, Wells, pp 101–193
Kobinger W, Walland A (1967) Investigation into the mechanism of hypotensive effect of 2, (2,6 dichlorophenylamino) 2-imidazoline HCl. Europ J Pharmacol 2:155–162
Kontos HA, Richardson DW, Norwell JE (1975) Mechanisms of circulatory dysfunction in orthostatic hypotension. Trans Am Clin Climatol Assoc 87:26–35
Korner PI (1971) Integrative neural cardiovascular control. Physiol Rev 51:312–367
Kostozewa RM, Jacobowitz DM (1974) Pharmacological actions of 6-hydroxydopamine. Pharmacol Rev 26:200–288
Lake CR, Zeigler MG, Coleman MD, Kopin IJ (1977) Age adjusted plasma norepinephrine levels are similar in normotensive and hypertensive subjects. N Engl J Med 296:208–208
Langer SZ (1974) Presynaptic regulation of catecholamine release. Biochem Pharmacol 23:1793–1800
Langer SZ (1977) Presynaptic receptors and their role in the regulation of transmitter release. Br J Pharmacol 60:481–498
Laubie M, Delbarre B, Bogaievsky Y, Bogaievsky D, Tsoucaris Kupfer D, Senon D, Schmitt H, Schmitt H (1976) Pharmacological evidence for a central sympathomimetic mechanism controlling blood pressure and heart rate. Circ Res 38 Suppl 2:35–41
Laubie M, Schmitt H, Vincent M, Remond G (1977) Central cardiovascular effects of morphinometic peptides in dogs. Eur J Pharmacol 46:67–71
Le Quan Bui KH, Elghozi J-L, Devynck M-A, Meyer P (1980) Early changes in noradrenaline content of some brain nuclei in spontaneously hypertensive rats. Clin Sci 59 Suppl 6:243–246
Lewis P (1976) The essential actions of propranolol in hypertension. Am J Med 60:837–852
Lewis PJ, Dargie H, Dollery CT (1975) Role of saline consumption in the prevention of deoxycorticosterone hypertension in rats by central 6-hydroxydopamine. Clin Sci Mol Med 48:327–330
Limas C, Limas CJ (1978) Reduced number of β-adrenergic receptors in the myocardium of spontaneously hypertensive rats. Biochem Biophys Res Comm 83:710–714

Loewy AD, McKellar S, Swensson EE, Panneton WM (1980) Onset of hypertension in spontaneously hypertensive rats despite the depletion of spinal cord catecholamines. Brain Res 185:449-454

Louis WJ, Krauss KR, Kopin IJ, Sjoerdsma A (1970) Catecholamine metabolism in hypertensive rats. Circ Res 27:589-594

Louis WJ, Doyle AE, Anavekar SN (1973) Plasma norepinephrine levels in essential hypertension. N Engl J Med 288:599-601

Lovenberg W, Yamabe H, deJong W, Hansen CT (1973) Genetic variation of the catecholamine biosynthetic enzyme activities in various strains of rats including the spontaneously hypertensive rat. In: Usdin E, Snyder S (Eds) Frontiers in Catecholamine Research. Pergamon Press Oxford, pp 891-895

McAllen RM (1976) Inhibition of the baroreceptor input to the medulla by stimulation of the hypothalamic defence area. J Physiol (Lond) 257:45p

McAllen RM, Spyer KM (1976) The location of cardiac preganglionic motoneurones in the medulla of the cat. J Physiol (Lond) 258:187-204

McCarty R, Chiueh CC, Kopin IJ (1978) Spontaneously hypertensive rats: adrenergic hyperresponsitivity to anticipation of electric shock. Behav Biol 23:180-188

McCubbin JW, Page IH (1963) Renal pressor system and neurogenic control of arterial pressure. Circ Res 12:553-559

Maling TJB, Ferrana A, Mucklow JC, Reid JL, Hamilton CA, Dollery CT (1979) Blood presure and plasma noradrenaline during single high dose beta adrenoceptor blockade. Eur J Clin Pharmacol 15:375-379

Manger WM, Gifford RW (1977) Phaeochromocytoma. New York, Springer Verlag

Mellander S, Johansson B (1968) Control of resistance, exchange and capacitance function in the peripheral circulation. Pharmacol Rev 20:117-196

Meyer P, Schmitt H (1978) Nervous System and Hypertension. John Wiles & Sons, New York

Miura M, Reis DJ (1969a) Termination and secondary projections of carotid sinus nerve in the cat brain stem. Am J Physiol 217:142-153

Miura M, Reis DJ (1969b) Cerebellum: a pressor response elicited from the fastigial nucleus and its efferent pathway in the brianstem. Brain Res 13:595-599

Miura M, Reis DJ (1970) A blood pressure response from fastigial nucleus and its relay pathway in the brainstem. Am J Physiol 219:1330-1336

Miura M, Reis DJ (1972) Role of the solitary and paramedian reticular nuclei in mediating cardiovascular reflex responses from carotid baro- and chemoreceptors. J Physiol (Lond) 223:525-548

Modlin IM, Farndon JR, Shephers A, Johnston IDA, Kennedy TL, Montgomery DAA, Welboum RB (1979) Phaeochromocytoma in 72 patients, clinical and diagnostic features, treatment and long term results. Br J Surg 66:456-465

Mroczek WJ, Davidov M, Finnerty FA (1973) Intravenous clonidine in hypertensive patients. Clin Pharm Ther 14:847-951

Mueller RA, Thoenen H, Axelrod J (1969) Adrenal tyrosine hydroxylase: compensatory increase in activity after chemical sympathectomy. Science. 163:468-469

Murray S, Jones DH, Davies DS, Dollery CT, Reid JL (1977) Free and total 3 methoxy 4 hydroxy phenylethylene glycol in human cerebrospinal fluid. Relationship to blood pressure in hypertensives and patients with neuropsychiatric disease. Clin Chim Acta 79:63-68

Myers MG, Lewis PJ, Reid JL, Dollery CT (1975) Brain concentration of propranolol in relation to hypotensive effect in the rabbit with observations on brain propranolol levels in man. J Pharmacol Exp Ther 192:327-335

Nagatsu T, Ikuta K, Numata Y, Kato T, Sano M, Nagatsu I, Takeuchi (1976) Vascular

and brain dopamine β-hydroxylase in genetic hypertensive rats. Science. 191:290-291
Naik RB, Mathias CJ, Wilson CA, Reid JL, Warren DJ (1981) Cardiovascular and autonomic reflexes in haemodialysis patients. Clin Sci 60:165-170
Nakamura K, Nakamura K (1978) Role of brain stem and spinal noradrenergic and adrenergic neurons in the development and maintenance of hypertension in spontaneously hypertensive rats. Nannyn-Schmiedeberg's Arch Pharmacol 305:127-133
Nakamura K, Gerald M, Thoenen H (1971) Experimental hypertension in the rat: reciprocal changes in norepinephrine turnover in the heart and brain stem. Nannyn-Schmiedeberg's Arch Pharmacol 268:125-135
Nakaoka A, Lovenberg W (1976) Plasma norepinephrine and dopamine beta hydroxylase in genetic hypertension. Life Sci 19:29-34
Nakaoka A, Lovenberg W (1977) Regional changes in the activities of aminergic biosynthetic enzymes in the brains of hypertensive rats. Eur J Pharmacol 43:292-306
Nanda RN, Johnson RH, Keogh HJ (1976) Treatment of neurogenic orthostatic hypotension with a monoamine oxidase inhibitor and tyramine. Lancet ii:1164-1167
Nathan MA, Reis DJ (1977) Chronic labile hypertension produced by lesions of the nucleus tractus solitarii in the cat. Circ Res 40:72-81
Natrajan PEG, McGarrigle HHG, Lawrence DM, Lachelin GCL (1982) Plasma noradrenaline and adrenaline levels in normal pregnancy and in pregnancy-induced hypertension. Br J Obstet Gynaecol 89:1041-1045
Neubauer B, Christensen NJ (1976) Norepinephrine, epinephrine and dopamine contents of the cardiovascular system in long term diabetes. Diabetes 25:6-10
Neumayr RJ, Hare BD, Franz DN (1974) Evidence for bulbospinal control of sympathetic preganglionic neurones by monoaminergic pathways. Life Sci 14:793-806
Nickerson M (1957) Nonequilibrium drug antagonism. Pharmacol Rev 9:246-259
Nickerson M, Goodman LS (1947) Pharmacological properties of a new adrenergic blocking agent. J Pharmacol Exp Ther 89:167-185
Nickerson M, Gump WS (1949) The chemical basis for adrenergic blocking activity in compounds related to dibenamine. J Pharmacol Exp Ther 97:25-47
Nijkamp FP, de Jong W (1977) Centrally induced hypotension by methyldopa and methylnoradrenaline in normotensive and renal hypertensive rats. Prog Brain Res 47:349-368
Norberg KA (1967) Transmitter histochemistry of the sympathetic adrenergic nervous system. Brain Res 5:125-170
Okamato K, Aoki K (1963) Development of a strain of spontaneously hypertensive rats. Jap Circ J 27:282-293
Okamoto K, Nosaka S, Yamori Y, Matsamoto M (1967) Participation of a neural factor in the pathogenesis of hypertension in the spontaneous hypertensive rat. Jap Heart J 8:168-180
Okamoto K, Yamon Y, Nagaoka A (1974) Establishment of the stroke prone spontaneously hypertensive rat (SHR). Circ Res 34, 35, Suppl 1:143-153
Oliver G, Schafer EA (1895) The physiological effects of extracts of the suprarenal capsules. J Physiol (Lond) 18:230-276
Palkovits M, Zaborsky L (1977) Neuroanatomy of central cardiovascular control. Prog Brain Res 47:9-34
Peach MJ, Bumpus FM, Khairallah PA (1969) Inhibition of norepinephrine uptake in hearts by angiotensin II and analogs. J Pharmacol Exp Ther 167:291-288
Pedersen E, Christiansen N (1975) Catecholamines in plasma and urine in patients with essential hypertension determined by double isotope derivative techniques. Acta Med Scand 198:373-377

Pedersen EB, Rasmussesn AB, Christensen NJ (1982) Plasma noradrenaline and adrenaline in pre-eclampsia, essential hypertension in pregnany and normotensive pregnant control subjects. Acta Endocrinologica 99:594-600

Petty MA, Reid JL (1977) Changes in noradrenaline concentration in brain stem and hypothalamic nuclei during the development of renovascular hypertension. Brain Res 136:376-380

Petty MA, Reid JL (1979) Catecholamine synthesizing enzyme in brain stem and hypothalamus during the development of renovascular hypertension. Brain Res 163:277-288

Philipp T, Distler A, Dordes U (1978) Sympathetic nervous system and blood pressure control in essential hypertension. Lancet 2:959-963

Philippu A, Roensberg W, Przuntek H (1973) Effects of adrenergic drugs on pressor responses to hypothalamic stimulation. Nannyn-Schmiedeberg's Arch Pharmacol 278:373-386

Pickering GW (1955) High Blood Pressure. Churchill London

Popper CW, Chiueh CC, Kopin IJ (1977) Plasma catecholamine concentrations in unanaesthetized rats during sleep, wakefulness, immobilization and âfter decapitation. J Pharmacol Exp Ther 202:144-148

Potter WP, Chubb IW, Put A, Schaepdryver AF (1971) Facilitation of the release of noradrenaline and dopamine-β-hydroxylase at low stimulation frequencies by α-blocking agents. Arch Int Pharmacodyn Ther 193:191-197

Pritchard BNC, Gillam PMS (1969) Treatment of hypertension with propranolol. Br Med J 1, 7

Provoost AP, de Jong W (1978) Differential development of renal, doca-salt and spontaneous hypertension in the rat after neonatal sympathectomy. Clin Exp Hypertension 1:177-189

Rascher W, Dietz R, Schömig A, Weber J Gross F (1980) Plasma catecholamine levels and vascular response in deoxycorticosterone salt hypertension of rats. Clin Sci 59 Symph 6:315-318

Reid JL (1979) Central action of antihypertensive drugs. In: Melmon K (Ed) Drug therapeutics: concepts for physicians. Elsevier, New York pp 135-150

Reid JL, Lewis PJ, Myers MG, Dollery CT (1974) Cardiovascular effects of intracerebroventricular d, l, and dl propranolol in the conscious rabbit. J Pharmacol Exp Ther 188:294-299

Reid JL, Zivin JA, Kopin IJ (1975) Central and peripheral adrenergic mechanisms in the development of DOCA saline hypertension in the rat. Circ Res 37:569-579

Reid JL, Dargie HJ, Franklin SS, Fraser B (1976) Plasma noradrenaline and renovascular hypertension in the rat. Clin Sci Mol Med 51:439-442S

Reid JL, Wing LMH, Mathias CJ, Frankel JH, Neill E (1977a) The central hypotensive effect of clonidine studies in tetraplegic subjects. Clin Pharmacol Ther 21:375-381

Reid JL, Wing LMH, Dargie HJ, Hamilton CA, Davies DS, Dollery CT (1977b) Clonidine withdrawal in hypertension. Lancet 1:1171-1174

Reid JL, Tangri KK, Wing LMH (1977c) The central hypotensive action of clonidine and propranolol in animals and man. Prog Brain Res 47:369-384

Reid JL, Mathias CJ, Jones DH, Wing LMH (1980) The contribution of central and peripheral adrenoceptors to the action of clonidine and alpha methyldopa in man. In: Fuxe K, Goldstein M, Hökfelt B, Hökfelt T (Eds) Central adrenaline neurons: basic aspects and their role in cardiovascular functions. Wennergren Symposium 33. Pergamon Press Oxford

Reid JL, Rubin PC, Petty MA (1984) Opioid peptides and central control of blood pressure. Clin Exp Hypertension A6:107-120

Reis DJ, Doba N (1974) The central nervous system and neurogenic hypertension. Prog Cardiovasc Dis 17:51-69
Reis DJ, Fuxe K (1968) Adrenergic innervation of the carotid sinus. Am J Physiol 215:1054-1057
Riley CM, Day RL, Greeley DM, Langford WS (1949) Central autonomic dysfunction with defective lacrimation. Report of five cases. Paediatrics 3:468-478
Robinson RL (1967) Stimulation of the catecholamine output of the isolated perfused adrenal gland of the dog by angiotensin and bradykinin. J Pharmacol Exp Ther 156:252-257
Roizen MF, Weise V, Grobecker H, Kopin IJ (1976) Plasma catecholamines and dopamine β-hydroxylase activity in spontaneously hypertensive rats. Life Sci 17:283-288
Rubin PC (1983) Hypertension occurring during pregnancy. In Birkenhager WH, Reid JL (Eds) Handbook of Hypertension, Vol. 2. Clinical Aspects of Secondary Hypertension Elsevier Science Publishers BV pp 304-318
Rubin PC (1984) Opioid peptides in blood pressure regulation in man. Clin Sci 66:625-630
Rubin P, Blaschke T (1980) Studies on the clinical pharmacology of prazosin. I. Cardiovascular, endocrine and catecholamine response to a single dose. Br J Clin Pharmacol 10:23-32
Rubin PC, Reid JL (1984) Strategies for dynamic assessment of sympathetic activity from plasma catecholamine measurements. In: Ziegler MG, Lake CR (Eds) Norepinephrine Williams and Wilkins, Baltimore, pp 209-216
Rubin PC, Blaschke T, Guilleminault C (1981) The influence of naloxone, a specific opioid antagonist, on blood pressure changes during sleep. Circulation 63:117-121
Rubin PC, Butters L, Reid JL (1983a) Plasma noradrenaline in pregnancy-associated hypertension. Clin Exp Hypertension B2:421-428
Rubin PC, McLean K, Reid JL (1983b) Endogenous opioids and baroreflex control in man. Hypertension 5:535-538
Saavedra JM (1979) Brain catecholamines during development of doca salt hypertension in rats. Brain Res 179:121-127
Saavedra JM, Grobecker H, Axelrod J (1976) Adrenaline forming enzyme in brainstem: elevation in genetic and experimental hypertension. Science 191:483-484
Saavedra JM, Grobecker H, Axelrod J (1978) Changes in central catecholaminergic neurons in the spontaneously (genetic) hypertensive rat. Circ Res 42:529-534
Saavedra JM, del Carmine R, Awai J, Alexander N (1979) Catecholamines in discrete areas of the rat brain in different forms of genetic and experimental hypertension. In: da Prada M, Albertini, Peskar BA (Eds) Radioimmunoassay of drugs and hormones in cardiovascular medicine. Elsevier, North Holland
Saran RK, Sahuja RC, Gupta NN, Hasan M, Bhargava KP, Shanker K, Kishor K (1978) 3-Methoxy-4-hydroxyphenylglycol in cerebrospinal fluid and vanillyl mandelic acid in urine of humans with hypertension. Science 200:317-318
Sattler RW, van Zwieten PA (1967) Acute hypotensive action of ST155 after infusion into the cat's vertebral artery. Eur J Pharmacol 2:9-13
Schanberg SM, Schildkraut JJ, Breese GR, Kopin IJ (1968) Metabolism of normetanephrine H^3 in rat brain—Identification of conjugated 3-methoxy-4-hydroxyphenylglycol as the major metabolite. Biochem Pharmacol 17:247-254
Scher AM (1977) Carotid and aortic regulation of arterial blood pressure. Circulation 56:521-528
Schmitt H, Schmitt H (1969) Localization of the hypotensive effect of 2(2,6-dichlorophenylamino)-2-imidazoline hydrochloride. Eur J Pharmacol 6:8-12
Schmitt H, Schmitt H, Fenard S (1973) Action of α-adrenergic blocking drugs on the

sympathetic centres and their interactions with the central sympatho-inhibitory effect of clonidine. Arzneim Forsch (Drug Res) 23:40-45

Scriabine A, Constantine JW, Hess HJ, McShare WC (1968) Pharmacological studies with some new antihypertensive aminoquinazolines. Experientia 24:1150-1152

Seller H, Illert M (1969) The localisation of the first synapse in the carotid sinus baroreceptor reflex pathway and its alteration of the afferent input. Pflügers Arch 306:1-19

Sever PS, Birch M, Osikowska B, Tunbridge RDG (1977) Plasma noradrenaline in essential hypertension. Lancet 1:1078-1081

Shoulson I, Ziegler M, Lake R (1976) Huntingdon's disease: determination of plasma norepinephrine and dopamine-β-hydroxylase. Neuroscience 2:800

Shy GM, Drager GA (1960) A neurological syndrome associated with postural hypotension: a clinical-pathological study. Arch Neurol 2:511-527

Sinha JN, Atkinson JM, Schmitt H (1973) Effect of clonidine and L-dopa on spontaneous and evoked splanchnic nerve discharges. Eur J Pharmacol 24:113-119

Sjöqvist B, Änggård E, Widerlov E, Lewander T (1979) Approaches to study catecholamine turnover in man using in vivo deuterium labelling. In Usdin E, Kopin IJ Barchas J (Eds) Catecholamines: Basic and Clinical Frontiers. Pergamon Press, Oxford

Smith AA, Dancis J (1964) Exaggerated response to infused norepinephrine in familial dysautonomia. N Engl J Med 270:704-707

Snyder, DW, Nathan MA, Reis DJ (1978) Chronic lability of arterial pressure produced by selective destruction of the catecholamine innervation of the nucleus tractus solitarii in the rat. Circ Res 43:662-671

Starke K, Endo T, Taube HD (1975) Relative pre- and postsynaptic potencies of alpha adrenoceptor agonists in rabbit pulmonary artery. Naunyn-Schmiedeberg's Arch Pharmacol 291:55-78

Stone EA (1976) Central noradrenergic activity and the formation of glycol sulphate metabolites of brain norepinephrine. Life Sci 19:1491-1498

Struyker-Boudier HAJ, Beker A (1975) Adrenaline-induced cardiovascular changes after intrahypothalamic administration to rats. Eur J Pharmacol 31:153-155

Struyker-Boudier HAJ, Smeets GWM, Brouwer GM, van Rossum JM (1975) Central nervous system adrenergic mechanisms and cardiovascular regulation in rats. Arch Int Pharmacodyn 213:285-293

Sundlöf G, Wallin BG (1978) Human muscle nerve sympathetic activity at rest. Relationship to blood pressure and age. J Physiol (Lond) 274:621-637

Taylor AA, Pool JL, Lake CR, Ziegler MG, Rosen RR, Rollins DE, Mitchel JR (1978) plasma norepinephrine concentration. No differences among normal volunteers and low, high or normal renin hypertensives. Life Sci 22:1499-1510

Thomas DM, Kaufman REP, Sprague JM (1956) Experimental studies of the vermal cerebellar projection in the brainstem of the cat. J Anat (Lond) 90:371-385

Tunbridge RDG, Donnai P (1981) Plasma noradrenaline in normal pregnancy and in hypertension of late pregnancy. Br J Obstet Gynaecol 88:105-108

van Ameringen MA, de Champlain J, Imbeanet S (1977) Participation of central noradrenergic neurons in experimental hypertension. Can J Physiol Pharmacol 55:1246-1251

van der Gugten J, Palkovits M, Wijnen HLJM, Versteeg DHG (1976) Regional distribution of adrenaline in rat brain. Brain Res. 107:171-175

van Zwieten PA (1975) Antihypertensive drugs with a central action. Progress in Pharmacology 1:1-63

Versteeg DHG, Palkovits M, van der Gugten J, Wijnen HLMLM, Smeets GWM, de

Jong W (1976a) Catecholamine content of individual brain regions of spontaneously hypertensive rats. Brain Res 112:429-434
Versteeg DHG, van der Gugten J, de Jong W, Palkovits M (1976b) Regional concentration of noradrenaline and dopamine in the rat brain. Brain Res 113:563-174
Volicer L, Scheer E, Hilse H, Visweswaram D (1968) The turnover of norepinephrine in the heart during experimental hypertension in rats. Life Sci 7:525-532
von Euler US (1946) A specific sympathomimetic ergone in adrenergic nerve fibres and its relation to adrenaline and noradrenaline. Acta Physiol Scand 12:73-97
Wallin BG, Delius W, Hagbarth KE (1973) Comparison of sympathetic activity in normo and hypertensive subjects. Circ Res 33:9-21
Watson RDS, Hamilton CA, Reid JL, Littler WA (1979) Changes in plasma norepinephrine blood pressure and heart rate during physical activity in hypertensive man. Hypertension 1:341-346
Wijnen HJLM, Versteeg DHG, Palkovits M, de Jong W (1977) Incereased adrenaline content of individual nuclei of the hypothalamus and the medulla oblongata of genetically hypertensive rats. Brain Res 135:180-185
Wijnen HJLM, Spierenburg HA, de Kloe R, de Jong W, Versteeg DHG (1980) Decrease in noradrenergic activity in hypothalamic nuclei during the development of spontaneous hypertension. Brain Res. 184:163-162
Willard PW, Fuller RW (1969) Functional significance of the sympathetic nervous system in production of hypertension. Nature 223:417-418
Williams LT, Snyderman R, Lefkowitz RJ (1976) Identification of adrenergic receptors in human lymphocytes by (−)-(^{3}H)-alprenolol binding. J Clin Invest 57:149-155
Wing LMH, Reid JL, Hamilton CA, Sever P, Davies DS, Dollery CT (1977) Effects of clonidine on biochemical indices of sympathetic function and plasma renin activity in normotensive man. Clin Sci Mol Med 53:45-53
Woodstock EA, Funder JW, Johnston CI (1979) Decreased cardiac adrenergic receptors in Doca salt and renal hypertensive rats. Circ Res 45:660-665
Yamaguchi J, de Champlain J, Nadeau RA (1977) Regulation of norepinephrine release from cardiac sympathetic fibres in the dog by presynaptic alpha and beta receptors. Circ Res 41:108-117
Zaimis E, Hanington E (1969) A possible pharmacological approach to migraine. Lancet 3:298-300
Zandberg P, de Jong W, de Wied D (1979) Effect of catecholamine receptor stimulating agents on blood pressure after local application in the nucleus tractus solitarii of the medulla oblongata. Eur J Pharmacol 55:43-56
Ziegler MG, Lake CR, Kopin IJ (1976) Deficient sympathetic nervous response in familial dysautonomia. N Engl J Med 294:630-633
Ziegler MG, Lake CR, Foppen FH, Shoulson I, Kopin IJ (1976) Norepinephrine in cerebrospinal fluid. Brain Res 108:436-440
Ziegler MG, Lake CR, Kopin IJ (1977) The sympathetic nervous system defect in primary orthostatic hypotension. N Engl J Med 296:293-297
Ziegler M, Lake CR, Wood JH, Brooks B (1980) Relationship between cerebrospinal fluid, norepinephrine and blood pressure in neurologic patients. Clin Exp Hypertension 2:995-1008
Zuspan FP (1979) Catecholamines—their role in pregnancy and the development of pregnancy induced hypertension. J Reprod Med 23:143-150

CHAPTER 18

Sympathomimetic Amines, β-Adrenoceptors and Bronchial Asthma

M. E. CONOLLY

A. Introduction

The sympathetic and parasympathetic nervous systems play a major, but still imperfectly understood, role in regulating the respiratory system. This fact is reflected in the widespread therapeutic use of sympathomimetic bronchodilators. If one includes ephedrine, adrenoceptor agonists have been used to treat asthma for thousands of years, and these agents continue to be the cornerstone of therapy for many patients. Following the recognition of the structure of adrenaline, a series of analogues have been developed over the last four decades, resulting in agents which are longer acting, orally active, and relatively selective for the β-2-adrenoceptor (though selectivity, of course, is never absolute). In this chapter the value of sympathomimetic bronchodilators as therapeutic agents, and the relative merits of the newer β-2-selective derivatives will be reviewed. Consideration will also be given to the role such agents may have played in the sharp rise in asthma deaths recorded in England and Wales as well as some other countries in the 1960's.

Much attention has been focussed on concepts invoking defects in adrenoceptor function in the etiology and pathogenesis of asthma. Despite intensive study, no consensus of opinion has so far emerged. However the weight of evidence is against the notion of a primary β-adrenoceptor defect, and rather more in favor of the concept that prolonged stimulation of the adrenoceptors by endogenous and exogenous β-agonists, or some other process, such as heterologous desensitization (v.i.) is responsible for any apparent loss of β-adrenoceptor response. The role (if any) of β-adrenoceptor antibodies awaits clarification. There have also been reports which indicate an increase in α-adrenoceptor activity in asthmatics. The nature of this is not known, however, nor is it clear what might be the magnitude of its contribution to the disease.

Surprisingly, there have been very few reliable measurements of endogenous catecholamines at any stage of the disease, so that little is known of the adequacy of the adrenergic response to bronchoconstriction in asthmatics.

In a number of studies, steroids have been shown to alter β-adrenoceptor number and function, but the importance of this in determining the efficacy of steroids in the treatment of asthma is unknown. Nor is there much information on what effect they may exert on the α-adrenoceptor population in asthmatics.

When investigating so complex a situation as that presented by a patient with bronchial asthma, it is often exceedingly difficult to dissect out the contribution made by endogenous catecholamines, exogenous β-adrenoceptor agonists and the functional competence of the adrenoceptors themselves. The situation is made more complex by potent nonadrenergic factors which play a major role in determining the end result which we observe clinically. Among these are inflammatory processes and mucosal edema, mucus secretion and mucus plugging, together with cholinergic and possibly purinergic neuronal modulation of bronchomotor tone. For these reasons and despite all its limitations, in order to study β-adrenoceptor function in health and disease, the leukocyte has been widely used as a source of readily accessible human β-adrenoceptors. They can be studied repeatedly regardless of the severity of the patient's illness, with few ethical restraints. The limitations and utilization of this model and the data it has furnished will be discussed in this chapter.

B. Catecholamines

In normal subjects, the usual response to stress includes an increase in sympathetic activity, which finds expression in an increase in heart rate and blood pressure, hyperglycemia, pupillary dilatation and bronchodilatation. The fact that propranolol uniformly leads to exacerbation of the symptoms and signs of asthma suggests that adrenergic activity truly is important in maintaining airway patency in asthmatics. It is unfortunate therefore that there have been very few reports on the adrenergic (catecholamine) response to asthma under any circumstances, and the data that do exist contain apparent anomalies and contradictions which are due, at least in part, to unreliable analytical techniques. No studies have been undertaken to determine the adrenergic response to induced bronchospasm in normal subjects.

MORRIS et al. (1972) reported a series of asthmatic children in whom they measured urinary catecholamines fluorometrically both while the disease was in remission and also during periods of relapse. They found that during the periods of quiescence, adrenaline excretion rates were significantly higher in asthmatics than in normal controls, the levels being 4.9 and 2.6 ng/min respectively, but, paradoxically, the excretion rate failed to increase significantly during exacerbations of their asthma which were severe enough to cause moderate distress. In sharp contrast, insulin-induced hypoglycemia in these same patients provoked a 7 to 10 fold increase in urinary adrenaline excretion rates, and the noradrenaline excretion rate rose by 40% ($p < 0.001$) at the same time. Interestingly, it was observed that during the hypoglycemic episode, presumably as a result of the increased catecholamine secretion, the asthmatic symptoms cleared and remained absent for some time after the test.

Several groups have examined various biochemical aspects of the sympathetically mediated response to exercise-induced asthma. BARBORIAK et al. (1973) found that the rise in free fatty acids was less in asthmatics than in normals, and that only a third of the asthmatics studied showed a rise in free fatty acids when adrenaline was infused. HARTLEY and DAVIES (1978) found that plasma cyclic AMP levels rose only 5% in asthmatics compared with 26% in normal subjects in response to exercise. They also noted that 18 days of administration of salbutamol aerosol in a dose of 1600 mg/day did not blunt the rise in plasma cyclic AMP in normal subjects.

All the early studies of noradrenaline relied upon fluorimetric assays. GRIFFITHS et al. (1972) found exercise-induced increases in plasma adrenaline and noradrenaline in asthmatic patients. The increases in normal subjects were significantly less than those seen in the asthmatics. However, despite their apparently greater catecholamine levels, the asthmatics showed a modest fall in the forced expiratory volume in one second (FEV_1) from 2.6 to 2.4 litres during exercise. ANDERSON et al. (1976) measured plasma noradrenaline in normal subjects and in patients with exercise-induced asthma. They could show no difference between the two groups either at rest or during exercise. The noradrenaline rose equally and significantly in both groups. Five minutes after exercise, when the peak expiratory flow rate had fallen by 47% in the asthmatics, plasma noradrenaline levels were found to be higher in the asthmatics than in the normal subjects, though both declined after the cessation of exercise, and the difference between the two was not significant. BEIL et al. (1977) reported a rise in noradrenaline greater in asthmatics than in normals, though surprisingly adrenaline levels did not change in either group. CHRYSSANTHOPOULOS et al. (1978) examined plasma noradrenaline and adrenaline in normal subjects and in patients with exercise-induced asthma. Neither at rest nor during exercise did the values obtained in asthmatics differ significantly from those seen in the controls.

Recently, BARNES et al (1980a) used a radioenzymatic assay to measure adrenaline and histamine levels in the plasma of patients with nocturnal asthma and in normal subjects. In both groups, an identical circadian rhythm of adrenaline concentrations in plasma was seen, with peak levels at 4 p.m. of 0.11 ng/ml, falling to around 0.04 ng/ml at 4 a.m. In asthmatics, as the adrenaline fell, the plasma histamine rose almost three hundred percent, and bronchoconstriction increased correspondingly. In contrast, normal subjects showed only a 50% increase in plasma histamine levels, and did not, of course, develop bronchospasm.

BARNES et al. (1981) also examined patients with exercise-induced asthma. They found a blunted rise in plasma noradrenaline in asthmatics where the level increased from 0.37 ng/ml to 0.76 ng/ml. In normal subjects, noradrenaline levels rose from 0.31 ng/ml to 1.67 ng/ml. The differences with adrenaline were even more striking, for levels which rose from 0.08 ng/ml to 0.25 ng/ml in normals did not change at all in asthmatics, in whom the peak expiratory flow rate fell by about 30%. Hyperventilation without exercise also caused a 29% fall in the peak expiratory flow rate in the asthmatics, but the bronchoconstriction did not provoke a rise in plasma catecholamines.

In a group of stable asthmatics, BARNES et al. (1982) found plasma histamine to be consistently 2-3 fold higher in asthmatics than in normal subjects (though the histamine levels did not correlate with the severity of their asthma), while plasma adrenaline and noradrenaline levels were the same as in normals.

At the present time, with little of the available data being based on reliable analytical techniques, it is difficult to reach firm conclusions. Intuitively one would expect asthmatics to secrete at least as much and possibly more catecholamines than normal subjects, even if this drive did not prove enough to overcome the bronchoconstriction, and the data of Griffiths et al. (1972), Anderson et al. (1976), and Beil et al. (1977) would support this view. On the other hand, the data of Morris et al. (1972) and more particularly Barnes et al. (1980a, 1981, 1982) with their superior methodology, seem to point in the opposite direction. It seems extraordinary that, in patients with nocturnal asthma, the plasma adrenaline should continue doggedly on with its downward circadian swing in the face of increasing bronchoconstriction. Similarly, the lack of an adrenergic response to exercise and to elevated histamine levels runs contrary to what one would intuitively predict. If confirmed, this might suggest either a defect of some peripheral sensing mechanism related to airway function, or a failure of some central mechanism to promote peripheral sympathetic activity. If this be the case it would seem to be a very specific defect, since Morris et al. (1972) showed a good adrenergic response to hypoglycemia, and asthmatics are not, as a matter of common observation, particularly subject to other manifestations of sympathetic failure such as postural hypotension.

I. β-Adrenoceptor Agonists

The modern use of sympathomimetics in the treatment of asthma began with the use of adrenaline sprays introduced in 1909. A wide range of synthetic agonists have been produced since isoprenaline first appeared in the 1940's (Konzett 1941a, b). It is beyond the scope of this chapter to examine all the information that has been published on these drugs and attention will be confined to those papers which deal with their clinical use. From the clinical point of view, these drugs are either essentially non-selective, such as isoprenaline and metaproterenol (orciprenaline), or show clinically significant degress (tenfold or more) of selectivity for the β_2-adrenoceptor, as do salbutamol, terbutaline, fenoterol, rimiterol, salmefamol and clenbuterol. No synthetic agonist which exhibits significant selectivity for the β_1-adrenoceptor has yet entered clinical use.

II. Structure Activity Relationship (Fig. 1)

The relationship between molecular configuration and pharmacological activity for adrenoceptor agonists in general is well known (Weiner 1980).

Loss of the catechol (3,4-dihydroxy) structure leads to diminution of both α- and β-agonist potency. Clearly, however, this is not absolute, since many of the drugs described here are resorcinols, and one is a saligenin. The side chain is a crucial feature. Optimal activity resides in those molecules whose side chain is a 2-C unit. The presence of the hydroxyl group on the β-carbon atom is generally regarded as essential, although dopamine which lacks this group still does have appreciable adrenergic activity. Isomerism occurs at this point, and biological activity resides predominantly in the (−)-isomer.

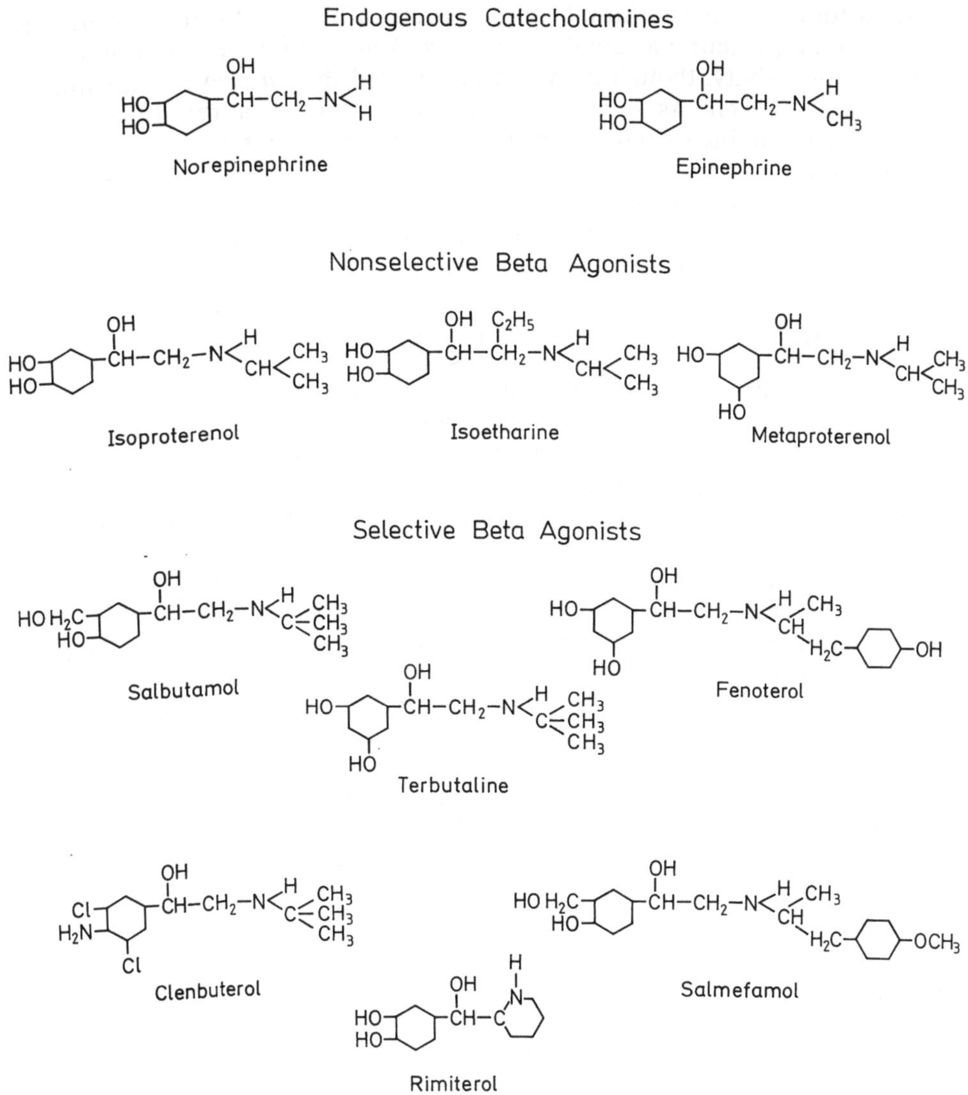

Fig. 1. Structures of β-adrenoceptor agonists. Dopamine, adrenaline and noradrenaline are naturally occurring compounds. The catechol structure is preserved in isoprenaline, isoetharine and rimiterol. Isoprenaline is the prototype non-β_2-selective agent. Isoetharine and the resorcinol derivative orciprenaline (metaproterenol) show no more than 2:1 selectivity for the β_2-receptor. Agents with larger substituent groups on the terminal amine moiety exhibit larger (10:1) and clinically useful degrees of β_2-selectivity

Substitution on the terminal amine group influences biological activity. Noradrenaline (if neuronal uptake is not blocked) exhibits predominantly α-adrenoceptor activity, though it should be stressed that *in vivo* it is an important β-agonist as well. As substitution increases through adrenaline (methyl group) to isoprenaline (isopropyl group), β-agonism increases. As substitution increases beyond this point to a tertiary butyl group (salbutamol, terbutaline) significant selectivity for the β-adrenoceptor appears. The substitution of an aromatic group between the amine group and the α-carbon atom (e.g. rimiterol) is a newer modification which is also associated with β_2-selectivity.

III. The Metabolism of β-Adrenoceptor Agonists

The metabolic fate of these compounds depends on their molecular configuration and the route of administration employed. Those drugs which are catecholamines share some of the metabolic fates of the endogenous compounds adrenaline and noradrenaline. However, the increased substitution on the terminal nitrogen either renders them immune to attack by monoamine oxidase or prevents neuronal uptake which might expose them to monoamine oxidase. Either way they give rise to no acidic urinary metabolites (CONOLLY et al. 1972). By implication, it follows that they are not dealkylated, since this would lead to the production of noradrenaline from which acidic metabolites could be formed. Where the catechol structure is preserved, such compounds can act as a substrate for catechol-O-methyltransferase. Thus a proportion of a dose of isoprenaline and isoetharine is metabolized to the 3-O-methyl derivatives. Many of the β-agonists in clinical use are not catecholamines; for instance, metaproterenol, terbutaline and fenoterol are resorcinols, while salbutamol [albuterol] is a saligenin. However, all of these compounds have phenolic hydroxyl groups which may be conjugated, most often with glucuronic or sulfuric acid—a fate which may also befall the 3-O methylated metabolites of the catecholamines. The route of administration may have important effects on the metabolism of a variety of drugs (DOLLERY et al. 1971). With respect to β-adrenoceptor agonists, this is particularly important. Following oral administration, a large fraction of the dose of drugs with the catechol structure is inactivated by conjugation in the gut wall (GEORGE et al. 1974), most commonly with sulfuric acid, before it ever reaches the systemic circulation. This has been reported for isoprenaline (CONOLLY et al. 1972), isoetharine (WILLIAMS et al. 1974), rimiterol (EVANS et al. 1974) and terbutaline (DAVIES et al. 1974). Salbutamol also undergoes conjugation during absorption from the gut, but the conjugating moiety does not appear to be glucuronic acid, sulfuric acid or β-glycoside, since only acid (but not enzymic) hydrolysis will liberate free salbutamol (LIN et al. 1972; EVANS et al. 1973). Gut wall conjugation is of major importance with respect to the clinical use of catecholamines, since approximately 90% of a dose administered by aerosol is deposited in the mouth and then swallowed (PATERSON et al. 1968). Administration of large doses of these drugs by mouth may also cause large, even dangerous, amounts of active drug to enter the systemic circulation by overloading the capacity of the conjugating enzymes (CONOLLY et al. 1972). It has been suggested (BENNETT et al.

1975) that competition by other drugs which are also substrates for the sulfate conjugation mechanism might deplete stores of available sulfate and lead to dangerous amounts of the β-agonist entering the systemic circulation. This remains, however, only a theoretical hazard having never been observed (or at least recognized) in clinical practice. The extent to which this conjugation takes place differs for each drug. Isoprenaline, for example, which is completely absorbed in normal doses, is virtually totally conjugated (CONOLLY et al. 1972) whereas, of that fraction of a dose of oral terbutaline which is absorbed (about 23 %), 75–90 % appears in the urine as a conjugate, the rest being excreted as the unchanged drug (DAVIES et al. 1974; HORNBLAD et al. 1976). An attempt has been made to prevent the conjugation of orally administered terbutaline by giving it in the form of a pro drug, the acyl diester, the hope being that the ester would be absorbed unchanged and then broken down by plasma diesterase to yield free terbutaline. Diisobutyryl terbutaline (ibuterol) has been studied in man (HORNBLAD et al. 1976; POULSÉN and PETERSEN 1976). This chemical modification was shown to increase the bioavailability only 1.6 fold. The peak clinical response (increase in FEV_1) occurred 30 minutes after dosing as opposed to 90 minutes when native terbutaline was used. However, for both preparations, the sulfate conjugate represented the major fraction of the urinary radioactivity. Although ibuterol was more consistently absorbed and gave a higher peak plasma level of terbutaline, the duration of action was shorter. As an oral preparation, therefore, the benefits of this new formulation appear to be marginal, and much the same conclusion appears to apply also to the use of this compound by inhalation (JOHNSON and CLARKE 1977).

The pattern of metabolism after intrabronchial administration is different. The lung has no conjugating mechanism but possesses catechol-O-methyl transferase (AXELROD 1959). Consequently, when catechols are given directly into the bronchial tree, a substantial fraction appears as the 3-O-methylated derivative (BLACKWELL et al. 1974) whereas the non-catechols are absorbed unaltered. Following absorption through the bronchial epithelium, conjugation with sulfuric acid or glucuronic acid (presumably in the liver), may then follow, prior to elimination in the urine (BRIANT et al. 1973).

IV. Clinical Pharmacology

The purpose of this section is to describe the clinical uses of β-adrenoceptor agonists, and, to that end, animal data and studies in normal subjects have been considered only where they serve to clarify a clinical situation. However, references are given to enable the reader to find information excluded from this presentation.

1. The Use of β-Adrenoceptor Agonists in Asthma

The mode of action of β-adrenergic bronchodilators is twofold. Firstly by direct action on bronchial smooth muscle they cause relaxation of the airways. Secondly, release of bronchoconstrictor mediators from mast cells, either within the bronchial lumen (PATTERSON et al. 1977), or in the lung parenchyma, is inhibited by β-receptor stimulation (SCHILD 1937; ORANGE et al. 1971). For many years these drugs have been the mainstay in the treatment of asthma, beginning in 1909 when MATTHEWS (1909) administered adrenaline as a nasal spray, and EPHRAIM (1910) instilled adrenaline into the bronchial tree via a bronchoscope. Delivery by oxygen powered nebulizer followed 20 years later (CAMPS 1929). Subsequently GRAESER and ROWE (1935) reported the beneficial use of a variety of hand operated nebulizers without significant side effects. NEILSEN (1936) reported further success (allegedly with negligible side effects) following inhalation of a 10% solution of adrenaline.

In 1940 the isopropyl analogue of adrenaline was synthesized (KONZETT 1941 a,b). The powerful bronchodilator properties of isoprenaline were soon recognized and it was introduced into the treatment of asthma, being given both parenterally and by aerosol inhalation. Its lack of α-adrenoceptor activity led to its rapid adoption as the standard adrenergic bronchodilator although adrenaline has continued in use to a limited extent even up to the present time, chiefly given subcutaneously in emergency situations.

Recent years have seen introduction of a variety of other β-adrenergic bronchodilators, namely metaproterenol [orciprenaline, "Alupent", (ENGELHARDT et al. 1961)], isotharine ["Bronkosol" (SIEGMUND et al. 1947)], albuterol [salbutamol (Syn. albuterol), "Ventolin" (CULLUM et al. 1969)] and terbutaline ["Bricanyl", "Brethine" (PERSSON and OLSSON 1970)], all of which are now in clinical use. More recent additions which are still in the clinical trial or early marketing phase include the catecholamine rimiterol (WG 253, "Pulmadil" (GRIFFIN and TURNER 1971; EVANS et al. 1974)], the resorcinol derivative fenoterol [Th 1165 a, "Berotec" (O'DONNELL 1970)] and salmefamol (AH 3923), an analogue of salbutamol (KENNEDY and DASH 1972).

These compounds differ from each other in various details but they also have in common several important properties which isoprenaline lacks. The non-catechols are active by mouth since they do not undergo as much sulfate conjugation in the gut wall. Likewise they are not subject to removal from the circulation by the uptake$_2$ mechanism [and it should be stressed that it is this mechanism and *not* metabolism by catechol-O-methyltranferase which determines the brief duration of effect of isoprenaline (LEVY and AHLQUIST 1971; CONOLLY et al. 1972)] so their action is relatively prolonged. Alone among the newer drugs, rimiterol has retained the catechol structure and it has a short duration of action whether given by inhalation (COOKE et al. 1974) or intravenously (MARLIN and TURNER 1975). Many of these compounds, particularly those which have groups larger than the isopropyl moiety substituted into the terminal amine group (Fig. 1), exhibit a degree of selectivity for the β_2-adrenoceptor and have appreciably less (at least 10 fold) potency with respect to those biological effects mediated via β_1-adrenoceptors, e.g. heart rate and force of contraction (LANDS et al. 1967).

The apparent selectivity of these agents may be further enhanced by giving them by inhalation (SPIRO et al. 1975; HETZEL and CLARK 1976) rather than orally or parenterally, for in this way a relatively high concentration is achieved in the respiratory tract, the effects of the swallowed dose being delayed and negligible (in the case of catecholamines it is virtually nonexistent). The convenience and acceptability of inhalation therapy has been greatly enhanced by the advent of pocket-sized pressurized aerosol dispensers, which use a mixture of halogenated hydrocarbons (freons) to propel an aerosol suspension of microfine crystals of the drug into the airstream during inhalation.

Pressurized aerosols are inadequate when asthma is severe and in this situation parenteral administration or inhalation therapy via a nebulizer with intermittent positive pressure ventilation (IPPV) is superior (CHOO-KANG and GRANT 1975).

It is not possible to provide more than a brief summary of the clinical use of these compounds in this chapter; the interested reader is referred to the papers cited below for more detail.

Isoprenaline has a pronounced action on all β-adrenoceptors, but is almost devoid of α-adrenoceptor activity. Its non selective β-agonist activity is its principal disadvantage when used to treat asthma, since effective bronchodilatation is always accompanied by marked cardiovascular side effects. The tachycardia is produced mainly by a direct action on the heart, although if the dose used is large enough to produce peripheral vasodilation with a fall in blood pressure, there will be a reflex withdrawal of vagal inhibitory tone. Its action on skeletal muscle gives rise to tremor, which even after aerosol inhalation can be quite pronounced, and, in combination with the perceptible tachycardia, may distress some patients. There is no clinically significant central nervous system stimulation. The metabolic consequences of isoprenaline in clinical situations are not important.

Isoprenaline is only suitable for inhalation or parenteral therapy because of sulfate conjugation in the gut wall (see above). When it is infused, bronchodilatation and increase in heart rate develop in a dose related manner simultaneously (PATERSON et al. 1971). However, even for isoprenaline, administration by inhalation achieves a modest degree of apparent selectivity. A number of pressurized aerosols delivering between 75 and 100 μg isoprenaline sulfate per metered dose are currently marketed, and are of established value in the routine control of asthmatic symptoms. For the more severely ill, administration of 0.5 % isoprenaline by nebulizer with IPPV is often effective, and is well established (WEINER 1980).

Metaproterenol (Orciprenaline) was the first alternative to isoprenaline to enter clinical use. Though often described as a β_2-selective agonist, this is open to question. Many of the clinical reports which claim to show a lesser cardiac effect have given small doses by inhalation. It is quite possible that, with its non-catechol structure, less of this small dose would be transported across the bronchial mucosa to affect the heart. Careful in-vitro studies do not support the contention that it has more than marginal β-selectivity (O'DONNELL 1970; BRITTAIN 1971; O'DONNELL 1972; JACK 1973; APPERLY et al. 1976). Rather, it appears that it has less effect on heart rate only because it is in all respects less potent than isoprenaline. Nevertheless, metaproterenol has been widely used as a bronchodilator, being effective whether given orally, by injection, or by inhalation. This effect is long lasting and is well maintained during prolonged therapy with recommended doses (DREWITT 1962; HURST 1973; KERR and GEBBIE 1973; SINHA et al. 1973; MCEVOY et al. 1973; KOIVIKKO 1974; REILLY et al. 1974; SOBOL and REED 1974; CHERVINSKY and CHERVINSKY 1975; BRANDON 1976; SACKNER et al. 1976).

Salbutamol was the first agent with noteworthy (more than 10 to 1) β_2-selectivity to be introduced into clinical practice. The dominant characteristics of salbutamol in clinical practice are its efficacy when given orally, parenterally or by inhalation, its prolonged effect and its relative selectivity for β_2-receptors. When given by intravenous infusion it is at least 10 times less active than isoprenaline in increasing the heart rate, though equipotent as a bronchodilator (WARRELL et al. 1970). Animal studies (CULLUM et al. 1969) suggest an even lower β_1-potency than that. Its use in the treatment of asthma has been widely described (CHOO-KANG et al. 1969; PARKER et al. 1971; SPITZER et al. 1972; WALKER et al. 1972; SINHA et al. 1973; CHOO-KANG et al. 1974; IGRAM et al. 1975; MAY et al. 1975). Much of the early data have been usefully summarized by FLETCHER et al. (1971).

When cumulative dose-response curves for the bronchodilator effect of pressurized aerosols of salbutamol and isoprenaline were examined, the response remained linear up to the maximum dose administered (500 μg) indicating that even at this dose the maximum bronchodilator response had probably not been achieved (WARRELL et al. 1970). Used in conventional doses of 100-200 μg by inhalation, increases of 40-50% in FEV_1 and FEV are commonly achieved (CHOO-KANG et al. 1969). Similar increases in FEV_1 have been reported after intravenous infusions (4 μg/min) or injections (200 μg over 5 minutes) (MAY et al. 1975; SPIRO et al. 1975). Hypokalaemia has been reported after parenteral salbutamol. LEITCH et al. (1976) observed a fall from 3.9 to 3.1 mmol/l after a 60 min infusion. The clinical importance of this is uncertain.

Although tolerance (tachyphylaxis) has been reported (CHOO-KANG et al. 1969) this does not appear to be clinically important when the drug is used correctly. Salbutamol has now become a reference compound against which newer drugs are compared, although, because of the development of benign smooth muscle tumors in rats, notwithstanding a decade of use in man elsewhere, its release in the USA was long delayed.

Terbutaline is a more recently developed agent with a pharmacological spectrum very similar to that of salbutamol (LEGGE et al. 1971). Since it does not possess the catechol structure it is effective when given by mouth. Parenteral preparations are also available where a fast response is sought (FREEDMAN 1971; CHERNACK et al. 1974; JOHANSEN 1974; SLY et al. 1975). Its duration of action is 4-6 hours, comparable to that of salbutamol. As with salbutamol, it has been found that delivery by inhalation (up to 8 μg) gives at least as good a bronchodilator response (52% increase in FEV_1) as intravenous infusions up to 0.8 μg kg/min (40% increase) while causing minimal cardiovascular side effects (THIRINGER and SVEDMYR 1976). Giving 5 mg by IPPV, SIMONSON et al.

(1976) observed a comparable improvement (47%) in ventilatory function (FEV_1) with negligible change in heart rate.

Several other selective β_2-agonists have undergone clinical evaluation. These include fenoterol (REBUCK and SAUNDERS 1972; DA COSTA and GOH 1973; MCLEOD and SELMAN 1973; POWLES 1975; ANDERSON et al. 1979; REBUCK and MARCUS 1979; TREMBATH et al. 1979; MILLER and RICE 1980; SASAKI et al. 1980; TANDON 1980), rimiterol (MACKAY and AXFORD 1977; MARLIN et al. 1977; PATERSON et al. 1977; PINDER et al. 1977; BLACKHALL et al. 1978; PHILLIPS et al. 1980), salmefamol (SILLETT et al. 1976a,b; DYSON and CAMPBELL 1977; MARMO et al. 1979) and clenbuterol (SALORINNE et al. 1975; ANDERSON and WILKINS 1977; KAMBUROFF et al. 1977; DELBONO et al. 1979; BARONTI et al. 1980). The published reports vary in their estimates of the relative potency of these newer agents, and the prominence or absence of side effects (muscle tremor being a clinically obvious feature of β_2-agonist activity). Under certain circumstances the short duration of action of rimiterol and the prolonged duration of action of fenoterol are claimed to be advantageous. However, if these newer agents have any major advantage over the bronchodilators already in place, it has yet to be demonstrated.

Caveat. Numerous factors contribute to airway narrowing. Smooth muscle contraction is an important and reversible element but wall thickening caused by inflammation and edema of the mucosa, (and, to a lesser extent, hypertrophy of the muscle and mucus glands), and occlusion of the lumen by plugs of mucus are other important factors. While β-adrenergic bronchodilators may reduce bronchospasm by several mechanisms (see above), the enthusiastic use of adrenergic bronchodilators must not lead to neglect of these other local factors, nor of an underlying respiratory tract infection, nor of systemic factors, such as dehydration. Neither must the physician let his reliance on these agents cause him to neglect adequate administration of steroids (see next section).

Footnote. In those clinical situations in which bronchospasm is precipitated by the inappropriate administration of β-adrenoceptor antagonists to patients with active or latent asthma, the clinician has several therapeutic options open to him. The first and most obvious is too withdraw the offending drug. Secondly, most physicians would give a theophylline preparation. Thirdly, since all clinically available β-adrenoceptor antagonists are competitive antagonists their effect could be overcome by administering a β-agonist bronchodilator by one of the routes described above. Additionally, in the treatment of hypertension and ischemic heart disease, the recent advent of angiotensin converting enzyme inhibitors such as captopril and enalapril, and of calcium channel blocking drugs such as nifedipine or verapamil is of far greater benefit than the development of the relatively β_1-selective β-adrenoceptor blockers (atenolol, metoprolol).

V. Reports of Death Associated with Sympathomimetic Bronchodilators

Adrenergic agents are of proven value in the treatment of bronchial asthma. However, a substantial number of reports have indicated that, if misused, these agents may produce a deterioration in the patient's asthmatic state. For the past 45 years, there has been a growing suspicion that there might be a link between overdosing with a sympathomimetic agent and death in asthma (Graeser 1939; Benson and Perlman 1948; Lowell et al. 1949).

Serious concern over the rising asthma mortality and its possible relation to therapeutic measures began to be felt from about 1964. In that year McMannis reported three patients who died suddenly following adrenaline injections given after excessive self medication with inhaled isoprenaline had failed to bring relief. The Australian Drug Evaluation Committee (Refshauge 1965) drew attention to the potential dangers of bronchodilator aerosols after reviewing the circumstance of death of five patients who had taken excessive doses of adrenaline, isoprenaline or metaproterenol.

Clinical and epidemiological reports by Smith (1966), Keighley (1966) Greenberg and Pines (1967), Van Metre (1967, 1969) and Gandevia (1968) all pointed to a continuously rising asthma mortality rate, although they did not specifically link this to misuse of sympathomimetic bronchodilators.

The most detailed investigation of the changing asthma death rate came from the United Kingdom, where this trend was most marked (Speizer et al. 1968a). Speizer and his colleagues noted that, beginning in 1961, there was a rise in the asthma death rate, particularly in the age range 5–34 years, the increase being greatest in patients aged 10–14 years, amongst whom the mortality rose from 0.33 to 2.46 per 100,000, an increase of 830%. When asthma death rates in other countries for the age range 5–34 were examined, an upward trend was seen in Japan (40%) and Australia (100%), but in other countries, which had a lower death rate to begin with (0.5 or less per 100,000), there was no increase. However, when the figures for the age range 10–14 years and 15–19 years were examined, a more pronounced pattern emerged. For Japan, Australia, New Zealand and Western Europe the death rate was found to have risen by 85–103% among the 10–14 year olds, and by 23–45% for patients aged 15–19 years. Increases in the United States were smaller, and evidence of a similar upward trend was lacking. Speizer and his colleagues eliminated the possibility that the increase in death rate was an artifact brought about by changing diagnostic criteria. Furthermore the number of cases classified as asthmatic on the basis of autopsy findings showed an even sharper increase than the cases so classified by clinicians. Morbidity figures relating to asthma indicated that the incidence of the disease in the population as a whole had not increased, so they concluded that a *true* increase in the case fatality rate of asthmatics had occurred. Although they were unable to test this postulate directly, Speizer and his colleagues put forward as supporting evidence the fact that among patients aged 5–34, the proportion of asthma deaths certified by coroners had increased from 35% in 1959 to 55% in 1966, which suggested that the mode of death from asthma among younger patients had in some way altered, becoming more often sudden and unexpected.

An explanation for this alteration was sought. Changes in atmospheric pollution did not seem to provide an adequate explanation since recent legislation had reduced smoke pollution, and in any case, the increase in asthma death rates in rural areas was found to be as marked as that in the urban population.

In considering the methods used to treat asthma, these authors observed that steroids had been introduced in 1952, whereas pressurized aerosols of sympathomimetic bronchodilators had been introduced only in 1960, and their use had increased four fold over the period of time that the asthma death rate had been increasing. This apparent correlation led to an investigation into the circumstances of death in 184 asthmatic patients, and the drugs used by these patients in their terminal illness (SPEIZER et al. 1968b). The investigators were able to obtain detailed information in 177 cases. Autopsy had been performed in 124 cases, and data from 113 (91%) of these were available to them. Death was regarded as sudden and unexpected in 80% of cases (137 out of 171). In 133 of these cases, where it was possible to estimate the duration of the terminal episode, 28% (37 out of 133) died within 24 hours of the onset of the asthmatic attack. In 110 of the cases with autopsy data (97%) the lungs were found to be over-distended, and in 106 cases (94%) the bronchi were extensively plugged with mucus.

Investigations into the numbers of patients receiving steroids or ACTH in the terminal illness showed that 50 out of 173 (29%) had never received steroids, and that 31 (18%) had had steroids previously but not in the fatal attack. A further 52 patients (30%) were kept on a moderate to low dose. That is to say, out of 173 patients, 77% appeared to have had little or no steroid therapy in the course of the fatal episode, even though 114 (66%) had been judged severe enough to receive steroid therapy at some earlier point in their history.

Inhaled adrenergic bronchodilators were taken by 150 out of 174 (86%), all but 4 using pressurized aerosols of which 108 (62%) contained isoprenaline, 33 (19%) contained metaproterenol, and 9 (5%) contained adrenaline. Out of 102 fatal cases previously classified as severe, 93 (91%) were known to have taken bronchodilators by inhalation, and in 85 of these death was sudden. Among 63 patients previously graded as not severe, 54 had taken bronchodilators by inhalation, and 34 had died suddenly. Approximately 60% of patients had taken ephedrine by mouth, either alone or in combination with theophylline and a sedative.

The fact that 66% of the cases reviewed in this study had received steroids at some time in their history led SPEIZER and his colleagues to conclude that as a group they had asthma of considerable severity, though the fact that 77% received little or no steroids in their final illness, and the observation that so many died suddenly at home suggested that these patients had been living closer to the limits of their ventilatory reserve than either they or their physician had realized. The close correlation between the rise in death rate and the increased use of pressurized aerosols was noted, although these authors felt that this correlation alone was insufficient evidence definitely to establish a causal link between the two.

During the period 1966–1969 the potential dangers of the misuse of sympathomimetic bronchodilators were widely recognized, and in the United Kingdom a warning notice was circulated to all doctors by the Commitee on Safety of Medicines, and the problem received extensive publicity. The rate of hospital admissions for patients with refractory asthma increased. Also, in December 1968, the sale of bronchodilators without a prescription was prohibited.

INMAN and ADELSTEIN (1969) again reviewed the asthma death figures for England and Wales, extending their observations to include the first quarter of 1969. It was confirmed by these authors that the previously reported rise in death rate was a genuine phenomenon, and that the methods of data collection had, if anything, led to an under-estimation of the size of this increase. They calculated that between 1961 and 1967, 3500 more patients had died from asthma than would have been predicted on the basis of the mortality figures for the preceding 10 years. Reviewing the trends in the the asthma death rate, and in consumption of bronchodilator aerosols since 1966, they noted a 25% fall in aerosol usage, and a much sharper fall in the death rate. This fell from an overall 400% increase over the 1952–1959 baseline, to only 50% above that level by the beginning of 1969. They considered it unreasonable to suggest that any factor, unrelated to therapy, and which could have caused the increase in the severity of asthma between 1961 and 1966, would suddenly have ceased to operate at exactly the same moment as excessive use of pressurized aerosols was discontinued, thereby producing a fall in the asthma death rate, coincident with, but unrelated to, the declining aerosol usage between 1966 and 1969. The only new therapeutic factor to appear in that time had been cromolyn sodium (disodium cromoglycate), the sales of which were then too small to have had any observable effect on the asthma death rate. Thus, it was argued that even if initially (at the time when SPEIZER and his colleagues first presented their data), the relationship between the increasing use of β-stimulants in aerosols and the rising death rate in asthmatics was not clear cut, the simultaneous down-turn in both aerosol usage and the asthma death rate between 1966 and 1969 constituted strong evidence for such a link. It is, however, important to acknowledge that, concomitant with the decline in use of bronchodilator aerosols (and use of these preparations did not decline as rapidly as the asthma death rate fell), there developed a much greater willingness on the part of physicians to prescribe steroids and to admit patients to hospital if an asthmatic attack proved resistant to bronchodilators. For this reason the decline in the asthma death rate cannot be considered as being due entirely to the more judicious use of sympathomimetic bronchodilators. Nevertheless, these preparations are a convenient and useful form of treatment which will continue to be widely used. It is therefore important to understand the mechanisms by which they might have contributed to the rise in death rate.

At one time it was suggested that the rise in the asthma death rate was confined to those countries which had allowed the high dose (500 µg isoprenaline/dose) pressurized aerosols on the market (STOLLEY 1972). However there appear to be too many exceptions to this postulate for it to be correct (HERXHEIMER 1972).

The aerosol propellants, which are chemically similar to halogenated hydrocarbon anesthetic agents (which are known to sensitize the heart to the arrhythmia-producing properties of catecholamines) also came under suspicion. However, even though these agents can be detected in the blood following inhalation (DOLLERY et al. 1970), it would be necessary for a patient to take a dose with each of 12–24 successive breaths for sensitizing blood levels of the freons to occur (DOLLERY et al. 1974), a most improbable circumstance. It seems likely therefore that, insofar as the bronchodilators *can* be implicated, it must be through some mechanism related to the receptors they stimulate, and this will be considered in the following section.

C. Adrenoceptors

I. The β-Adrenoceptor

Attention was first focused on the possible role of β-adrenoceptors in the pathogenesis of bronchial asthma by SZENTIVANYI (1968). He argued that asthma, atopic disease and several other pathological processes could be explained by assuming a primary defect in the β-adrenoceptor. Initial studies in asthmatics, relying on indirect measurements such as changes in free fatty acid levels, blood glucose or cardiopulmonary responses to infused β-agonists (COOKSON and REED 1963; INOUE 1967; KIRKPATRICK and KELLER 1967; LOCKEY et al. 1967) appeared to support this suggestion. As FURCHGOTT (1972) has pointed out however, studies designed to define and characterize receptors must be done in isolated tissues, not in intact subjects. In man this presents peculiar ethical difficulties, particularly if repeated studies are contemplated in sick patients. The only practical resolution of this problem stems from the discovery by SCOTT (1970) that human leukocytes contain an adenylate cyclase system which can be activated by several agents including β-adrenoceptor agonists, histamine and prostaglandin E_1(PGE_1), each working through specific cell surface receptors. As a model for studying disease of the respiratory system, the leukocyte (granulocyte or lymphocyte) prompts several obvious questions. Clearly, in anatomical terms, it bears no relationship to bronchial smooth muscle. However, other cells in the lung such as the mast cell are also to varying degrees involved in the asthmatic process, and leukocytes may more closely model such cells.

The cell surface β-receptor of the lymphocyte, of bronchial smooth muscle and of lung parenchyma characterized by response to a range of agonists and antagonists all appear to conform to the β_2 classification proposed in 1967 by LANDS et al. (CONOLLY and GREENACRE 1977; DAVIS et al. 1980). Assuming that all receptors of a particular type (e.g. β_2) would share equally in any inherent defect, and assuming that systemically administered drugs would affect all these receptors equally, the leukocyte would seem reasonable to use as a model for study of the β-adrenoceptor in asthma. There are some data which suggest that a reduction in receptor density in polymorphonuclear leukocytes may be truly drug induced, while that in lymphocytes may not be wholly drug related (SANO et al. 1979). As yet, this question has not been resolved and there is no agreement on whether granulocytes, lymphocytes or a mixed leukocyte population should be utilized. Most workers have opted for the lymphocyte in the belief that a more uniform cell population can be obtained. Being moreover a longer lived cell than the granulocyte, the lymphocyte would seem a better choice for studying the impact of a prolonged diseased state, or of the drugs administered for it, on receptor number and function. Most workers have used a modification of BOYUM'S (1968) density gradient centrifugation technique which yields 75–95% lymphocytes (a mixed T and B lymphocyte population), the rest being monocytes. Further separation into T and B fractions is time consuming, may yield cells of uncertain viability, and is probably not necessary since the pharmacological responsiveness of the two fractions has been reported to be very similar (BISHOPRIC et al. 1980). It has been our experience that platelet-generated prostaglandins may cause considerable problems where measurement of cyclic AMP is contemplated, since PGE_1 is so much more potent as an adenylate cyclase stimulant than isoprenaline. If platelets are traumatized during cell handling, or are allowed to aggregate and thereby to liberate PGE_1, background levels of cyclic AMP become so high that no response to isoprenaline can be detected. In our view it is necessary to guard against this by carefully removing as many platelets as possible in an initial centrifugal separation, and to maintain an anticoagulant concentration of EDTA throughout the entire procedure. Even with careful handling, the cyclic AMP response may show considerable variation within and between subjects (Fig. 2; CONOLLY and GREENACRE 1977). This variability may be overcome to some extent by expressing the changes as a percent of baseline values, but this does not eliminate all the variability. That which persists inevitably limits the precision of the lymphocyte as a model for studying receptor function, and diminishes its usefulness in measuring day by day changes in an individual patient. Fortunately, direct study of the cell surface β-receptors by competitive binding studies utilizing radiolabelled β-adrenoceptor antagonists such as ^{3}H-dihydroalprenolol (DHA) or ^{125}iodocyanopindolol is less capricious. In our experience, inter-assay variability of receptor density is usually no more than 15%.

Early reports describing the use of the lymphocyte or leukocyte model appeared before receptor binding studies were widely used, and most focused therefore on the production of cyclic AMP in response to β-adrenoceptor agonists. The majority of the intial reports showed an impaired cyclic AMP response in cells from asthmatic subjects (LOGSDON et al. 1972; PARKER and SMITH 1973; ALSTON et al. 1974), although GILLESPIE et al. (1974) were unable to confirm this. None of these studies made adequate allowances for the possible effect of prolonged antiasthmatic medication on the leukocyte β-receptor, and this problem was not specifically addressed until CONOLLY and GREENACRE (1976) took two groups of asthmatics of comparable severity, who differed only in respect of therapy. One group was using β-adrenergic bron-

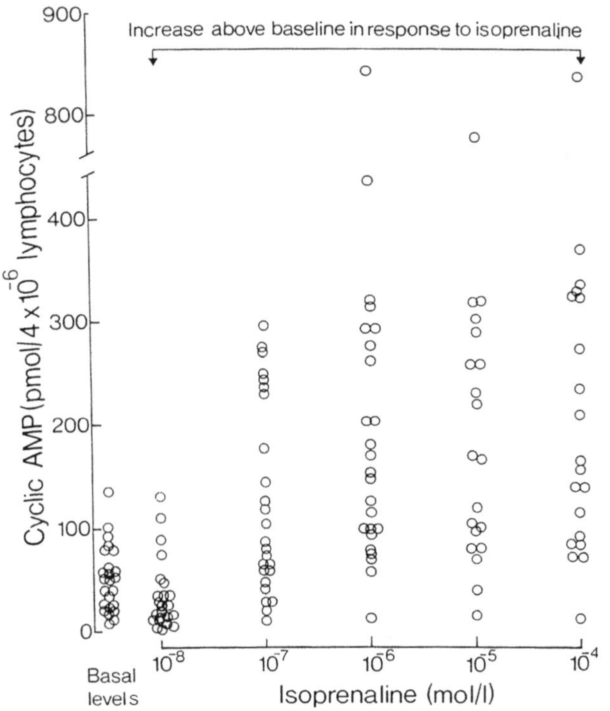

Fig. 2. Normal lymphocytes show a marked variability in the cyclic AMP response to adrenoceptor stimulation. The data represent basal levels of cyclic AMP, and increases above baseline seen at the end of a 15 min incubation at 37 °C, with the indicated concentrations of isoprenaline. The data were obtained from 8 normal subjects (on no medication) who were studied on 1 to 5 separate occasions. (Reproduced from Br. J. Pharmacol., 59:17-23, 1977)

chodilator therapy extensively, while the other was using exclusively non-adrenergic medication, namely beclomethasone dipropionate and cromolyn sodium. When compared with lymphocytes from normal subjects, the cyclic AMP response (percent increase above baseline) in lymphocytes taken from asthmatics using β-adrenergic bronchodilators showed a marked depression, while that from the group of non-adrenergic drugs was indistinguishable from normal (Fig. 3). Weaning patients off adrenergic drugs onto nonadrenergic agents resulted in the restoration of a normal response (Fig. 4).

A number of studies both *in vivo* and *in vitro* have been carried out to explore the development of desensitization of the β-receptor and to examine the underlying mechanism(s). *In vitro* studies have consistently demonstrated the development of impaired β-adrenoceptor function in a variety of tissues from several species including human lymphocytes and tracheobronchial smooth muscle after exposure to β-agonists for varying periods of time (Makman 1971; Franklin and Foster 1973; Mickey et al. 1975; Mukherjee et al. 1976; Williams et al. 1976; Greenacre et al. 1978; Davis and Conolly 1980). Stud-

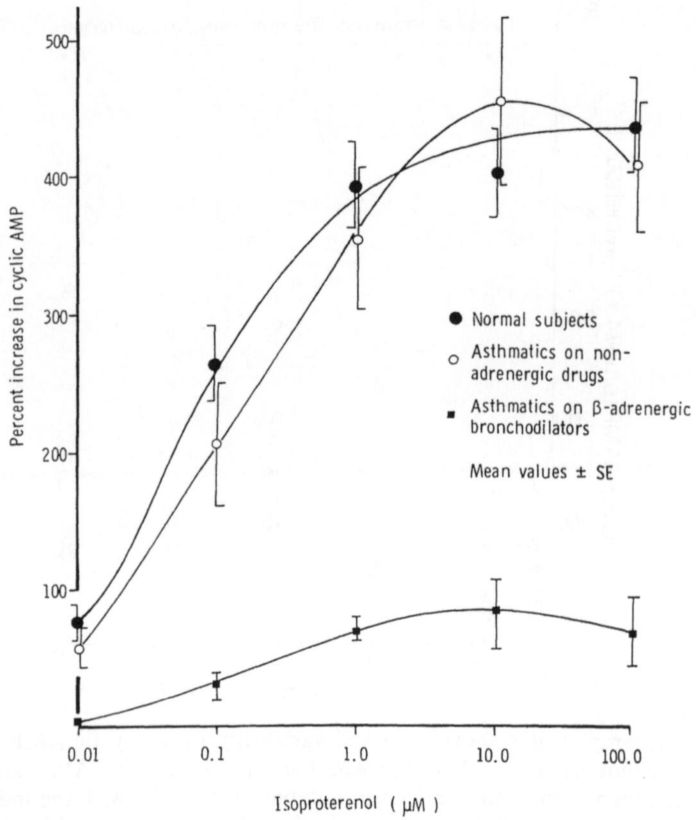

Fig. 3. Production of cyclic AMP in lymphocytes in response to a range of concentrations of isoprenaline. The cells were derived from normal subjects (●——●), or from asthmatics whose disease was controlled exclusively with non-adrenergic drugs (○——○) or from asthmatics of comparable severity but who were using excessive doses of β-adrenergic bronchodilators (■——■). (Reproduced from the Journal of Clinical Investigation *58* 1307–1316, 1976)

ies in intact subjects and animals are broadly in agreement with these findings. The early clinical reports of Keighley (1966) and Van Metre (1969) suggesting the development of resistance to β-adrenoceptor agonists find support in some of the more recent studies. Choo-Kang et al. (1969) described diminution in the response to inhaled salbutamol following repeated administration. Paterson et al. (1971) observed a marked rebound bronchoconstriction in 4 out of 15 patients following prolonged infusions of isoprenaline. Holgate et al. (1977) have demonstrated a waning bronchodilator response to inhalations of salbutamol during a four week period of exposure of normal subjects to that drug. Jenne et al. (1977), Nelson et al. (1977), Plummer (1978) and Van Arsdel et al. (1978) have reported impaired bronchodilator response in asthmatics exposed to β-adrenergic bronchodilators. Galant et al. (1978) using polymorphonuclear leukocytes (PMN), Greenacre and Conolly

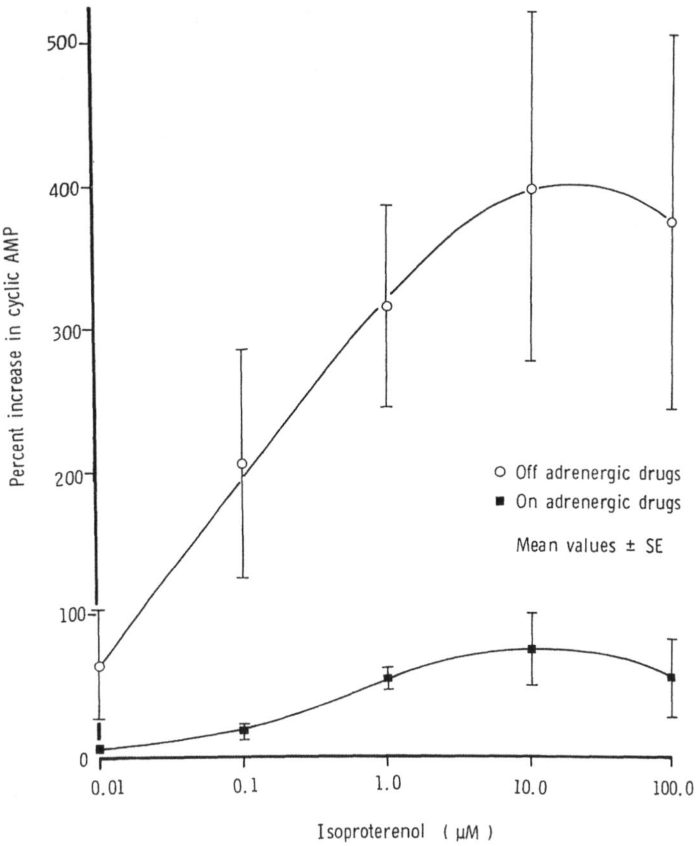

Fig. 4. Reversion of the cyclic AMP response to isoprenaline in lymphocytes taken from asthmatic subjects firstly while consuming excessive amounts of a β-adrenergic bronchodilator (■——■) and subsequently when weaned off these drugs and on to non-adrenergic medication (○——○). (Reproduced from the Journal of Clinical Investigation *58* 1307–1316, 1976)

(1978) and TASHKIN et al. (1982) using lymphocytes have reported reduced β-receptor numbers and function in subjects receiving β-adrenoceptor agonists. GALANT et al. (1980) examined PMN β-adrenoceptor numbers, binding affinity and cyclic AMP response to isoprenaline in normals, asthmatics on nonadrenergic medication and asthmatics receiving adrenergic bronchodilators. They found results consonant with those of CONOLLY and GREENACRE (1976) in that normal subjects and asthmatics on non-adrenergic therapy were indistinguishable from each other with respect to adrenoceptor number and responsiveness, whereas marked reductions therein were associated with the use of drugs acting on adrenoceptors. BUSSE et al. (1979) have also demonstrated reduced PMN responses (inhibition of release of β-glucuronidase) to isoprenaline during treatment with adrenergic aerosols. Interestingly this deficit persisted for several weeks after the cessation of therapy. Whether, in response to bronchospasm, the body might maintain a high level of catecholamine secretion leading thereby to β-adrenoceptor down regulation independent of treatment is at present unknown. The finding (LEMANSKE et al. 1980) of impaired β-receptor responsiveness in PMN's (inhibition of release of β-glucuronidase by isoprenaline) from non-asthmatic patients with chronic obstructive pulmonary disease might favor such a concept. Another mechanism whereby β-adrenoceptor dysfunction might come about is heterologous desensitization. Asthma may often be associated with increased circulating levels of bronchoconstrictor substances such as histamine (BARNES et al. 1980a) and it has been shown that prolonged exposure to one agonist capable of activating adenylate cyclase can lead to a reduced cyclic AMP response to heterologous agonists Thus GREENACRE et al. (1978) demonstrated a reduced response to isoprenaline after pretreatment with PGE_1; SAFKO and HANIFIN (1980) showed a reduced response to isoprenaline after exposure to histamine; TUCK et al. (1980) demonstrated desensitization to PGE_2 by isoprenaline, and BUSSE et al. (1979) showed a reduced response to histamine in PMN's from patients previously treated with isoprenaline or salbutamol. Perhaps the demonstration of a decline in PMN β-receptor responsiveness in association with upper respiratory tract infection (BUSSE 1977) is indicative of a similar process. Whether such heterologous desensitization finds expression at the cell surface receptor as it does intracellularly is at present unknown.

Finally, loss of DHA binding sites could be because of interaction between β-adrenoceptors and antibodies specifically directed towards them (VENTER et al. 1980). Such antibodies are well recognized as being clinically important in Graves's disease and in myasthenia gravis. The significance of the β-adrenoceptor antibodies in asthma is not known, as they can also be found in non-asthmatic patients, though an anecdotal report of apparently successful treatment of severe asthma by plasmapheresis (GARTMANN et al. 1978) gives added interest to it. Clearly this is an area urgently in need of further investigation.

The nature of the changes which underlie desensitization have been extensively investigated in recent years using a variety of amphibian, avian and mammalian cells. Understanding these data will be facilitated by a brief review of our current concepts of hormonally mediated adenlyate cyclase activation. It is believed that during occupancy by an agonist, but not by an antagonist (LIMBIRD et al. 1980), the receptor becomes coupled to adenylate cyclase by a regulatory protein referred to as the G or N protein because an essential component of its activity is the binding of guanine nucleotides (RODBELL et al. 1974a,b; KASLOW et al. 1980; DELEAN et al. 1980; JEFFERY et al. 1980). Indeed it has been suggested that the only purpose of the hormone receptor interaction is in some way to permit the displacement of GDP from this N protein by GTP (CASSEL and SELINGER 1976, 1977; CASSEL and PFEUFFER 1978; PFEUFFER 1979). Binding of GTP to the N protein appears to increase its affinity for the catalytic unit of adenylate cyclase and leads to its activation.

Among the earliest investigators, ROMERO and AXELROD (1975) demonstrated a circadian fluctuation of β-adrenoceptor density in the rat pineal gland, which varied inversely as the prevailing level of noradrenaline altered with the light-dark cycle. Similarly MUKHERJEE et al. (1975) reported a loss of β-adrenoceptor binding sites in frog erythrocytes following prolonged exposure to isoprenaline. Other workers have since shown down-regulation in β-receptor numbers following prolonged exposure to β-adrenoceptor agonists (SHEAR et al. 1976; WOLFE et al. 1977; TASHKIN et al. 1982). The cause of this loss of binding sites has been investigated by several groups. It was at one time suggested on the basis of broken cell studies that it might be caused by persistence of a receptor-agonist complex which failed to dissociate (WILLIAMS and LEFKOWITZ 1977), but more recent data from the same laboratory (LEFKOWITZ et al. 1978) using whole cell preparations have indicated that such a mechanism cannot adequately account for the loss of cell surface receptors. A more likely explanation is that (like other hormone receptors) the β-adrenoceptors become internalized within the cell (CHUANG and COSTA 1979). This loss of binding sites can be prevented by low temperatures or by inhibition of protein synthesis (HOMBURGER et al. 1980). This process is slow to resolve. In

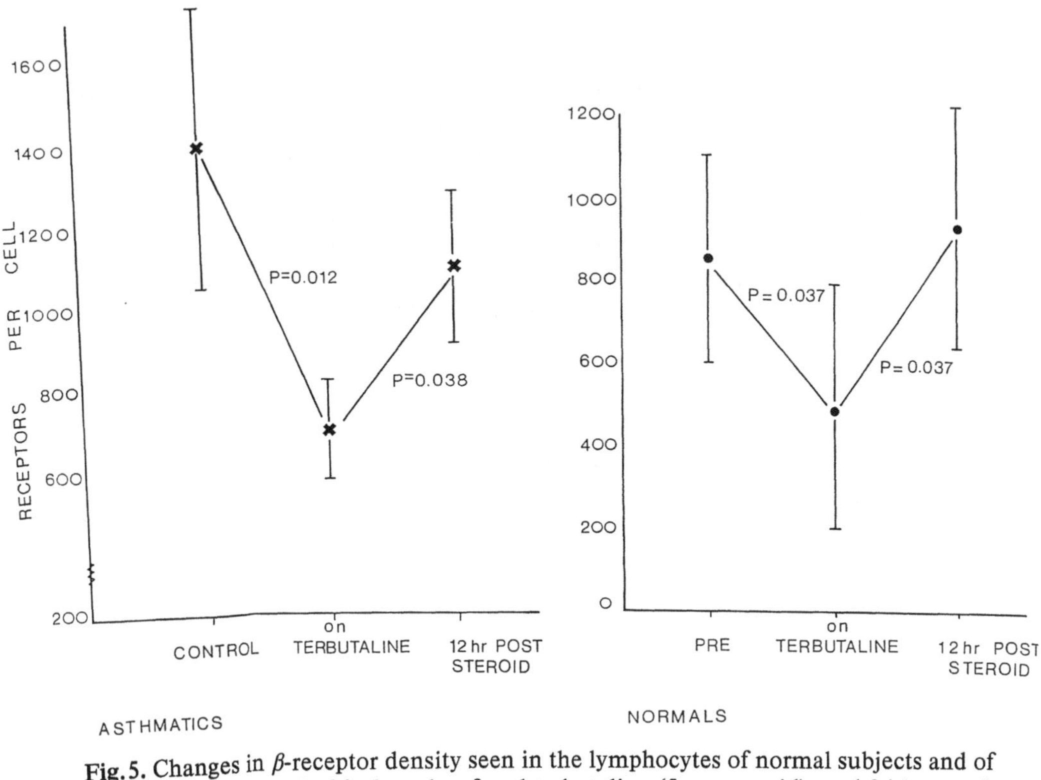

Fig. 5. Changes in β-receptor density seen in the lymphocytes of normal subjects and of asthmatics at the end of 3–5 weeks of oral terbutaline (5 mg p.o. tid), and 24 hours after a single dose of methyl prednisolone (2 mg/kg i.v.). Normalization of receptor density was seen within 24 hours after steroid administration. In the absence of steroids this normalization takes 1–2 weeks to be achieved. (Reproduced from the Journal of Allergy and Clinical Immunology *32*, 566–571, 1982)

the patients observed by CONOLLY and GREENACRE (1976) full recovery of the cyclic AMP response took up to a month to occur. Recovery of leukocyte β-receptor numbers in patients treated less intensively with β-agonists has been reported to take a week (GALANT et al. 1978) or longer (SANO et al. 1979). Recently, it has been observed that steroids accelerate the recovery so that within 16 hours of a single dose of methylprednisolone (2 mg/kg) β-adrenoceptor density is back to normal levels (Fig. 5) (HUI et al. 1982).

Careful studies by several groups (SU et al. 1980; KRALL et al. 1980; HOMBURGER et al. 1980) have shown that loss of cell surface receptors is not a completely adequate explanation for the observed desensitization. A loss of responsiveness of adenylate cyclase to β-adrenoceptor agonists occurs before a demonstrable loss of receptors, and unlike loss of receptors, is not prevented by low temperatures or cycloheximide. This process is moreover fairly rapidly reversible, and (of possible relevance to the asthmatic situation) can be induced by very low (0.01 nanomolar) concentrations of the agonist. This form of desensitization is thought to represent uncoupling of the receptor from the N protein-adenylate cyclase mechanism.

II. The α-Adrenoceptor

In contrast to β-receptors, the α-adrenoceptors have received relatively little attention in the context of bronchial asthma. A number of clinical studies have alluded to the usefulness of α-receptor blockade in asthmatics (BIANCO et al. 1974; BEIL and DEKOCK 1978; BLACK et al. 1978). More recently there have been reports of increased pupillary α-adrenoceptor responsiveness in asthmatics (HENDERSON et al. 1979). We have found α-adrenoceptor responsiveness (pressor response to phenylephrine) to be increased after prolonged exposure to β-agonists, but unlike the β-adrenoceptor desensitization it was not reversed by steroids (CONOLLY et al. in press). The nature of the interrelationship between α- and β-adrenoceptors is not known. The postulate that α- and β-adrenoceptors may, under appropriate circumstances, be interconvertible seems not to be true (BENFEY 1980).

Whether or not α-adrenoceptors may in some way modulate β-adrenoceptor numbers or vice versa is not known, but the increase in α-adrenoceptor density and a decrease in that of β-adrenoceptors in an animal model of asthma (BARNES et al. 1980b) suggest that there may be some dynamic interrelation between the two receptor populations.

III. Consequences of Adrenoceptor Imbalance

Regardless of whether imbalance between α- and β-adrenoceptors occurs because of increased α-adrenoceptor number or responsiveness, or because of loss of numbers or responsiveness of β-adrenoceptors, the consequences of such an imbalance in an asthmatic are potentially calamitous. As noted earlier, β-agonist activity in asthma accomplishes two things. Firstly, by direct action, it promotes bronchodilation. Secondly, it inhibits the release of bronchoconstrictor mediators from the mast cell. Conversely α-adrenoceptor

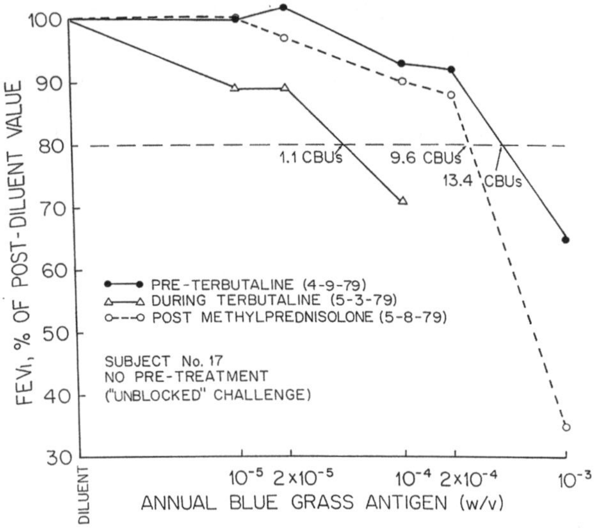

Fig. 6. The development of increased susceptibility to antigen-induced bronchospasm during prolonged treatment with terbutaline (5 mg p.o., tid for 3-5 weeks) and its rapid reversal by a single dose of methyl prednisolone (2 mg/kg i.v.). Prior to terbutaline, 13.4 cumulative breath units (CBU) were required to produce a 20 % fall in FEV_1. After terbutaline treatment the same degree of bronchospasm was caused by 1.1 CBU. Twenty four hours after the single dose of methyl prednisolone, 9.6 CBU were required to achieve the same degree of bronchoconstriction, thus indicating a substantial reversal of the terbutaline effect. (Reproduced from the Amer. Rev. Resp. Dis. *125* 185-193, 1982)

stimulation produces bronchoconstriction and enhances mediator release (Orange et al. 1971). Loss of β-adrenoceptor-related functions is clearly undesirable in an asthmatic, and exaggeration of this loss by increased α-agonist activity doubly so. Illustrations of such consequences have been provided by the demonstrations of increased susceptibility of asthmatics to exercise induced bronchospasm after prolonged exposure to conventional doses of β-adrenergic bronchodilators (Gibson et al. 1978), and by finding that some asthmatics so treated may become more susceptible to the bronchoconstrictor effect of inhaled allergen (Fig. 6) (Tashkin et al. 1982).

Perhaps not every asthmatic will be equally susceptible to the hazard of adrenoceptor imbalance. It has been argued that one reason for the transience of the rise in asthma deaths in the United Kingdom could have been due to the elimination of a particularly susceptible subset of the asthmatic population (Inman 1974). Davis and Conolly (1980) demonstrated a definite interindividual variability in the ease with which β-receptor desensitization could be produced in isolated bronchial smooth muscle. It is conceivable that in susceptible individuals, a vicious circle of events could have led to their deaths in the following way. In response to a severe asthmatic attack, large doses of β-agonist bronchodilators may have been taken. β-adrenoceptor desensitization could arise from this, leading to worsening of the asthma, and an increased endogenous catecholamine output. The β-adrenoceptors having become unresponsive, the endogenous compounds would predominantly stimulate α-adrenoceptors [which may paradoxically have come to develop increased responsiveness (see above)], leading to further deterioration of the patient's state, further consumption of bronchodilators, and further β-adrenoceptor desensitization, death eventually ensuing. Whether or not this was the case will now never be known with certainty. However, the concept is plausible enough to enjoin considerable restraints on the use of β-adrenergic bronchodilators in severe asthma and recent attempts to restore these agents to the status of non-prescription items should be deplored. Further, the marked restorative effect of glucocorticoids on down-regulated β-adrenoceptors should serve as a reminder of the crucial importance of steroids in treating this potentially lethal disease.

D. References

Alston WC, Patel KR, Kerr JW (1974) Response of leucocyte adenyl cyclase to isoprenaline and effect of alpha blocking drugs in extrinsic bronchial asthma. Brit Med J 1:90-93

Anderson G, Wilkins E (1977) A trial of clenbuterol in bronchial asthma. Thorax 32:717-719

Anderson G, Wilkins E, Jariwalla AG (1979) Fenoterol in Asthma. Brit J Dis Chest 73:81-84

Anderson SD, Pojer R, Smith ID, Temple D (1976) Exercise-related changes in plasma levels of 15 keto-13,14-dihydroprostaglandin F_2 and noradrenaline in asthmatic and normal subjects. Scand J Resp Dis 57:41-48

Apperly GH, Daly MJ, Levy GP (1976) Selectivity of adrenoceptor agonists and antagonists on bronchial skeletal vascular and cardiac muscle in the anesthetized cat. Brit J Pharmacol 57:235-246

Axelrod J (1959) Metabolism of epinephrine and other sympathomimetic agents. Physiol Rev 39:751-766

Barboriak JJ, Sosman AJ, Fink NJ, Maksud MG, McConnell LH, Hamilton LH (1973) Metabolic changes in exercise induced asthma. Clin Allergy 3:83-89

Barnes P, Fitzgerald G, Brown M, Dollery C (1980a) Nocturnal asthma and changes in circulating epinephrine, histamine and cortisol. New Engl J Med 303:263-267

Barnes PJ, Dollery CT, MacDermot J (1980b) Increased pulmonary alpha adrenergic and reduced beta adrenergic receptors in experimental asthma. Nature 285:569-571

Barnes PJ, Brown MJ, Silverman M, Dollery CT (1981) Circulatory catecholamines in exercise and hyperventilation induced asthma. Thorax 36:435-440

Barnes PJ, Ind PW, Brown MJ (1982) Plasma histamine and catecholamines in stable asthmatic subjects. Clin Sci 62:661-665

Baronti A, Grieco A, Vibelli C (1980) Oral NAB 365 (clenbuterol) and terbutaline in chronic obstructive lung disease: a double blind two week study. Int J Clin Pharmacol Ther Toxicol 18:21-25

Beil M, deKock MA (1978) Role of alpha adrenergic receptors in exercise induced bronchoconstriction. Respiration 35:78-86

Beil M, Brecht HM, Rasche B (1977) Plasma catecholamines in exercise induced bronchoconstriction. Klin Wschr 55:577-581
Benfey BG (1980) The evidence against interconversion of α- and β-adrenoceptors. Trends in Pharmacological Sciences 1:193-194
Bennett PN, Blackwell EW, Davies DS (1975) Competition for sulphate during detoxification in the gut wall. Nature (Lond) 258:247-248
Benson RL, Perlman F (1948) Clinical effects of epinephrine by inhalation. J Allergy 19:120-140
Bianco S, Griffin JP, Kamburoff P, Prime FJ (1974) Prevention of exercise induced asthma by indoramin. Brit Med J 4:18-20
Bishopric NH, Cohen HJ, Lefkowitz RJ (1980) β-adrenergic receptors in lymphocyte subpopulation. J Allergy Clin Immunol 65:29-40
Black JL, Temple DM, Anderson SD (1978) Long term trial of an alpha adrenoreceptor blocking drug (indoramin) in asthma. Scand J Resp Dis 59:307-312
Blackhall MI, Macartney B, O'Donnell SR (1978) Acute effects of the administration of rimiterol aerosol in asthmatic children. Brit J Clin Pharmacol 6:59-62
Blackwell EW, Briant RH, Conolly ME, Davies DS, Dollery CT (1974) Metabolism of isoprenaline after aerosol and direct intrabronchial administration in man and dog. Brit J Pharmacol 50:587-591
Boyum A (1968) Isolation of mononuclear cells and granulocytes from human blood. Scand J Clin Lab Invest 21(suppl 97):77-89
Brandon ML (1976) Long term metaproterenol therapy in asthmatic children. J Amer Med Assoc 235:736-737
Briant RH, Blackwell EW, Williams FM, Davies DS, Dollery CT (1973) The metabolism of sympathomimetic bronchodilator drugs by the isolated perfused dog lung. Xenobiotica 3:787-799
Brittain RT (1971) A comparison of the pharmacology of salbutamol with that of isoprenaline, orciprenaline and tremetoquinol. Postgrad Med J 47(suppl):11-16
Busse WW (1977) Decreased granulocyte response to isoproterenol in asthma during upper respiratory infections. Am Rev Resp Dis 115:783-791
Busse WW, Bush RK, Cooper W (1979) Granulocyte response in vitro to isoproterenol, histamine and prostaglandin E_1 during treatment with beta adrenergic aerosols in asthma. Am Rev Resp Dis 120:377-384
Camps PWL (1929) A note on the inhalation treatment of asthma. Guy's Hosp Reports 79:496
Cassel E, Pfeuffer T (1978) Mechanism of cholera toxin action: covalent modification of the guanyl nucleotide-binding protein of the adenylate cyclase system. Proc Natl Acad Sci 75:2669-2673
Cassel D, Selinger Z (1976) Catecholamine stimulated GTPase activity in turkey erythrocyte membranes. Biochem Biophys Acta 452:538-551
Cassel D, Selinger Z (1977) Mechanism of adenylate cyclase activation by cholera toxin: inhibition of GTP hydrolysis at the regulatory site. Proc Natl Acad Sci USA 74:3307-3311
Chernack WJ, Davis WJ, Pang LM, Rodriguez-Martinez F, Bierstuempel HC, Mellins RB (1974) Terbutaline in the treatment of acute childhood asthma. Amer Rev Resp Dis 109:741
Chervinsky P, Chervinsky G (1975) Metaproterenol tablets: their duration of effect by comparison with ephedrine. Curr Ther Res 17:507-518
Choo-Kang YFJ, Grant IWB (1975) Comparison of two methods of administering bronchodilator aerosol to asthmatic patients. Brit Med J 2:119-120
Choo-Kang YFJ, Simpson WT, Grant IWB (1969) Controlled comparison of the bron-

chodilator effects of 3 beta-adrenergic stimulant drugs administered by inhalation to patients with asthma. Brit Med J 2:287-292

Choo-Kang YFJ, Tribe AE, Grant IWB (1974) Salbutamol by intermittent positive pressure ventilation in status asthmaticus. Scot Med J 19:191-195

Chryssanthopoulos C, Barboriak JJ, Fink JN, Stekiel WJ, Maksud MG (1978) Adrenergic responses of asthmatic and normal subjects to submaximal and maximal work levels. J Allerg Clin Immunol 61:17-22

Chuang DM, Costa E (1979) Evidence for internalization of the recognition site of beta adrenergic receptors during receptor subsensitivity induced by (−)-isoproterenol. Proc Natl Acad Sci USA 76:3024-3028

Conolly ME, Greenacre JK (1976) The lymphocyte beta-adrenoceptor in normal subjects and patients with bronchial asthma: effect of different forms of treatment. J Clin Invest 58:1307-1316

Conolly ME, Greenacre JK (1977) The beta-adrenoceptor of the human lymphocyte and human lung parenchyma. Brit J Pharmacol 59:17-23

Conolly ME, Davies DS, Dollery CT, Morgan CD, Paterson JW, Sandler M (1972) Metabolism of isoprenaline in dog and man. Brit J Pharmacol 46:458-472

Conolly ME, Tashkin DP, Ertle A, Hui KK, Borst S, Bricknell KS, Lee E, Dauphinee B (in press) The effect of chronic treatment with β-adrenergic bronchodilators on α- and β-adrenoceptor number and function in mild asthmatics. Bulletin Européen de Physiopathologie Respiratoire

Cooke NJ, Kerr JA, Willey RF, Hoare MW, Grant IWB, Crompton GK (1974) Response to rimiterol and salbutamol aerosols administered by intermittent positive pressure ventilation. Brit Med J 2:250-252

Cookson DU, Reed CE (1963) A comparison of the effects of isoproterenol in normal and asthmatic subjects. Am Rev Resp Dis 88:636-643

Cullum VA, Farmer JB, Jack D, Levy GP (1969) Salbutamol: a new, selective β_2-adrenoceptive receptor stimulant. Brit J Pharmacol 35:141-151

DaCosta JL, Goh BK (1973) A comparative trial of subcutaneous terbutaline (Th 1165 a) and adrenaline in bronchial asthma. Med J Aust 2:588-591

Davies DS, George CF, Blackwell EW, Conolly ME, Dollery CT (1974) Metabolism of terbutaline in man and dog. Br J Clin Pharmacol 1:129-136

Davis C, Conolly ME (1980) Tachyphylaxis to beta adrenoceptor agonists in human bronchial smooth muscle: studies in vitro. Brit J Clin Pharmacol 10:417-423

Davis C, Conolly ME, Greenacre JK (1980) Beta adrenoceptors in human lung bronchus and lymphocytes. Brit J Clin Pharmacol 10:425-432

DelBono N, Quarteieri F, Vibelli C (1979) Intravenous NAB 365 (clenbuterol) and terbutaline in exercise induced bronchospasm (EIB). Clin Allergy 9:277-282

DeLean A, Stadel JM, Lefkowitz RJ (1980) A ternary complex model explains the agonist-specific binding properties of the adenylate cyclase coupled beta adrenergic receptor. J Biol Chem 255:7108-7117

Dollery CT, Davies DS, Draffan GH, Williams FM, Conolly ME (1970) Blood concentrations in man of fluorinated hydrocarbons after inhalation of pressurized aerosols. Lancet II:1164-1166

Dollery CT, Davies DS, Conolly ME (1971) Differences in the metabolism of drugs depending upon their route of administration. Ann NY Acad Sci 179:108-112

Dollery CT, Williams FM, Draffan GH, Wise G, Sahyoun H, Paterson JW, Walker SR (1974) Arterial blood levels of fluorocarbons in asthmatic patients following use of pressurized aerosols. Clin Pharm Ther 15:59-66

Drewitt AH (1962) First clinical experiences with Alupent—a new bronchodilator. Brit J Clin Pract 16:549-551

Dyson AJ, Campbell IA (1977) Interaction between choline theophyllinate and salmefamol in patients with reversible airways obstruction. Brit J Clin Pharmacol 4:677-682

Englehardt A, Hoefke W, Wick H (1961) Zur Pharmakologie des Sympathomimeticums 1-(3,5-dihydroxphenyl)-1-hydroxy-2-isoprophyaminoäthan. Arzn Forsch 11:521-525

Ephraim A (1910) Über endobronchiale Therapie. Berliner Klin Wochenschr 47:1317-1320

Evans ME, Walker SR, Brittain RT, Paterson JW (1973) The metabolism of salbutamol in man. Xenobiotica 3:113-120

Evans ME, Shenfield GM, Thomas N, Walker SR, Paterson JW (1974) The pharmacokinetics of rimiterol in man. Xenobiotica 4:681-692

Fletcher CM, Herxheimer H, Howell JBL, Lewis AAG, Jack D (1971) Salbutamol-proceedings of an international symposium. Postgrad Med J (suppl) 47:1-133

Franklin TJ, Foster SJ (1973) Hormone induced desensitization of hormonal control of cyclic AMP levels in human diploid fibroblasts. Nature (New Biol) 246:146-148

Freedman BJ (1971) Trial of new bronchodilator, terbutaline in asthma. Brit Med J 1:633-636

Furchgott RF (1972) The classification of adrenoceptors (adrenergic receptors). An evaluation from the standpoint of receptor theory. In: Blaschko H, Muscholl E (eds) Catecholamines. Springer, Berlin Heidelberg New York Tokyo (Handbook of Experimental Pharmacology, vol 33) pp 283-335

Galant SP, Duriseti L, Underwood S, Insel PA (1978) Decreased beta-adrenergic receptors on polymorphonuclear leukocytes after adrenergic therapy. New Engl J Med 299:933-936

Galant SP, Duriseti L, Underwood S, Allred S, Insel PA (1980) Decreased Beta-adrenergic receptors of polymorphonuclear particulates in bronchial asthma. J Clin Invest 65:577-585

Gandevia B (1968) The changing pattern of mortality from asthma in Australia. 2. Mortality and modern therapy. Med J Aust 1:884-891

Gartmann J, Grob P, Frey M (1978) Plasmapheresis in severe asthma. Lancet 2:40

George CF, Blackwell EW, Davies DS (1974) Metabolism of isoprenaline in the intestine. J Pharm Pharmacol 26:265-267

Gibson GJ, Greenacre JK, Konig P, Conolly ME, Pride NB (1978) Use of exercise challenge to investigate possible tolerance to beta adrenoceptor stimulation in asthma. Brit J Dis Chest 72:199-206

Gillespie E, Valentine MD, Lichtenstein L (1974) Cyclic AMP metabolism in asthma: studies with leukocytes and lymphocytes. J Allergy Clin Immunol 53:27-33

Graeser JB (1939) Inhalation therapy of bronchial asthma. JAMA 112:1223-1226

Graeser JB, Rowe AH (1935) Inhalation of epinephrine for the relief of asthmatic symptoms. J Allergy 6:415-420

Greenacre JK, Conolly ME (1978) Desensitization of the beta adrenoceptor of lymphocytes from normal subjects and patients with phaeochromocytoma: studies in vivo. Brit J Clin Pharmacol 5:191-197

Greenacre JK, Schofield P, Conolly ME (1978) Desensitization of the beta adrenoceptor of lymphocytes from normal subjects and asthmatic patients: studies in vitro. Brit J Clin Pharmacol 5:199-206

Greenberg MJ, Pines A (1967) Pressurized aerosols in asthma—a letter. Brit Med J 1:563

Griffin JP, Turner P (1971) Preliminary studies of a new bronchodilator (WG 253) in man. J Clin Pharmacol 11:280-287

Griffiths J, Leung FY, Grzybowski S, Chan-Yeung MMW (1972) Sequential estimation of plasma catecholamines in exercise induced asthma. Chest 62:527-533

Hartley JPR, Davies CT (1978) Plasma cyclic nucleotides in exercise induced asthma. Thorax 33:668

Henderson WR, Shelhamer JM, Reingold DB, Smith LJ, Evans R, Kaliner M (1979) Alpha adrenergic hyperresponsiveness in asthma. N Engl J Med 300:642-647

Herxheimer H (1972) Asthma deaths. Brit Med J 4:795

Hetzel MR, Clark TJH (1976) Comparison of intravenous and aerosol salbutamol. Brit Med J 4:795

Holgate ST, Baldwin CJ, Tattersfield AE (1977) Beta adrenergic agonist resistance in normal human airways. Lancet 2:375-377

Homburger V, Lucas M, Cantan B, Barabe J, Penit J, Bockaert J (1980) Further evidence that desensitization of beta adrenergic-sensitive adenylate cyclase proceeds in two steps. J Biol Chem 255:10436-10444

Hornblad Y, Ripe E, Magnusson PO, Tegner K (1976) The metabolism and clinical activity of terbutaline and its prodrug ibuterol. Eur J Clin Pharmacol 10:9-18

Hui KKP, Conolly ME, Tashkin DP (1982) Reversal of human lymphocyte beta adrenoceptor desensitization by glucocorticoids. Clin Pharmacol Ther 32:566-571

Hurst A (1973) Metaproternol, a potent and safe bronchodilator. Ann Allergy 31:460-466

Ingram J, Gaddie J, Skinner C, Palmer KNV (1975) The effect of intramuscular salbutamol in asthmatics. Brit J Clin Pharmacol 2:263-266

Inman WHW (1974) Recognition of unwanted drug effects with special reference to pressurized bronchodilator aerosols. In: Burley DM, Clarke SW, Cuthbert MF, Paterson JW, Shelly JH (Eds) Evaluation of Bronchodilator Drugs. Trust for Education ~ Research in Therapeutics, London, pp 191-200

Inman WHW, Adelstein AM (1969) Rise and fall of asthma mortality in England and Wales in relation to use of pressurized aerosols. Lancet 2:279-285

Inoue S (1967) Effects of epinephrine on asthmatic children. J Allergy 40:355-348

Jack D (1973) Selectively acting beta adrenoceptor stimulants in asthma. In: Austen KF, Lichenstein LM (eds) Asthma: Physiology, Immunopharmacology and Treatment. Academic Press, New York, pp 251-266

Jeffery DR, Charlton RR, Venter JC (1980) Reconstruction of turkey erythrocyte beta-adrenergic receptors into human erythrocyte acceptor membranes. J Biol Chem 255:5015-5018

Jenne JW, Chick TW, Strickland RD, Wall FJ (1977) Subsensitivity of beta responses during therapy with a long acting beta-2-preparation. J Allergy Clin Immunol 59:383-390

Johansen S (1974) Clinical comparison of intramuscular terbutaline and subcutaneous adrenaline in bronchial asthma. Europ J Clin Pharmacol 7:163-167

Johnson N McI, Clarke SW (1977) Asthma: ibuterol versus terbutaline inhalation. Brit Med J 1:1006

Kamburoff PL, Prime FJ, Schmidt OP (1977) Bronchodilator effect of NAB 365. Brit J Clin Pharmacol 4:67-71

Kaslow HR, Johnson GL, Brothers VM, Bourne HR (1980) A regulatory component of adenylate cyclase from human erythrocyte membrane. J Biol Chem 255:3736-3741

Keighley JF (1966) Iatrogenic asthma associated with adrenergic aerosols. Ann Int Med 65:985-995

Kennedy MCS, Dash CH (1972) The bronchodilator effect of a new adrenergic aerosol -salmefamol. Acta Allerg (Koh) 27:22-26

Kerr A, Gebbie T (1973) Comparison of orciprenaline, ephedrine and methoxyphenamine as oral bronchodilators. NZ Med J 77:320-322

Kirkpatrick CH, Keller C (1967) Impaired responsiveness to epinephrine in asthma. Am Rev Resp Dis 96:692-699

Koivikko A (1974) A comparison of the effects of subcutaneous orciprenaline salbutamol and terbuline in asthmatic children. Ann Clin Res 6:99-104

Konzett H (1941a) Neue broncholytisch hochwirksame Körper der Adrenalinreihe. Arch Exp Path und Pharmakol 197:27-40

Konzett H (1941b) Zur Pharmakologie neuer Adrenalin verwandter Körper. Arch Exp Path und Pharmakol 197:41-56

Krall FJ, Connelly M, Tuck ML (1980) Acute regulation of beta adrenergic catecholamine sensitivity in human lymphocytes. J Pharmacol Exp Ther 214:544-560

Lands AM, Arnold A, McAuliff JP, Ludeuna FP, Brown TG (1967) Differentiation of receptor systems activated by sympathomimetic amines. Nature (Lond) 241:597-598

Lefkowitz RJ, Mullikin D, Williams LT (1978) A desentitized state of the beta adrenergic receptor not associated with high affinity agonist occupancy. Mol Pharmacol 14:376-380

Legge JS, Gaddie J, Palmer KNV (1971) Comparison of two oral selective beta-2-adrenergic stimulant drugs in bronchial asthma. Brit Med J 1:637-639

Leitch AG, Clancy LJ, Costello JF, Flenley DC (1976) Effect of intravenous infusion of salbutamol on ventilatory response to carbon dioxide and hypoxia and on heart rate and plasma potassium in normal men. Brit Med J 1:365-367

Lemanske RF, Anderson C, Braun S, Skatrud J, Busse WW (1980) Impaired in vitro beta-adrenergic granulocyte response in chronic obstructive pulmonary disease. Am Rev Rep Dis 122:213-219

Levy B, Ahlquist RP (1971) Andrenergic drugs. In: DiPalma JR (ed) Drill's Pharmacology in Medicine. McGraw-Hill Book Co. New York pp. 627-674

Limbird LL, Gill DM, Lefkowitz JR (1980) Agonist promoted coupling of the beta-adrenergic receptor with the guanine nucleotide regulatory protein of the adenylate cyclase system. Proc Natl Acad Sci USA 77:775-779

Lin C, Magat J, Calesnick B, Synchowicz S (1972) Absorption, excretion and urinary metabolic pattern of ^{3}H-albuterol aerosol in man. Xenobiotica 2:506-515

Lockey SD, Glennon JA, Reed CE (1967) Comparison of some metabolic responses in normal and asthmatic subjects to epinephrine and glucagon. J Allergy 40:349-354

Logsdon PA, Middleton E, Coffey RG (1972) Stimulation of leukocyte adenyl cyclase by hydrocortisone and isoproterenol in asthmatic and non asthmatic subjects. J Allergy Clin Immunol 50:45-56

Lowell FC, Curry JJ, Schiller IW (1949) A clinical and experimental study of isuprel in spontaneous and induced asthma. New Engl J Med 240:45-51

MacKay AD, Axford AT (1977) Bronchodilator activity of rimiterol (Pulmadil) in patients with chronic airways obstruction. Comparison with salbutamol (Ventolin). Clin Trials J 14:7379

Makman MH (1971) Properties of adenylate cyclase of lymphoid cells. Proc Natl Acad Sci USA 68:885-889

Marlin GE, Turner P (1975) Intravenous treatment with rimiterol and salbutamol in asthma. Brit Med J 2:715-719

Marlin GE, Harnett BJS, Berend N (1977) Comparative potencies and beta-2-adrenoceptor selectivities of rimiterol and salbutamol aerosols. Brit J Clin Pharmacol 4:77-79

Marmo E, Rossi F, Giordans L, Lampa E, Rosalti G, DiMezza G (1979) Salmefamol and salbutamol. An experimental comparative study of their action on tracheal and bronchial smooth muscles. Respiration 38:257-265

Matthews C (1909) Adrenaline in asthma—a letter. Brit Med J 2:1323

May CS, Spiro SG, Johnson AJ, Paterson JW (1975) Intravenous infusion of salbutamol in the management of asthma. Thorax 30:236

McEvoy JDS, Vall-Spinosa A, Paterson JW (1973) Assessment of orciprenaline and isoproterenol infusions in asthmatic patients. Amer Rev Resp Dis 108:490-500

McLeod J, Selman J (1973) A comparison of Berotec (Th 1165 a) and salbutamol aerosols. NZ Med J 78:209-210

McMannis AG (1964) Adrenaline and isoprenaline: a warning—a letter. Med J Aust 2:76

Mickey JV, Tate R, Lefkowitz RJ (1975) Subsensitivity of adenylate cyclase and decreased beta adrenergic receptor binding after chronic exposure to (1)-isoproterenol in vitro. J Biol Chem 250:5727-5729

Miller WC, Rice DL (1980) A comparison of oral terbutaline and fenoterol in asthma. Ann Allergy 44:15-18

Morris HG, DeRoche G, Earle MR (1972) Urinary excretion of epinephrine and norepinephrine in asthmatic children. J Allergy Clin Immunol 50:138-145

Mukherjee C, Caron MG, Lefkowitz RJ (1975) Catecholamine induced subsensitivity of adenylate cyclase associated with loss of beta adrenergic receptor binding sites. Proc Natl Acad Sci USA 72:1945-1949

Mukherjee C, Caron MG, Lefkowitz RJ (1976) Regulation of adenylate cyclase coupled beta adrenergic receptors by beta adrenergic catecholamines. Endocrinol 99:347-357

Neilsen NA (1936) Treatment of asthmatic attacks by inhalation of adrenaline. Lancet 2:848

Nelson HS, Raine D, Doner C, Posey WC (1977) Subsensitivity to the bronchodilator action of albuterol produced by chronic administration. Am Rev Resp Dis 116:871-878

O'Donnell SR (1970) A selective beta adrenoceptor stimulant (Th 1165 a) related to orciprenaline. Eur J Pharmacol 12:35-43

O'Donnell SR (1972) An examination of some beta-adreno-receptor stimulants for selectivity using the isolated trachea and atria of the guinea pig. Eur J Pharmacol 19:371-379

Orange RP, Kaliner MA, Laraia PJ, Austen KF (1971) Immunological release of histamine and slow reacting substance of anaphylaxis from human lung II. Influence of cellular levels of cyclic AMP. Fed Proc 30:1725-1729

Parker CW, Smith JW (1973) Alterations in cyclic adenosine monophosphate metabolism in human bronchial asthma. I. Leukocyte responsiveness to beta adrenergic agonist. J Clin Invest 52:48-59

Parker SS, Choo Kang YFG, Cooper EJ, Cameron SJ, Grant IWB (1971) Bronchodilator effect of oral salbutamol in asthmatics treated with corticosteroids. Brit Med J 4:139-142

Paterson IC, Willey RF, Shotter MV, Crompton GK (1977) Further studies of rimiterol and salbutamol administered by intermittent positive pressure ventilation and an important observation on the technique of using the Bennett ventilator. Brit J Clin Pharmacol 4:605-609

Paterson JW, Conolly ME, Davies DS, Dollery CT (1968) Isoprenaline resistance and the use of pressurized aerosols in asthma. Lancet II:426-429

Paterson JW, Courtnay Evans RJ, Prime FJ (1971) Selectivity of bronchodilator action of salbutamol in asthmatic patients. Brit J Dis Chest 65:21-28

Patterson R, McKenna JM, Suszko IM, Solliday NH, Purzansky JJ, Roberts M, Kehow TJ (1977) Living histamine-containing cells from the bronchial lumens of humans. J Clin Invest 59:217-225

Persson H, Olsson T (1970) Some pharmacological properties of terbutaline, (1-3,5-dihydroxyphenyl)-2-(t-butylamino)-ethanol, a new sympathomimetic beta-receptor stimulating agent. Acta Med Scand Suppl 512:11-19

Pfeuffer T (1979) Guanine nucleotide controlled interaction between components of adenylate cyclase. FEBS Letters 101:85-89

Phillips PJ, Vedig AE, Jones PL, Chapman MG, Collins M, Edwards JB, Smeaton TC, Duncan BM (1980) Metabolic and cardiovascular side effects of the beta-2-adrenoceptor agonists salbutamol and rimiterol. Brit J Clin Pharmacol 9:483-491

Pinder RM, Brogen RN, Speight TM, Avery GS (1977) Rimiterol: a review of its pharmacological properties and therapeutic efficacy in asthma. Drugs 14:81-104

Plummer AL (1978) Development of drug tolerance to beta-2-adrenergic agents in asthmatics. Chest 73 (suppl):949-957

Poulsen SS, Petersen BN (1976) Ibuterol hydrochloride and terbutaline in asthma. Brit Med J 1:835

Powles ACP (1975) The bronchodilator effects of fenoterol (Berotec). NZ Med J 81:249-251

Rebuck AS, Marcus JI (1979) ScH 1000 in psychogenic asthma. Scand J Resp Dis (Suppl): 103:186-191

Rebuck AS, Saunders NA (1972) A new bronchodilator, TH 1165a (Berotec). Med J Aust 1:225-227

Refshauge WD (1965) Sympathomimetic drugs and bronchial asthma—a letter from the Australian Drug Evaluation Comm. Med J Aust 1:93

Reilly EB, Rodgers JM, Bickerman HA (1974) A comparison of the onset of bronchodilator activity of metaproterenol and isoproterenol aerosols. Curr Ther Res 16:759-765

Rodbell M, Lin MC, Salomon Y, Londos C, Harwood JP, Martin BR, Rendell M, Berman M (1974a) The role of adenine and guanine nucleotides in the activity and response of adenylate cyclase to hormones: evidence for multisite transition states. Acta Endocr 77 (suppl 191):11-37

Rodbell M, Lin MC, Salomon Y (1974b) Evidence for interdependent action of glucagon and nucleotides on the hepatic adenylate cyclase system. J Biol Chem 249:59-65

Romero JA, Axelrod J (1975) Regulation of sensitivity to beta adrenergic stimulation in induction of pineal N-acetyltransferase. Proc Natl Acad Sci USA 72:1661-1665

Sackner MA, Silva G, Marks MB (1976) Long term effects of metaproterenol in asthmatic children. Amer Rev Resp Dis 133 (suppl):130

Safko MJ, Hanifin JM (1980) Receptor desensitization as a possible basis for the beta-adrenergic theory of atopic disease. Clin Res 28:22 A

Salorinne Y, Stenius B, Tukiainen P, Poppins H (1975) Double blind cross over comparison of clenbuterol and salbutamol tablets in asthmatic outpatients. Eur J Clin Pharmacol 8:189-195

Sano Y, Ruprecht H, Mano K, Begley M, Bewtra A, Townley R (1979) Leukocyte beta adrenergic receptor assay in normals and asthmatics. Clin Res 27:403A

Sasaki T, Takishima T, Siegwra R (1980) Comparison of fenoterol and orciprenaline with regard to bronchodilating action and beta-2-selectivity. J Int Med Res 8:205-216

Schild HO (1937) Histamine release and anaphylatic shock in isolated lungs of guinea pigs. Quart J Exp Physiol 26:165-179

Scott RE (1970) Effect of prostaglandins, epinephrine and sodium fluoride on human leukocyte, platelet and liver adenylate cyclase. Blood 35:514

Shear M, Insel PH, Melmon KL, Coffino P (1976) Agonist specific refractoriness induced by isoproterenol. J Biol Chem 251:7572-7576
Siegmund OH, Granger HR, Lands AM (1947) The bronchodilator action of compounds structurally related to epinephrine. J Pharmacol Exp Ther 90:254-259
Sillett RW, Dash CH, McNicol MW (1976a) Comparison of salmefamol with salbutamol aerosols in asthmatics. Eur J Clin Pharmacol 9:277-280
Sillett RW, Dash CH, McNicol MW (1976b) Salmefamol orally in asthmatics—two doses compared. Eur J Clin Pharmacol 9:281-284
Simonson BG, Svedenblad H, Strom B (1976) Bronchodilatory and circulatory doses of a beta-2-agonist (terbutaline) inhaled with IPPB in patients with reversible airways obstruction. Scand J Resp Dis 57:252-258
Sinha BN, Allan GW, Lees AW (1973) Oral orciprenaline, salbutamol and isoetharine compared. Brit J Dis Chest 67:61-63
Sly RM, Badiei B, Faciane J (1975) Treatment of acute asthma in children with subcutaneous terbutaline. J Allergy 55:97-98
Smith JM (1966) Death from asthma—a letter. Lancet 1:1042
Sobol B, Reed A (1974) The rapidity of onset of bronchodilatation: a comparison of alupent and isoproterenol. Ann Allergy 32:137-141
Speizer FE, Doll R, Heaf P (1968a) Observations on recent increase in mortality from asthma. Brit Med J 1:335-339
Speizer FE, Doll R, Heaf P, Strange LB (1968b) Investigation into use of drugs preceding death from asthma. Brit Med J 1:339-343
Spiro SG, May CS, Johnson AJ, Paterson JW (1975) Intravenous injections of salbutamol in the management of asthma. Thorax 30:236
Spitzer SA, Goldschmidt Z, Dubrawsky C (1972) The bronchodilator effect of salbutamol administered by IPPB to patients with asthma. Chest 62:273-276
Stolley PD (1972) Asthma mortality: why the United States was spared an epidemic of deaths due to asthma. Amer Rev Resp Dis 105:883-890
Su YF, Harden TK, Perkins JP (1980) Catecholamine-specific desensitization of adenylate cycylase. J Biol Chem 225:7410-7419
Su YF, Harden TK, Perkins JP (1979) Isoproterenol-induced desensitization of adenylate cyclase in human astrocytoma cells. J Biol Chem 254:38-41
Szentivanyi A (1968) The beta adrenergic theory of the atopic abnormality in bronchial asthma. J Allergy 42:203-232
Tandon MK (1980) Cardiopulmonary effects of fenoterol and salbutamol aerosols. Chest 77:429-431
Tashkin DP, Conolly ME, Deutsch R, Hui KK, Litner M, Scarpace P, Abrass I (1982) Subsensitization of beta adrenoceptors in airways and lymphocytes of healthy and asthmatic subjecets. Amer Rev Resp Dis 125:185-193
Thiringer G, Svedmyr N (1976) Comparison of infused and inhaled terbutaline in patients with asthma. Scand J Resp Dis 57:1724
Trembath PW, Greenacre JK, Anderson M, Dimmock S, Mansfield L, Wadsworth J, Green M (1979) Comparison of four weeks' treatment with fenoterol and terbutaline aerosols in adult asthmatics. J Allergy Clin Immunol 63:395-400
Tuck ML, Fittingoff D, Connelly M, Krall JF (1980) Beta adrenergic catecholamine regulation of lymphocyte sensitivity: heterologous desensitization to prostaglandin E_2 by isoproterenol. J Clin Endocr Metab 51:1-6
Van Arsdel PP, Schaffrin RM, Rosenblatt J, Sprenkle AC, Altman LC (1978) Evaluation of oral fenoterol in chronic asthmatic patients. Chest 73: (suppl):997-999
Van Metre TE (1967) Death in asthmatics. Trans Amer Clin Climatological Assoc 78:58-69

Van Metre TE (1969) Adverse effects of inhalation of excessive amounts of nebulized isoproterenol in status asthmaticus. J Allergy 43:101–113

Venter JC, Fraser CM, Harrison LC (1980) Autoantibodies to beta-2-adrenergic receptors: a possible cause of adrenergic hyporesponsiveness in allergic rhinitis and asthma. Science 207:1361–1363

Walker SR, Evans ME, Richards AJ (1972) The clinical pharmacology of oral and inhaled salbutamol. Clin Pharmacol Ther 13:861–867

Warrell DA, Robertson DC, Newton Howes JA, Conolly ME, Paterson JW, Beilin LJ, Dollery CT (1970) Comparison of the cardiorespiratory effects of isoprenaline and salbutamol in patients with bronchial asthma. Brit Med J 1:65–70

Weiner N (1980) Norepinephrine, epinephrine and the sympathomimetic amines. In: Gilman A, Goodman LS, Gilman A (eds) The pharmacological Basis of Therapeutics. Macmillan, New York, pp 138–175

Williams FM, Briant RH, Dollery CT, Davies DS (1974) The influence of the route of administration on urinary metabolites of isoetharine. Xenobiotica 4:343–353

Williams LT, Lefkowitz RJ (1977) Slowly reversible binding of catecholamine to a nucleotide-sensitive state of the beta adrenergic receptor. J Biol Chem 252:7207–7213

Williams LT, Snyderman R, Lefkowitz RJ (1976) Identification of beta adrenergic receptors in human lymphocytes by (1)-(^{3}H)-alprenolol binding. J Clin Invest 57:149–155

Wolfe BB, Harden TK, Molinoff PB (1977) In vitro study of beta adrenergic receptors. Ann Rev Pharmacol Toxicol 17:575–604

CHAPTER 19

Catecholamine Biochemical Genetics

R. M. WEINSHILBOUM[1]

A. Introduction

Catecholamine biochemical genetics has advanced significantly during the past decade. Biochemical genetic studies have already contributed to our understanding of the regulation of catecholamine biosynthesis and metabolism and promise to help clarify the biological basis of individual variation of human adrenergic function. The most extensive catecholamine biochemical genetic data relate to the effects of inheritance on the enzymes that catalyze catecholamine biosynthesis and degradation. Before these data are reviewed, the research strategies and analytical techniques most commonly used in biochemical genetic experiments will be described briefly.

I. Biochemical Genetic Research Strategy

When BATESON coined the term "genetics" to describe the science of heredity, he emphasized that it involved the study of both inheritance and variation (BATESON 1907). The potential contribution of studies of inheritance to the understanding of biological variation has not always been appreciated. Although genetic information is extremely important in its own right, genetic techniques also provide powerful probes by which the biochemical mechanisms responsible for natural variation may be studied. The goals of biochemical genetic studies in experimental animals and man are identical: to determine whether inheritance might play a role in the regulation of the variation of a trait and, if so, to explore the biological basis of that regulation. Although their goals are similar, the research strategies used in experiments with laboratory animals and man differ.

Virtually all catecholamine biochemical genetic studies performed with experimental animals have used inbred rodents, and the strategy employed in all of these experiments has been similar. Enzyme activities or other enzyme properties have been compared among several inbred strains. If large strain variations were found, it was assumed that these differences were most likely due to inheritance. Breeding studies have then been performed to determine whether strain variations were inherited in a monogenic (mendelian) or polygenic fashion. Determination of the mechanism of inheritance is not just of

---

[1] Supported in part by NIH grants NS 11014, HL 17487 and GM 28157. Dr. Weinshilboum is a Burroughs Wellcome Scholar in Clinical Pharmacology.

academic importance. This is true in part because it is potentially easier to elucidate the biochemical basis of a monogenically inherited trait, one due primarily to the effect of a single genetic locus, than that of a polygenically inherited trait, one due to the effects of several loci. The term "locus" used in this context refers to the position of a gene on a chromosome and is often used as a synonym for gene. Alternative forms of a gene at a specific locus are referred to as alleles.

Breeding studies performed with inbred animals have usually included matings between members of the parental strains to produce F1 (first filial or hybrid) animals. Hybrid offspring have then been mated with each other to obtain F2 (second filial) animals. F1 hybrids have also been mated to the original parental strains in so-called backcross experiments. When a trait is inherited in a monogenic fashion, it is expected that both F2 animals and the offspring of backcross experiments will show segregation of the trait, distribution according to Mendel's famous ratios (MENDEL 1865). Even if segregation does not occur, it is possible on the basis of such breeding experiments to estimate the number of genes responsible for the variance of any trait that can be expressed quantitatively (BRUELL 1962; FALCONER 1963). Unfortunately, traditional breeding studies often involve large numbers of animals and require long periods of time to perform. Therefore, newer approaches such as those involving "recombinant inbred" strains have been developed that simplify the use of inbred animals for certain biochemical genetic experiments (BAILEY 1971; SWANK and BAILEY 1973). It is beyond the scope of this discussion to describe the analysis of data from breeding experiments with laboratory animals, but such discussions may be found in many standard genetic textbooks (e.g. GARDNER 1984).

Research strategies used to perform human biochemical genetic studies differ from those used to study laboratory animals. The difficulties of genetic studies in man are compounded in the case of catecholamines by the fact that tissue samples from the organs of greatest interest, the nervous system and the adrenal medulla, cannot be obtained from large numbers of subjects and are impossible to obtain from multiple members of families and kindreds. One approach that has been used as a first step in such studies is measurement in an easily obtainable tissue of the activities of enzymes involved in catecholamine biosynthesis and degradation. The most commonly used tissues have been blood and skin fibroblasts. The major advantage of blood elements such as plasma, erythrocytes and platelets is that blood may be obtained from large numbers of subjects and from multiple family members. Fibroblasts are more difficult to obtain because a skin biopsy is required. In addition, long time periods are required to culture skin fibroblasts. However, this approach also has significant advantages. Conditions of cell culture can be controlled and systematically altered to study enzyme regulation, and fibroblasts may be used to form hybrids with cells of other species. Study of these hybrid cells may make it possible to assign genes to specific chromosomes (see the subsequent discussion of MAO). The relative advantages and disadvantages of blood elements and fibroblasts for biochemical genetic studies of catecholamines have been reviewed elsewhere (WEINSHILBOUM 1979b; BREAKEFIELD et al. 1981).

Experiments performed with blood elements and fibroblasts involve the tacit assumptions that the enzymes in these tissues are biochemically identical with those in the nervous system and adrenal medulla, and that genetic regulation in blood or fibroblasts is similar to that in the adrenal medulla and in nervous tissue. These assumptions must eventually be tested experimentally.

Included among commonly used human biochemical genetic research strategies are measurements of enzyme activities or other enzyme properties in large population samples, in twins or in first-degree relatives within families. Data from twin and family studies may be used to calculate estimates of the "heritability" of a trait. For example, if the variance of a polygenically inherited trait is due entirely to the effects of inheritance (heritability of 1.0), monozygotic twins would be expected to have a correlation coefficient for the trait of 1.0, while dizygotic twins and siblings, subjects who share on the average half of their genes, would have a correlation coefficient of 0.5 (Cavalli-Sforza and Bodmer 1971).

Even if the heritability of a trait is high, the question of whether inheritance is monogenic or polygenic remains unanswered. The shapes of frequency distribution histograms of enzyme activities or other enzyme characteristics from large population samples can give a clue to the possibility of monogenic inheritance. A bimodal or trimodal frequency distribution raises the possibility of monogenic inheritance, although a distribution that is apparently Gaussian does not eliminate that possibility (Hopkinson et al. 1964). Monogenically inherited traits detected in human population samples of reasonable size usually represent genetic polymorphisms rather than rare variants. Polymorphism refers to a situation in which there are two alleles at a locus, and the least common of the alleles has a gene frequency of one percent or greater (Cavalli-Sforza and Bodmer 1971). This definition is not arbitrary since a gene frequency of one percent or greater usually cannot be maintained in a population by spontaneous mutation alone. The very existence of a polymorphism may imply that selective evolutionary pressure is involved in the maintenance of multiple alleles in the population (Cavalli-Sforza and Bodmer 1971; Lewontin 1974). However, that interpretation is the subject of intense controversy (Kimura and Ohta 1974). Ultimately, family studies are required to define the mechanism of inheritance of a trait. Data from family studies can be used for segregation analysis to determine whether inheritance might be monogenic (e.g. autosomal recessive, sex-linked, etc.). A detailed discussion of methods for the genetic analysis of data from family and population studies is also beyond the scope of this review, but such discussions may be found in many standard textbooks of human genetics (Li 1961; Cavalli-Sforza and Bodmer 1971; Vogel and Motulsky 1979).

II. Biochemical Genetic Analytical Techniques

A variety of analytical techniques have been used to study the genetic regulation of enzymes. These include, among others, measurement of total enzyme activity, measurement of electrophoretic mobility, and measurement of thermal stability. Electrophoresis has probably been used more often than any other technique for the detection of inherited enzyme variation (HARRIS 1980). However, electrophoresis has found only limited application in catecholamine biochemical genetic studies. This is true at least in part because of the sensitivity of the radiochemical enzymatic assays usually used to measure the activities of catecholamine biosynthetic and metabolic enzymes. The most commonly used technique in studies of the effects of inheritance on the enzymes of catecholamine biosynthesis and metabolism has been measurement of enzyme activity under optimal conditions. Determinations of enzyme thermal stability have also been used frequently (WEINSHILBOUM 1981). Studies of thermal stability have proven useful in the detection of inherited variations in the structures of many proteins. However, differences in thermal stability are not necessarily due to variation in the structure of the enzyme under study. These differences might represent variations in the thermal stability of enzyme activators, inhibitors or other constituents of tissue homogenates. Precautions required for the proper control of thermal stability experiments and for the interpretation of the results of such studies have been described in detail elsewhere (WEINSHILBOUM 1981). Other enzyme properties that have been measured in the course of biochemical genetic studies have included variations in the effects of enzyme inhibitors and in apparent Michaelis-Menten constants for substrates. There are many examples of genetic variation in which only a single property of an enzyme might differ among variant forms while all other properties are identical (HARRIS 1980). An ideal biochemical genetic study would include simultaneous measurement of as many enzyme properties as possible.

III. Mechanisms of Gene Effects

There are several different mechanisms by which inheritance might affect enzyme activity. Genetic effects might result in variations in either the structure of the enzyme or in the quantity of enzyme protein. They might also alter the cellular environment in such a way as to result in a change in enzyme activity. Gene effects on the "realization" of an enzyme activity, the final functional level of activity in a tissue, have been classified as those due to the actions of four major hypothetical gene types: structural, regulatory, processing, and temporal genes (PAIGEN 1971, 1979; PAIGEN et al. 1975). Structural genes determine the primary structure of the enzyme protein. Regulatory genes modulate the rate of enzyme synthesis and often respond to physiologic signals such as hormones. Processing genes regulate, for example, the cellular apparatus that determines the subcellular localization and metabolic degradation of enzyme molecules. Temporal genes determine the program for the expression of enzyme activity during the course of growth and development. Inherited variations in each of these gene types have been found to play a role in the realization of many enzyme activities in eukaryotic systems (PAIGEN et al. 1975; PAIGEN 1979). Although this classification is undoubtedly overly simplified, it will be useful to keep it in mind as the results of catecholamine biochemical genetic studies are reviewed.

B. Biochemical Genetics of Catecholamine Biosynthesis

I. Introduction

A great deal is now known with regard to the genetic regulation of TH, DBH, and PNMT, three of the four enzymes involved in catecholamine biosynthesis. Most of this information is based on studies of inbred rodents. Unfortunately, even though TH catalyzes the reaction that is thought to be the rate-limiting step in catecholamine biosynthesis (LEVITT et al. 1965), and even though PNMT catalyzes the formation of adrenaline from noradrenaline (AXELROD 1962), no human biochemical genetic data are available for these enzymes. This is true because neither TH nor PNMT is present in an easily obtainable human tissue such as blood or skin fibroblasts. DBH is the only one of the four catecholamine biosynthetic enzymes present in an easily obtainable human source, serum (WEINSHILBOUM and AXELROD 1971a). Extensive human biochemical genetic data with regard to serum DBH are available. In the subsequent discussion the results of studies of each enzyme performed with experimental animals will be reviewed followed by human biochemical genetic data whenever human information is available.

II. Tyrosine Hydroxylase (Tyrosine 3-Monooxygenase, EC 1.14.16.2, TH)

1. Experimental Animal Biochemical Genetics

a. Adrenal TH

Biochemical genetic studies of adrenal TH in laboratory animals have proceeded in a stepwise fashion. First, inbred strains of mice were surveyed and wide strain variations in TH activity were reported. Breeding experiments were then performed to characterize the mode of inheritance of the level of enzyme activity. Finally, attempts were made to define the biochemical basis of inherited variations in enzyme activity.

Early studies showed wide strain variations in adrenal TH activities among inbred mice (Table 1; CIARANELLO et al. 1972a; KESSLER et al. 1972). These inbred strains were mated to yield F1 or hybrid animals (Table 1). The results of the breeding studies were most compatible with the conclusions that the gene or genes controlling adrenal TH activity in BALB/cJ mice were "recessive" to those in the other two strains studied, and that the gene or genes regulating adrenal TH in CBA/J mice were "dominant" to those in the C57BL/Ka strain (Table 1; KESSLER et al. 1972). The terms "dominant" and "recessive" must be understood to be relative in their meanings (FALCONER 1963). It should also be emphasized that since these experiments did not include F2 or backcross animals, the mode of inheritance could not really be determined, and it was not possible to estimate the number of genes responsible for the strain differences in adrenal TH activity.

Table 1. TH, PNMT and DBH activities in mice. Data for BALB/cJ, CBA/J, C57BL/Ka and their F1 hybrids are from KESSLER et al. (1972). Data for BALB/cJ, BALB/cN, and their F1 and F2 generation descendants are from CIARANELLO et al. (1974). TH activities are expressed as nmoles of dihydroxyphenylalanine formed per hour per pair of adrenals or per g of brain. PNMT activity is expressed as nmoles of N-methyl-phenylethanolamine formed per hour per pair of adrenals. DBH activity is expressed as nmoles of phenylethanolamine formed per hour per pair of adrenals

Mouse Strain	Tyrosine Hydroxylase			Phenylethanolamine N-Methyltransferase		Dopamine-β-Hydroxylase	
	N	Adrenal	Brain	N	Adrenal	N	Adrenal
BALB/cJ	8	5.28 ± 0.81	14.6 ± 1.14	8	0.198 ± 0.015	—	—
CBA/J	8	1.99 ± 0.14	9.4 ± 0.66	8	0.124 ± 0.007	—	—
C57BL/Ka	8	1.40 ± 0.09	11.6 ± 0.99	8	0.085 ± 0.008	—	—
BALB/cJ × CBA/J, (F1)	16	2.24 ± 0.10	12.5 ± 0.66	16	0.119 ± 0.008	—	—
CBA/J × C57BL/Ka, (F1)	16	1.82 ± 0.16	11.7 ± 0.51	16	0.143 ± 0.010	—	—
C57BL/Ka × BALB/cJ, (F1)	16	2.17 ± 0.10	10.4 ± 0.70	16	0.118 ± 0.006	—	—
BALB/cJ	20	7.88 ± 0.70	—	43	0.367 ± 0.001	20	29.72 ± 2.98
BALB/cN	22	3.78 ± 0.37	—	42	0.194 ± 0.006	23	16.97 ± 1.11
BALB/cJ × BALB/cN, (F1)	48	5.62 ± 0.34	—	43	0.255 ± 0.006	48	21.13 ± 1.28
F1 × F1, (F2)	161	5.28 ± 0.27	—	163	0.263 ± 0.005	160	23.00 ± 0.92

More complete genetic studies were performed subsequently with two BALB substrains, BALB/cJ and BALB/cN (CIARANELLO et al. 1974). The adrenals of BALB/cN animals had only about half the TH activity present in the adrenals of BALB/cJ mice (Table 1). The BALB/cN animals also had only about half the adrenal activities of DBH and PNMT present in the BALB/cJ substrain (Table 1; CIARANELLO et al. 1974). The average adrenal TH activities in F1 and F2 generation animals resulting from matings between BALB/cN and BALB/cJ mice were intermediate to those in the inbred parental lines (Table 1). In addition, there was Mendelian segregation of the level of enzyme activity in F2 animals. Of these mice, 20% were included in a high enzyme activity subgroup, 54% were included in an intermediate activity subgroup, and 26% were included in a low activity subgroup (CIARANELLO et al. 1974). These percentages approximated the expected 1:2:1 ratios for monogenic inheritance by an autosomal codominant mechanism. Even more intriguing was the observation that adrenal DBH and PNMT activity levels in F2 mice were also inherited in a monogenic fashion and that they cosegregated with the levels of TH activity, i.e. an F2 mouse with low adrenal TH activity also had low DBH and PNMT activities. The same was true for animals with high or intermediate levels of activity. These results raised the possibility that a single genetically determined system might control the levels of activity of all three enzymes. That possibility will be discussed further when the data for PNMT are described.

b. Brain TH

The initial studies in which TH activity was determined in the adrenal glands of inbred mice included measurements of brain TH levels (Table 1). There were also strain variations in brain TH activities, but the magnitudes and directions of variation did not correlate directly with strain differences in adrenal TH (KESSLER et al. 1972). Variations in mouse brain TH activity were subsequently investigated with a series of biochemical and immunohistochemical experiments. The decreased level of TH activity in the brains of CBA/J mice compared with the brains of BALB/cJ mice was found to be due entirely to differences in enzyme activities in areas of the brain in which neurons contained dopamine rather than noradrenaline (Ross et al. 1976). TH activity in the brains of CBA/J mice in areas containing dopamine nerve cell bodies or terminals was only about 70–80% of that in the same brain areas of BALB/cJ animals. Immunoprecipitation with anti-TH antibodies demonstrated that the decrease in TH enzymatic activity in CBA/J mouse brain was due to a decrease in the quantity of TH protein (Ross et al. 1976). Immunohistofluorescent staining of TH-containing neurons showed that the total number of those neurons in areas of CBA/J mouse brain in which dopamine was the neurotransmitter was only about 70–80% of the number of TH-containing neurons in the same areas of BALB/cJ mouse brain (Ross et al. 1976; BAKER et al. 1980). However, the number of neurons per unit volume was the same in the two strains. Therefore the total volume of brain areas containing dopamine neurons or terminals was greater in BALB/cJ than in CBA/J mice. The caudate nucleus, for example, was significantly larger in the brains of BALB/cJ than in the brains of CBA/J animals. There were no differences between the two strains in brain TH activities at birth (REIS et al. 1981). Variations appeared only during the course of growth and development. Overall, these observations were best explained by strain-specific differences in the number of dopamine neurons in the brains of the two strains, differences that appeared during growth and development. Determination of the mode of inheritance of this presumed genetic trait awaits the completion of breeding studies. Breeding experiments will also help to clarify the possible relationship of variations in brain TH activity to behavioral and pharmacologic differences between these two mouse strains.

BALB/cJ mice display significantly more spontaneous locomotor activity and exploratory behavior than do CBA/J animals (REIS et al. 1981). They are also more sensitive than are CBA/J mice to drugs such as (+)-amphetamine, butyrophenones and phenothiazines that can alter dopamine neurotransmission (REIS et al. 1981). It has been speculated that there might be a relationship between strain variations in behavior and in drug sensitivity and strain variations in brain TH activity. If behavioral and pharmacologic differences co-segregate with brain TH levels during breeding experiments, that would provide strong support for a relationship between these phenomena. This example demonstrates one potential advantage of genetic experiments. The use of genetic techniques makes it possible to determine whether biochemical variables such as enzyme activities and functional variables such as drug response co-segregate in the course of breeding experiments.

III. Aromatic L-Amino Acid Decarboxylase (EC 4.1.1.28, AADC)

1. Experimental Animal Biochemical Genetics

Biochemical genetic studies of "dopa decarboxylase" activity have been performed in the fruit fly (HODGETTS 1975). Although the fruit fly data are of interest, the relationship of those results to the regulation of the enzyme activity in mammals is not clear. Therefore, studies of dopa decarboxylase in *Drosophila* will not be discussed further here. However, there has been at least one systematic attempt to study strain variations in the biochemical characteristics of AADC in inbred mice (CAVALLI-SFORZA et al. 1974). Electrophoresis of brain homogenates from 5 inbred mouse strains was performed and AADC activity was measured. No differences were found in the Rf values for the peak of AADC activity, but there were wide strain variations in total brain AADC activity. Breeding studies and attempts to detect inherited variations in other properties of AADC among inbred strains of laboratory animals have not been reported.

2. Human Biochemical Genetics

Biochemical genetic studies of human AADC have been hampered by inability to measure the enzyme activity in an easily obtainable tissue. Although it was reported that human erythrocytes contained AADC activity (TATE et al. 1971), later experiments indicated that the apparent erythrocyte activity was probably an artifact (DAIRMAN and CHRISTENSON 1973). AADC activity in preparations of human lymphocytes or platelets cannot be detected even with sensitive radiochemical enzymatic assays (WEINSHILBOUM, unpublished observation). Nevertheless, at least one attempt has been made to perform biochemical genetic studies of AADC in man. A search for electrophoretic variants of AADC in homogenates of human liver obtained at autopsy failed to detect significant variation in Rf values among the 20 samples tested (CAVALLI-SFORZA 1976). However, there were wide variations in the thermal stability of the enzyme. Heating of human liver homogenates at 55 °C resulted in a 90% inactivation of AADC after 3 minutes in some homogenates but only after 20 minutes in other samples (CAVALLI-SFORZA 1976). It was not possible to determine whether these variations in enzyme thermal stability might segregate within families. Thus, even though techniques for the detection of variations in Rf values of AADC after electrophoresis were developed expressly to evaluate the possibility of genetic polymorphism, little progress has been made in human biochemical genetic studies of this enzyme. The example provided by AADC illustrates the importance of the presence of the enzyme activity in an easily obtainable tissue for the population and family studies that are the first step in human biochemical genetic experiments. The lack of an easily obtainable source of the enzyme has hindered human biochemical genetic studies of AADC and has made it impossible to study TH and PNMT in man.

IV. Dopamine-β-Hydroxylase (Dopamine-β-Monooxygenase, EC 1.14.17.1, DBH)

1. Introduction

Many biochemical genetic studies of DBH have been performed in experimental animals and in man. The enzyme has been measured in both solid tissues and in serum of laboratory animals. Only serum DBH data are available for man. Although DBH is localized to catecholamine-containing vesicles in the adrenal medulla and adrenergic nerves, is released with catecholamines and is found in blood, the relationship of serum DBH to tissue levels of the enzyme is complex and controversial (WEINSHILBOUM 1978c). Therefore, the genetic data on serum DBH described below cannot be extrapolated to the situation in the adrenal medulla and nervous system. Finally, the fact that DBH is a tetrameric glycoprotein (WALLACE et al. 1973) may complicate the interpretation of the genetic data since several different structural genes may be involved in the regulation of this complex enzyme.

2. Experimental Animal Biochemical Genetics

a. Adrenal DBH

The BALB/cJ and BALB/cN mouse sublines with genetic differences in adrenal TH activities also had very different adrenal DBH activities (Table 1). Adrenal DBH activity in the BALB/cN strain was only about 60% of that present in BALB/cJ mice (CIARANELLO et al. 1974). Average adrenal DBH activities in F1 and F2 generation animals were intermediate to those of the parental strains (Table 1). Of more importance was the fact that there was segregation in F2 mice of the level of enzyme activity. Of the F2 animals, 27% were included in a high DBH activity subgroup, 48% were included in an intermediate activity subgroup and 25% were included in a low activity subgroup (CIARANELLO et al. 1974). These results, like those for adrenal TH, were compatible with the monogenic inheritance of the level of adrenal DBH activity by an autosomal codominant mechanism. As discussed previously, there was also co-segregation of the level of adrenal DBH activity with the levels of adrenal TH and PNMT activities.

b. Serum DBH

There are significant variations of serum DBH activity among inbred strains of rat (STOLK et al. 1979; Table 2). The results of breeding experiments performed with inbred rats were compatible with the autosomal codominant in-

Table 2. Serum DBH activity in inbred rats. Data are from STOLK et al. (1979). DBH activity is expressed as nmoles of synephrine formed per hour per ml of serum. All values are mean ± SEM

Strain	N	Male	N	Female
Lewis	37	12.47 ± 1.17	11	11.10 ± 0.73
Buffalo	13	10.32 ± 0.63	6	13.07 ± 0.63
Fischer-344	60	9.99 ± 0.83	39	12.90 ± 1.24
Wistar-Furth	35	7.62 ± 1.00	30	6.94 ± 0.91

heritance of the level of serum DBH activity in female rats. Inheritance was also monogenic in males, but low enzyme activity behaved as a dominant trait. The investigators reporting these data interpreted them as indicating the existence of sex-related modifiers of serum DBH activity in the rat (STOLK et al. 1979). The biochemical basis of monogenically inherited variations in rat serum DBH activity is not understood, and the relationship, if any, of these observations to the genetic regulation of human serum DBH activity to be described subsequently is unclear.

3. Human Biochemical Genetics

Biochemical genetic studies of human serum DBH have involved a much different research strategy than that used in experiments performed with inbred laboratory animals. These human experiments began with studies of siblings and twins, progressed to measurements of serum DBH activities in large randomly selected population samples, were expanded to include family studies, and finally involved evaluations of enzyme properties other than just the level of DBH activity.

Human serum DBH genetic studies began with the observation of a significant familial aggregation of the level of enzyme activity. Sibling-sibling correlation coefficients of approximately 0.5 were reported (WEINSHILBOUM et al. 1973; OGIHARA et al. 1975), and in three studies of monozygotic twins correlation coefficients of 0.96, 0.99, and 0.92 were found (ROSS et al. 1973; LEVITT and MENDLEWICZ 1975; WINTER et al. 1978). Based on the analysis of

▶

Fig. 1. Frequency distribution of human serum DBH activity. *TOP:* 48 siblings of probands with very low serum DBH activity (<50 units/ml). *MIDDLE:* 554 unrelated children. This distribution does not differ from that for randomly selected unrelated adults. *BOTTOM:* 32 parents of probands. Only data from 16 families in which both parents had activity of greater than 50 units/ml (i.e. both were presumed to be heterozygous at the locus *DBH* with genotype DBH^{L} DBH^{H}) are included in the top and bottom panels. DBH activity is expressed as nmoles of phenylethanolamine formed per hour per ml of serum. (Reproduced with permission of the Mayo Clinic Proceedings; WEINSHILBOUM 1977)

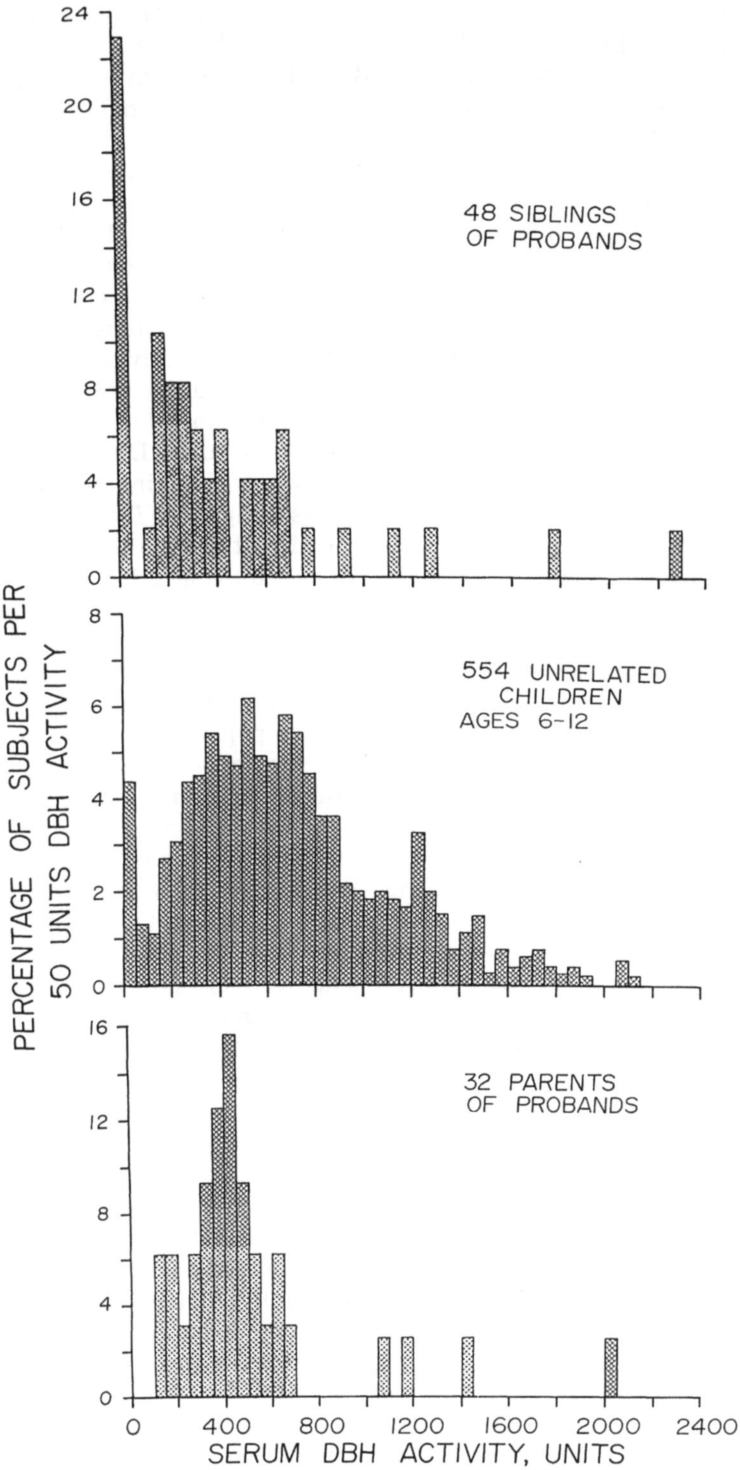
48 SIBLINGS
OF PROBANDS
554 UNRELATED
CHILDREN
AGES 6-12
32 PARENTS
OF PROBANDS
PERCENTAGE OF SUBJECTS PER
50 UNITS DBH ACTIVITY
SERUM DBH ACTIVITY, UNITS
0
400
800
1200
1600
2000
2400

data from one twin study, the heritability of serum DBH was estimated to be greater than 0.9 (LEVITT and MENDLEWICZ 1975). Although these results indicated that the heritability of serum DBH activity was high, they shed no light on the mechanism of inheritance. However, when serum DBH activity was measured in blood obtained from large randomly selected population samples, the frequency distribution histogram was skewed, and there was a subgroup of approximately 3-4% of subjects with very low enzyme activity (< 50 units/ml, Fig. 1, middle panel; WEINSHILBOUM et al. 1973, 1975; WEINSHILBOUM 1977). The existence of this subgroup has been confirmed independently during another large population study in which a completely different DBH assay technique was used (HEISS et al. 1980). The results of pedigree and segregation analysis of data from first-degree relatives of subjects with very low enzyme activity (< 50 units/ml) were compatible with the genetic regulation of human serum DBH activity by the autosomal codominant inheritance of two alleles, one for low and one for high serum DBH activity (WEINSHILBOUM et al. 1975). The allele for low activity was originally referred to as *d* and the alternative allele for high activity was designated *D* (WEINSHILBOUM et al. 1975). It has subsequently been proposed that these alleles be designated DBH^{L} and DBH^{H} (WEINSHILBOUM 1978c, 1979b), respectively, to conform to the recommendations of the Committee on Nomenclature of the Third International Workshop on Human Gene Mapping (1976).

The gene frequency of DBH^{L} was estimated to be approximately 0.2 and that of DBH^{H} was estimated to be approximately 0.8 (WEINSHILBOUM et al. 1975). About one-third of a randomly selected population was expected to be heterozygous at the locus *DBH*, and obligate heterozygotes (e.g. parents of offspring with low activity who themselves did not demonstrate the trait) had an average serum DBH activity significantly below that found in a randomly selected population sample but much higher than that in blood of subjects homozygous for DBH^{L} (Fig. 1, lower panel; WEINSHILBOUM et al. 1975). This finding was compatible with the autosomal codominant hypothesis. When matings of 16 pairs of obligate heterozygotes were studied, 22% of the siblings of probands with low activity themselves had very low enzyme activity (Fig. 1, upper panel; WEINSHILBOUM et al. 1975). Probands are the "index subjects" who are used to identify families for study. In this case all probands had low DBH activity. The 22% value for siblings of probands who also had low activity was much higher than the 3-4% of subjects with very low DBH found in a randomly selceted population sample. This 22% figure was similar to the 25% expected on the basis of Mendelian (monogenic) inheritance. Several subsequent studies of families and kindreds have confirmed monogenic inheritance of the basal level of human serum DBH enzymatic activity (ELSTON et al. 1979; GERSHON et al. 1980). It has been estimated that the locus *DBH* accounts for 50-75% of the total variance of serum DBH activity (ELSTON et al. 1979). Although the chromosomal location of the locus *DBH* is unknown, linkage studies have raised the possibility that *DBH* is linked to the *ABO* locus in man (GOLDIN et al. 1982).

Biochemical genetic studies of human serum DBH have included measurement of circulating DBH protein by both immunotitration and radioimmunoassay. A significant positive correlation between DBH enzymatic activity and DBH immunoreactive protein was found in samples from randomly selected subjects and in blood from first-degree relatives of subjects with very low enzymatic activity (< 50 units/ml; DUNNETTE and WEINSHILBOUM 1976, 1977). For example, in one radioimmunoassay study the correlation between enzymatic activity and the level of immunoreactive DBH protein was 0.84 ($n = 134$, $P < 0.001$; DUNNETTE and WEINSHILBOUM 1977). Subjects with very low levels of circulating enzyme activity also had very low levels of DBH immunoreactive protein (< 100 ng/ml), and the trait of low DBH immunoreactive protein co-segregated with the trait of low enzymatic activity (DUNNETTE and WEINSHILBOUM 1977). These results indicated that the locus *DBH* regulated not only the level of enzymatic activity but also the quantity of DBH protein in blood.

In the course of radioimmunoassay studies of serum DBH, one family was discovered in which the mother and three of five children had much lower ratios of DBH enzymatic activity to DBH immunoreactive protein than were found in any other of more than 200 subjects tested (DUNNETTE and WEINSHILBOUM 1977). The "dissociation" between enzymatic and immunoreactive DBH values in this family raised the possibility of a rare familial form of DBH that was enzymatically less active than the usual enzyme. The relative rarity of this trait contrasted with the frequency of the common polymorphism represented by the locus *DBH*.

As was discussed previously, genetic effects on the realization of an enzyme activity have been classified as those due to the effects of structural, regulatory, processing or temporal genes (PAIGEN 1971; PAIGEN et al. 1975). Studies of the effect of growth and development on human serum DBH have provided evidence that the locus *DBH* might be a temporal gene. At birth all children studied had very low umbilical cord blood levels of both DBH enzymatic activity (< 50 units/ml) and immunoreactive DBH protein (< 100 ng/ml; WEINSHILBOUM et al. 1978). These values were similar to those in the serum of adults homozygous for the allele DBH^{L}. In most subjects there was a rapid increase in serum DBH activity during the first few years of life (WEINSHILBOUM and AXELROD 1971b; FREEDMAN et al. 1972), an increase that apparently did not occur in the blood of subjects homozygous for the allele DBH^{L}. Therefore, the effect of the allele DBH^{H} is apparently that of a temporal gene, but the biochemical mechanism responsible for the effect is unknown.

Attempts to study the biochemical basis of the genetic regulation of human serum DBH also led to experiments in which variations in the thermal stability of the enzyme were detected. Approximately 8% of a randomly selected population sample was found to have a thermolabile form of serum DBH (DUNNETTE and WEINSHILBOUM 1979, 1982a; BARON et al. 1982a). Thermolability was a characteristic of the DBH molecule itself (DUNNETTE and WEINSHILBOUM 1979; BARON et al. 1982a), and depended on an interaction of DBH with oxygen (DUNNETTE and WEINSHILBOUM 1981). The trait of thermolability showed a significant familial aggregation (DUNNETTE and WEINSHILBOUM 1979, 1982a), and subjects with thermolabile serum DBH had basal enzyme activities that averaged only about 55% of those in the blood of subjects with thermostable enzyme (DUNNETTE and WEINSHILBOUM 1979, 1982a). However, this trait did not co-segregate with the allele *DBH*^{L} (DUNNETTE and WEINSHILBOUM 1979, 1982a). Therefore, the familial trait of thermolabile serum DBH is associated with and explains a part of the residual variance in basal levels of the serum enzyme activity not regulated by the locus *DBH* itself (DUNNETTE and WEINSHILBOUM 1979, 1982a). The biochemical basis for the trait of thermolability is not known. Interest in this familial enzyme characteristic is heightened by the possibility that it may represent a genetic polymorphism of a structural gene for human DBH, a polymorphism separate and distinct from the polymorphism at the locus *DBH*.

Finally, biochemical genetic studies of serum DBH have potential phylogenetic and evolutionary importance. Until recently, the high level of serum DBH activity that is associated with the allele *DBH*^{H} in man had not been found in any non-human species (WEINSHILBOUM 1978c). All other species previously studied had levels of activity one to two orders of magnitude lower than those in the blood of humans who carry the allele *DBH*^{H}. In an attempt to determine how recently in the course of evolution the allele *DBH*^{H} arose, DBH activity was measured in serum samples from five species of ape and two species of monkey (DUNNETTE and WEINSHILBOUM 1982b, 1983). Samples were obtained from 33 gorillas, 45 chimpanzees, 4 pygmy chimpanzees, 9 orangutans, 13 gibbons, 10 rhesus monkeys and 10 squirrel monkeys. Only gorillas had levels of serum DBH activity comparable to those in humans with the allele *DBH*^{H}. It cannot be determined with certainty whether high serum DBH activity in gorillas is due to a polymorphism at the locus *DBH*. However, whether the genetic polymorphism is present only in man, or only in man and the gorilla, the allele *DBH*^{H} and its manifestation of high serum enzyme activity are of very recent origin in the course of primate evolution. Even though it appeared only recently, *DBH*^{H} has become the predominant allele in man with a gene frequency of 80%. The possible selective advantage of *DBH*^{H}, if any, is unknown.

Biochemical genetic studies of human serum DBH demonstrate several ways in which genetic information may help to increase our understanding of catecholamine enzymes. Initially, it was suggested that serum DBH activity might be a simple and accurate measure of the exocytotic release of catecholamines (WEINSHILBOUM and AXELROD 1971a). However, the results of subsequent clinical studies of serum DBH were often confusing and difficult to interpret (WEINSHILBOUM 1978c). It is now known that 50–75% of the variance in basal human serum DBH activity results from the effects of the genetic polymorphism at the locus *DBH*, and that much of the remaining variance may be related to the familial trait of thermolabile DBH. The relationship of these factors to the exocytotic release of catecholamines is unclear. However, the biochemical genetic results will now make it possible to evaluate measurements of serum DBH in a more intelligent fashion than was previously possible.

V. Phenylethanolamine-N-Methyltransferase (Noradrenaline-N-Methyltransferase, EC 2.1.1.28, PNMT)

1. Introduction

PNMT catalyzes the N-methylation of noradrenaline to form adrenaline (AXELROD 1962). It is a "cytoplasmic" enzyme localized to the adrenal medulla and to specific nuclei in the brain. Both humoral and neural factors are involved in the regulation of adrenal PNMT (WURTMAN and AXELROD 1966; CIARANELLO et al. 1972b; CIARANELLO 1978), but hormones are usually considered of greatest regulatory importance. Adrenal PNMT in rats and mice decreases after hypophysectomy and increases in response to treatment of hypophysectomized animals with adrenocorticotrophic hormone or glucocorticoids (WURTMAN and AXELROD 1966; CIARANELLO et al. 1972b). Much is now known about the genetic regulation of adrenal PNMT in rodents. Unfortunately, nothing is known of a possible role of inheritance in the regulation of this important enzyme in man since there is no easily obtainable source of human PNMT.

2. Experimental Animal Biochemical Genetics

The same studies in which TH activity was measured in the adrenal glands and brains of inbred mice included determinations of adrenal PNMT activity. Greater than 2-fold strain variations in adrenal PNMT activity were found (CIARANELLO et al. 1972a,b; KESSLER et al. 1972; Table 1). F1 hybrids from mating among these strains gave results similar to those found for adrenal TH measured in the same strains (Table 1). The gene or genes controlling BALB/cJ adrenal PNMT activity were apparently recessive to those in the other two strains studied, and the gene or genes regulating adrenal PNMT in CBA/J mice were dominant to those of the C57BL/Ka strain. Unfortunately, since these breeding experiments did not include F2 or backcross animals, it was not possible to estimate either the number of genes responsible for the presumed inherited differences in PNMT activities or to determine the mode of inheritance.

Later it was reported that there were large differences in adrenal PNMT activities between the BALB/cJ and BALB/cN substrains in which there were also differences in adrenal TH and DBH activities (CIARANELLO and AXELROD 1973; CIARANELLO et al. 1974). PNMT activity was approximately twice as great in BALB/cJ as it was in BALB/cN mice (Table 1). Breeding studies showed that average enzyme activities in F1 and F2 animals were intermediate to those of the parental strains and that there was segregation of the level of adrenal PNMT activity in F2 mice (Table 1). Of these animals, 27% were included in a low activity subgroup, 49% were included in an intermediate activity subgroup, and 24% had high enzyme activity (CIARANELLO et al. 1974). These results were compatible with the monogenic inheritance of the level of PNMT activity by an autosomal codominant mechanism. PNMT activity cosegregated with the levels of adrenal TH and DBH activities. Immunoprecipitation performed with anti-bovine adrenal PNMT antibodies showed that differences in PNMT activities between the BALB/cJ and BALB/cN substrains were due to differences in the quantity of PNMT protein (CIARANELLO and AXELROD 1973). Several biochemical properties of adrenal PNMT including thermal stability, stability to proteolytic digestion, apparent K_m constants for substrate and electrophoretic mobility were tested in the two substrains and were found to be identical, an observation that made structural gene variation unlikely as an explanation for the substrain differences in levels of enzyme activity (CIARANELLO and AXELROD 1973). Rates of adrenal PNMT synthesis and degradation were then measured in these mice. PNMT protein was labeled with radioactive leucine and immunoprecipitation with anti-PNMT antibodies was performed at different times after *in vivo* labeling. The results of these experiments were compatible with the conclusion that the rate of degradation of PNMT was twice as rapid in BALB/cN as in BALB/cJ mice (CIARANELLO and AXELROD 1973). The coordinate genetic regulation in these two substrains of the activities of three adrenal catecholamine biosynthetic enzymes, TH, DBH, and PNMT, led to speculation that the degradation of all three enzymes was controlled by a common inherited mechanism (CIARANELLO et al. 1974). That hypothesis remains unproven.

Possible genetic regulation of the response of adrenal PNMT activity to pharmacologic treatment and "stress" has also been studied. Wide strain variations in the magnitude of the response of adrenal PNMT to treatment with glucocorticoids, ACTH, phenoxybenzamine and cold exposure in inbred mice have been described (CIARANELLO et al. 1972b). Although it was assumed that these strain differences were inherited, no breeding studies were performed and nothing is known of the biochemical basis for possible genetic variations in the response of PNMT to drugs, hormones or environmental stress.

Studies of the inheritance of mouse adrenal PNMT activity provide an example of an additional potential benefit of genetic experiments. Because these studies focussed attention on the degradation of the enzyme, they led to a series of experiments the results of which indicated that the degradation of adrenal PNMT may be partially regulated by "stabilization" of the enzyme by S-adenosyl-L-methionine, the methyl donor for the PNMT-catalyzed reaction (CIARANELLO 1978). Although these observations raised the possibility of inherited variations in regulation of levels of S-adenosyl-L-methionine that might in turn result in variations in the regulation of PNMT, that possibility remains to be tested experimentally.

C. Biochemical Genetics of Catecholamine Metabolism

I. Introduction

The biochemical genetics of the enzymes involved in catecholamine metabolism have been studied more extensively in man than in experimental animals. This is true in part because the two major enzymes of catecholamine catabolism, COMT and MAO, can be measured in human blood elements and in skin fibroblasts. In addition, interest in a possible role of these enzymes in the pathophysiology of neuropsychiatric disease has served as a stimulus for human biochemical genetic studies. Extensive genetic studies of COMT have also been performed with laboratory animals, and these inbred animals can be used for experiments that cannot be performed in man. In addition to MAO and COMT, other enzymes such as phenol sulfotransferase may play a role in the metabolism of catecholamines. Little is known about the possible role of inheritance in the regulation of these other catecholamine catabolic enzyme activities.

II. Catechol-O-Methyltransferase (EC 2.1.1.6, COMT)

1. Introduction

COMT catalyzes the ring O-methylation of catecholamines and of catechol drugs such as isoprenaline (Axelrod and Tomchick 1958; Conolly et al. 1972). S-Adenosyl-L-methionine is the methyl donor for the reaction and magnesium is required to activate the enzyme (Axelrod and Tomchick 1958). COMT is present in a variety of tissues including the erythrocyte (Axelrod and Tomchick 1958; Axelrod and Cohen 1971). Erythrocyte COMT is biochemically and immunologically similar to COMT in other tissues (Assicot and Bohuon 1969; Creveling et al. 1973; Quiram and Weinshilboum 1976). A small amount of the COMT activity in erythrocytes and other tissues is membrane-associated (Assicot and Bohuon 1971; Roffman et al. 1976). All of the biochemical genetic studies described below involved measurements of only the soluble enzyme activity. The existence of isozymes of COMT in some experimental animal tissues has also been reported (Axelrod and Vesell 1970; Huh and Friedhoff 1979). The relationship of these observations to the biochemical genetic data reviewed below has not been studied.

2. Experimental Animal Biochemical Genetics

An early study of COMT activity in inbred mice showed that C57BL/6J animals had higher brain enzyme activity than did 6 other strains including DBA/2J animals (Tunnicliff et al. 1973). These results were subsequently confirmed when COMT activity was measured in brains and livers of C57BL/6J and DBA/2J mice (Schlesinger et al. 1975; Table 3). Strain variations in the enzyme activity were organ-specific since COMT activities in the livers of the two strains were virtually identical (Table 3). F1 offspring from matings between C57BL/6J and DBA/2J mice had brain enzyme activities between those of the two parental strains, but closer to activities in the brains of C57BL/6J animals (Schlesinger et al. 1975; Table 3). It was not possible to evaluate the number of loci involved in the presumed genetic regulation of mouse brain COMT activity since the breeding studies were only carried through an F1 generation. There were no differences between these strains in apparent K_m constants of the enzyme for the catechol substrate, 3,4-dihydroxybenzoic acid, or for the methyl donor, S-adenosyl-L-methionine (Schlesinger et al. 1975). Optimal magnesium concentrations for brain COMT were

Table 3. COMT activity in mice. Activity in C57BL/6J and DBA/2J mice was measured with 3,4-dihydroxybenzoic acid as substrate (SCHLESINGER et al. 1975). Results of those experiments are expressed as nmoles of 3-methoxy-4-hydroxybenzoic acid formed per hour per mg protein. Data for C57BL and C3HF mice were obtained with 3,4-dihydroxyphenylacetic acid as substrate (GERSHON and JONAS 1976). The results of those experiments are expressed as nmoles of 3-methoxy-4-hydroxyphenylacetic acid formed per hour per mg protein. All values are mean ± SEM

Mouse Strain	Tissue	N	COMT Activity
C57BL/6J	Brain	15	2.13 ± 0.29
DBA/2J	Brain	15	1.29 ± 0.23
C57BL/6J × DBA/2J, (F1)	Brain	10	2.02 ± 0.24
C57BL/6J	Liver	5	13.23 ± 1.50
DBA/2J	Liver	5	14.20 ± 1.53
C57BL	Brain	—	26.03 ± 0.47
C3HF	Brain	—	18.22 ± 0.33
C57BL × C3HF, (F1)	Brain	—	24.47 ± 0.41
C57BL	Erythrocyte	—	46.44 ± 1.27
C3HF	Erythrocyte	—	55.85 ± 0.95
C57BL × C3HF, (F1)	Erythrocyte	—	40.88 ± 0.58

also identical in the two strains. An independent study of COMT activity in inbred mice showed that C57BL animals had higher brain enzyme activity than did C3HF mice (GERSHON and JONAS 1976; Table 3). F1 hybrids from matings between these strains had levels of brain enzyme activity similar to those in the C57BL parental strain (GERSHON and JONAS 1976). There was also evidence of tissue specificity for the regulation of enzyme in these animals since C3HF mice had higher erythrocyte COMT activity than did C57BL animals. These breeding experiments were also carried only through the F1 generation. However, the results of both of these experiments in mice were compatible with the dominance or "partial" dominance of the gene or genes regulating brain COMT in C57BL over those regulating enzyme activity in the brains of DBA/2J and C3HF mice.

COMT activity has also been measured in the tissues of several strains of rat with the goal of performing biochemical genetic experiments. Variations in liver COMT activities among 1 outbred and 8 inbred rat strains are shown in Fig. 2 (WEINSHILBOUM and RAYMOND 1977b). Fischer-344 rats had much lower liver and kidney COMT activities than did the other 7 inbred strains studied. For example, Fischer-344 animals had only about 60% of the liver and kidney activities found in Wistar strains such as Wistar-Furth (WEINSHILBOUM and RAYMOND 1977b). Variations in rat COMT activity, like the activity in mice, showed organ specificity. Liver and kidney contained the highest COMT levels of any organs studied in these animals. Heart, brain, lung, intestine, and erythrocyte contained less than 10% as much activity as the liver

(WEINSHILBOUM and RAYMOND 1977b). There was little difference between Fischer-344 and Wistar-Furth animals in heart, brain, lung, intestine, and erythrocyte COMT activities. The results of breeding experiments performed with Fischer-344 and Wistar-Furth rats were compatible with the inheritance of the level of liver and kidney COMT activities in these rats in a monogenic fashion (WEINSHILBOUM et al. 1979). The trait of low enzyme activity was inherited as an autosomal recessive characteristic, and 24.5% (36/147) of F2 animals had a low activity phenotype. This result was close to the expected Mendelian 1:3 ratio. There was only incomplete dominance of the high activity phenotype since both liver and kidney values in F1 animals were closer to, but not identical with, those in the high activity than those in the low activity parental animals. The locus responsible for inherited variation of COMT activity in these rats was designated *Co* with two alleles, Co^{l} and Co^{h} for low and high activities, respectively (WEINSHILBOUM et al. 1979).

An attempt was made to determine whether the locus *Co* might represent the structural gene for COMT. Several biochemical properties of the enzyme were measured in tissue homogenates from the two strains. No differences were found in any of the following properties: thermal stability; apparent Michaelis-Menten constants for catechol substrate, S-adenosyl-L-methionine or magnesium; IC50 values for three different types of COMT inhibitors; and electrophoretic mobility (WEINSHILBOUM et al. 1979). Immunotitration with anti-rat COMT antibody demonstrated that the differences in liver and kidney enzyme activity in these strains were due to differences in the quantity of COMT protein (GOLDSTEIN et al. 1980). Although these results did not eliminate the possibility that the locus *Co* might be the structural gene for COMT in the rat, they made it much less likely and raised the possibility that regula-

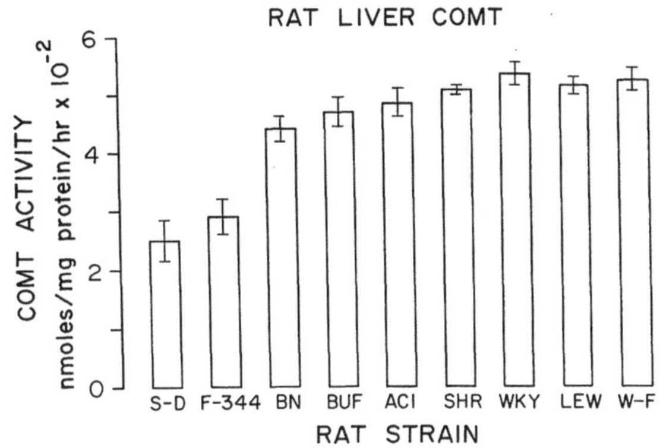

Fig. 2. Rat liver COMT. COMT activities in livers of male rats 9 to 11 weeks old are shown. Abbreviations for strain names include the following: S-D, Sprague-Dawley; F-344, Fischer-344; BUF, Buffalo; SHR, Spontaneously hypertensive rats of Okamoto; WKY, Wistar Kyoto; LEW, Lewis Wistar; W-F, Wistar-Furth. All values are mean ± SEM for 5 animals. COMT activity is expressed as nmoles of 3-methoxy-4-hydroxybenzoic acid formed per hour per mg protein. (Reproduced with the permission of Pergamon Press; WEINSHILBOUM and RAYMOND 1977b)

tory, processing or temporal gene effects might be responsible for differences in enzyme activity between Wistar-Furth and Fischer-344 animals.

To test the possibility of temporal gene effects, the patterns of change of COMT during growth and development were studied. There was little change in organs such as the heart and brain with relatively low enzyme activities, but there were 5-10 fold increases in liver and kidney activities in both strains between birth and 12 weeks of age (GOLDSTEIN et al. 1980). However the patterns of change in hepatic and renal COMT activity during growth and development were identical in the two strains. This observation made temporal gene effects much less likely as an explanation for the strain differences in COMT activity. It remains to be determined whether differences in the rates of COMT synthesis and/or degradation are responsible for inherited differences in the enzyme activity in Fischer-344 and Wistar-Furth rats. However, it is clear that the genetic regulation of COMT in these animals is much different from the genetic regulation of COMT in man that is described subsequently. Although these inbred rats may not be an adequate genetic model for the human situation, they may be useful for pharmacologic experiments that cannot be performed in man.

3. Human Biochemical Genetics

Biochemical genetic studies of human erythrocyte COMT, like those of human serum DBH, began with reports of a significant familial aggregation of the level of enzyme activity. Sibling-sibling correlation coefficients of approximately 0.5 for erythrocyte COMT activity were reported (WEINSHILBOUM et al. 1974; GERSHON and JONAS 1975). Two independent groups of investigators found correlation coefficients of 0.37 and 0.65 for dizygotic twin pairs while those for monozygotic twin pairs were 0.90 and 0.95 (GRUNHAUS et al. 1976; WINTER et al. 1978). Estimates of heritability calculated on the basis of the data from one twin study ranged from 0.68 to 1.0 (GRUNHAUS et al. 1976).

The possibility of monogenic inheritance of erythrocyte COMT activity was raised when large randomly selected population samples were found to include a subgroup of approximately 25-30% of subjects with low COMT activity (< 8 units/ml; Figs. 3 and 4A; WEINSHILBOUM et al. 1974). This subgroup was identified consistently in all large studies in which care was taken to remove calcium from red cell lysates (WEINSHILBOUM et al. 1974; WEINSHILBOUM and RAYMOND 1977b; WETTERBERG et al. 1979; SCANLON et al. 1979; FLODERUS and WETTERBERG 1981). Calcium is a potent COMT inhibitor (RAYMOND and WEINSHILBOUM 1975; WEINSHILBOUM and RAYMOND 1976). The results of pedigree and segregation analysis of data from families selected because they included a proband from the "low" activity subgroup were compatible with monogenic inheritance of the trait of low COMT activity (< 8 units/ml) by an autosomal recessive mechanism (WEINSHILBOUM and RAYMOND 1977a; WEINSHILBOUM 1978a). Subsequent analysis of these data showed that the overall level of enzyme activity, as contrasted with the "trait" of low activity, was inherited in an autosomal codominant fashion (SPIELMAN and WEINSHILBOUM 1979, 1981). Subjects homozygous for an allele for low ac-

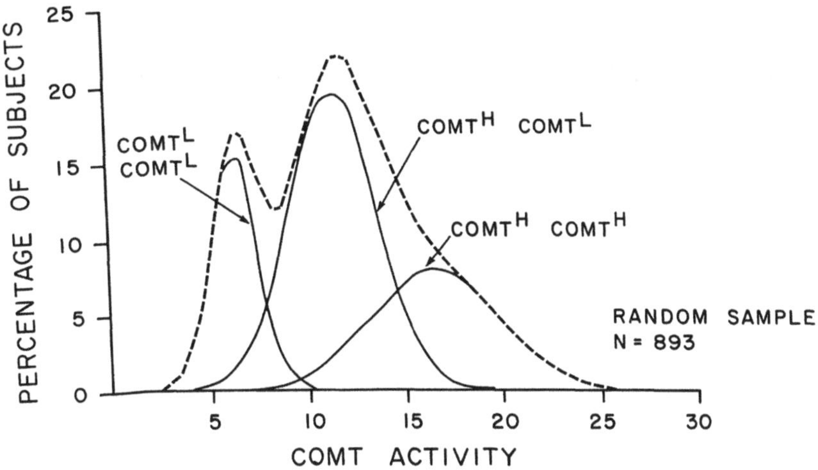

Fig. 3. Schematic representation of the frequency distribution of erythrocyte COMT activities and of *COMT* genotypes in a randomly selected population sample. The dashed line represents the distribution of levels of erythrocyte COMT activity in blood samples from 893 randomly selected subjects. The solid lines represent calculated distributions of genotypes at the locus *COMT*. COMT activity is expressed as nmoles of 3-methoxy-4-hydroxybenzoic acid formed per hour per ml packed erythrocytes. (Modified from SPIELMAN and WEINSHILBOUM 1981. Reproduced with permission of Alan R. Liss, Inc.)

tivity, $COMT^L$, had levels of enzyme activity of less than 8 units/ml while subjects homozygous for an alternative allele for high activity, $COMT^H$, had high erythrocyte COMT (Fig. 3). About half of a randomly selected population was heterozygous and had intermediate activity. It was estimated that this polymorphism accounted for at least 60% of the total variance in human erythrocyte COMT activity (GERSHON et al. 1980). The gene frequencies of the two alleles at the locus *COMT* were approximately equal in a white population (WEINSHILBOUM and RAYMOND 1977 a; WEINSHILBOUM 1978 a), but preliminary data indicated that the gene frequencies of $COMT^L$ and $COMT^H$ might vary widely among different racial groups (REILLY and RIVERA-CALIMLIM 1979). Subsequent studies by other investigators have confirmed the autosomal codominant inheritance of the level of activity of human erythrocyte COMT (GERSHON et al. 1980; FLODERUS and WETTERBERG 1981; FLODERUS et al. 1981).

The next step in biochemical genetic studies of erythrocyte COMT involved attempts to determine whether inherited regulation of this enzyme might result from structural gene variation. If the polymorphism were due to the effects of a structural gene, the biochemical properties of the enzyme might differ in blood from subjects with different genotypes at the locus *COMT*. A series of biochemical characteristics of the human erythrocyte enzyme were tested including pH optimum, apparent Michaelis-Menten constants for substrate and magnesium, effect of COMT inhibitors and thermal

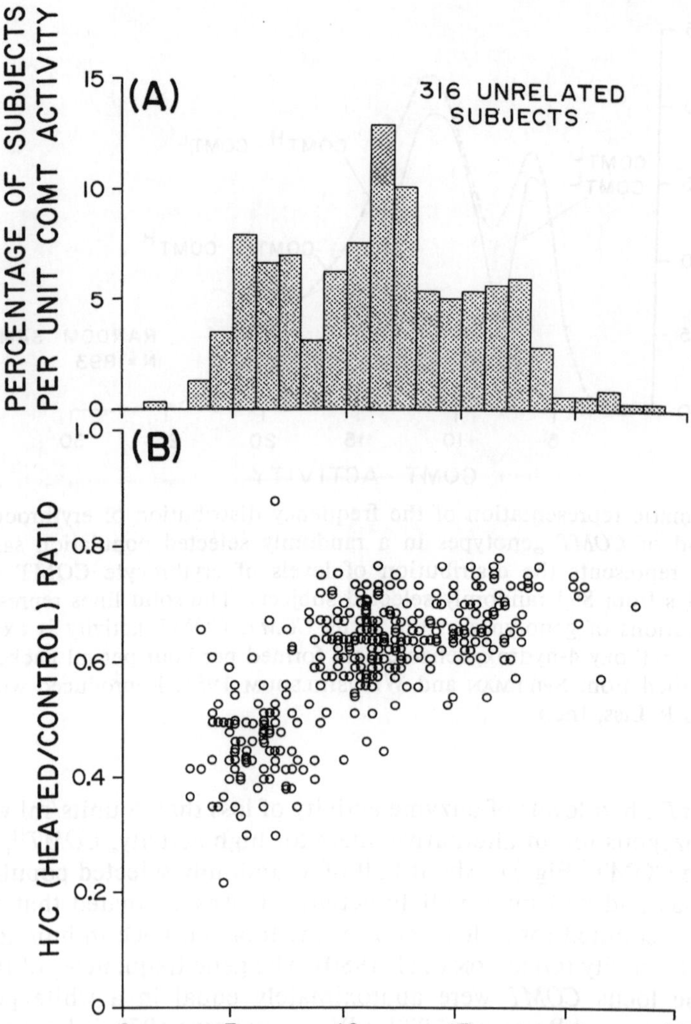

Fig. 4. Erythrocyte (RBC) COMT activity and thermal stability of erythrocyte COMT. (A) Frequency distribution of erythrocyte COMT values in blood samples from 316 randomly selected adult subjects. COMT activity is expressed as nmoles of 3-methoxy-4-hydroxybenzoic acid formed per hour per ml packed erythrocytes. **(B)** Thermal stability of RBC COMT as measured by heated divided by control (H/C) ratios is plotted against basal enzyme activity for the same 316 subjects for whom data are shown in (A). See text for details. (Reproduced with permission of the American Association for the Advancement of Science; SCANLON et al. 1979)

stability. Only thermal stability was found to differ among samples from subjects with different genotypes at the locus *COMT* (SCANLON et al. 1979; WEINSHILBOUM 1979a). The thermal stability of COMT was measured in erythrocyte lysates from 316 randomly selected subjects by heating the lysates at 48 °C for 15 minutes. The fraction of enzyme activity that remained after heating, a heated (H) divided by control (C) or H/C ratio, was used as a measure of thermal stability. The 84 subjects who were presumed to be homozygous for the allele $COMT^L$ (< 8 units/ml) had a significantly lower average H/C ratio, 0.47 ± 0.01 (mean ± SEM), than did the 232 subjects with higher enzyme activity, 0.65 ± 0.01 ($P < 0.001$; Fig. 4B; SCANLON et al. 1979; WEINSHILBOUM 1979a). Variations in thermal stability were due to differences in COMT itself and were not due to the effects of heating on other constituents of the lysates (SCANLON et al. 1979; WEINSHILBOUM 1979a). These differences in erythrocyte COMT thermal stability have subsequently been independently confirmed (BARON 1982b). The fact that the biochemical properties of COMT differ in erythrocytes of subjects with different genotypes at the locus *COMT* suggested that this locus was the structural gene for human COMT. However, genetically determined post-translational modification of the enzyme cannot be excluded as a possible explanation for the findings.

If *COMT* is the structural gene for the human enzyme, a genetic polymorphism at this locus might be expressed in many different tissues. To test this possibility, human lung and kidney tissue were obtained during clinically indicated surgical procedures and COMT activity was measured. There was a significant positive correlation between relative erythrocyte COMT activity and that in renal cortical tissue ($r = 0.81$, $P < 0.005$) from 12 subjects and in lung tissue ($r = 0.62$, $P < 0.001$) from 29 subjects (WEINSHILBOUM 1978b). The genetically determined level of erythrocyte COMT also correlated significantly with lymphocyte COMT activity ($r = 0.733$, $n = 23$, $P < 0.001$; SLADEK-CHELGREN and WEINSHILBOUM 1981). However, it cannot be assumed that the genetic polymorphism for COMT activity that is expressed in the erythrocyte is the only source of regulation of the enzyme. The results of at least one study of cultured skin fibroblasts showed no significant correlation between fibroblast COMT activity and the enzyme activity in erythrocytes from the same subjects (GROSHONG et al. 1977).

The fact that genetically determined levels of erythrocyte COMT reflect levels of enzyme activity in other tissues and organs increased the possibility that this common genetic polymorphism might result in functionally significant variations in the metabolism of endogenous catecholamines and/or of catechol drugs. Data which support that possibility have appeared. There was a significant positive correlation between erythrocyte COMT activity and the proportion of L-dopa converted to 3-O-methyldopa in patients with Parkinson's disease who were treated with L-dopa (REILLY et al. 1980). There was also a significant correlation between erythrocyte COMT activity and resistance to L-dopa in the same patients. Thus, there is evidence that the common genetic polymorphism at the locus *COMT* is related to the regulation of the enzyme activity in other tissues and is functionally significant—at least in a "pharmacogenetic" sense. It remains to be determined whether individual variations in the metabolism of endogenous catecholamines in man are also affected by inherited variations in COMT activity. Finally, the genetic polymorphism for COMT in man is, like that for serum DBH, a balanced or common polymorphism. The possibility exists that these polymorphisms are maintained in the human population by selective evolutionary pressures. The nature of these selective pressures and the possible significance of these genetic polymorphisms for the evolution of man remain unanswered questions.

III. Monoamine Oxidase (Amine Oxidase [Flavin-containing], EC 1.4.2.4, MAO)

1. Introduction

MAO is a membrane-bound mitochrondrial enzyme (BLASCHKO 1974). MAO is thought to be a dimer composed of two subunits of approximately equal molecular weight, one of which has a covalently attached flavin cofactor (MINAMIURA and YASUNOBU 1978). Mitochondrial MAO activity is classified as either type A or type B on the basis of differential substrate specificity and sensitivity to inhibitors (JOHNSON 1968; FUENTES and NEFF 1975; MURPHY 1978). Type A MAO preferentially catalyzes the oxidation of noradrenaline and 5-HT, while phenylethylamine and benzylamine are relatively specific substrates for type B MAO. Clorgyline is a more potent inhibitor of type A than of type B MAO, while deprenyl is a more potent inhibitor of the type B enzyme (JOHNSON 1968; FUENTES and NEFF 1975; MURPHY 1978). Controversy has existed as to whether types A and B MAO represent separate enzymes (HOUSLAY et al. 1976). Evidence based on limited proteolysis and peptide mapping favors the view that the enzymes are truly different (CAWTHON and BREAKEFIELD 1979). Human platelet MAO, the source of the enzyme for many human biochemical genetic studies, behaves as a type B enzyme (MURPHY 1978), while the MAO in human fibroblasts is primarily type A (ROTH et al. 1976). Because of the biochemical heterogeneity of MAO it is important to consider both the tissue source and the specific substrates used in biochemical genetic studies of this enzyme. Although there is a plasma MAO activity, that enzyme differs biochemically from mitochondrial MAO (BLASCHKO 1974). Plasma MAO activity will not be discussed further here.

2. Experimental Animal Biochemical Genetics

Very few biochemical genetic studies of MAO have been performed with experimental animals. However, significant strain variations in brain MAO activity among inbred mice and rats have been reported. In one study BALB/cJ mice were found to have only about 64% as much brain MAO activity as did CBA/CaJ mice (TUNNICLIFF et al. 1973). Another group of investigators reported that Fawn Hooded rats had about twice the brain MAO activity present in Long-Evans rats (BELLIN and SORRENTINO 1974). Neither of these reports included systematic breeding studies, so nothing is really known of the possible role of inheritance in the regulation of MAO activity in inbred rodents.

At least one attempt was made to study the role of inheritance in the regulation of platelet MAO in the rhesus monkey *(Macaca mulatta)*. The correlation coefficient for platelet MAO activity in 29 pairs of half sibling monkeys (fathers unknown) was 0.26 and the correlation coefficient for randomly selected unrelated pairs was 0.04 ($P < 0.05$; MURPHY et al. 1978). The investigators who reported these results interpreted them as an indication of significant "genetic influences" on platelet MAO activity in the rhesus monkey.

3. Human Biochemical Genetics

a. Platelet MAO

Biochemical genetic studies of human platelet MAO began with the observation of a significant positive correlation ($r = 0.94$) of platelet enzyme activity in blood samples from 9 pairs of normal monozygotic twins (WYATT et al. 1973). Tryptamine was used as a substrate for these experiments. Simultaneously, a separate group of investigators reported a correlation coefficient of 0.76 for platelet MAO in monozygotic twins, a correlation coefficient of 0.39 in dizygotic twins and a correlation coefficient of −0.16 in pairs of control subjects (NIES et al. 1973). Benzylamine was used as substrate. The heritability of platelet MAO was estimated to be 0.83 on the basis of these latter data. The same authors reported a bimodal frequency distribution of platelet MAO activites in 80 subjects. Approximately equal numbers of subjects were included in "low" and "high" enzyme activity subgroups (NIES et al. 1973). Another group of investigators studying a relatively small number of subjects reported an apparent bimodal distribution of platelet MAO activity when phenylethylamine was used as substrate (SANDLER et al. 1974). However, when platelet MAO activity in blood samples from 680 randomly selected subjects was measured with benzylamine as substrate, the frequency distribution was slightly skewed, but there was no evidence of bimodality (MURPHY et al. 1976). The authors of other studies involving large population samples have reported skewed frequency distributions with no apparent subgroups (PANDEY et al. 1979; GERSHON et al. 1980).

The high heritability of platelet MAO activity has been verified repeatedly. For example, when 255 normal adult members of 112 families were studied, the correlation coefficient for platelet MAO activity measured with tyramine as a substrate was 0.27 for 45 parent-offspring pairs and was 0.32 for 75 sibling-sibling pairs (PANDEY et al. 1979). In the same study correlation coefficients of 0.54 and 0.55 for parent-offspring and sibling-sibling pairs, respectively, were found when benzylamine was used as substrate. Heritability was estimated to be 0.54 for the tyramine data and 1.0 for the data obtained with benzylamine as substrate (PANDEY et al. 1979). No evidence of monogenic inheritance of platelet MAO activity was found in this study or in another large study of 1,125 relatives of 162 patients with primary affective disorders and 10 patients with schizo-affective disease (GERSHON et al. 1980). Data from the latter study, one in which benzylamine was used as a substrate, was evaluated by the computer pedigree analysis method developed by ELSTON and STEWART (1971).

The possible inheritance of properties of platelet MAO other than enzyme activity has also been investigated. Wide individual variations in the thermal stability of human platelet MAO have been reported (WISE et al. 1979; BRIDGE et al. 1981). The possibility of monogenic inheritance of this characteristic of the enzyme was suggested on the basis of the shapes of frequency distributions and on the basis of very limited family data (BRIDGE et al. 1981). In the absence of larger family or twin studies, it is not yet possible to decide whether MAO thermal stability might be an inherited trait. However, further experiments of this type offer one possible avenue through which additional biochemical genetic information on human platelet MAO might be obtained.

In summary, there has been much interest in the biochemical genetics of platelet MAO. This interest was generated in part because of the association of low platelet enzyme activity with psychiatric disease such as chronic schizophrenia (WYATT et al. 1979). Inheritance clearly plays an important role in the regulation of human platelet MAO activity. The mode of inheritance, monogenic or polygenic, is not known, but there is presently no evidence for monogenic inheritance. The biochemical basis for the genetic regulation of the enzyme activity is also unclear. Finally, it is not known whether the genetic regulation of MAO B in the human platelet reflects the regulation of MAO B activity in other tissues and organs.

b) Fibroblast MAO

Studies of MAO activity in cultured skin fibroblasts have resulted in the assignment of the structural gene for the flavin-polypeptide of human MAO A to the X chromosome. These experiments were initiated when it was observed that mouse neuroblastoma cells had very low MAO activity when they were cultured in media designed to allow growth of only cells that lacked the enzyme hypoxanthine phosphoribosyltransferase (EC 2.4.2.8, HPRT; BREAKEFIELD et al. 1976). The structural gene for HPRT is located on the X chromosome, and a deficiency in HPRT activity is found in patients with an X-linked inherited neurologic disorder, the Lesch-Nyhan syndrome (SEEGMILLER 1976). The simultaneous loss of HPRT and of MAO activities in cultured cells was pursued in a series of experiments which demonstrated that MAO activity is significantly decreased in skin fibroblasts cultured from patients with the Lesch-Nyhan syndrome (EDELSTEIN et al. 1978; CASTRO COSTA et al. 1980). In the course of these studies it was noted that there was a significant correlation of the levels of MAO A activity in fibroblasts from monozygotic twins (BREAKEFIELD et al., 1980), an indication that the heritability of the level of enzyme activity might be high. The frequency distribution of MAO activities in fibroblasts cultured from control subjects was also found to be "bimodal"—an observation which led to the suggestion that there might be a genetic polymorphism for human MAO A activity (BREAKEFIELD et al. 1979). That suggestion awaits confirmation by data obtained from family studies.

Hybrid cell culture techniques were then used to study the question of the chromosomal localization of the structural gene for MAO A. These experiments were made possible by the development of a method for the electrophoresis of MAO after exposure to radioactively labeled pargyline, an irreversible MAO inhibitor (Castro Costa and Breakefield 1979). When this procedure was combined with limited proteolysis of the radioactively labeled protein and a second eloctrophoretic step, a distinctive peptide "map" resulted that made it possible to distinguish murine from human MAO. The ability to distinguish murine from human MAO was critical for the success of attempts to use cell hybridization techniques to assign a chromosomal location to the structural gene for human MAO A. This was true because hybrid cells lose human chromosomes as they are cultured. Chromosomal assignment of genes is based on the simultaneous loss of the human enzyme and of a specific human chromosome. When hybrid cells derived from human fibroblasts and murine neuroblastoma cells were cloned and were grown in culture, the human flavin polypeptide of MAO A was lost from clones that had lost the human X chromosome. This observation made it possible to assign the structural gene for the protein to the X chromosome (Pintar et al. 1981). The assignment of the structural gene for the flavoprotein polypeptide of MAO A to the X chromosome represented an important advance in catecholamine biochemical genetics.

IV. Other Catecholamine Metabolic Enzymes

Although the reactions catalyzed by MAO and COMT are the major pathways for catecholamine catabolism, other enzymes also play a role in the metabolism of these neurotransmitter compounds. For example, almost all of the dopamine and most of the noradrenaline in human plasma is conjugated (Kuchel et al. 1979, 1980). Phenol sulfotransferase (EC 2.8.2.1, PST) catalyzes the sulfate conjugation of both of these compounds as well as that of a large number of catechol and phenolic drugs (Dodgson 1977; Renskers et al. 1980). The presence of potent endogenous inhibitors of PST in tissue homogenates made it difficult in the past to measure this enzyme activity in tissue homogenates accurately (Anderson and Weinshilboum 1979). The recent development of assay procedures which negate the effects of tissue PST inhibitors will make biochemical genetic studies of this enzyme possible (Anderson and Weinshilboum 1980). Experiments with laboratory animals have already shown wide variations among 9 inbred strains of rat in kidney, liver, and brain PST activities (Maus et al. 1982). These observations open the way for biochemical genetic studies of PST in inbred rodents.

PST activity is also present in an easily obtainable human tissue, the blood platelet (Hart et al. 1979; Anderson and Weinshilboum 1980; Anderson et al. 1981). Human platelets contain at least two independently regulated forms of PST, forms that differ in substrate specificity, sensitivity to inhibitors, and physical properties (Rein et al. 1981; Reiter and Weinshilboum 1982). The presence of PST in the platelet will make it possible to perform biochemical genetics studies of this enzyme activity in man.

Finally, the implications of the simultaneous and independent genetic regulation of each of several catecholamine metabolic enzymes in an individual subject must not be overlooked. There might be great differences in adrenergic function in subjects with genetically low levels of, for example, COMT, MAO, and PST when compared with that in subjects with inherited high levels of all three enzyme activities.

D. Conclusion

An important beginning has been made in biochemical genetic studies of catecholamine biosynthetic and metabolic enzymes. Much has been learned about the genetic regulation of these enzymes in experimental animals. Laboratory animal models have already helped to increase our understanding of the regulation of catecholamine synthesis and degradation by inheritance. The activities of serum DBH, erythrocyte COMT, and platelet MAO, three enzymes that can be studied easily in man, are strongly influenced by inheritance, and common genetic polymorphisms have been described for serum DBH and erythrocyte COMT. The possible implications of these observations with respect to individual variations in neural function and variations in response to drugs have only begun to be explored. The question of whether selective evolutionary pressure is involved in the maintenance of these polymorphisms in the human population has not been addressed. The possible relationship of these biochemical genetic observations to the pathophysiology of diseases for which heritability is high, diseases such as schizophrenia and the affective disorders, is only beginning to be studied (Heston 1977; Gershon 1979). All of these advances, however, represent only a beginning. Additional laboratory animal experiments are needed to define genetic mechanisms for the regulation of catecholamine enzymes, mechanisms that may then be studied in man. New and imaginative approaches must be developed to study the possible role of inheritance in the regulation of TH, AADC, and PNMT in man. Genetic studies of adrenoceptors and catecholamine uptake mechanisms are only beginning to appear. Eventually, catecholamine biochemical genetics may not only increase our understanding of the biological regulation of one class of neurotransmitter compounds, but may also serve as a model to be used in the study of the effects of inheritance on many other neurotransmitter systems.

Acknowledgements. The author thanks Luanne Wussow for her help in the preparation of this manuscript.

E. References

Anderson RJ, Weinshilboum RM (1979) Phenolsulphotransferase: enzyme activity and endogenous inhibitors in the human erythrocyte. J Lab Clin Med 94:158-171

Anderson RJ, Weinshilboum RM (1980) Phenolsulphotransferase in human tissue: radiochemical enzymatic assay and biochemical properties. Clin Chim Acta 103:79-90

Anderson R, Weinshilboum R, Phillips S, Broughton D (1981) Human platelet phenol sulphotransferase: assay procedure, substrate and tissue correlations. Clin Chim Acta 110:157-167

Assicot M, Bohuon C (1969) Production of antibodies to catechol-O-methyltransferase in rat liver. Biochem Pharmacol 18:1893-1898

Assicot M, Bohuon C (1971) Presence of two distinct catechol-O-methyltransferase activities in red blood cells. Biochimie 53:871-874

Axelrod J (1962) Purification and properties of phenylethanolamine N-methyl-transferase. J Biol Chem 237:1657-1660

Axelrod J, Cohen CK (1971) Methyltransferase enzymes in red blood cells. J Pharmacol Exp Ther 176:650-654

Axelrod J, Tomchick R (1958) Enzymatic O-methylation of epinephrine and other catechols. J Biol Chem 233:702-705

Axelrod J, Vesell ES (1970) Heterogeneity of N- and O-methyltransferases. Mol Pharmacol 6:78-84

Bailey DW (1971) Recombinant-inbred strains. Transplantation 11:325-327

Baker H, Joh TH, Reis DJ (1980) Genetic control of the number of midbrain dopaminergic neurons in inbred strains of mice: relationship to size and neuronal density of the striatum. Proc Nat Acad Sci USA 77:4369-4373

Baron M, Kreuz D, Levitt M, Gruen R, Asnis L (1982a) Variation in thermal stability of human plasma dopamine-beta-hydroxylase. Biol Psychiat 17:621-626

Baron M, Levitt M, Hunter C, Gruen R, Asnis L (1982b) Thermolabile catechol-O-methyltransferase in human erythrocytes: a confirmatory note. Biol Psychiat 17:265-270

Bateson W (1907) The progress of genetic research. In: Spottiswoode London, Report of the Third International Conference 1906 on Genetics. pp 90-97

Bellin JW, Sorrentino JM (1974) Kinetic characteristics of monoamine oxidase and serum cholinesterase in several related rat strains. Biochem Genet 11:309-317

Blaschko H (1974) The natural history of amine oxidases. Rev Physiol Biochem Pharmacol 70:81-148

Breakefield XO, Castiglione CM, Edelstein SB (1976) Monoamine oxidase activity decreased in cells lacking hypoxanthine phosphoribosyltransferase activity. Science 192:1018-1020

Breakefield XO, Edelstein SB, Castro Costa MR (1979) Genetic analysis of neurotransmitter metabolism in cell culture: studies on the Lesch-Nyhan syndrome. In Breakefield XO (Eds) Neurogenetics: Genetic Approaches to the Nervous System. Elsevier, New York, pp 197-234

Breakefield XO, Giller EL, Nurnberger JI, Castiglione CM, Buchsbaum MS, Gershon ES (1980) Monoamine oxidase type A in fibroblasts from patients with bipolar depressive illness. Psychiat Res 2:307-314

Breakefield XO, Edelstein SB, Grossman MH (1981) Variations in MAO and NGF in cultured human skin fibroblasts. In: Gershon ES, Matthysse S, Breakefield XO, Ciaranello RD, (Eds) Genetic Strategies in Psychobiology and Psychiatry. Boxwood Press, Pacific Grove, pp 129-142

Bridge TP, Wise CD, Potkin SG, Phelps BH, Wyatt RJ (1981) Platelet monoamine oxidase: studies of activity and thermolability in a general population. In: Gershon ES, Matthysse S, Breakefield XO, Ciaranello RD (Eds), Genetic Strategies in Psychobiology and Psychiatry. Boxwood Press, Pacific Grove, pp 95-104

Bruell JH (1962) Dominance and segregation in the inheritance of quantitative behavior in mice. In: Bliss EL (Ed) Roots of Behavior. Harper and Brothers, New York pp 48-67

Castro Costa MR, Breakefield XO (1979) Electrophoretic characterization of monoamine oxidase by [^{3}H]pargyline binding in rat hepatoma cells with A and B activity. Mol Pharmacol 16:242-249

Castro Costa MR, Edelstein SB, Castiglione CM, Chao H, Breakefield XO (1980) Properties of monoamine oxidase in control and Lesch-Nyhan fibroblasts. Biochem Genet 18:577-590

Cavalli-Sforza LL (1976) Enzyme polymorphism. Neurosci Res Prog Bull 14:56-58

Cavalli-Sforza LL, Bodmer WF (1971) The Genetics of Human Populations. San Francisco: Freeman

Cavalli-Sforza LL, Santachiara SA, Wang L (1974) Electrophoretic study of 5-hydroxytryptophan decarboxylase from brain and liver in several species. J Neurochem 23:629-634

Cawthon RM, Breakefield XO (1979) Differences in A and B forms of monoamine oxidase revealed by limited proteolysis and peptide mapping. Nature (Lond) 281:692-694

Ciaranello RD (1978) Regulation of phenylethanolamine N-methyltransferase. Biochem Pharmacol 27:1895-1897

Ciaranello RD, Axelrod J (1973) Genetically controlled alterations in the rate of degradation of phenylethanolamine N-methyltransferase. J Biol Chem 248:5616-5623

Ciaranello RD, Barchas R, Kessler S, Barchas JD (1972a) Catecholamines: strain differences in biosynthetic enzyme activity in mice. Life Sci 11:565-572

Ciaranello RD, Dornbusch JN, Barchas JD (1972b) Regulation of adrenal phenylethanolamine N-methyltransferase activity in three inbred mouse strains. Mol Pharmacol 8:511-520

Ciaranello RD, Hoffman HF, Shire JGM, Axelrod J (1974) Genetic regulation of the catecholamine biosynthetic enzyme. II. Inheritance of tyrosine hydroxylase, dopamine-beta-hydroxylase and phenylethanolamine N-methyltransferase. J Biol Chem 249:4528-4536

Committee on Nomenclature (1976) Report of Committee on Nomenclature. Birth Defects. Original Articles 12:65-74

Conolly ME, Davies DS, Dollery CT, Morgan CD, Paterson JW, Sandler M (1972) Metabolism of isoprenaline in dog and man. Br J Pharmacol 46:458-472

Creveling CR, Borchardt RT, Isersky C (1973) Immunological characterization of catechol-O-methyltransferase. In: Usdin E, Snyder S (Eds) Frontiers in Catecholamine Research Pergamon Press New York, pp 117-119

Dairman W, Christenson JG (1973) Properties of human red blood cell L-3,4-dihydroxyphenylalanine decarboxylase activity. Eur J Pharmacol 22:135-140

Dodgson KS (1977) Conjugation with sulfate. In: Parke DV, Smith RL (Eds) Drug Metabolism from Microbe to Man. Francis London, pp 91-104

Dunnette J, Weinshilboum R (1976) Human serum dopamine-beta-hydroxylase: correlation of enzymatic activity with immunoreactive protein in genetically defined samples. Am J Hum Genet 28:155-166

Dunnette J, Weinshilboum R (1977) Inheritance of low immunoreactive human plasma dopamine-beta-hydroxylase: radioimmunoassay studies. J Clin Invest 60:1080-1087

Dunnette J, Weinshilboum R (1979) Human plasma dopamine-beta-hydroxylase: variation in thermal stability. Mol Pharmacol 15:649-660

Dunnette J, Weinshilboum R (1981) Human plasma dopamine-beta-hydroxylase: oxygen and thermal stability. Experientia 37:115-117

Dunnette J, Weinshilboum R (1982a) Family studies of plasma dopamine-beta-hydroxylase thermal stability. Am J Hum Genet 34:84-99

Dunnette J, Weinshilboum R (1982b) Serum dopamine-beta-hydroxylase activity in non-human primates. The Pharmacologist 24:243

Dunnette J, Weinshilboum R (1983) Serum dopamine-beta-hydroxylase activity in non-human primates: phylogenetic and genetic implications. Comp Biochem Physiol 75:85-91

Edelstein SB, Castiglione CM, Breakefield XO (1978) Monoamine oxidase activity in normal and Lesch-Nyhan fibroblasts. J Neurochem 31:1247-1254

Elston RC, Stewart J (1971) A general model for the genetic analysis of pedigree data. Hum Heredity 21:523-542

Elston RC, Namboodiri KK, Hames CG (1963) Segregation and linkage analysis of dopamine-beta-hydroxylase activity. Hum Heredity 29:284-292

Falconer DS (1963) Quantitative inheritance. In: Burdette MJ (Ed) Methodology in Mammalian Genetics. Holden-Day, San Francisco, pp 193-216

Floderus Y, Wetterberg L (1981) The inheritance of human erythrocyte catechol-O-methyltransferase (COMT) activity. Clin Genet 19:392-395

Floderus Y, Ross SB, Wetterberg L (1981) Erythrocyte catechol-O-methyltransferase (COMT) activity in a Swedish population. Clin Genet 19:389-392

Freedman LS, Ohuchi T, Goldstein M, Axelrod F, Fish I, Dancis J (1972) Changes in human serum dopamine-beta-hydroxylase activity with age. Nature (Lond) 236:310-311

Fuentes JA, Neff NH (1975) Selective monoamine oxidase inhibitor drugs as aids in evaluating the role of type A and B enzymes. Neuropharmacol 14:819-825

Gardner EJ (1984) Principles of Genetics. New York: John Wiley and Sons

Gershon ES (1979) Genetics of the affective disorders. Hosp Pract 14:117-122

Gershon ES, Jonas WZ (1975) Erythrocyte soluble catechol-O-methyltransferase activity in primary affective disorder. Arch Gen Psychiat 32:1351-1356

Gershon ES, Jonas WZ (1976) Inherited differences in brain and erythrocyte soluble catechol-O-methyltransferase activity in two mouse strains. Biol Psychiat 11:641-645

Gershon ES, Goldin LR, Lake CR, Murphy DL, Guroff JJ (1980) Genetics of plasma dopamine-beta-hydroxylase, erythrocyte catechol-O-methyltransferase and plasma monoamine oxidase in pedigrees of patients with affective disorders. In: Usdin E, Sourkes P, Youdim MBH (Eds) Enzymes and Neurotransmitters in Mental Disease. New York: Wiley and Sons pp 281-299

Goldin LR, Gershon ES, Lake CR, Murphy DL, McGinniss M, Sparkes RS (1982) Segregation and linkage studies of plasma dopamine-beta-hydroxylase (DBH), erythrocyte catechol-O-methyltransferase (COMT), and platelet monoamine oxidase (MAO): possible linkage between ABO locus and a gene controlling DBH activity. Am J Hum Genet 34:250-262

Goldstein DJ, Weinshilboum RM, Dunnette JH, Creveling CR (1980) Developmental patterns of catechol-O-methyltransferase in genetically different rat strains: enzymatic and immunochemical studies. J Neurochem 34:153-162

Groshong R, Gibson DA, Baldessarini RJ (1977) Monoamine oxidase activity in cultured human skin fibroblasts. Clin Chim Acta 80:113-120

Grunhaus L, Ebstein R, Belmaker R, Sandler SG, Jonas W (1976) A twin study of human red blood cell catechol-O-methyltransferase. Br J Psychiat 128:494-498

Harris H (1980) The Principles of Human Biochemical Genetics. Third edition. American Elsevier, New York

Hart RF, Renskers KJ, Nelson EB, Roth JA (1979) Localization and characterization of phenol sulfotransferase in human platelets. Life Sci 24:125-139

Heiss G, Tyroler HA, Gunnells JC, McGuffin WL, Hames CG (1980) Dopamine-beta-hydroxylase in a biracial community: demographic, cardiovascular and familial factors. J Chron Dis 33:301-310
Heston LL (1977) Schizophrenia: genetic factors. Hosp Pract 12:43-49
Hodgetts RB (1975) The response of dopa decarboxylase activity to variations in gene dosage in Drosophila: a possible location of the structural gene. Genetics 79:45-54
Hopkinson DA, Spencer N, Harris H (1964) Genetical studies on human red cell acid phosphatase. Hum Genet 16:141-154
Houslay MD, Tipton KF, Youdim MBH (1976) Multiple forms of monoamine oxidase: fact and artifact. Life Sci 19:467-478
Hum MMO, Friedhoff AJ (1979) Multiple molecular forms of catechol-O-methyltransferase. J Biol Chem 254:299-308
Johnson JP (1968) Some observations upon a new inhibitor of monoamine oxidase in brain tissue. Biochem Pharmacol 17:1285-1297
Kessler S, Ciaranello RD, Shire JGM, Barchas JD (1972) Genetic variation in activity of enzymes involved in synthesis of catecholamines. Proc Nat Acad Sci USA 69:2448-2450
Kimura M, Ohta T (1974) On some principles governing molecular evolution. Proc Nat Acad Sci USA 71:2848-2852
Kuchel O, Buu NT, Unger T (1979) Free and conjugated dopamine; physiological and clinical implications. In Imbs JL, Schwartz J (Eds) Peripheral Dopaminergic Receptors. Pergamon Press, New York, pp 15-27
Kuchel O, Buu NT, Fontaine A, Hamet P, Beroniade V, Larochelle P, Genet J (1980) Free and conjugated catecholamines in hypertensive patients with and without pheochromocytoma. Hypertension 2:177-186
Levitt M, Mendlewicz J (1975) A genetic study of plasma dopamine-beta-hydroxylase in affective disorder. Mod Probl Pharmacopsychiat 10:89-98
Levitt M, Spector S, Sjoerdsma A, Udenfriend S (1965) Elucidation of the rate limiting step in norepinephrine biosynthesis on the perfused guinea-pig heart. J Pharmacol Exp Ther 148:1-8
Lewontin RC (1974) The Genetic Basis of Evolutionary Change. Columbia University Press.
Li CC (1961) Human Genetics, Principles and Methods. McGraw-Hill, New York
Maus TP, Pearson RK, Anderson RJ, Woodson LC, Reiter C, Weinshilboum RM (1982) Rat phenol sulfotransferase: assay procedure, developmental changes and glucocorticoid regulation. Biochem Pharmacol 31:849-856
Mendel G (1865) Versuche über Pflanzen-Hybride. Verhandlungen des naturforschenden Vereines in Brünn 4:3-47
Minamiura N, Yasumobu KT (1978) Bovine liver monoamine oxidase: a modified purification procedure and preliminary evidence for two subunits and one FAD. Arch Biochem Biophys 189:481-489
Murphy DL (1978) Substrate-selective monoamine oxidases—inhibitor, tissue, species and functional differences. Biochem Pharmacol 27:1889-1893
Murphy DL, Wright C, Buchsbaum M, Nichols A, Costa JL, Wyatt RJ (1976) Platelet and plasma amine oxidase activity in 680 normals: sex and age differences and stability over time. Biochem Med 16:254-265
Murphy DL, Redmond DE Jr, Bauler J, Donnelly CH (1978) Platelet monoamine oxidase activity in 116 normal rhesus monkeys: relations between enzyme activity and age, sex, and genetic factors. Comp Biochem Physiol 60C:105-108
Nies A, Robinson DS, Lamborn KR, Lampert RP (1973) Genetic control of platelet and plasma monoamine oxidase activity. Arch Gen Psychiat 28:834-838

Ogihara T, Nugent CA, Shen SW, Goldfein S (1975) Serum dopamine-beta-hydroxylase activity in parents and children. J Lab Clin Med 85:566-573

Paigen K (1971) The genetics of enzyme realization. In: Rechcigl M Jr (Ed) Enzyme Synthesis and Degradation in Mammalian Systems. University Park Press, Baltimore pp 1-44

Paigen K (1979) Acid hydrolases as model of genetic control. Ann Rev Genet 13:417-466

Paigen K, Swank RT, Tomino S, Ganschow RE (1975) The molecular genetics of mammalian glucuronidase. J Cell Physiol 85:379-392

Pandey GN, Dorus E, Shaughnessy R, Davis JM (1979) Genetic control of platelet monoamine oxidase activity: studies on normal families. Life Sci 25:1173-1178

Pintar JE, Barbosa J, Francke U, Castiglione CM, Hawkins MH Jr, Breakefield XO (1981) Gene for monoamine oxidase type A assigned to the human X chromosome. J Neurochem 1:166-175

Quiram DR, Weinshilboum RM (1976) Catechol-O-methyltransferase in rat erythrocyte and three other tissues: comparison of biochemical properties after removal of inhibitory calcium. J Neurochem 27:1197-1203

Raymond FA, Weinshilboum RM (1975) Microassay of human erythrocyte catechol-O-methyltransferase: removal of inhibitory calcium with chelating resin. Clin Chim Acta 58:185-194

Reilly DK, Rivera-Calimlim L (1979) Racial difference in catechol-O-methyltransferase activity? A comparison of Filipinos with Caucasians in the United States. Clin Pharmacol Ther 25:244

Reilly DK, Rivera-Calimlim L, Van Dyke D (1980) Catechol-O-methyltransferase activity: a determinant of levodopa response. Clin Pharmacol Ther 28:278-286

Rein G, Glover V, Sandler M (1981) Phenolsulphotransferase in human tissue: evidence for multiple forms. In Sandler M, Usdin E (Eds) Phenolsulfotransferase in Mental Health Research MacMillan, London, pp 98-126

Reis DJ, Baker H, Fink JS, Joh TH (1981) A genetic control of the number of dopamine neurons in mouse brain: its relationship to brain morphology, chemistry and behavior. In: Genetic Strategies in Psychobiology and Psychiatry. Gershon ES, Matthysse S, Breakefield XO, Ciaranello RD (Eds) Boxwood Press, Pacific Grove, pp 215-229

Reiter C, Weinshilboum R (1982) Acetaminophen and phenol: substrates for both a thermostable and a thermolabile form of human platelet phenol sulfotransferase. J Pharmacol Exp Ther 221:43-51

Renskers KJ, Feor KD, Roth JA (1980) Sulfation of dopamine and other biogenic amines by human brain phenol sulfotransferase. J Neurochem 34:1362-1368

Roffman M, Reigle TG, Orsalak PJ, Schildkraut JJ (1976) Properties of catechol-O-methyltransferase in soluble and particulate preparations from rat red blood cells. Biochem Pharmacol 25:208-209

Ross SB, Wetterberg L, Myrhed M (1973) Genetic control of plasma dopamine-beta-hydroxylase. Life Sci (I) 12:529-532

Ross RA, Judd AB, Pickel VM, Joh TH, Reis DJ (1976) Strain-dependent variations in number of midbrain dopaminergic neurons. Nature (Lond) 264:654-656

Roth JA, Breakefield XO, Castiglione CM (1976) Monoamine oxidase and catechol-O-methyltransferase activities in cultured human skin fibroblasts. Life Sci 19:1705-1710

Sandler M, Youdim MBH, Harrington E (1974) A phenylethylamine oxidizing defect in migraine. Nature (Lond) 250:335-337

Scanlon PD, Raymond FA, Weinshilboum RM (1979) Catechol-O-methyltransferase:

thermolabile enzyme in erythrocytes of subjects homozygous for the allele for low activity. Science 203:63-65

Schlesinger K, Harkins J, Deckard BS, Paden C (1975) Catechol-O-methyltransferase and monoamine oxidase activities in brains of mice susceptible and resistant to audiogenic seizures. J Neurobiol 6:587-596

Seegmiller JE (1976) Inherited deficiency of hypoxanthine-guanine phosphoribosyltransferase in X-linked uric aciduria (the Lesch-Nyhan syndrome and its variants). Advan Human Genet 6:75-163

Sladek-Chelgren S, Weinshilboum RM (1981) Catechol-O-methyltransferase biochemical genetics. Biochem Genet 19:1037-1053

Spielman RS, Weinshilboum RM (1979) Family studies of low red cell COMT activity. Am J Hum Genet 31:63A

Spielman RS, Weinshilboum RM (1981) Genetics of red cell COMT activity: analysis of thermal stability and family data. Am J Med Genet 10:279-290

Stolk JM, Hurst JH, Van Riper DA, Harris PQ (1979) Genetic analysis of serum dopamine-beta-hydroxylase activity in rats. Mol Pharmacol 16:922-931

Swank RT, Bailey DW (1973) Recombinant inbred lines: value in the genetic analysis of biochemical variants. Science 181:1249-1252

Tate SS, Sweet R, McDowell RH, Meister A (1971) Decrease of the 3,4-dihydroxyphenylalanine (DOPA) decarboxylase activities in human erythrocytes and mouse tissues after administration of DOPA. Proc Nat Acad Sci USA 68:2121-2123

Tunnicliff G, Wimer CC, Wimer RE (1973) Relationships between neurotransmitter metabolism and behaviour in seven inbred strains of mice. Brain Res 61:428-434

Vogel F, Motulsky AG (1979) Human Genetics, Problems and Approaches. Springer, New York

Wallace EF, Krantz MJ, Lovenberg W (1973) Dopamine-beta-hydroxylase: a tetrameric glycoprotein. Proc Nat Acad Sci USA 70:2253-2255

Weinshilboum RM (1977) Serum dopamine-beta-hydroxylase activity and blood pressure. Mayo Clinic Proc 52:374-378

Weinshilboum RM (1978a) Human biochemical genetics of plasma dopamine-beta-hydroxylase and erythrocyte catechol-O-methyltransferase. Hum Genet Suppl I 101-112

Weinshilboum RM (1978b) Human erythrocyte catechol-O-methyltransferase: correlation with lung and kidney activity. Life Sci 22:625-630

Weinshilboum RM (1978c) Serum dopamine-beta-hydroxylase. Pharmacol Rev 30:133-166

Weinshilboum RM (1979a) Genetic regulation of catechol-O-methyltransferase. Soc Neurosci Symp 4:67-82

Weinshilboum RM (1979b) Catecholamine biochemical genetics in human populations. In: Breakefield XO (Ed) Neurogenetics: Genetic Approaches to the Nervous System. Elsevier, New York, pp 257-282

Weinshilboum RM (1981) Enzyme thermal stability and population genetic studies: application to erythrocyte catechol-O-methyltransferase and plasma dopamine-beta-hydroxylase. In: Gershon ES, Matthysse S, Breakefield XO, Ciaranello RD (Eds) Genetic Strategies in Psychobiology and Psychiatry. Boxwood Press Pacific Grove, pp 79-94

Weinshilboum R, Axelrod J (1971a) Serum dopamine-beta-hydroxylase. Cir Res 28:307-315

Weinshilboum RM, Axelrod J (1971b) Reduced plasma dopamine-beta-hydroxylase activity in familial dysautonomia. New Engl J Med 285:938-942

Weinshilboum RM, Raymond FA (1976) Calcium inhibition of rat liver catechol-O-methyltransferase. Biochem Pharmacol 25:573-579

Weinshilboum RM, Raymond FA (1977a) Inheritance of low erythrocyte catechol-O-methyltransferase activity in man. Am J Hum Genet 29:125-135

Weinshilboum RM, Raymond FA (1977b) Variations in catechol-O-methyltransferase activity in inbred strains of rats. Neuropharmacol 16:703-706

Weinshilboum RM, Raymond FA, Elveback LR, Weidman WH (1973) Serum dopamine-beta-hydroxylase activity: sibling-sibling correlation. Science 181:943-945

Weinshilboum RM, Raymond FA, Elveback LR, Weidman WH (1974) Correlation of erythrocyte catechol-O-methyltransferase activity between siblings. Nature (Lond) 252:490-491

Weinshilboum RM, Schrott HG, Raymond FA, Weidman WH, Elveback LR (1975) Inheritance of very low serum dopamine-beta-hydroxylase activity. Am J Hum Genet 27:573-585

Weinshilboum RM, Dunnette J, Raymond F, Kleinberg F (1978) Erythrocyte catechol-O-methyltransferase and plasma dopamine-beta-hydroxylase in human umbilical cord blood. Experientia 34:310-311

Weinshilboum RM, Raymond FA, Frohnauer M (1979) Monogenic inheritance of catechol-O-methyltransferase in the rat: biochemical and genetic studies. Biochem Pharmacol 28:1239-1248

Wetterberg L, Book JA, Floderus Y, Ross SB (1979) Genetics and biochemistry of schizophrenia in a defined poulation. In: Usdin E, Kopin IJ, Barchas J (Eds) Catecholamines: Basic and Clinical Frontiers. Pergamon Press, New York, pp 1857-1859

Winter H, Herschel M, Propping P, Friedl W, Vogel F (1978) A twin study on three enzymes (DBH, COMT, MAO) of catecholamine metabolism. Psychopharmacol 57:63-69

Wise CD, Bridge P, Potkin SB, Wyatt RJ (1979) Platelet monoamine oxidase: studies on the rate of heat inactivation in normal and in paranoid and nonparanoid chronic schizophrenic groups. Psychiat Res 1:187-190

Wurtman RJ, Axelrod J (1966) Control of enzymatic synthesis of adrenaline in the adrenal medulla by adrenal cortical steroids. J Biol Chem 241:2301-2305

Wyatt RJ, Murphy DL, Belmaker R, Cohen S, Donnelly CH, Pollin W (1973) Reduced monoamine oxidase activity in platelets: a possible genetic marker for vulnerability to schizophrenia. Science 173:916-918

Wyatt RJ, Potkin SG, Murphy DL (1979) Platelet monoamine oxidase activity in schizophrenia: a review of the data. Am J Psychiat 136:4A:377-385

Weinshilboum RM, Raymond FA (1976) Calcium inhibition of rat liver catechol-O-methyltransferase. Biochem Pharmacol 25:573–579
Weinshilboum RM, Raymond FA (1977a) Inheritance of low erythrocyte catechol-O-methyltransferase activity in man. Am J Hum Genet 29:125–135
Weinshilboum RM, Raymond FA (1977b) Variations in catechol-O-methyltransferase activity in inbred strains of rats. Neuropharmacol 16:703–706
Weinshilboum RM, Raymond FA, Elveback LR, Weidman WH (1973) Serum dopamine-beta-hydroxylase activity: sibling-sibling correlation. Science 181:943–945
Weinshilboum RM, Raymond FA, Elveback LR, Weidman WH (1974) Correlation of erythrocyte catechol-O-methyltransferase activity between siblings. Nature (Lond) 252:490–491
Weinshilboum RM, Schrott HG, Raymond FA, Weidman WH, Elveback LR (1975) Inheritance of very low serum dopamine-beta-hydroxylase activity. Am J Hum Genet 27:573–585
Weinshilboum RM, Dunnette J, Raymond F, Kleinberg F (1978) Erythrocyte catechol-O-methyltransferase and plasma dopamine-beta-hydroxylase in human umbilical cord blood. Experientia 34:310–311
Weinshilboum RM, Raymond FA, Frohnauer M (1979) Monogenic inheritance of catechol-O-methyltransferase in the rat: biochemical and genetic studies. Biochem Pharmacol 28:1239–1248
[illegible]
White HL, [illegible] (1978) A twin study on three enzymes (DBH, COMT, MAO) of catecholamine metabolism. [illegible]
[illegible]
[illegible]
[illegible]

CHAPTER 20

The Role of Tyrosine Hydroxylase in the Regulation of Catecholamine Synthesis

J. M. MASSERANO, P. R. VULLIET, A. W. TANK and N. WEINER

A. Introduction and Enzymology of Tyrosine Hydroxylase

Tyrosine-3-monoxygenase (EC 1.14.16.2) (tyrosine hydroxylase) is the rate limiting enzyme in the pathway for the synthesis of catecholamines. This enzyme was first demonstrated in bovine adrenal medulla (NAGATSU et al. 1964a; BRENNEMAN and KAUFMAN 1964). Tyrosine hydroxylase is a mixed function oxidase, requiring molecular oxygen (DALY et al. 1968) and a reduced pterin (NAGATSU et al. 1964a; BRENNEMAN and KAUFMAN 1964) as cosubstrates. The putative natural pterin cosubstrate is believed to be tetrahydrobiopterin and was first demonstrated in adrenal medulla cells (LLOYD and WEINER 1971). The cofactor is also present in brain tissue and the concentration of tetrahydrobiopterin in various brain regions is correlated with the distribution of catecholamines (GAL et al. 1976; BULLARD et al. 1978; LEVINE et al. 1979; MANDELL et al. 1980). Biopterin can be syntesized from guanosine in mouse neuroblastoma clones (BUFF and DAIRMAN 1974) and from guanosine triphosphate (GTP) in rat brain (GAL and SHERMAN 1976). Dihydropteridine reductase, the enzyme that catalyzes the reduction of 7,8-dihydropterins to the active 5,6,7,8-tetrahydroform, has been demonstrated in sheep liver (KAUFMAN 1964; NIELSEN et al.1969; CRAINE et al. 1972), beef adrenal medulla (MUSACCHIO 1969; MUSACCHIO et al. 1971) and brain (TURNER et al. 1974; SPECTOR et al. 1977). The enzyme requires a reduced pyridine nucleotide for activity. Either NADH or NADPH serves as a cofactor for dihydropteridine reductase, but the former pyridine nucleotide exhibits a higher affinity for the enzyme (NIELSEN et al. 1969; CRAINE et al. 1972).

A number of synthetic analogues of tetrahydrobiopterin are available for the study of tyrosine hydroxylase and other aromatic amino acid hydroxylases. Pterin compounds with an alkyl substituent on the six-position of the pterin ringsystem are now regarded as the preferred synthetic cosubstrates for use in studies involving aromatic amino acid hydroxylases. Of these, one of the most commonly employed is 6-methyltetrahydropterin. In addition, biopterin is available commercially and can be readily reduced to the tetrahydroform (LOVENBERG et al. 1975) and should be considered the preferred substrate.

Ferrous ion appears to be required for optimal activity of tyrosine hydroxylase (NAGATSU et al. 1964a; PETRACK et al. 1968; WAYMIRE et al. 1972). Iron chelating reagents, such as α,α-dipyrdyl, have been shown to completely inhibit the enzyme (TAYLOR et al. 1969). HOELDTKE and KAUFMAN (1977) found that tyrosine hydroxylase purified from solubilized bovine adrenal medulla has approximately 0.5 to 0.75 mole of iron per mole of enzyme monomer. They concluded that, like other hydroxylases, such as phenylalanine hydroxylase, which is an iron containing enzyme (Fisher et al. 1972) and DBH which contains copper (Goldstein et al. 1968; Craine et al. 1973), tyrosine hydroxylase is a metallo-enzyme.

I. Mechanism for Tyrosine Hydroxylation

The mechanism for the hydroxylation of tyrosine by tyrosine hydroxylase appears to be analogous to that which occurs with hepatic phenylalanine hydroxylase (Kaufman and Fisher 1974). On the basis of kinetic studies with both solubilized and particulate bovine adrenal tyrosine hydroxylase Kaufman and Fisher (1974) proposed that the hydroxylation of tyrosine involves a sequential mechanism in which a quaternary complex, consisting of enzyme, substrate, tetrahydropterin and oxygen, is an intermediate in the catalysis. Recently, this has been verified using bovine striatal tyrosine hydroxylase. Bullard and Capson (1983) performed a comprehensive kinetic analysis of striatal tyrosine hydroxylase using L-tyrosine, 5,6,7,8-tetrahydrobiopterin, and oxygen as substrates and dopa as product, a deazapterin as a substrate analogue and 3-iodo-L-tyrosine and dopamine as product analogue inhibitors. They report that the model provides evidence for a sequential mechanism with partially ordered sequences for substrate addition and product release.

The mechanism of oxygen utilization by tyrosine hydroxylase has been evaluated by Dix et al. (1987). These authors report that a putative hydroxylating intermediate, peroxytetrahydropterin, is formed by the combination of oxygen and the cofactor in the presence of the Fe^{2+}-containing enzyme tyrosine hydroxylase. The outermost oxygen atom of the peroxytetrahydropterin can be transferred onto tyrosine, forming dopa and 4a-carbinolamine, with this latter species spontaneously dehydrating to form dihydrobiopterin.

II. Similarities of Tyrosine Hydroxylase to Phenylalanine Hydroxylase

Tyrosine hydroxylase and phenylalanine hydroxylase have many features in common and it has been suggested that these enzymes share a common reaction mechanism (Kaufman and Fisher 1974; Ikeda et al. 1965, 1967; Shiman et al. 1971). Tyrosine hydroxylase is able to hydroxylate both phenylalanine and tyrosine (Ikeda et al. 1965, 1967; Shiman et al. 1971). Using bovine adrenal enzyme, Shiman and coworkers found that phenylalanine was a relatively better substrate than tyrosine in the presence of tetrahydrobiopterin, the putative natural cofactor, compared with 6,7-dimethyltetrahydropterin. Furthermore, in the presence of the natural cofactor, high concentrations of tyrosine were associated with substrate inhibition of the enzyme. These workers concluded that, under circumstances which may exist *in vivo* (presence of tetrahydrobiopterin and relatively high concentrations of tyrosine and phenylalanine), phenylalanine may be the preferred substrate. However, in the intact vas deferens preparation, hydroxylation of tyrosine to dopa is preferred over the conversion of phenylalanine to dopa, when both are present in the medium in physiological concentrations (WEINER et al. 1974). These *in situ* studies, however, may not reflect the true rate of phenylalanine hydroxylation, since the newly formed labelled tyrosine presumably dissociates from the enzymes and mixes with the much larger pool of unlabelled tyrosine in the neuron. Thus, the specific activity of the tyrosine formed from phenylalanine would be markedly reduced and the rate of conversion of labelled tyrosine to dopa would be expected to be small in proportion to this dilution.

KATZ et al. (1976a) compared the rate of hydroxylation of phenylalanine and tyrosine to dopa by soluble rat striatal enzyme and by the intact P_2 fraction of synaptosomes from this tissue. With soluble enzyme, much more exogenous tyrosine than phenylalanine was converted to dopa, although there was a brisk conversion of phenylalanine to tyrosine. In contrast, the synaptosomal preparations converted a considerable amount of the phenylalanine to tyrosine which was further transformed to dopa. These results suggest that the synaptosomal pool of tyrosine formed from phenylalanine is preferentially utilized for dopa formation. This probably reflects the existence of a small pool of the precursor near the enzyme, which is maintained by the structural integrity of the synaptosomal membrane.

HAAVIK and FLATMARK (1987) have isolated and characterized the oxidative products of tetrahydropterin generated in the tyrosine hydroxylase reaction. They report that, similar to phenylalanine hydroxylase, the first product of the tetrahydropterin cofactor detected in the tyrosine hydroxylase-catalyzed reaction was 4a-hydroxytetrahydropterin and provides evidence for a similar catalytic mechanism for the two enzymes. LEDLEY et al. (1985) have examined the molecular similiarities of phenylalanine hydroxylase and tyrosine hydroxylase by amino acid and cDNA sequencing. They find a large degree of homology at the amino acid and nucleic acid levels. There is little homology between the two proteins at the amino terminus, but considerable homology is apparent in the central region and carboxyl terminus of the two proteins. Similarily, no homology is apparent in the extreme 5′ sequences of the two proteins but there is considerable homology between the central regions of the genes. Based on studies using the active proteolytic fragment of tyrosine hydroxylase and phenylalanine hydroxylase, it is suggested that the highly homologous domain at the center and carboxyl terminus of each enzyme represents the catalytic core of the enzymes and the nonhomologous domain at the amino terminus of the enzymes represent regulatory determinants that contribute to substrate specificity and regulation of the two enzymes (LEDLEY et al. 1985; GRIMA et al. 1985; VIGNY and HENRY 1981; HOELDTKE and KAUFMAN 1977).

III. Oxygen Requirements of Tyrosine Hydroxylase

The interaction of oxygen with tyrosine hydroxylase is also critically dependent on the cofactor employed. With tetrahydrobiopterin, the apparent K_m of the brain and adrenal enzyme for O_2 is approximately 1% (solubilized bovine adrenal enzyme) or less than 1% (bovine brain enzyme). Enzyme inhibition was seen at oxygen concentrations of 40% for the former enzyme and 8% for the latter enzyme. The apparent K_m of O_2 for the enzyme is much higher when 6,7-dimethyltetrahydropterin is employed as cofactor, and the enzyme is much more resistent to inhibition by high O_2 concentrations (FISHER and KAUFMAN 1972). The physiological relevance of the inhibition of the enzyme by high oxygen concentrations is dubious, however, since WEINER et al. (1973) have shown that, with the isolated vas deferens preparation, the hydroxylation of tyrosine to dopa increases progressively as the oxygen concentration in the medium is altered from 20% to 95%. Similar results have been obtained with brain tyrosine hydroxylase *in situ* in rats exposed to different concentrations of oxygen in the inspired air (DAVIS and CARLSON 1973; DAVIS 1976). Of course, it is possible that the high oxygen concentrations in the incubation medium or in the inspired air do not reach the site of the enzyme. Because of diffusion limitations, the oxygen concentration in the isolated organ or in the brain may be very much lower and it may not be possible to reach levels at the site of the enzyme which are inhibitory.

IV. Inhibition of Tyrosine Hydroxylase by Substrate and Cofactor End-Products

Tyrosine hydroxylase is inhibited by catechol compounds. For most catechol compounds the inhibition of tyrosine hydroxylase appears to be competitive with the pterin cofactor (NAGATSU et al. 1964a; UDENFRIED et al. 1965). However, KAUFMAN and coworkers have claimed that the inhibition of tyrosine hydroxylase by the immediate product of the hydroxylation reaction, dopa, is noncompetitive with pterin cofactor (SHIMAN et al. 1971; KAUFMAN and FISHER 1974). In addition to the substrate end-product inhibition of tyrosine hydroxylase, KUHN and LOVENBERG (1983) found that tyrosine hydroxylase is inactivated by incubation with its reduced pterin cofactors. L-erythrotetrahydrobiopterin was found to inhibit the enzyme resulting in a reduction of the V_{max} for both tetrahydrobiopterin and tyrosine. Each of the two diastereoisomers of L-erythrotetrahydrobiopterin inactivate tyrosine hydroxylase but the natural (6 R) form of the pterin is much more effective than the unnatural (6 S) form at equimolar concentrations. The 6-methyl-5-deazatetrahydropterin form can inactivate tyrosine hydroxylase but has little cofactor activity, suggesting that pterin binding to the enzyme and not cofactor activity underlies the pterin-induced inactivation of tyrosine hydroxylase.

V. Kinetics of Tyrosine Hydroxylase for Cofactor

The affinity of the enzyme for tyrosine appears to vary with the cofactor employed and the source of the enzyme. SHIMAN et al. (1971) have demonstrated that the affinity of tyrosine for bovine adrenal tyrosine hydroxylase is considerably increased in the presence of tetrahydrobiopterin, exhibits an intermediate value in the presence of 6-methyltetrahydropterin, and is lowest in the presence of 6,7-dimethyltetrahydropterin. Furthermore, substrate inhibition by excess tyrosine occurs at lower concentrations of tyrosine in the presence of the putative natural cofactor.

Estimates of the affinity of pterin cofactor for tyrosine hydroxylase vary considerably, depending upon the nature of the cofactor employed, the source of the enzyme and the method of assay. In general, using tissue obtained from the striatum (KUCZENSKI and MANDELL 1972b; LOVENBERG and BRUCKWICK 1975; ZIVKOVIC et al. 1975), nucleus accumbens, hypothalamus, brain stem (ZIVKOVIC and GUIDOTTI 1974; ZIVKOVIC et al. 1975), and vas deferens (WEINER et al. 1978), K_m values for the pterin cofactor (6-methyltetrahydropterin or 6,7-dimethyltetrahydropterin) for control tissues were within the range of 0.47 to 0.80 mM. Various treatments, including nerve stimulation (WEINER et al. 1978), heparin sulfate (KUCZENSKI and MANDELL 1972a), antipsychotic drugs (LOVENBERG and BRUCKWICK 1975; ZIVKOVIC et al. 1975; KAPATOS and ZIGMOND 1979), melanin (NAGATSU et al. 1978) and a cyclic-AMP-dependent protein phosphorylation system (LOVENBERG et al. 1975; MORGENROTH et al. 1975a; MORITA and OKA 1977; WEINER et al. 1978) appear to be associated with shifts in the apparent K_m for the cofactor to values which are within the range of 0.11 to 0.29 m*M*. MASSERANO and WEINER (1979) found that tyrosine hydroxylyse, obtained from rat adrenal glands removed in a nonstressful manner, exhibited both the high and low K_m forms of the enzyme. This was suggested from the curvilinear Lineweaver-Burk kinetics seen in the adrenal medulla supernatants prepared from the non-stressed animals. In these animals the adrenal tyrosine hydroxylase was approximately 25% in the activated (low K_m) form and 75% in the less active (high K_m) form. Following various stresses including decapitation (MASSERANO and WEINER 1979), electroconvulsive shock (MASSERANO et al. 1981) or subcutaneous formaldehyde injection (MASSERANO and WEINER 1981), approximately 60% of the enzyme was found in the active (low K_m) form, suggesting that a considerable fraction of the less active form of the enzyme is rapidly transformed to the active form during stress. These data indicate that, in the rat adrenal medulla, two forms of tyrosine hydroxylase coexist and the proportions of these two enzymes forms depend upon the manner in which the animals are manipulated prior to and during killing. A similar, although more complete, transformation of the less active form of the enzyme to the active form can be produced if the soluble enzyme from either stressed or nonstressed rats is incubated in the presence of a cyclic AMP-dependent protein phosphorylating system (MASSERANO and WEINER 1979). Employing an indirect phosphorylation procedure, MELIGENI et al. (1981) developed evidence to suggest that tyrosine hydroxylase in rat adrenal is either phosphorylated during decapitation stress or undergoes a conformational change which renders it less susceptible to phosphorylation by cyclic AMP-dependent protein kinase. The enzyme in the guinea-pig vas deferens preparation appears to be modified in a similiar manner consequent to nerve stimulation (WEINER et al. 1981).

It is interesting to note that areas of the brain that contain predominantly dopamine (striatum, nucleus accumbens) contain tyrosine hydroxylase that appears to be in the less active form and can be readily activated by drug treatments (ZIVKOVIC and GUIDOTTI 1974; LOVENBERG and BRUCKWICK 1975; ZIVKOVIC et al. 1974, 1975). This is in contrast to areas of the brain that contain predominantly noradrenaline (cortex and hippocampus). In these areas tyrosine hydroxylase appears to exists in a more activated form ($K_m = 0.3$ mM) and is only marginally activated by drug treatments ($K_m = 0.2$ mM) (FRENCH et al. 1983). These studies indicate that tyrosine hydroxylase in noradrenaline containing brain regions is maintained in a higher state of activation than tyrosine hydroxylase in dopamine-containing brain regions. It is possible that decapitation of animals, prior to removal of the brains, can preferentially activate tyrosine hydroxylase in noradrenaline-containing brain areas to a greater extent than that seen in the predominantly dopamine-containing brain areas.

B. Assay of Tyrosine Hydroxylase

Many methods for the assay of tyrosine hydroxylase in homogenates and in the purified state have been employed by various investigators. Although fluorimetric assays measuring either dopa or catecholamines from tyrosine can be employed to estimate tyrosine hydroxylase activity, these assay procedures are generally restricted to purified enzyme of relatively high activity. In most studies of tyrosine hydroxylase, radiometric assays have been utilized. The precursor 3,5-ditritiotyrosine has been commonly employed in the assay of tyrosine hydroxylase. During the hydroxylation reaction, a tritium atom is displaced from the tyrosine and ultimately is conjugated with OH^- to form ^3HOH as a product of the reaction. The ^3HOH can then be assayed by separation of the radioactive water from precursor and catechol products by Dowex-50 column chromatography (NAGATSU et al. 1964b; SHIMAN et al. 1971). A simplified version of this assay has been developed using the direct extraction of the ^3HOH into isoamyl alcohol, thus avoiding the ion-exchange column separation (LEVINE et al. 1984). Alternatively, ^3H-dopa may be assayed following chromatographic separation of the catechol from tyrosine by absorption on and elution from alumina or alumina and Dowex-50 columns (COYLE 1972). Tyrosine hydroxylase may also be conveniently measured by the use carboxyl-labelled tyrosine. The initial reaction entails the conversion of ^{14}C-tyrosine into ^{14}C-dopa. In the presence of excess L-aromatic amino acid decarboxylase, purified from hog kidney (WAYMIRE et al. 1971), the ^{14}C-dopa product is converted to dopamine with the liberation of $^{14}CO_2$, and then collected in a basic trapping agent and counted by liquid scintillation spectrometry (WAYMIRE et al. 1971, ZIVKOVIC and GUIDOTTI 1974; KAPOTOS and ZIGMOND 1979; SHEN et al. 1986). The purification of decarboxylase from hog kidney for use in this assay can be avoided by utilizing ferricyanide to non-enzymatically decarboxylate the dopa (OKUNO and FUJISAWA 1983). Other methods of measuring the product of the reaction have also been employed, including derivatization and either thin layer or gas chromatographic separation of the derivatized products and mass spectrometry for identification and quantitation of the product (COSTA et al. 1972).

The assay system most commonly employed to measure tyrosine hydroxylase activity in intact tissues is the dopa accumulation procedure. NSD-1015 or NSD-1055, inhibitors of L-aromatic amino acid decarboxylase, are either administered parenterally or tissues are incubated with these compounds, and the dopa formed from endogenous tyrosine is measured using high-performance liquid chromatography (HPLC) with electrochemical detection (CARLSSON et al. 1972; HIRATA et al. 1973; FRENCH and WEINER 1984)

C. Localization of Tyrosine Hydroxylase

Tyrosine hydroxylase appears to be exclusively localized either in the adrenal medulla or noradrenergic neurons of the central or peripheral nervous systems (PICKEL et al. 1975, 1977, 1981; STEPHENS et al. 1981). Sympathetic denervation of various target organs or destruction of sympathetic neurons by 6-hydroxydopamine results in the complete disappearance of tyrosine hydroxylase activity from the target tissue (POTTER et al. 1965). It is generally believed that tyrosine hydroxylase is primarily or exclusively localized in the axoplasm of adrenergic neurons in soluble form. However, a considerable fraction of the tyrosine hydroxylase in bovine adrenal medulla appears to be associated with the chromaffin granules (PETRACK et al. 1968). The possibility that the association of tyrosine hydroxylase with chromaffin granules may play a role in the regulation of catecholamine synthesis has been suggested by MORITA et al. (1987) and STEPHENS et al. (1981). MORITA et al. (1987) found that the incubation of soluble tyrosine hydroxylase with chromaffin granule membranes in a mixture that may approximate the intracellular environment of the resting cell, produced an increase in enzyme activity recovered in the membrane fraction and a decrease in activity of the soluble enzyme. STEPHENS et al. (1981), utilizing immunohistochemical techniques, found that, in the rat adrenal medulla, tyrosine hydroxylase was associated with the cytosol of chromaffin cells. In these studies the percentage of chromaffin granules which appeared to contain tyrosine hydroxylase was lower in animals subjected to severe stress (electroconvulsive shock plus decapitation) as compared with animals whose adrenal glands were removed non-stressfully under pentobarbital anesthesia. These studies suggest that interactions between tyrosine hydroxylase and the chromaffin granule may play a possible role in the regulation of catecholamine biosynthesis in the adrenal medulla.

In homogenates of sympathetic ganglia and of bovine splenic nerves, tyrosine hydroxylase appears to be largely in the soluble supernatant fraction (STJÄRNE 1966; MUELLER et al. 1969a). Some of the brain tyrosine hydroxylase, notably in the striatum, is associated with particles (IYER et al. 1963; MCGEER et al. 1965; PATRICK and BARCHAS 1974). Although much of the enzyme present in the particulate fraction of brain tissue may actually be present in solution in the axoplasm which is trapped within synaptosomes, a significant fraction does appear to be membrane-bound (KUCZENSKI and MANDELL 1972b). Two types of tyrosine hydroxylase may exist which exhibit different kinetic properties and may be localized in different cell fractions, at least in bovine adrenal medulla (IKEDA et al. 1966; JOH et al. 1969) and in rat striatum (KUCZENSKI and MANDELL 1972b; KUCZENSKI 1973a, 1973b, 1983). KUCZENSKI (1983) reports that membrane-bound tyrosine hydroxylase from the rat striatum exhibits lower K_{ms} for both tyrosine (7 μ*M*) and pterin cofactor (110 μ*M*) as compared to the soluble enzyme (47 μ*M* and 940 μ*M*, respectively). Treatment of membrane-bound tyrosine hydroxylase with *V.* russelli phospholipase A_2 increased the K_m for tyrosine to 48 μ*M*, increased the V_{max} for tyrosine by 274%, and increased the K_m for cofactor to 560 μ*M*. In contrast, treatment of membrane-bound tyrosine hydroxylase with *C.* perfringens phospholipase *C* increased the K_m of the enzyme for tyrosine to 27 μ*M* and the V_{max} by 63% without changing the K_m for cofactor. It appears that the activity of membrane-bound tyrosine hydroxylase may be regulated by the membrane-associated enzymes, phospholipase *C* and phospholipase *A*. The importance of this membrane-bound tyrosine hydroxylase to the *in vivo* regulation of catecholamine synthesis has not been determined.

D. Short-Term Regulation of Tyrosine Hydroxylase

I. The Regulation of Catecholamine Synthesis

Tyrosine hydroxylase is generally regarded to be the rate limiting enzyme in the biosynthesis of catecholamines in brain as well as peripheral adrenergic tissue. Employing the isolated, perfused heart, LEVITT et al. (1965) demonstrated that the rate of noradrenaline synthesis from tyrosine was relatively low and did not increase when physiological concentrations of tyrosine in the perfusate were exceeded. In contrast, noradrenaline synthesis increased markedly when the perfusate contained either dopa or dopamine as precursors. The rate of noradrenaline synthesis from either dopa or dopamine increased in proportion to the concentration of the precursor in the perfusate. Their results suggested that, in situ, noradrenaline synthesis is limited by the hydroxylation of tyrosine. Based on the studies of NAGATSU et al. (1964a) and UDENFRIEND et al. (1965), which demonstrated inhibition of tyrosine hydroxylase activity by noradrenaline and other catechols, UDENFRIEND and coworkers suggested that the rate limiting step in the biosynthesis of catecholamines in tissues may be regulated by end-product feedback inhibition.

For many years it has been known that noradrenaline synthesis in adrenergic neurons and in the adrenal medulla is enhanced as a consequence of stimulation of these tissues. This conclusion was based on the observation that acute or chronic stresses to intact animals are associated with increased excretion of catecholamine metabolites in the urine without an associated decline in tissue catecholamine levels, unless either biosynthesis or storage of the biogenic amine is impaired (VON EULER et al. 1955). In more direct studies, HOLLAND and SCHÜMANN (1956), BUTTERWORTH and MANN (1957) and BYGDEMAN and VON EULER (1958) demonstrated that stimulation of the isolated perfused adrenal gland is associated with the release of amounts of catecholamines from that organ which greatly exceed the difference between the remaining catecholamine content in the stimulated gland and that in the unstimulated contralateral preparation.

ALOUSI and WEINER (1966) and ROTH and coworkers (1967) demonstrated directly that nerve stimulation is associated with an increase in noradrenaline synthesis from tyrosine. In the studies of ALOUSI and WEINER (1966), the isolated hypogastric nerve-vas deferens preparation of the guinea pig was stimulated in the presence of ^{3}H-tyrosine and the production of ^{3}H-catecholamines in the stimulated preparation was compared with that in the contralateral control. Intermittent nerve stimulation was associated with an approximately 50% increase in catecholamine synthesis. This effect of nerve stimulation appeared to be at the tyrosine hydroxylation step, since no enhanced synthesis of catecholamines was demonstrated when ^{3}H-dopa was employed as substrate (WEINER and RABADJIJA 1968). WEINER and RABADJIJA (1968) observed that the effect of nerve stimulation on catecholamine synthesis could be either reduced or blocked by the addition of catecholamines to the medium. These workers concluded that reduced end-product feedback inhibition was responsible for the accelerated noradrenaline synthesis associated with nerve stimulation. Since no difference in tissue catecholamine content was demonstrable between stimulated and contralateral control vasa deferentia, it was proposed that a small pool of noradrenaline located in solution in the axoplasm of the adrenergic neuron regulated the activity of tyrosine hydroxylase. Presumably, during nerve stimulation this small intracytosolic pool of soluble noradrenaline was reduced in concentration and, as a consequence, the normal tonic suppression of tyrosine hydroxylase activity was diminished and enhanced noradrenaline synthesis ensued.

This attractive theory for the regulation of tyrosine hydroxylase activity during nerve stimulation was supported by a variety of pharmacological studies. Indirectly acting sympathomimetic amines and monoamine oxidase inhibitors which exhibit an ability to release catecholamines from storage sites profoundly reduce the activity of tyrosine hydroxylase in intact tissues (WEINER and SELVARATNAM 1968; WEINER et al. 1972; BJUR and WEINER 1975). These amines possess no direct inhibitory effect on tyrosine hydroxylase isolated from the same preparations. It was postulated that these agents were able to release stored noradrenaline from intraneuronal vesicles into the axoplasm of the neuron where the catecholamine could interact with tyrosine hydroxyase and inhibit the enzyme. Similarly, acute reserpine administration, which would inhibit the uptake of noradrenaline and dopamine into adrenergic vesicles, and should, by this mechanism, lead to an increase in free cytosolic catecholamine until tissue stores are severely depleted, exerts an inhibitory effect on tyrosine hydroxylase *in situ* (WEINER et al. 1972; PFEFFER et al. 1975). Again, the drug possesses no direct inhibitory effect on soluble tyrosine hydroxylase. The inhibitory effect of reserpine on tyrosine hydroxylase activity in intact tissue is profoundly enhanced by the concomitant administration of a monoamine oxidase inhibitor (PFEFFER et al. 1975). This result is consistent with the concept that elevated levels of free intraneuronal noradrenaline do indeed regulate tyrosine hydroxylase activity in the intact tissue.

Several years ago, BJUR and WEINER (1975; WEINER and BJUR 1972) were able to demonstrate that exogenous reduced pterin cofactor could stimulate tyrosine hydroxylase activity in intact tissue. Presumably, the pterin is able to penetrate the adrenergic nerve terminal and serve as a cofactor for the enzymatic hydroxylation of tyrosine. These results suggested that either the concentration of pterin cofactor in the adrenergic nerve terminal is not adequate to saturate the enzyme or free intracytosolic catecholamines are present at levels which are able to antagonize competitively the interaction with the enzyme and otherwise saturating concentrations of cofactor. BJUR and WEINER (1975) and PFEFFER et al. (1975) demonstrated that the inhibitory effects of indirectly acting sympathomimetic amines and reserpine on tyrosine hydroxylase activity in either intact vasa deferentia or adrenal slices were overcome competitively by addition of synthetic tetrahydropterin cofactor to the incubation system. In contrast, the inhibition of tyrosine hydroxylase was not antagonized by addition of higher concentrations of tyrosine.

The ability of exogenous pterin cofactor to overcome the inhibitory effects of noradrenaline on tyrosine hydroxylase activity in intact tissue enabled CLOUTIER and WEINER (1973) to test the hypothesis that reduced end-product feedback inhibition could account for the accelerated synthesis of noradrenaline associated with nerve stimulation. It was argued that, if the enhancement of tyrosine hydroxylase activity associated with nerve stimulation was due to reduced end-product feedback inhibition, the differential effect of nerve stimulation should be overcome by the presence of high concentrations of pterin cofactor in the medium, since sufficiently high concentrations of exogenous cofactor should eliminate any end-product feedback inhibition in both stimulated and contralateral control preparations. However, CLOUTIER and WEINER (1973) demonstrated that the enhancement of noradrenaline synthesis associated with nerve stimulation was even further heightened by pterin cofactor and the difference in tyrosine hydroxylase activity between control and stimulated preparations was at least as pronounced as that seen in the absence of exogenous pterin cofactor. This finding was in contrast to the inhibitory effect of exogenous noradrenaline on tyrosine hydroxylase activity in intact vasa deferentia, which could be competitively overcome by added pterin cofactor. These results suggested that the effect of nerve stimulation on tyrosine hydroxylase activity in the intact vas deferens preparation is not due to simple reduction of end-product feedback inhibition by a mechanism involving competitive interaction with reduced pterin.

However, evaluating tyrosine hydroxylase activity in a retinal preparation, IUVONE et al. (1985) found no additivity between activated tyrosine hydroxylase and exogenously added cofactor. IUVONE et al. (1978) reported that the exposure of dark-adapted rats to light increased tyrosine hydroxylase activity in the retina. Kinetic analysis indicated that the increase was associated with a decrease in the K_m of tyrosine hydroxylase for pteridine cofactor with no effect on the V_{max}. Utilizing this technique IUVONE et al. (1985) found that the intravitreal injection of tetrahydrobiopterin increased the accumulation of dopa in retinas of dark-adapted rats, similar to that which occurs following light exposure. In contrast, the intravitreal injection of tetrahydrobiopterin had no significant effect on dopa accumulation in retinas of light-exposed rats. These data indicate that tyrosine hydroxylase is not saturated with endogenous tetrahydrobiopterin in the retinas of dark-adapted rats, but becomes saturated when the affinity of the enzyme for its cofactor is increased following light exposure. The relationship of this retinal effect to the feedback inhibition of catecholamines on tyrosine hydroxylase has not been determined.

The possibility that the levels of endogenous pterin cofactor can be regulated and may play a role in the regulation of catecholamine synthesis has been addressed by a number of laboratories (Abou-Donia and Viveros 1981; Viveros et al. 1981; Lee and Mandell 1985; Abou-Donia et al. 1986). There is evidence to suggest that tetrahydrobiopterin is actively synthesized in the adrenal medulla (Viveros et al. 1981) and in the central nervous system from the precursor, GTP (Gal and Sherman 1976). Abou-Donia and Viveros (Abou-Donia and Viveros 1981; Viveros et al. 1981) have shown that tetrahydrobiopterin in adrenal medulla increases following insulin administration or reserpine treatment and that these effects can be blocked by adrenal denervation. In addition, Viveros et al. (1981) have shown that, following adrenal stimulation, guanosine triphosphate cyclohydrolase activity (the rate-limiting enzyme in the biosynthesis of tetrahydrobiopterin) is increased in the medulla and is significantly reduced after denervation. More recently, Abou-Donia et al. (1986) reported that treatment of chromaffin cell cultures with compounds that increase the levels of, or mimic the actions of, cyclic AMP (forskolin, cholera toxin, theophylline, dibutyryl and 8 bromo cyclic AMP) not only increase tyrosine hydroxylase activity in these cells, but also increase guanosine triphosphate-cyclohydrolase activity. Tetrahydrobiopterin levels and intact cell tyrosine hydroxylase were both increased after 8-bromocyclic AMP treatment. These data indicate that intracellular levels of tetrahydrobiopterin can be regulated by stress, sympathetic nerve stimulation and intracellular cyclic AMP, and provide an additional mechanism for the control of the rate of tyrosine hydroxylation. In contrast, Nelson and Kaufman (1987) have shown that tetrahydropterins can activate a tyrosine hydroxylase phosphatase which can dephosphorylate tyrosine hydroxylase, thus decreasing the activity of the enzyme. These authors isolated three peaks of phosphatase activity from rat caudate nucleus by chromatography on DEAE-cellulose. The major peak exhibited activity that could dephosphorylate tyrosine hydroxylase previously phosphorylated by cyclic AMP-dependent protein kinase. The two minor peaks could dephosphorylate tyrosine hydroxylase which had been phosphorylated by calcium-calmodulin-dependent protein kinase. The tyrosine hydroxylase phosphatase, which was active on the cyclic AMP-dependent site, could be activated maximally at concentrations of 5 to 10 μM tetrahydrobiopterin with about a 2.7 fold activation occuring at a 1 μM concentration. Earlier work by Pollock et al. (1981) indicates that the activation of tyrosine hydroxylase by phosphorylation decreases the K_m of the enzyme for the pterin cofactor from 30 μM to 0.9 μM, similar to the known levels of tetrahydrobiopterin in the brain, approximately 1.5 μM. Therefore, the phosphorylated enzyme would increase in activity as tetrahydrobiopterin levels were increased, but as higher levels of tetrahydrobiopterin were achieved, this would offset the increase in tyrosine hydroxylase activity by producing a dephosphorylation of the enzyme. These authors conclude that the overall effect of tetrahydrobiopterin on the activities of tyrosine hydroxylase and tyrosine hydroxylase phosphatase would be to maintain relatively constant levels of catecholamine synthesis in spite of major changes in the levels of tetrahydrobiopterin.

II. Other Possible Allosteric Regulators of Tyrosine Hydroxylase Activity

A considerable number of putative allosteric modulators of rat striatal tyrosine hydroxylase have been proposed and one or more of these substances may be responsible for the enhanced activity of this enzyme which is associated with neural activity. MANDELL et al. (1972) noted that administration of either amphetamine, reserpine or α-methyl-p-tyrosine to rats is associated with a shift of caudate tyrosine hydroxylase from the soluble to the particulate state, presumably as a consequence of membrane binding. This was associated with a modest activation of the enzyme. However, stresses that presumably enhance central neural activity did not reproduce the effects of the drugs on this enzyme.

KUCZENSKI and MANDELL (1972b) observed that sulfate ions and heparin activated rat caudate soluble tyrosine hydroxylase *in vitro*. Both the maximal velocity and the affinity of the enzyme for cofactor were enhanced by heparin. In addition, the K_i for dopamine was increased by this sulfated polysaccharide. The changes produced by heparin resulted in an enzyme whose kinetic properties resembled closely those of the particulate enzyme. KUCZENSKI and MANDALL (1972a) proposed that physiological stimuli may modify the soluble enzyme or its environment in a manner which allows the molecule to become membrane-bound. As a consequence, the enzyme is activated. However, the relevance of this effect of heparin to neurally induced enzyme activation is questionable (WEINER et al. 1974).

MORITA et al. (1986a) found that the basic polypeptides, polyserine and polyarginine, enhance the ATP activation of tyrosine hydroxylase obtained from the bovine adrenal medulla. Kinetic studies showed that the basic polypeptides caused an increase in the V_{max} of the ATP-activated enzyme for the cofactor without any change in the K_m. The polyamines, putrescine, spermidine and spermine, failed to produce any effect on the ATP activation of tyrosine hydroxylase, indicating that the effect of the basic polypeptides on tyrosine hydroxylase is not due simply to their positive charge, but may also be related to their molecular size. These data suggest that macromolecular cell components, such as basic polypeptides, may be involved in the regulation of the activity of tyrosine hydroxylase through their modulating effects on the sensitivity of the enzyme to ATP within the cell.

Other anions, such as phosphatidyl-L-serine and polyglutamic acid, are also able to activate rat caudate tyrosine hydroxylase by enhancing the affinity of the enzyme for cofactor and shifting the optimal pH of the enzyme toward the more physiological pH region (LLOYD and KAUFMAN 1974; KATZ et al. 1976b). RAESE et al. (1977) have demonstrated activation of rat striatal tyrosine hydroxylase also by lysolecithin and phosphatidyl-L-serine. The activation in volves a three to four-fold reduction in the K_m for 6-methyltetrahydropterin or tetrahydrobiopterin. No change in the apparent affinity constant for L-tyrosine or in the maximal velocity of the reaction is demonstrable. The K_i for dopamine inhibition is increased approximately four-fold. It is of interest that both heparin and phosphatidyl-L-serine have been reported to be without effect on the activity of tyrosine hydroxylase from peripheral adrenergic neurons (NUMATA and NAGATSU 1975). Recently, MORITA et al. (1986b) reported that the membrane lipids, phosphatidylinositol (PI), phosphatidylinositol 4-phosphate (PIP) and phosphatidylinositol 4,5,-bisphosphate $(PIP)_2$ caused a rapid and concentration-dependent activation of adrenal tyrosine hydroxylase assayed *in vitro*. PIP_2 produced the greatest effect. These authors suggest that the activation of tyrosine hydroxylase by inositol phospholipids may be due to their electrostatic action.

In 1974, MORGENROTH et al. reported changes in the activity of soluble tyrosine hydroxylase prepared from vasa deferentia after field stimulation, as compared with the enzyme prepared from unstimulated organs. They demonstrated a marked reduction in the K_m for both 6,7-dimethyltetrahydropterin and tyrosine and a substantial increase in the K_i for noradrenaline. They also demonstrated similar kinetic changes in supernatant enzyme prepared from unstimulated vasa deferentia when 10–50 μM calcium was added to the assay medium. Similarily, MORGENROTH et al. (1974, 1975a) and ROTH et al. (1975a) reported that the addition of calcium to soluble tyrosine hydroxylase prepared from control vasa deferentia and from rat hippocampi results in an enhanced affinity of the enzyme for both substrate and cofactor and a reduced affinity of the enzyme for noradrenaline. These results are akin to those which they observed following field stimulation of the vas deferens (MORGENROTH et al. 1974) or electrical stimulation of the locus coeruleus (ROTH et al. 1975a; MORGENROTH et al. 1975a). They therefore proposed that the mechanism of activation of tyrosine hydroxylase following nerve stimulation involves uptake of calcium during depolarization and direct activation of the enzyme by the divalent cation. OSBORNE and NEUHOFF (1976) also have reported activation of soluble tyrosine hydroxylase from the circumesophageal ganglion of the snail (Helix pomatia) with $CaCl_2$. Definite effects, consisting of approximately a two to three fold decrease in K_m for both substrate (tyrosine) and cofactor (6-methyltetrahydropterin) occurred in the presence of 1 mM $CaCl_2$. At lower concentrations, the effects of calcium were unimpressive or absent.

More recently, IUVONE (1984) reported that rat striatal tyrosine hydroxylase subjected to Sephadex G-25 gel filtration could be activated by the addition of calcium, ATP and magnesium. The activation was reversible and was characterized by a large decrease of the apparent K_m for the pteridine cofactor and a small decrease in V_{max}. The activation was greatest at sub-optimal pH values and did not appear to be due to a calcium-activated protease or a cyclic-AMP-dependent protein kinase.

Several other laboratories, however, have been unable to duplicate these effects of calcium on soluble tyrosine hydroxylase. KNAPP et al. (1975) and LERNER et al. (1977) failed to demonstrate an activation of brain or adrenal tyrosine hydroxylase with calcium.

The influence of calcium on tyrosine hydroxylase has been suggested to be through the two calcium-dependent protein kinases, calcium-calmodulin-dependent and calcium-phospholipid-dependent protein kinases. Both these kinases have been shown to phosphorylate and in some cases activate tyrosine hydroxylase. The inconsistencies reported in the earlier work of the direct activation of tyrosine hydroxylase by calcium may be related to the method of preparation of the enzyme or to the assay procedures, which in some cases may allow for the calcium-dependent kinases to be active and to play a role in the subsequent activation of the enzyme.

In contrast to the purported stimulatory effects of calcium on tyrosine hydroxylase prepared from central and peripheral noradrenergic neurons (MORGENROTH et al. 1974; 1975b), GOLDSTEIN et al. (1970) found that calcium ions inhibited tyrosine hydroxylase in rat striatal slices. ROTH et al. (1974a) obtained analogous results with a high-speed supernatant preparation of striatal enzyme. Futhermore, ROTH et al. (1974a) observed that ethylene glycol bis (α-aminoethyl-ether)-N-N′-tetraacetic acid (EGTA) activates striatal tyrosine hydroxylase by enhancing its affinity for substrate and cofactor and reducing its affinity for dopamine.

The simulatory effects of EGTA and inhibitory effects of calcium on striatal tyrosine hydroxylase have been used to explain the peculiar observation that reducing impulse flow in the dopamine neurons of the striatum, either by administration of γ-hydroxybutyrolactone or by lesioning the nigrostriatal tract, is associated with increased tyrosine hydroxylase activity in the striatum (ROTH et al. 1974b). This enhanced activity of tyrosine hydroxylase can be antagonized by administration of dopamine agonists, such as apomorphine or piribedil (WALTERS and ROTH 1974). These agonists, like γ-hydroxybutyrolactone, suppress firing of dopamine neurons of the nigrostriatal tract (WALTERS et al. 1975). The tentative explanation proposed for these observations is that there exist dopamine presynaptic receptors which, when activated by dopamine or dopamine agonists in the presynaptic space, initiate a negative feedback effect on presynaptic tyrosine hydroxylase. Dopamine receptor blockers are able to prevent this effect of dopamine agonists. Thus, administration of haloperidol, chlorpromazine or other neuroleptics results in activation of tyrosine hydroxylase in the striatum by reversing this presynaptic inhibition (ZIVKOVIC and GUIDOTTI 1974; ZIVKOVIC et al. 1974, 1975; COSTA et al. 1974; LOVENBERG and BRUCKWICK 1975; ROTH et al. 1975b). ZIVKOVIC and GUIDOTTI (1974), ZIVKOVIC et al.(1974, 1975), COSTA et al. (1974) and LOVENBERG and BRUCKWICK (1975) have observed that administration of neuroleptics to rats results in an activation of soluble striatal tyrosine hydroxylase which is characterized by a reduced K_m for pterin cofactor and an increased K_i for dopamine. In contrast to these studies in which an increase in cofactor affinity alone was reported, ROTH et al. (1975b) also found that the enzyme exhibited a reduced K_m for tyrosine. The reason for this discrepancy is unclear. It is interesting that enhanced nigrostriatal neuronal activity, which presumably occurs following neuroleptic administration, and which is assumed to be associated with calcium uptake, results in activation of striatal tyrosine hydroxylase, in view of the reported effects of EGTA and calcium which are summarized above.

The unique activation of rat brain striatal tyrosine hydroxylase by both reduced and increased impulse flow in the nigrostriatal system remains unexplained. It is possible that the enzyme in this region is different from that in peripheral and central noradrenergic neurons. The enzyme from striatal dopamine neurons does appear to exhibit special regulatory properties, such as polyanion sensitivity. As noted above, NUMATA and NAGATSU (1975) were unable to demonstrate changes in the kinetic properties of bovine mandibular nerve tyrosine hydroxylase in the presence of either heparin or lysolecithin, suggesting the isoenzymes in adrenergic and dopaminergic neurons may be different. However, the physical properties of central nervous system and peripheral nervous system tyrosine hydroxylase appear to be identical, as described in the subsequent section.

III. Purification and Physical Properties of Tyrosine Hydroxylase

Tyrosine hydroxylase activity was first demonstrated by BRENNMAN and KAUFMAN (1964) and NAGATSU et al. (1964a). Since discovery of this enzyme, much effort has been devoted to purifying and characterizing this key regulatory protein. Initial attempts to purify tyrosine hydroxylase utilized bovine adrenal medullary tissue. These studies suggested that tyrosine hydroxylase might be associated with the cell membrane, since greater activity was found in the particulate fraction of the cell. However, tyrosine hydroxylase is now generally regarded as a soluble enzyme. Subsequent studies demonstrated that tyrosine hydroxylase will irreversibly aggregate with an unidentified membrane component in bovine adrenal medullary homogenates (WURTZBURGER and MUSACCHIO 1971). This aggregation can be prevented by partial proteolytic digestion of the enzyme with trypsin or chymotrypsin (PETRACK et al. 1968). Using proteolytic digested tyrosine hydroxylase, a catalytically active fragment of tyrosine hydroxylase was purified to homogeneity with a specific activity of 360 nanomoles of dopa formed per mg protein per minute (HOELDTKE and KAUFMAN 1977). SDS gel electrophoresis and density gradient sedimentation studies established that this proteolytic fragment migrated as a protein with a molecular weight of 34,000 Daltons. Stoichiometric amounts of iron (0.5 to 0.75 mole of iron per mole of enzyme subunit) were found to be present in this preparation confirming that tyrosine hydroxylase is a metallo-enzyme containing iron. Since the activity of this fragment could no longer be increased upon exposure to phospholipids or conditions optimal for protein phosphorylation, this suggested that only the catalytic portion of the molecule had been purified and that the regulatory domains of tyrosine hydroxylase might be located in a different region of the enzyme than the region involved in its catalytic activity (HOELDTKE and KAUFMAN 1977).

In the past few years, native tyrosine hydroxylase has been purified to homogeneity from rat pheochromocytoma (VULLIET et al. 1980; MARKEY et al. 1980), human pheochromocytoma (PARK and GOLDSTEIN 1975), rat and bovine adrenal medulla (OKUNO and FUJISAWA 1982; OKA et al. 1983), and rat caudate nucleus (RICHTAND et al. 1985). The rat pheochromocytoma homogenate has the highest initial starting specific activity of tyrosine hydroxylase of all the tissues examined, thus this tissue was initially used to purify tyrosine hydroxylase to homogeneity in several laboratories (VULLIET et al. 1980; MARKEY et al. 1980). Two different techniques for the purification of tyrosine hydroxylase from this tissue have been developed. VULLIET et al. (1980, 1984) employed differential ammonium sulfate fractionation, DEAE chromatography and heparin sepharose affinity chromatography to purify tyrosine hydroxylase 100-fold to a specific activity of approximately 500 nanomoles of dopa formed per mg protein per min at 37 °C. The purified enzyme from pheochromocytoma tumors migrated as a single band with a molecular weight of 60,000 Daltons. Purity was judged to be greater than 90% by SDS gel electrophoresis. Inclusion of proteolytic inhibitors throughout the purification was essential to prevent proteolysis of the N-terminal region of tyrosine hydroxylase. MARKEY et al. (1980) employed differential ammonium sulfate fractionation, gel permeation chromatography and sucrose density sedimentation to purify the enzyme from PC 12 cells. Enzyme prepared in this manner had a specific activity of approximately 370 nanomoles of dopa formed per mg protein per min.

Native tyrosine hydroxylase has been purified from the bovine (OKA et al. 1983), rat (OKUNO and FUJISAWA 1982) and human adrenal glands (KOJIMA et al. 1984). Most of these investigators have employed ammonium sulfate fractionation, DEAE chromatography, gel permeation chromatography, and/or heparin sepharose affinity chromatography to yield a preparation that was homogeneous as judged by SDS polyacrylamide gel electrophoresis. The purified enzymes from rat and bovine adrenal medulla had specific activities of 1600 and 1880 nanomoles dopa formed per mg protein per min, respectively, when assayed at 30 °C using a direct measurement of dopa production by either fluorometric or HPLC techniques. The purified human adrenal enzyme had a specific activity of 310 nanomoles of dopa formed per mg protein per minute at 37 °C when assayed using HPLC methods for quantitating dopa production.

Tyrosine hydroxylase from the central nervous system has been purified to homogeneity from bovine (OKA et al. 1983) and rat caudate nucleus (RICHTAND et al. 1985). The physical properties of brain tyrosine hydroxylase appear to be very similar to the physical properties of the enzyme isolated from the peripheral nervous system. RICHTAND et al. (1985) reported a specific activity of 406 nanomoles of dopa formed per mg protein per minute at 37 °C, which is in agreement with the values reported for enzyme purified from peripheral neuronal tissues. The enzyme migrates on SDS gel electrophoresis as a protein with a subunit molecular weight of 61,300 Daltons.

Tyrosine hydroxylase purified from various tissues of each species appears to exhibit similar physical properties. For the most part the subunits of purified tyrosine hydroxylase migrate as a single polypeptide with a molecular mass of approximately 60,000 Daltons (Table 1). These values agree with the 55,903 Dalton molecular weight, that was calculated for pheochromocytoma tyrosine hydroxylase by sequencing the complementary DNA coding for tyrosine hydroxylase (GRIMA et al. 1985). The Stokes radius ranges from 68.2 A for the human pheochromocytoma form of the enzyme to 54.3 A for the rat adrenal enzyme. Variations in the reported subunit size of tyrosine hydroxylase may result from the use of different molecular weight standards used to calibrate the SDS gel, the production of proteolytic artifacts during the purification of the enzyme, or the expression of variant forms of the enzyme with different sequences in the N-terminal region (GRIMA et al. 1987). Other physical properties of tyrosine hydroxylase purified from various tissues are also presented in Table 1.

An important question in studying the regulation of catecholamine biosynthesis is whether there is a difference between tyrosine hydroxylase from the central nervous system and the peripheral nervous system. Several studies that have investigated the physical properties of enzyme isolated from each source suggest that only slight physical differences exist (see Table 1). ROSENBERG and LOVENBERG (1983) found that the rat adrenal gland enzyme had a Stokes radius of 60.9 A and a sedimentation coefficient of 9.10 S compared to 54.3 A and 9.38 S for the striatal form of the enzyme. OKUNO and FUJISAWA (1985) performed a comparative study using tyrosine hydroxylase isolated from the adrenal medulla, substantia nigra, and striatum of rats. These authors found no major difference in the physical properties of the enzyme isolated from these sources. Using a monoclonal antibody prepared against purified rat adrenal tyrosine hydroxylase, these investigators were able to purify tyrosine hydroxylase from both brain regions and adrenal medulla, and then compare the physical properties of the enzymes by SDS gel electrophoresis

Table 1. Molecular Parameters of Tyrosine Hydroxylase

Subunit Source	Species	m.w.	Stokes Radius	S20,W	f/f_0	Reference
Adrenal	Bovine	60,000	—	—	—	OKA et al. (1983)
	rat	60,000	60.9	9.1	1.39	ROSENBERG and LOVENBERG (1983)
	rat	59,000	—	—	—	OKUNO and FUJISAWA (1985)
	human	60,000	—	—	—	KOJIMA et al. (1984)
Caudate	bovine	60,000	—	—	—	OKA et al. (1982)
	bovine	62,000	—	—	—	LAZAR et al. (1982)
	rat	61,300	57.4	9.58	—	RICHTAND et al. (1985)
	rat	59,000	—	—	—	OKUNO and FUJISAWA (1985)
	rat	—	54.3	9.38	1.28	ROSENBERG and LOVENBERG (1983)
Pheochromocytoma	rat	60,000	59	9.22	1.47	VULLIET and WEINER (1981)
	rat	62,000	—	7.6	—	PARK and GOLDSTEIN (1975)
	human	—	68.2	9.08	1.5	ROSENBERG and LOVENBERG (1983)

Abbrevations: m.w. = molecular weight; S20,W = sedimentation constant (Martin and Ames, 1961); f/f_0 = Functional coefficient (Siegel and Monty, 1966).

and gel permeation chromatography. Identical molecular weights were obtained by SDS gel electrophoresis. However, gel permeation chromatography demonstrated that the striatal enzyme eluted as a protein with a slightly larger molecular weight than adrenal tyrosine hydroxylase. Since tyrosine hydroxylase from each of these tissues has an identical subunit molecular weight, it is likely that slight differences in tyrosine hydroxylase isolated from the different tissues, such as lipid content, may account for the observed chromatographic differences. Thus, it seems likely that no major physical differences exist between tyrosine hydroxylase from the central and the peripheral nervous systems.

Comparisons of the specific activities of enzyme preparations of tyrosine hydroxylase from different species or tissues are difficult because many investigators utilize different conditions for assay of the enzyme. Ionic strength, pH, and buffering agents are known to affect the activity of tyrosine hydroxylase. Since phosphorylated tyrosine hydroxylase is thermally labile (Lazar et al. 1981; Vrana and Roskoski 1983), many laboratories utilize 30 °C to assay this enzyme. Purified tyrosine hydroxylase, when assayed using a radiometric assay, exhibits a specific activity between 300 and 500 nanomoles of dopa formed per mg protein per minute in most of the species and tissues examined.

In general the specific activity of tyrosine hydroxylase purified from the central nervous system is lower than tyrosine hydroxylase purified from the peripheral nervous system (Oka et al. 1983; Okuno and Fujisawa 1985). This may result from a less pure preparation of tyrosine hydroxylase from central nervous system tissue or from the presence of inhibitory factors in this tissue. Several investigators have noted that during purification of tyrosine hydroxylase, there is an increase in tyrosine hydroxylase activity over what would have been predicted from the prior purification step after accounting for recovery. Since this occurs in the presence of protease inhibitors, certain purification steps may be removing inhibitory factors. Okuno and Fujisawa (1985) observed significant activity differences between tyrosine hydroxylase purified from the adrenal and striatum when assayed at physiological pH in the presence of proteolytic inhibitors. If an inhibitor of tyrosine hydroxylase is present and serves to regulate the activity of the enzyme, it would be especially abundant in the brain. Although no major physical differences have been observed between central and peripheral tyrosine hydroxylase, differences in kinetic properties may exist.

It is generally accepted that tyrosine hydroxylase exists as a tetramer of identical subunits in the native state. It is thought to be an asymmetric molecule in its holoenzymatic form, as indicated by the fractional ratio (f/f_0) of greater than one (Table 1). Evidence for the existence of identical subunits comes from the observation that preparations of purified tyrosine hydroxylase migrate as a single band during SDS gel electrophoresis. Phosphorylation of the purified material by the catalytic subunit of cyclic AMP-dependent protein kinase occurs at a single site with a ratio of one molecule of phosphate per molecule of enzyme, further demonstrating that all of the subunits contain this phosphorylation site. Investigations using PC 12 cells have demonstrated that these cells produce only one sequence of mRNA coding for tyrosine hydroxylase (GRIMA et al. 1987). When taken together, these data strongly suggest that tyrosine hydroxylase in the native form is a tetramer of identical subunits.

IV. In Vitro Phosphorylation of Tyrosine Hydroxylase

Many key enzymes in intermediary metabolism and in biosynthetic reactions are regulated by reversible phosphorylation-dephosphorylation reactions, as well as allosteric effectors (KREBS and BEAVO 1979; COHEN 1980). It has become increasingly apparent in recent years that the phosphorylation of these enzymes may be at multiple sites and may be catalyzed by different protein kinases. In addition, a large number of protein kinases have been described over the past several years (GREENGARD 1975, 1978; KREBS and BEAVO 1979; COHEN 1980). These protein kinases can be currently grouped into four major classes (a) cyclic nucleotide-dependent protein kinases (LANGAN 1968; CORBIN et al. 1973; BECHTEL et al. 1977; HUTTNER and GREENGARD 1979); (b) calcium-calmodulin-dependent protein kinases (HUTTNER and GREENGARD 1979; O'CALLAGHAN et al. 1980); (c) calcium-phospholipid-dependent protein kinase (TAKAI et al. 1979; WRENN et al. 1980); and (d) protein kinases which are dependent upon neither cyclic AMP nor calcium for activity (RAESE et al. 1979; ANDREWS et al. 1983). All of these protein kinases have been demonstrated to modulate tyrosine hydroxylase activity. The physiological role of these protein kinases in regulating tyrosine hydroxylase activity and thus affecting the rate of catecholamine biosynthesis is an active area of research.

The possibility that tyrosine hydroxylase activity might be regulated by protein phosphorylation mechanisms was first suggested by the experiments of GOLDSTEIN and coworkers in which dibutyryl cyclic AMP was added to caudate slices and an increase in the production of dopa was observed (GOLDSTEIN et al. 1973). Other workers found that incubating extracts of caudate tissue under conditions that were optimal for phosphorylation by cyclic AMP-dependent protein kinase (e.g. inclusion of ATP, sodium fluoride and cyclic AMP) produced an increase in tyrosine hydroxylase activity due to an increase in the affinity of the enzyme for its reduced pterin cofactor (LOVENBERG et al. 1975). MORGENROTH et al. (1975b) demonstrated that this activation persisted following chromatography on G-25 Sephadex and suggested that the enzyme might be activated by a covalent incorporation of a phosphate group. However, to demonstrate that a cyclic AMP-dependent phosphorylation of tyrosine hydroxylase actually occurs, requires the availability of purified tyrosine hydroxylase and protein kinases.

Using a partially purified preparation of tyrosine hydroxylase from the rat caudate, JOH et al. (1978) reported that purified cyclic AMP-dependent protein kinase phosphorylated and activated tyrosine hydroxylase. This cyclic AMP-dependent protein kinase-mediated activation of tyrosine hydroxylase was reversible (YAMAUCHI and FUJISAWA 1979a, b). VULLIET et al. (1980) demonstrated that highly purified rat pheochromocytoma tyrosine hydroxylase incorporated stoichiometric amounts of phosphate in the presence of cyclic AMP-dependent protein kinase, ATP and magnesium. The amount of phosphate incorporation was highly correlated with the degree of activation of the enzyme, thus demonstrating that the direct incorporation of a phosphate group was the mechanism by which tyrosine hydroxylase was activated in the presence of cyclic AMP-dependent protein kinase.

As described previously, calcium has been shown to activate soluble tyrosine hydroxylase in numerous tissues (MORGENROTH 1974, 1975a, b; ROTH et al. 1975a; OSBORNE and NEUHOFF 1976). In addition, depolarization-dependent activation of tyrosine hydroxylase requires the presence of extracellular calcium (WEINER et al. 1978; CHALFIE et al. 1978). Recently, EL-MESTIKAWY et al. (1983) and IUVONE (1984) demonstrated an ATP and calcium-dependent activation of tyrosine hydroxylase in striatal homogenates and suggested that a calcium-dependent protein phosphorylation may be involved in regulating catecholamine biosynthesis in this tissue.

A calcium-dependent protein phosphorylation of tyrosine hydroxylase has been illustrated in a number of purified preparations. YAMAUCHI and FUJISAWA (1981) have demonstrated three distinct calcium-dependent protein kinases in rat cerebral cortex which they have designated kinases 1, 2 and 3. Kinase 1 appears to be similar to or identical with phosphorylase kinase. Kinase 3 phosphorylates casein or myosin light chain and may be identical with myosin light chain kinase. Kinase 2 exhibits a broad substrate specificity and appears to be involved in the regulation of tyrosine hydroxylase and tryptophan hydroxylase. These investigators have shown that this calcium-calmodulin-dependent protein kinase (kinase 2) directly phosphorylates both tyrosine hydroxylase and tryptophan hydroxylase and activates these enzymes when a specific activator protein is included. This observation was recently confirmed by ATKINSON et al. (1987) using the rat striatal enzyme. These authors report that tyrosine hydroxylase purified from rat striatum could be phosphorylated by calcium-calmodulin-dependent protein kinase 2. This kinase catalyzed the incorporation of approximately 0.8 moles $^{32}PO_4$ per mole of tyrosine hydroxylase subunit and this occurred at a different serine residue than was phosphorylated by the cyclic AMP-dependent protein kinase. Tyrosine hydroxylase activity was increased following phosphorylation by calcium-calmodulin-dependent protein kinase 2, only in the presence of an activator protein.

This activator protein has been purified to near homogeneity and has a molecular weight of 70,000 Daltons (YAMAUCHI et al. 1981; YAMAUCHI and FUJISAWA 1981). The activator protein has recently been characterized by ICHIMURA et al. (1987). These authors report that the tyrosine hydroxylase activator protein first reported by YAMAUCHI et al. (1981) is identical to brain 14-3-3 protein described by MOORE and PEREZ (1967). This activator protein is present in brain tissue at high concentrations (13.3 μg per mg soluble protein) (BOSTON et al. 1982) and will activate maximally tryptophan 5-monooxygenase in the presence of calcium-calmodulin-dependent protein kinase 2, at concentrations ranging from 10–20 μg per ml (ICHIMURA et al. 1987). The distribution of the 14-3-3 protein does not appear to parallel the distribution of the monooxygenases, thus implying that this protein may be involved not only

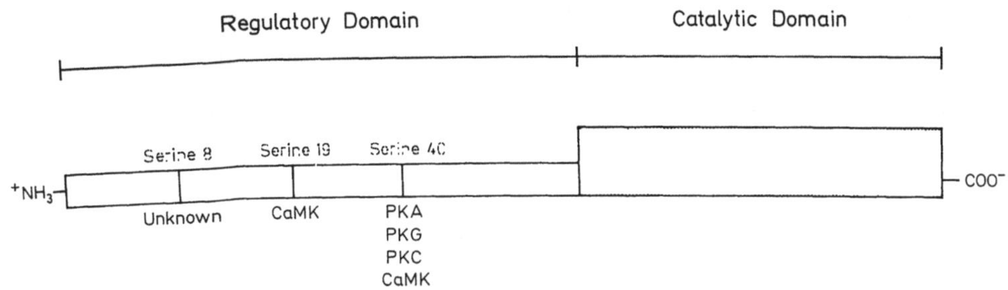

Fig. 1. Schematic Diagram of the Regulatory and Catalytic Domains of Tyrosine Hydroxylase Illustrating the Three Major Phosphorylation Sites. Abbreviations: PKA — cyclic AMP-dependent protein kinase; PKG — cyclic GMP-dependent protein kinase; PKC — calcium-phospholipid-dependent protein kinase; CaMK — calcium-calmodulin-dependent protein kinase.

in the regulation of 5-hydroxytryptamine and catecholamine biosynthesis but also in other biological processes.

CAMPBELL et al. (1986), using highly purified rat pheochromocytoma tyrosine hydroxylase, isolated and sequenced four specific phosphorylation sites on the enzyme. All of these sites were found to be present in the sequence of rat pheochromocytoma reported by GRIMA et al. (1985), thus providing direct verification that this sequence was coding for the tyrosine hydroxylase molecule. The phosphorylation sites identified were serine 8, serine 19, serine 40 and serine 153 (see Fig. 1). Serine 40 was found to be phosphorylated by cyclic AMP-dependent protein kinase, cyclic GMP-dependent protein kinase, calcium-phospholipid-dependent protein kinase and secondarily by calcium-calmodulin-dependent protein kinase. Serine 19 was preferentially phosphorylated by calcium-calmodulin-dependent protein kinase and was not observed to be phosphorylated by the other protein kinases examined (VULLIET et al. 1984). Serine 8 was phosphorylated by a novel protein kinase that is believed to mediate a phosphorylation associated with treatment of PC 12 cells with nerve growth factor (P. R. VULLIET, unpublished observation). Serine 153 appeared to be a secondary site phosphorylated by cyclic AMP-dependent protein kinase. This only occurred when tyrosine hydroxylase was purified in the absence of protease inhibitors, and was found to have lost about 20 residues from the N terminus, indicating that this site may only be phosphorylated when the enzyme is proteolytically degraded or denatured.

V. In Situ Phosphorylation of Tyrosine Hydroxylase

The *in situ* phosphorylation of tyrosine hydroxylase has been studied in a variety of catecholaminergic tissues that maintain a high degree of structural organization. These tissues include dissociated bovine adrenal medullary cells (HAYCOCK et al. 1982a, 1982b; POCOTTE and HOLZ 1986), PC 12 cells (TACHIKAWA et al. 1986, 1987), the superior cervical ganglia (CAHILL and PERLMAN 1984a, 1984b), and synaptosomes isolated from the rat striatum (HAYcock 1987). Generally, the approach has been to add a receptor agonist or an activator of a specific kinase system to cells that have been previously incubated with ^{32}P inorganic phosphate to label endogenous ATP. Tyrosine hydroxylase is then isolated by either immunoprecipitation or SDS gel electrophoresis and the radioactivity incorporated into the enzyme quantitated by scintillation counting or autoradiography. The specific sites labelled in tyrosine hydroxylase are then examined by proteolytic digestion of the enzyme and analysis of the peptides by either HPLC or two dimensional gel techniques. Since the specific sites phosphorylated *in vitro* by each of the kinases have been well characterized, it is then possible to determine whether these same sites are phosphorylated *in situ*. This analysis can determine the role that each of these kinases has in regulating catecholamine biosynthesis.

The regulation of tyrosine hydroxylase activity by cyclic AMP-dependent protein kinase has been extensively studied in chromaffin cells and pheochromocytoma PC 12 cells. The addition of 8-bromo cyclic AMP to bovine adrenal chromaffin cells resulted in an increase in the rate of catecholamine biosynthesis as a result of the phosphorylation and activation of tyrosine hydroxylase by cyclic AMP-dependent protein kinase (Meligeni et al. 1982). Addition of cyclic AMP analogues or compounds that cause an increase in intracellular cyclic AMP levels in PC 12 cells also results in the activation and phosphorylation of tyrosine hydroxylase (Yanagihara et al. 1986; Tachikawa et al. 1987). This activation is the result of increased phosphorylation on the same serine residue that is phosphorylated by cyclic AMP-dependent protein kinase *in vitro* (J. P. Mitchell and P. R. Vulliet, unpublished observation). The treatment of PC 12 cells with forskolin, an activator of adenylate cyclase, also produces an activation and phosphorylation of tyrosine hydroxylase in PC 12 cells (Tachikawa et al. 1987).

The phosphorylation of tyrosine hydroxylase by calcium-phospholipid-dependent protein kinase has been investigated using PC 12 cells (Tachikawa et al. 1987), dissociated bovine adrenal medullary cells (Pocotte and Holz 1986), and the superior cervical ganglia (Wang et al. 1986). Incubation of these tissues in the presence of phorbol esters, which selectively activate calcium-phospholipid-dependent protein kinase, results in an increase in tyrosine hydroxylase activity and an increase in the amount of radioactive phosphate incorporated into the tyrosine hydroxylase molecule. Interestingly, incubation of PC 12 cells with activators of calcium-phospholipid-dependent protein kinase results in a phosphorylation of a site different from the site phosphorylated *in vitro* (Tachikawa et al. 1987). *In vitro* studies have demonstrated that serine 40, the site phosphorylated by the catalytic subunit of cyclic AMP-dependent protein kinase, is also phosphorylated by purified calcium-phospholipid-dependent protein kinase (Vulliet et al. 1985). However, when phosphopeptide analysis is performed on cells treated *in situ* with phorbol esters, a site other than the cyclic AMP-dependent protein kinase site (serine 40) is found to be phosphorylated (Tachikawa et al. 1987). This observation suggests that the involvement of calcium-phospholipid-dependent protein kinase in the *in situ* regulation of catecholamine biosynthesis is more complicated than is indicated by the *in vitro* studies.

A calcium-dependent increase in tyrosine hydroxylase activity was observed following depolarization of cells with acetylcholine (HAYCOCK et al. 1982b). Treatment of chromaffin cells with acetylcholine resulted in an increase in phosphorylation of two sites; the E site and the C site. The E site was phosphorylated by treatment of the adrenal chromaffin cells with 8-bromo cyclic AMP and this site most likely represents the bovine site that is equivalent to rat pheochromocytoma serine 40. The phosphorylation of site C was dependent upon the presence of extracellular calcium (HAYCOCK et al. 1982b). This observation was further extended by MCTIGUE et al. (1985) when they identified four different tryptic phosphopeptides isolated from tyrosine hydroxylase following treatment of PC 12 cells with a variety of agonists. These tryptic phosphopeptides were labelled T1 through T4 and demonstrated a different degree of phosphorylation depending upon the agonist used to treat the cells. Treatment of the cells with acetylcholine caused a calcium-dependent increase in phosphorylation of peptides T2 and T3. These authors suggested that a calcium-dependent protein kinase other than protein kinase C may mediate the actions of this depolarizing agent. Following treatment with phorbol ester, the phosphorylation of peptide T1 was increased. Treatment of PC 12 cells with dibutyryl cyclic AMP resulted in the phosphorylation of peptide T3. Treatment of these cells with nerve growth factor (NGF) resulted in an increase in phosphorylation of peptides T1 and T3 and these authors suggested that both cyclic AMP-dependent protein kinase and calcium-phospholipid-dependent protein kinase were involved in mediating some of the actions of NGF on these cells.

Cytosolic tyrosine hydroxylase in adrenal medullary cells, PC 12 cells or superior cervical ganglia is also known to be phosphorylated following treatment with membrane depolarizing agents, such as acetylcholine and its analogs (HAYCOCK et al. 1982a, 1982b; CAHILL and PERLMAN 1984a), electrical stimulation (CAHILL and PERLMAN 1984b), peptide hormones, such as vasoactive intestinal peptide (CAHILL and PERLMAN 1984a) and growth factors (HALEGOUA and PATRICK 1980; GREENE et al. 1984; MCTIGUE et al. 1985). In most of these studies, the total incorporation of phosphate into tyrosine hydroxylase was investigated and the specific sites of phosphorylation were not determined. To elucidate which of the protein kinases are responsible for the *in situ* phosphorylation of tyrosine hydroxylase will require site-specific analysis of the phosphorylated enzyme with direct comparisons made to the known kinase-dependent phosphorylation sites on tyrosine hydroxylase phosphorylated *in vitro* using purified enzyme and kinases.

VI. Summary of the Short-Term Regulation of Tyrosine Hydroxylase

Based upon the similar effects that polyanions, anionic phospholipids, limited tryptic digestion and phosphorylation by specific protein kinases have on the activity of tyrosine hydroxylase, it is possible that the less active form of tyrosine hydroxylase may exist in a constrained conformation, wherein a regulatory strand on the polypeptide interacts coulombically with an anionic group

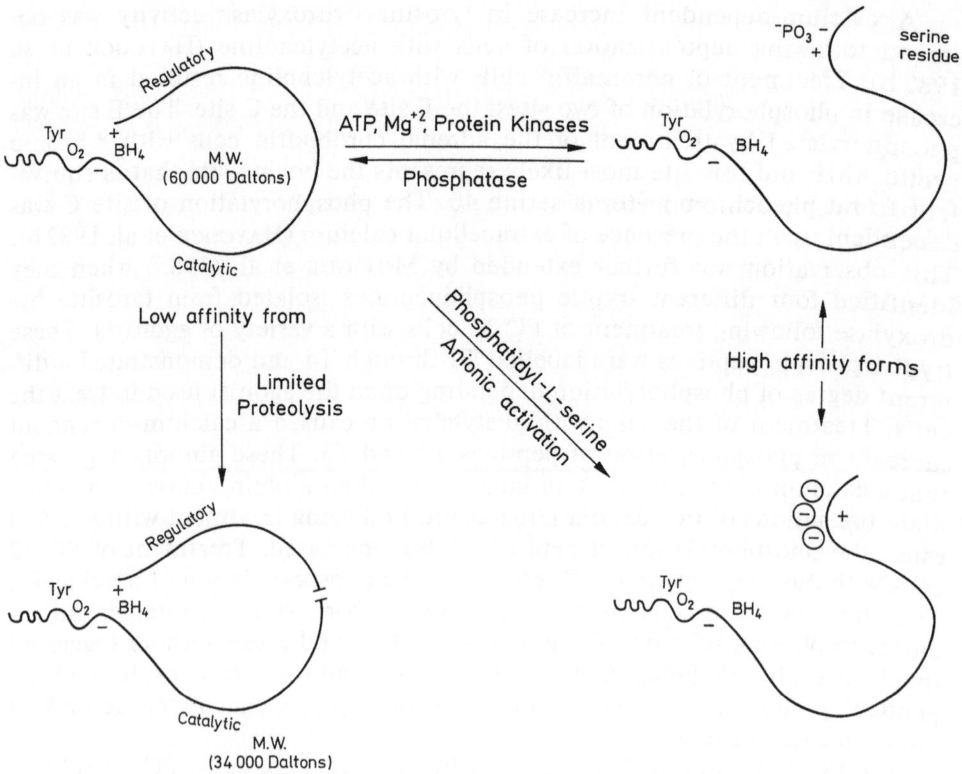

Fig. 2. Hypothetical Model for the Activation of Tyrosine Hydroxylase by Various Treatments. Abbreviations: Tyr — tyrosine; O_2 — oxygen; BH_4 — tetrahydrobiopterin

at the active site of the enzyme at which the pterin cofactor normally binds (Fig. 2). The constrained conformation is associated with a low affinity form of the enzyme for pterin cofactor. When a cationic group on the regulatory strand is neutralized by polyanionic interactions, interaction with phospholipids, or phosphorylation of an adjacent serine hydroxyl, the conformation of the enzyme is altered and the anionic site at which pterin cofactor binds is rendered more accessible to interaction with the cofactor. Limited proteolysis, which activates tyrosine hydroxylase, may result in a selective removal of this peptide strand or subunit from the region of the pterin binding site, thus resulting in irreversible activation of the enzyme.

The potential mechanisms by which the serine groups of tyrosine hydroxylase are phosphorylated, thus activating the enzyme, are illustrated in Fig. 3. The information illustrated in Fig. 3 was obtained from studies in the literature using isolated chromaffin cells, PC 12 cells and the superior cervical ganglia. Tyrosine hydroxylase is thought to be regulated *in vivo* by one of three possible mechanisms: a cyclic AMP-dependent protein phosphorylating system, a calcium-calmodulin-dependent protein phosphorylating system, or a calcium-phospholipid-dependent protein phosphorylating system. The two

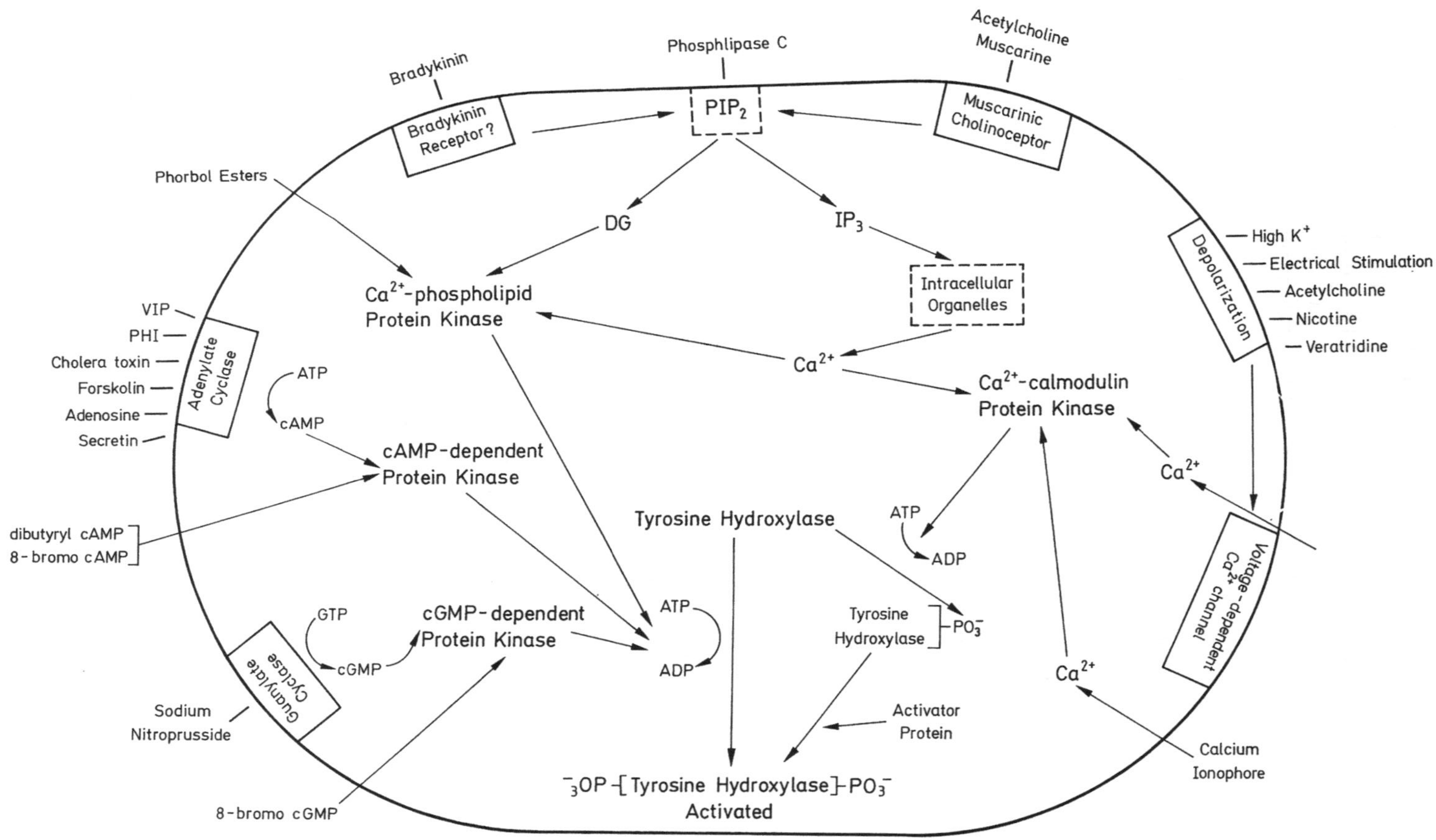

Fig. 3. Intracellular Components that Influence the Phosphorylation and Activation of Tyrosine Hydroxylase. Abbreviations: VIP — vasoactive intestinal peptide, PHI — a 27 amino acid peptide of the secretin family, PIP_2 — phosphatidylinositol 4,5-bisphosphate; DG — diacylglycerol; IP_3 — inositol trisphosphate.

major second messengers in neuronal tissues that enhance tyrosine hydroxylase activity through kinase activation are cyclic AMP and calcium. Cyclic AMP is known to activate a cyclic AMP-dependent protein kinase by attaching to the regulatory subunit of this kinase and releasing the catalytic subunit which is the active component. The catalytic subunit produces its effect on catecholamine synthesis by phosphorylating tyrosine hydroxylase and increasing the activity of the enzyme. Compounds that are known to activate tyrosine hydroxylase in intact cells, through a cyclic AMP-generating system, include secretin, vasopressin, PHI (peptide-histidine-isoleucine) (IP et al. 1982a, 1984, 1985; TISCHLER et al. 1985), cholera toxin (MCTIGUE et al. 1985; CAHILL and PERLMAN 1984a), forskolin (CAHILL and PERLMAN 1984; TACHIKAWA et al. 1987), and adenosine (ERNY et al. 1981). A similar mechanism of regulation of tyrosine hydroxylase has been reported to occur by an activation of guanylate cyclase in chromaffin cells by sodium nitroprusside, leading to an increase in cyclic GMP levels and an activation of tyrosine hydroxylase (ROSKOSKI and ROSKOSKI 1987). The addition of 8-bromo cyclic AMP (HAYCOCK et al. 1982b; ROSKOSKI and ROSKOSKI 1987), dibutyryl cyclic AMP (MCTIGUE et al. 1985; YANAGIHARA et al. 1986) or 8-bromo cyclic GMP (ROSKOSKI and ROSKOSKI 1987) to intact cells also activates and phosphorylates tyrosine hydroxylase.

Intracellular calcium is thought to act as a second messenger through an interaction with either a calcium-phospholipid-dependent protein kinase or a calcium-calmodulin-dependent protein kinase. The activation of these kinases leads to a phosphorylation of tyrosine hydroxylase, and an increase in the activity of the enzyme in the case of the calcium-phospholipid-dependent protein kinase, whereas the calcium-calmodulin-dependent protein kinase requires the presence of an activator protein to activate the enzyme.

Intracellular calcium can be elevated following an influx of extracellular calcium caused by membrane deopolarization leading to an activation of voltage-dependent calcium-channels. Treatments that are known to activate tyrosine hydroxylase in intact cells though an influx of extracellular calcium include, high potassium (YANAGIHARA et al. 1986; TACHIKAWA et al. 1986), electrical depolarization (CAHILL and PERLMAN 1984b), acetylcholine (HAYCOCK et al. 1982a, 1982b), nicotine (IP et al. 1982b; POCOTTE et al. 1986), veratridine (HORWITZ and PERLMAN 1984), and the calcium ionophore, A23187 (TACHIKAWA et al. 1986).

Intracellular calcium may also be elevated by calcium released from intracellular organelles. This is controlled by receptor-stimulated phospholipid hydrolysis. The breakdown of the membrane lipid, phosphatidylinositol 4,5-bisphosphate (PIP_2) produces two putative second messengers in the cell, inositol trisphosphate (IP_3) and diacylglycerol (DG). Diacylglycerol is thought to act by increasing the affinity of calcium-phospholipid-dependent protein kinase for calcium, thus enhancing the activity of this kinase. Inositol trisphosphate is thought to act by mobilizing calcium from an intracellular vesicular pool, such as the mitochondria or the endoplasmic reticulum. This intracellular calcium can bind to calmodulin and activate calcium-calmodulin-dependent protein kinase or can increase the affinity of diacylglycerol for calcium-phospholipid-dependent protein kinase, and therefore play a role in the phosphorylation and activation of tyrosine hydroxylase. Bradykinin (VAN CALKER and HEUMANN 1987) and muscarine (HORWITZ et al. 1984) have been shown to increase inositol trisphosphate in PC 12 cells and superior cervical ganglia, respectively, and to activate tyrosine hydroxylase (HORWITZ et al. 1984; HOUCHI et al. 1988). These compounds may activate tyrosine hydroxylase either by activating calcium-calmodulin-dependent protein kinase through an increase in intracellular calcium caused by IP_3 generation, or by increasing diacylglycerol intracellularly and thus increasing the affinity of calcium for calcium-phospholipid-dependent protein kinase. Phospholipase C, the membrane-bound enzyme that breaks down PIP_2, increases the turnover of phospholipids in superior cervical ganglia cells and also increases the activity of tyrosine hydroxylase and the incorporation of ^{32}P into the enzyme (HORWITZ et al. 1984; CAHILL and PERLMAN 1984a). Phorbol mesylate ester (PMA), a selective activator of calcium-phospholipid-dependent protein kinase, also increases the activity of tyrosine hydroxylase in PC 12 cells (McTIGUE et al. 1985; TACHIKAWA et al. 1987).

The activation of tyrosine hydroxylase appears to be under the control of the three major protein kinases, cyclic AMP-dependent protein kinase, calcium-phospholipid-dependent protein kinase and calcium-calmodulin-dependent protein kinase. The importance of each kinase in the regulation of catecholamine synthesis may depend upon the ability of these kinases to activate and phosphorylate tyrosine hydroxylase in neuronal tissues where they are co-localized.

E. Long Term Regulation of Tyrosine Hydroxylase

I. Regulation of Tyrosine Hydroxylase Enzyme Levels by Different Environmental Stimuli

Almost 20 years ago two laboratories demonstrated that tyrosine hydroxylase activity in the adrenal medulla and sympathetic ganglia increased in animals exposed to catecholamine-depleting drugs, such as reserpine or 6-hydroxydopamine, or to prolonged stress (MUELLER et al. 1969a; THOENEN et al. 1969; KVETNANSKY et al. 1971). A number of studies have indicated that these treatments elevate tyrosine hydroxylase activity by increasing the stimulation of these catecholaminergic cells for prolonged periods of time (THOENEN et al. 1969; ZIGMOND et al. 1974; IP and ZIGMOND 1985). Other early studies have indicated that the increase in activity is blocked by nicotinic receptor antagonists and by inhibitors of RNA and protein synthesis (MUELLER et al. 1969b, 1970). JOH et al. (1973) have demonstrated that this increase in enzyme activity is due to an increase in enzyme protein. Hence, this effect on tyrosine hydroxylase has been termed transsynaptic induction, and is thought to be elicited by prolonged occupation of nicotinic cholinoceptors by acetylcholine released from presynaptic nerves.

The intracellular steps that occur between the occupation of the nicotinic receptors and the elevation of tyrosine hydroxylase remain obscure. Early evidence has suggested that an increase in cyclic AMP levels is associated with this transsynaptic induction (GUIDOTTI and COSTA 1973). However, this association is controversial (OTTEN et al. 1973), and the mechanism by which nicotinic receptor activation leads to cyclic AMP elevation has not been elucidated. Calcium has been shown to play a role in the elevation of the enzyme in cultured sympathetic neurons (HEFTI et al. 1982); however, more work is required to determine whether calcium induces the enzyme in other catecholaminergic systems. In addition, more studies are needed to determine whether neurohumoral factors, other than acetylcholine, that are released from presynaptic nerves also play a role in the transsynaptic induction of tyrosine hydroxylase.

Glucocorticoids and nerve growth factor (NGF) have also been shown to influence the levels of tyrosine hydroxylase. Glucocorticoids induce the enzyme in a number of tissue culture model systems, such as rat sympathetic ganglia (OTTEN and THOENEN 1976), rat pheochromocytoma cells (LEWIS et al. 1983), and mouse neuroblastoma cells (TANK and WEINER 1982). In addition, glucocorticoids have been shown to modulate the transsynaptic induction of tyrosine hydroxylase observed *in vivo* during prolonged stress (OTTEN and THOENEN 1975). NGF is thought to be essential for the expression of tyrosine hydroxylase in developing and adult sympathetic neurons. Furthermore, concentrations of NGF greater than those necessary to maintain the basal expression of the enzyme further increase the enzyme levels in sympathetic neurons and adrenal chromaffin cells (HEFTI et al. 1982; ACHESON et al. 1984).

A number of other factors have been shown to influence tyrosine hydroxylase levels in developing catecholaminergic cells. Incubation of dissociated sympathetic neurons derived from immature rats in the presence of an elevated concentration of potassium results in an increase in the enzyme level (WALICKE et al. 1977). In contrast, a soluble factor synthesized and secreted from cardiac muscle decreases the expression of the enzyme in these cultured sympathetic neurons (WOLINSKI and PATTERSON 1983). Interaction with the extracellular matrix protein, laminin, and increased cell-cell contact have also been shown to elevate tyrosine hydroxylase in catecholaminergic cells (ACHESON and THOENEN 1983; ADLER and BLACK 1985; ACHESON et al. 1986).

II. Regulation of Tyrosine Hydroxylase mRNA Levels

Much recent work has focused on the effects of many of the inducing agents described in the preceding section on the levels of the RNA coding for tyrosine hydroxylase (TH-mRNA). The measurement of TH-mRNA has been made possible by the isolation of cloned cDNA's coding for this mRNA (LAMOUROUX et al. 1982; LEWIS et al. 1983). Using these cloned probes, a number of laboratories have shown that the major TH-mRNA transcript present in all catecholaminergic cells tested so far is approximately 1800–1900 nucleotides. In addition, the isolation of a full-length cDNA that codes for the complete tyrosine hydroxylase molecule has permitted the deduction of the amino acid sequence of the enzyme (GRIMA et al. 1985). Translation of the RNA for which this cDNA codes in Xenopus oocytes results in tyrosine hydroxylase activity, demonstrating that this single mRNA species codes for a functional enzyme (HORELLOU et al. 1986). This information, along with primer extension studies and partial characterization of the TH gene indicate that the TH-mRNA molecule consists of 1494 nucleotides of sequences coding for the enzyme, 35 nucleotides of 5′ untranslated sequences, and approximately 265 nucleotides of 3′ untranslated sequences (GRIMA et al. 1985, 1987; HARRINGTON et al, 1987). Furthermore, a comparison of the sequence of TH-mRNA with those of other mRNA's indicates that tyrosine hydroxylase is highly homologous to other enzymes that catalyze steps in the biosynthesis of the catecholamines (BAETGE et al. 1986; JOH et al. 1984), as well as to the liver enzyme, phenylalanine hydroxylase (LEDLEY et al. 1985), suggesting that tyrosine hydroxylase may be a member of a multi-gene family (JOH et al. 1985).

A number of laboratories have shown that the levels of TH-mRNA increase in rat adrenal medulla and superior cervical ganglion after reserpine administration or prolonged stress (TANK et al. 1985; BLACK et al. 1985; STACHOWIAK et al. 1986; FAUCON BIGUET et al. 1986). In the adrenal medulla this increase in the level of TH-mRNA occurs much more rapidly than the increase in the level of enzyme; however, the extent of the increases in these two parameters are equivalent, suggesting that the changes in TH-mRNA totally account for the subsequent changes in enzyme levels. The discrepancy between the time courses for the increases in these two parameters is presumably due to differences in their rates of degradation. Reserpine administration also induces TH-mRNA in the locus coeruleus of rat brain; this induction has been observed both by Northern blot analysis of RNA isolated from this brain region and by *in situ* hybridization of histological sections (FAUCON BIGUET et al. 1986; BEROD et al. 1987).

The induction of TH-mRNA has been further investigated using rat pheochromocytoma cells in culture. Cyclic AMP and glucocorticoids have both been shown to induce TH-mRNA in this model system (LEWIS et al. 1983; TANK et al. 1986b). Similar to the results obtained *in vivo*, the elevation of TH-mRNA elicited by either of these two inducing agents precedes, but is quantitatively equal to, the elevation of tyrosine hydroxylase enzyme levels (TANK et al. 1986a, 1986b). Recent studies using nuclear run-on assays have demonstrated that the elevation of TH-mRNA elicited by cyclic AMP and glucocorticoids in rat pheochromocytoma cells is at least partially due to an increase in the rate of transcription of the tyrosine hydroxylase gene (LEWIS et al. 1987; FOSSOM et al. 1987). Furthermore, LEWIS et al. (1987) have clearly demonstrated that a functional cyclic AMP response element exists in the 5′ flanking region of the tyrosine hydroxylase gene. Consensus sequences for cyclic AMP and glucocorticoid response elements have been identified in this 5′ flanking region; however, whether the glucocorticoid response element is functional in the intact cell remains to be determined (HARRINGTON et al. 1987; LEWIS et al. 1987). Interestingly, BLUM et al. (1987) have shown that the rate of transcription of the tyrosine hydroxylase gene is decreased *in vivo* in the dopaminergic neurons of the rat arcuate nucleus after the administration of estrogen to the rats. The mechanism responsible for this estrogen-dependent effect on the transcription rate is not yet clear. As yet, studies on the effects of other stimuli on the rate of transcription of the tyrosine hydroxylase gene have not been reported. Furthermore, regulation of the stability of TH-mRNA by different stimuli cannot be ruled out.

When rat pheochromocytoma cells are exposed to elevated levels of both cyclic AMP and glucocorticoids, the enzyme level and rate of synthesis of tyrosine hydroxylase increase additively (TANK et al. 1986a). Interestingly, under these conditions, the level of TH-mRNA does not increase above that observed in the presence of either inducing agent alone (TANK et al. 1986b). Recent results indicate that the translational activity of TH-mRNA increases additively in cells exposed to both inducing agents (BOWYER et al. 1988). Hence, cyclic AMP and glucocorticoids regulate TH-mRNA by two mechanisms: (1) an elevation of TH-mRNA levels, which is at least partially due to an enhancement of the rate of transcription of the TH gene; and (2) an increase in the translational activity per hybridizable molecule of TH-mRNA. The mechanism for this effect on translational activity remains to be determined.

Recently, a number of laboratories have reported the existence of multiple TH-mRNA transcripts present in humans, suggesting that multiple forms of tyrosine hydroxylase may exist in different tissues (GRIMA et al. 1987; KANEDA et al. 1987). These multiple forms of TH-mRNA differ by 12 to 93 nucleotides, and are due to alternate splicing of a nuclear RNA precursor derived from a single tyrosine hydroxylase gene. The alternate splicing occurs at the 3′ end of the first exon, and the insertion of these sequences does not alter the reading frame of the protein-coding region. The alternate splicing presumably results in the production of multiple forms of tyrosine hydroxylase molecules that differ in amino acid sequence in the region near their amino-termini. Since the catalytic and regulatory domains of the enzyme molecule are thought to reside in the carboxyl-terminus and amino-terminus, respectively, it has been suggested that the alternate forms of tyrosine hydroxylase may possess different regulatory properties (GRIMA et al. 1987; KANEDA et al. 1987). If these alternate forms exist in different neurons in the central and peripheral nervous systems, then differences in the regulation of the enzyme in these different neurons might be predicted. The existence of multiple forms of tyrosine hydroxylase in different neurons, the functional significance of these different forms, and the regulation of the alternate splicing requires further investigation.

Acknowledgement. The authors would like to acknowledge the technical assistance of Ellen Barnes in preparing this manuscript.

F. References

Abou-Donia MM, Viveros OH (1981) Tetrahydrobiopterin increases in adrenal medulla and cortex: a factor in the regulation of tyrosine hydroxylase. Proc Nat Acad Sci USA 78:2703–2706

Abou-Donia MM, Wilson SP, Zimmerman TP, Nichol CA, Viveros OH (1986) Regulation of guanosine triphosphate cyclohydrolase and tetrahydrobiopterin levels and the role of the cofactor in tyrosine hydroxylation in primary cultures of adrenomedullary chromaffin cells. J Neurochem 46:1190-1199

Acheson AL and Thoenen H (1983) Cell contact-mediated regulation of tyrosine hydroxylase synthesis in cultured bovine adrenal chromaffin cells. J Cell Biol 97:925-928

Acheson AL, Naujoks K, Thoenen H (1984) Nerve growth factor-mediated enzyme induction in primary cultures of bovine adrenal chromaffin cells: specificity and level of regulation. J Neuroscience 4:1771-1780

Acheson AL, Edgar D, Timpl R, Thoenen H (1986) Laminin increases both levels and activity of tyrosine hydroxylase in calf adrenal chromaffin cells. J Cell Biol 102:151-159

Adler JE, Black IB (1985) Sympathetic neuron density differentially regulates transmitter phenotypic expression in culture. Proc Nat Acad Sci USA 82:4296-4300

Alousi A, Weiner N (1966) The regulation of norepinephrine synthesis in sympathetic nerves. Effect of nerve stimulation, cocaine and catecholamine releasing agents. Proc Nat Acad Sci USA 56:1491-1496

Andrews DW, Langan TA, Weiner N (1983) Evidence for the involvement of a cyclic AMP-independent protein kinase in the activation of soluble tyrosine hydroxylase from rat striatum. Proc Nat Acad Sci USA 80:2097-2101

Atkinson J, Richtand N, Schworer C, Kuczenski R, Soderling T (1987) Phosphorylation of purified rat striatal tyrosine hydroxylase by Ca^{2+}/calmodulin-dependent protein kinase II: effect of an activator protein. J Neurochem 49:1241-1249

Baetge EE, Suh YH, Joh TH (1986) Complete nucleotide and deduced amino acid sequence of bovine phenylethanolamine N-methyltransferase: partial amino acid homology with rat tyrosine hydroxylase. Proc Nat Acad Sci USA 83:5454-5458

Bechtel PJ, Beavo JA, Krebs EG (1977) Purification and characterization of catalytic subunit of skeletal muscle adenosine 3′:5′-monophosphate-dependent protein kinase. J Biol Chem 252:2691-2697

Berod A, Faucon Biguet N, Dumas S, Bloch B, Mallet J (1987) Modulation of tyrosine hydroxylase gene expression in the central nervous system visualized by in situ hybridization. Proc Nat Acad Sci USA 84:1699-1703

Bjur RA, Weiner N (1975) The activity of tyrosine hydroxylase in intact adrenergic neurons of the mouse vas deferens. J Pharmacol Exp Ther 194:9-26

Black IB, Chikaraishi D, Lewis EJ (1985) Transsynaptic increase in RNA coding for tyrosine hydroxylase in a rat symapthetic ganglion. Brain Res 339:151-153

Blum M, McEwen BS, Roberts JL (1987) Transcriptional analysis of tyrosine hydroxylase gene expression in the tuberoinfundibular dopaminergic neurons of the rat arcuate nucleus after estrogen treatment. J Biol Chem 262:817-821

Boston PF, Jackson P, Thompson RJ (1982) Human 14-3-3 protein: radioimmunoassay, tissue distribution, and cerebrospinal fluid levels in patients with neurological disorders. J Neurochem 38:1475-1482

Bowyer J, Curella P, Tank AW Regulation of the translational activity of tyrosine hydroxylase mRNA by cyclic AMP and glucocortoids. Submitted for publication

Brenneman AR, Kaufman S (1964) The role of tetrahydropteridines in the enzymatic conversion of tyrosine to 3,4-dihydroxyphenylalanine. Biochem Biophys Res Comm 17:177-183

Buff K, Dairman W (1974) Biosynthesis of biopterin by two clones of mouse neuroblastoma. Mol Pharmacol 11:87-93

Bullard WP, Capson TL (1983) Steady-state kinetics of bovine striatal tyrosine hydroxylase. Mol Pharmacol 23:104-111

Bullard WP, Guthrie PB, Russo PV, Mandell AJ (1978) Regional and subcellular distribution and some factors in the regulation of reduced pterins in rat brain. J Pharmacol Exp Ther 206:4-20

Butterworth KR, Mann M (1957) The release of adrenaline and noradrenaline from the adrenal glands of the cat by acetylcholine. Brit J Pharmacol Chemother 12:422-426

Bygdeman S, Euler von US (1958) Resynthesis of catechol hormones in the cat's adrenal medulla. Acta Physiol Scand 44:375-383

Cahill AL, Perlman RL (1984a) Phosphorylation of tyrosine hydroxylase in the superior cervical ganglion. Biochem Biophys Acta 805:217-226

Cahill AL, Perlman RL (1984b) Electrical stimulation increases phosphorylation of tyrosine hydroxylase in superior cervical ganglion of rat. Proc Nat Acad Sci USA 81:7243-7247

Campbell DG, Hardie DG, Vulliet PR (1986) Identification of four phosphorylation sites in the N-terminal region of tyrosine hydroxylase. J Biol Chem 261:10489-10492

Carlsson A, Davis JN, Kehr W, Lindqvist M, Atack CV (1972) Simultaneous measurement of tyrosine and tryptophan hydroxylase activities in brain in vivo using an inhibitor of the aromatic amino acid decarboxylase. Naunyn-Schmiedeberg's Arch Pharmacol 275:153-168

Chalfie M, Settipani L, Perlman RL (1978) The role of cyclic adenosine 3′:5′-monophosphate in the regulation of tyrosine 3-monooxygenase activity. Mol Pharmacol 15:263-270

Cloutier G, Weiner N (1973) Further studies on the increased synthesis of norepinephrine during nerve stimulation of guinea-pig vas deferens preparations: effect of tyrosine and 6,7-dimethyltetrahydropterin. J Pharmacol Exp Ther 186:75-85

Cohen P (1980) Recently Discovered Systems of Enzyme Regulation by Reversible Phosphorylation. Elsevier-North Holland, Amsterdam

Corbin JD, Soderling TR, Park CR (1973) Regulation of adenosine 3′:5′-monophosphate-dependent protein kinase. I. Preliminary characterization of the adipose tissue enzyme in crude extracts. J Biol Chem 248:1813-1821

Costa E, Green AR, Koslow SH, LeFevre HF, Revuelta AV, Wang C (1972) Dopamine and norepinephrine in noradrenergic areas: a study in vivo of their precursor product relationship by mass fragmentography and radiochemistry. Pharmacol Rev 24:167-190

Costa E, Guidotti A, Zivkovic B (1974) Short- and long-term regulation of tyrosine hydroxylase. In: Usdin E (Ed) Neuropsychopharmacology of Monoamines and Their Regulatory Enzymes. Raven Press New York, pp 161-175

Coyle JT (1972) Tyrosine hydroxylase in rat brain – cofactor requirements, regional and subcellular distribution. Biochem Pharmacol 21:1935-1944

Craine JE, Hall ES, Kaufman S (1972) The isolation and characterization of dihydropteridine reductase from sheep liver. J Biol Chem 247:6082-6091

Craine JE, Daniels GH, Kaufman S (1973) Dopamine-beta-hydroxylase: the subunit structure and anion activation of the bovine adrenal enzyme. J Biol Chem 248:7838-7844

Daly J, Levitt M, Guroff G, Udenfriend S (1968) Isotope studies on the mechanism of action of adrenal tyrosine hydroxylase. Arch Biochem Biophys 126:593-598

Davis JN (1976) Brain tyrosine hydroxylation: alteration of oxygen affinity in vivo by immobilization or electroshock in the rat. J Neurochem 27:211-215

Davis JN, Carlsson A (1973) Effect of hypoxia on tyrosine and tryptophan hydroxylation in unanesthetized rat brain. J Neurochem 20:913-915

Dix TA, Kuhn DM, Benkovic SJ (1987) Mechanism of oxygen activation by tyrosine hydroxylase. Biochemistry 26:3354-3361

El-Mestikawy SE, Glowinski J, Hamon M (1983) Tyrosine hydroxylase activation in depolarized dopaminergic terminals: involvement of Ca^{2+}-dependent phosphorylation. Nature 302:830-832

Erny RE, Berezo MW, Perlman RL (1981) Activation of tyrosine 3-monooxygenase in pheochromocytoma cells by adenosine. J Biol Chem 256:1335-1339

Euler von US, Luft R, Sundin T (1955) The urinary excretion of noradrenaline and adrenaline in healthy subjects during recumbancy and standing. Acta Physiol Scand 34:169-174

Faucon Biguet N, Buda M, Lamouroux A, Samolyk D, Mallet J (1986) Time course of the changes of TH mRNA in rat brain and adrenal medulla after a single injection of reserpine. EMBO J 5:287-291

Fisher DB, Kaufman S (1972) The inhibition of phenylalanine and tyrosine hydroxylase by high oxygen levels. J Neurochem 19:1359-1365

Fischer DB, Kirkwood R, Kaufman S (1972) Rat liver phenylalanine hydroxylase, an iron enzyme. J Biol Chem 16:5161-5167

Fossom LH, Weiner N, Tank AW (1987) Increased transcription of the gene for tyrosine hydroxylase induced by cyclic AMP in a pheochromocytoma cell line. Society for Neuroscience 12:599

French TA, Weiner N (1984) Effect of ethanol on tyrosine hydroxylation in brain regions of long and short sleep mice. Alcohol 1:247-252

French TA, Masserano JM, Weiner N (1983) Activation of tyrosine hydroxylase in the frontal cortex by phentolamine and prazosin. J Pharm Pharmacol 35:618-620

Gal EM, Sherman AD (1976) Biopterin. II. Evidence for cerebral synthesis of 7,8-dihydrobiopterin in vivo and in vitro. Neurochem Res 1:627-639

Gal EM, Hanson G, Sherman A (1976) Biopterin. I. Profile and quantitation in rat brain. Neurochem Res 1:511-523

Goldstein M, Joh TH, Garvey III TQ (1968) Kinetic studies of the enzymatic dopamine-beta-hydroxylation reaction. Biochemistry 7:2724-2730

Goldstein M, Ohi Y, Backstrom T (1970) The effect of ouabain on catecholamine biosynthesis in rat brain cortex slices. J Pharmacol Exp Ther 174:77-82

Goldstein M, Agagnoste B, Shirron C (1973) The effect of trivastol, haloperidol and dibutyryl cyclic AMP on ^{14}C-dopamine synthesis in rat striatum. J Pharm Pharmacol 25:348-351

Greene LA, Seeley PJ, Rukenstein A, DiPiazza M, Howard A (1984) Rapid activation of tyrosine hydroxylase in response to nerve growth factor. J Neurochem 42:1728-1734

Greengard P (1975) Cyclic nucleotides, protein phosphorylation and neuronal function. In: Drummond GI, Greengard P, Robison GA (Eds) Advances in Cyclic Nucleotide Research. Raven Press, New York pp 585-601

Greengard P (1978) Phosphorylated proteins as physiological effectors, Science 199:146-152

Grima B, Lamouroux A, Blanot F, Faucon Biguet N, Mallet J (1985) Complete coding sequence of rat tyrosine hydroxylase mRNA. Proc Nat Acad Sci USA 82:617-621

Grima B, Lamouroux A, Boni C, Julien JF, Javoy-Agid F, Mallet J (1987) A single human gene encoding multiple tyrosine hydroxylases with different predicted functional characteristics. Nature 326:707-711

Guidotti A, Costa E (1973) Involvement of adenosine 3′,5′-monophosphate in the activation of tyrosine hydroxylase elicited by drugs. Science 179:902-904

Haavik J, Flatmark T (1987) Isolation and characterization of tetrahydropterin oxidation products generated in the tyrosine 3-monooxygenase (tyrosine hydroxylase) reaction. Eur J Biochem 168:21-26

Halegoua S, Patrick J (1980) Nerve growth factor mediates phosphorylation of specific proteins. Cell 22:571-581

Harrington CA, Lewis EJ, Chikaraishi D (1987) Identification and cell type specificity of the tyrosine hydroxylase gene promoter. Nucleic Acids Res 15:2363-2383

Haycock JW (1987) Stimulation-dependent phosphorylation of tyrosine hydroxylase in rat corpus striatum. Brain Res Bull 19:619-622

Haycock JW, Meligeni JA, Bennett WF, Waymire JC (1982a) Phosphorylation and activation of tyrosine hydroxylase mediate the acetylcholine-induced increase in catecholamine biosynthesis in adrenal chromaffin cells. J Biol Chem 257:12641-12648

Haycock JW, Bennett WF, George RJ, Waymire JC (1982b) Multiple site phosphorylation of tyrosine hydroxylase, differential regulation in situ by 8-bromo-cAMP and acetylcholine. J Biol Chem 257:13699-13703

Hefti F, Gnahn H, Schwab ME, Thoenen H (1982) Induction of tyrosine hydroxylase by nerve growth factor and by elevated potassium concentrations in cultures of dissociated sympathetic neurons. J Neurosci 2:1554-1560

Hirata Y, Togari A, Nagatsu T (1983) Studies on tyrosine hydroxylase system in rat brain slices using high-performance liquid chromatography with electrochemical detection. J Neurochem 40:1585-1589

Hoeldtke R, Kaufman S (1977) Bovine adrenal tyrosine hydroxylase: purification and properties. J Biol Chem 252:3160-3169

Holland WC, Schümann HJ (1956) Formation of catecholamines during splanchnic stimulation of the adrenal gland of the cat. Brit J Pharmacol Chemother 11:449-453

Horellou P, Guibert B, Leviel V, Mallet J (1986) A single RNA species injected in Xenopus oocytes directs the synthesis of active tyrosine hydroxylase. FEBS Lett 205:6-10

Horwitz J, Perlman RL (1984) Stimulation of DOPA synthesis in the superior cervical ganglion by veratridine. J Neurochem 42:384-389

Horwitz J, Tsymbalov S, Perlman RL (1984) Muscarine increases tyrosine 3-monooxygenase activity and phospholipid metabolism in the superior cervical ganglion of the rat. J Pharmacol Exp Ther 229:577-582

Houchi H, Masserano JM, Weiner N Bradykinin activates tyrosine hydroxylase in rat pheochromocytoma PC12 cells. Submitted for publication

Huttner WB, Greengard P (1979) Multiple phosphorylation sites in protein I and their differential regulation by cyclic AMP and calcium. Proc Nat Acad Sci USA 76:5402-5406

Ichimura T, Isobe T, Okuyama T, Yamauchi T, Fujisawa H (1987) Brain 14-3-3 protein is an activator protein that activates tryptophan 5-monooxygenase and tyrosine 3-monooxygenase in the presence of Ca^{2+}, calmodulin-dependent protein kinase II. FEBS Lett. 219:79-82

Ikeda M, Levitt M, Udenfriend S (1965) Hydroxylation of phenylalanine by purified preparation of adrenal and brain tyrosine hydroxylase. Biochem Biophys Res Comm 18:482-488

Ikeda M, Fahien LA, Udenfriend S (1966) A kinetic study of bovine adrenal tyrosine hydroxylase. J Biol Chem 241:4452-4456

Ikeda M, Levitt M, Udenfriend S (1967) Phenylalanine as substrate and inhibitor of tyrosine hydroxylase, Arch Biochem Biophys 120:420-427

Ip NY, Zigmond RE (1985) Long-term regulation of tyrosine hydroxylase activity in the superior cervical ganglion in organ culture: effects of nerve stimulation and dexamethasone. Brain Res 338:61-70

Ip NY, Ho CK, Zigmond RE (1982a) Secretin and vasoactive intestinal peptide acutely increase tyrosine 3-monooxygenase in the rat superior cervical ganglion. Proc Nat Acad Sci USA 79:7566-7569

Ip NY, Perlman RL, Zigmond RE (1982b) Both nicotinic and muscarinic agonists acutely increase tyrosine 3-monooxygenase activity in the superior cervical ganglion. J Pharmacol Exp Ther 223:280-283

Ip NY, Baldwin C, Zigmond RE (1984) Acute stimulation of ganglionic tyrosine hydroxylase activity by secretin, VIP and PHI. Peptides 5:309-312

Ip NY, Baldwin C, Zigmond RE (1985) Regulation of the concentration of adenosine 3',5'-cyclic monophosphate and the activity of tyrosine hydroxylase in the rat superior cervical ganglion by three neuropeptides of the secretin family. J Neurosci 5:1947-1954

Iuvone PM (1984) Calcium, ATP, and magnesium activate soluble tyrosine hydroxylase from rat striatum. J Neurochem 43:359-368

Iuvone PM, Galli CL, Garrison-Fund CK, Neff NH (1978) Light stimulates tyrosine hydroxylase activity and dopamine synthesis in retinal amacrine neurons. Science 202:901-902

Iuvone PM, Reinhard JF, Abou-Donia MM, Viveros OH, Nichol CA (1985) Stimulation of retinal dopamine biosynthesis in vivo by exogenous tetrahydrobiopterin: relationship to tyrosine hydroxylase activation. Brain Res 359:392-396

Iyer NT, McGeer PL, McGeer EG (1963) Conversion of tyrosine to catecholamines by rat brain slices. Canad J Biochem Physiol 41:1565-1570

Joh TH, Kapit R, Goldstein M (1969) A kinetic study of particulate bovine adrenal tyrosine hydroxylase. Biochem Biophys Acta 171:378-380

Joh TH, Geghman C, Reis D (1973) Immunochemical demonstration of increased accumulation of tyrosine hydroxylase protein in sympathetic ganglion and adrenal medulla elicited by reserpine. Proc Nat Acad Sci USA 70:2676-2681

Joh TH, Park DH, Reis DJ (1978) Direct phosphorylation of brain tyrosine hydroxylase by cyclic AMP-dependent protein kinase: mechanism of enzyme activation. Proc Nat Acad Sci USA 75:4744-4748

Joh TH, Baetge EE, Reis DJ (1984) Molecular biology of catecholamine neurons: similar gene hypothesis. Hypertension Suppl. II, 6:1-6

Joh TH, Baetge EE, Ross ME, Lai CY, Docherty M, Bradford H, Reis DJ (1985) Genes for neurotransmitter synthesis, storage and uptake. Fed Proc 44:2773-2779

Kaneda N, Kobayashi K, Ichinose H, Kishi F, Nakazawa A, Kurosawa Y, Fujita K, Nagatsu T (1987) Isolation of a novel cDNA clone for human tyrosine hydroxylase: alternative RNA slicing produces four kinds of mRNA from a single gene. Biochem Biophys Res Comm 146:971-975

Kapatos G, Zigmond MJ (1979) Effect of haloperidol on dopamine synthesis and tyrosine hydroxylase in striatal synaptosomes. J Pharmacol Exp Ther 208:468-475

Katz I, Lloyd T, Kaufman S (1976a) Studies on phenylalanine and tyrosine hydroxylation by rat brain tyrosine hydroxylase. Biochimica et Biophysica Acta 445:567-578

Katz I, Yamauchi T, Kaufman S (1976b) Activation of tyrosine hydroxylase by polyanions and salts. Biochimica et Biophysica Acta 429:84-95

Kaufman S (1964) Studies on the structure of the primary oxidation product formed from tetrahydropteridines during phenylalanine hydroxylation. J Biol Chem 239:332-338

Kaufman S, Fisher DB (1974) Pterin-requiring aromatic amino acid hydroxylases. In: Hayaishi O (Ed) Molecular Mechanism of Oxygen Activation. Academic Press, New York, pp 809-816

Knapp S, Mandell AJ, Bullard WP (1975) Calcium activation of brain tryptophan hydroxylase. Life Sci 10:1583-1594

Kojima K, Mogi M, Oka K, Nagatsu T (1984) Purification and immunological characterization of human adrenal tyrosine hydroxylase. Neurochem Int 6:475-480

Krebs EG, Beavo JA (1979) Phosphorylation-dephosphorylation of enzymes. Ann Rev Biochem 48:923-959

Kuczenski R (1973a) Soluble, membrane-bound, and detergent-solubilized rat striatal tyrosine hydroxylase. J Biol Chem 248:5074-5080

Kuczenski R (1973b) Striatal tyrosine hydroxylases with high and low affinity for tyrosine: implication for the multiple-pool concept of catecholamines. Life Sci 13:247-255

Kuczenski R (1983) Effects of phospholipases on the kinetic properties of rat striatal membrane-bound tyrosine hydroxylase. J Neurochem 40:821-829

Kuczenski RT, Mandell AJ (1972a) Allosteric activation of hypothalamic tyrosine hydroxylase by ions and sulphated mucopolysaccharides. J Neurochem 19:131-137

Kuczenski RT, Mandell AJ (1972b) Regulatory properties of soluble and particulate rat brain tyrosine hydroxylase. J Biol Chem 247:3114-3122

Kuhn KM, Lovenberg W (1983) Inactivation of tyrosine hydroxylase by reduced pterins. Biochem Biophys Res Comm 117:894-900

Kvetnansky R, Gerwitz G, Weise VK, Kopin I (1971) Catecholamine-synthesizing enzymes in the rat adrenal gland during exposure to cold. Amer J Physiol 220:928-931

Lamouroux A, Faucon Biguet N, Samolyk D, Privat A, Salomon JC, Pujol JF, Mallet J (1982) Identification of cDNA clones coding for rat tyrosine hydroxylase antigen. Proc Nat Acad Sci USA 79:3881-3885

Langan TA (1968) Histone phosphorylation: stimulation by adenosine 3':5'-monophosphate. Science 162:579-580

Lazar MA, Truscott RJW, Raese JD, Barchas JD (1981) Thermal denaturation of native striatal tyrosine hydroxylase: increased thermolability of the phosphorylated form of the enzyme. J Neurochem 36:677-682

Lazar MA, Lockfield AJ, Truscott RJW, Barchas JD (1982) Tyrosine hydroxylase from bovine striatum: catalytic properties of the phosphorylated and nonphosphorylated forms of the purified enzyme. J Neurochem 39:409-422

Ledley FD, DiLella AG, Kwok SCM, Woo LC (1985) Homology between phenylalanine and tyrosine hydroxylases reveals common structural and functional domains. Biochemistry 24:3389-3394

Lee EHY, Mandell AJ (1985) Relationships between drug-induced changes in tetrahydrobiopterin and biogenic amine concentration in rat brain. J Pharmacol Exp Ther 234:141-146

Lerner P, Ames MM, Lovenberg W (1977) The effect of ethylene glycol bis (b-aminoethylether)-N,N'-tetraacetic acid and calcium on tyrosine hydroxylase activity. Mol Pharmacol 13:44-49

Levine RA, Kuhn DM, Lovenberg W (1979) The regional distribution of hydroxylase cofactor in rat brain. J Neurochem 32:1575-1578

Levine RA, Pollard HB, Kuhn DM (1984) A rapid and simplified assay method for tyrosine hydroxylase. Anal Biochem 143:205-208

Levitt M, Spector S, Sjoerdsma A, Udenfriend S (1965) Elucidation of the rate-limiting step in norepinephrine biosynthesis in the perfused guinea-pig heart. J Pharmacol Exp Ther 148:1-8

Lewis EJ, Tank AW, Weiner N, Chikaraishi D (1983) Regulation of tyrosine hydroxylase mRNA by glucocorticoid and cyclic AMP in a rat pheochromocytoma cell line: isolation of a cDNA clone for tyrosine hydroxylase mRNA. J Biol Chem 258:14632-14637

Lewis EJ, Harrington CA, Chikaraishi D (1987) Transcriptional regulation of the tyrosine hydroxylase gene by glucocortcoid and cyclic AMP. Proc Nat Acad Sci USA 84:3550-3554

Lloyd T, Kaufman S: The stimulation of partially purified bovine caudate tyrosine hydroxylase by phosphatidyl-L-serine. Biochem Biophys Res Comm 59:1262-1269

Lloyd T, Weiner N (1971) Isolation and characterization of a tyrosine hydroxylase cofactor from bovine adrenal medulla. Mol Pharmacol 7:569-580

Lovenberg W, Bruckwick EA (1975) Mechanisms of receptor mediated regulation of catecholamine synthesis in brain. In: Usdin E, Bunney Jr WE (Eds) Pre- and Postsynaptic Receptors. Marcel Dekker, New York, pp 149-168

Lovenberg W, Bruckwick EA, Hanbauer I (1975) ATP, cyclic AMP and magnesium increase the affinity of rat striatal tyrosine hydroxylase for its cofactor. Proc Nat Acad Sci USA 72:2955-2958

Mandell AJ, Knapp S, Kuczenski RT, Segal DS (1972) Methamphetamine-induced alteration in the physical state of rat caudate tyrosine hydroxylase. Biochem Pharmacol 21:2737-2750

Mandell AJ, Bullard WP, Yellin JB, Russo PV (1980) The influence of D-amphetamine on rat brain striatal reduced biopterin concentration. J Pharmacol Exp Ther 213:569-574

Markey KA, Kondo S, Shenkman L, Goldstein M (1980) Purification and characterization of tyrosine hydroxylase from a clonal pheochromocytoma cell line. Mol Pharmacol 17:79-85

Martin R, Ames B (1961) A method for determining the sedimentation behavior of enzymes: application to protein mixtures. J Biol Chem 236:1372-1379

Masserano JM, Weiner N (1979) The rapid activation of adrenal tyrosine hydroxylase by decapitation and its relationship to a cyclic AMP-dependent phosphorylating mechanism. Mol Pharmacol 16:513-528

Masserano JM, Weiner N (1981) The rapid activation of tyrosine hydroxylase by the subcutaneous injection of formaldehyde. Life Sci 29:2025-2029

Masserano JM, Takimoto GS, Weiner N (1981) Electroconvulsive shock increases tyrosine hydroxylase activity in the brain and adrenal glands of the rat. Science 214:662-665

McGeer PL, Bagchi SP, McGeer EG (1965) Subcellular localization of tyrosine hydroxylase in beef caudate nucleus. Life Sci 4:1859-1867

McTigue M, Cremins J, Halegoua S (1985) Nerve growth factor and other agents mediate phosphorylation and activation of tyrosine hydroxylase, a convergence of multiple kinase activities. J Biol Chem 260:9047-9056

Meligeni J, Tank AW, Stephens JK, Dreyer E, Weiner N (1981) In vivo phosphorylation of rat adrenal tyrosine hydroxylase during acute decapitation stress. In: Rosen OM, Krebs EG (Eds) Cold Spring Harbor Conferences on Cell Proliferation, Vol 8. pp 1377-1389

Meligeni JA, Haycock JW, Bennett WF, Waymire JC (1982) Phosphorylation and activation of tyrosine hydroxylase mediate the cAMP-induced increase in catecholamine biosynthesis in adrenal chromaffin cells. J Biol Chem 257:12641-12648

Moore BW, Perez VJ (1967) Physiological and Biochemical Aspects of Nervous Integration. Prentice-Hall, Englewood, Cliff NJ, pp 343-359

Morgenroth III VH, Broadle-Biber M, Roth RH (1974) Tyrosine hydroxylase: activation by nerve stimulation. Proc Nat Acad Sci USA 71:4283-4287

Morgenroth III VH, Hegstrand LR, Roth RH, Greengard P (1975a) Evidence for involvement of protein kinase in the activation by adenosine 3′:5′-monophosphate of brain tyrosine 3-monooxygenase. J Biol Chem 250:1946-1948

Morgenroth III VH, Boadle-Biber MC, Roth RH (1975b) Activation of tyrosine hydroxylase from central noradrenergic neurons by calcium. Mol Pharmacol 11:427-435

Morita K, Oka M (1977) Activation by cyclic AMP of soluble tyrosine hydroxylase in bovine adrenal medulla. FEBS Lett 76:148-150

Morita K, Nakanishi A, Houchi H, Oka M, Teracaoka K, Minakuchi K, Hamano S, Murakumo Y (1986a) Modulation by basic peptides of ATP-induced activation of tyrosine hydroxylase prepared from bovine adrenal medulla. Arch Biochem Biophys 247:84-90

Morita K, Houchi H, Nakanishi A, Minakuchi K, Teracaoka K, Oka M (1986b) Inositol phospholipids cause the activation of adrenal tyrosine hydroxylase through their electrostatic action on the enzyme. Int J Biochem 18:857-860

Morita K, Teraoka K, Oka M (1987) Interaction of cytoplasmic tyrosine hydroxylase with chromaffin granule. In vitro studies on association of soluble enzyme with granule membranes and alteration in enzyme activity. J Biol Chem 262:5654-5658

Mueller RA, Thoenen H, Axelrod J (1969a) Increase in tyrosine hydroxylase activity after reserpine administration. J Pharmacol Exp Ther 169:74-79

Mueller RA, Thoenen H, Axelrod J (1969b) Inhibition of transsynaptically increased tyrosine hydroxylase activity by cycloheximide and actinomycin D. Mol Pharmacol 5:463-469

Mueller RA, Thoenen H, Axelrod J (1970) Inhibition of neurally induced tyrosine hydroxylase by nicotinic receptor blockade. Eur J Pharmacol 10:51-56

Musacchio JM (1969) Beef adrenal dihydropteridine reductase. Biochem Biophys Acta 191:485-487

Musacchio JM, D'Angelo GL, McQueen CA (1971) Dihydropteridine reductase: implication on the regulation of catecholamine biosynthesis. Proc Nat Acad Sci USA 68:2087-2091

Nagatsu T, Levitt M, Udenfriend S (1964a) Tyrosine hydroxylase: the initial step in norepinephrine biosynthesis. J Biol Chem 238:2910-2917

Nagatsu T, Levitt M, Udenfriend S (1964b) A rapid and simple radioassay for tyrosine hydroxylase activity. Anal Biochem 9:122-126

Nagatsu T, Numata Y, Kato T, Sugiyama K, Akino M (1978) Effects of melanin on tyrosine hydroxylase and phenylalanine hydroxylase. Biochim Biophys Acta 523:47-52

Nelson TJ, Kaufman S (1987) Activation of rat caudate tyrosine hydroxylase phosphatase by tetrahydropterin. J Biol Chem 262:16470-16475

Nielsen KH, Simonsen V, Lind KE (1969) Dihydropteridine reductase: a method for the measurement of activity and investigations of the specificity for NADH and NADPH. Eur J Biochem 9:497-502

Numata (Sudo) Y, Nagatsu T (1975) Properties of tyrosine hydroxylase in peripheral nerves. J Neurochem 24:317-322

O'Callaghan JP, Dunn L, Lovenberg W (1980) Calcium-regulated phosphorylation in synaptosomal cytosol: dependence on calmodulin. Proc Nat Acad Sci USA 77:5812-5816

Oka K, Ashiba G, Sugimoto T, Matsuura S, Nagatsu T (1982) Kinetic properties of tyrosine hydroxylase purified from bovine adrenal medulla and bovine caudate nucleus. Biochim Biophys Acta 706:188-196

Oka K, Kojima K, Nagatsu T (1983) Characterization of tyrosine hydroxylase from bovine adrenal medulla. Biochem Int 7:387-393

Okuno S, Fujisawa H (1982) Purification and some properties of tyrosine 3-monooxygenase from rat adrenal. Eur J Biochem 122:49-55

Okuno S, Fujisawa H (1983) Assay of tyrosine 3-monooxygenase using the coupled nonenzymatic decarboxylation of dopa. Anal Biochem 129:405–411

Okuno S, Fujisawa H (1985) A comparative study of tyrosine 3-monooxygenase from rat adrenal and brainstem. J Biochem 97:265–273

Osborne NN, Neuhoff V (1976) Activitation of snail (Helix pomatia) nervous tissue tyrosine monooxygenase by calcium in vitro. Hoppe-Seyler's J Physiol Chem 357:1271–1275

Otten U, Thoenen H (1975) Circadian rhythm of tyrosine hydroxylase induction by short-term cold stress: modulary action of glucocorticoids in newborn and adult rats. Proc Nat Acad Sci USA 72:1415–1419

Otten U, Thoenen H (1976) Selective induction of tyrosine hydroxylase and dopamine-beta-hydroxylase in sympathetic ganglia in organ culture: role of glucocorticoids as modulators. Mol Pharmacol 12:353–361

Otten U, Oesch F, Thoenen H (1973) Dissociation between the changes in cyclic AMP and subsequent induction of TH in the rat superior cervical ganglion and adrenal medulla. Naunyn-Schmiedeberg's Arch Pharmacol 285:233–242

Park DH and Goldstein M (1975) Purification of tyrosine hydroxylase from pheochromocytoma tumors. Life Sci 18:55–60

Patrick RL, Barchas JD (1974) Regulation of catecholamine synthesis in rat brain synaptosomes. J Neurochem 23:7–15

Petrack B, Sheppy F, Fetzer V (1968) Studies on tyrosine hydroxylase from bovine adrenal medulla. J Biol Chem 243:743–748

Pfeffer RI, Mosimann W, Weiner N (1975) Time course of the effect of reserpine administration on tyrosine hydroxylase activity in adrenal glands and vasa deferentia. J Pharmacol Exp Ther 193:533–548

Pickel VM, Joh TH, Reis DJ (1975) Ultrastructural localization of tyrosine hydroxylase in noradrenergic neurons of brain. Proc Nat Acad Sci USA 72:659–663

Pickel VM, Joh TH, Reis DJ (1977) Regional and ultrastructural localization of tyrosine hydroxylase by immunocytochemistry in dopaminergic neurons of the mesolimbic and nigroneostriatal system. In: Costa E, Gessa GL (Eds) Advances in Biochemical Psychopharmacology. Raven Press, New York pp 321–329

Pickel VM, Beckley SC, Joh TH, Reis DJ (1981) Ultrastructural immunocytochemical localization of tyrosine hydroxylase in the neostriatum. Brain Res 225:373–385

Pocotte SL, Holz RW (1986) Effects of phorbol ester on tyrosine hydroxylase phosphorylation and activation in cultured bovine adrenal chromaffin cells. J Biol Chem 261:1873–1877

Pocotte SL, Holz RW, Ueda T (1986) Cholinergic receptor-mediated phosphorylation and activation of tyrosine hydroxylase in cultured bovine adrenal chromaffin cells. J Neurochem 46:610–622

Pollock RJ, Kapatos G, Kaufman S (1981) Effect of cyclic AMP-dependent protein phosphorylating conditions on the pH dependent activity of tyrosine hydroxylase from beef and rat striata. J Neurochem 37:855–860

Potter LT, Cooper T, Willman VL, Wolfe DE (1965) Synthesis, binding, release, and metabolism of norepinephrine in normal and transplanted dog hearts. Circulation Res 16:468–481

Raese JD, Edleman AM, Lazar MA, Barchas JD (1977) Bovine striatal tyrosine hydroxylase: multiple forms and evidence for phosphorylation by cyclic AMP-dependent protein kinase. In Usdin E, Weiner N, Youdim MBH (Eds) Structure and Function of Monoamine Enzymes. Marcel Dekker, New York, pp 383–421

Raese JD, Edleman AM, Makk G, Bruckwick EA, Lovenberg W, Barchas JD (1979)

Brain striatal tyrosine hydroxylase: activation of the enzyme by cyclic AMP-independent phosphorylation. Comm Psychopharmacol 3:295-301
Richtand NM, Inagami T, Misono K, Kuczenski R (1985) Purification and characterization of rat striatal tyrosine hydroxylase: comparison of the activation of cyclic AMP dependent phosphorylation and by other effectors. J Biol Chem 260:8465-8473
Rosenberg RC, Lovenberg W (1983) Determination of some molecular parameters of tyrosine hydroxylase from rat adrenal, rat striatum, and human pheochromocytoma. J Neurochem 40:1529-1533
Roskoski R, Roskoski LM (1987) Activation of tyrosine hydroxylase in PC12 cells by the cyclic GMP and cyclic AMP second messenger systems. J Neurochem 48:236-242
Roth RH, Stjärne L, Euler von US (1967) Factors influencing the rate of norepinephrine biosynthesis in nerve tissue. J Pharmacol Exp Ther 158:373-377
Roth RH, Salzman PM, Morgenroth III VH (1974a) Noradrenergic neurons: allosteric activation of hippocampal tyrosine hydroxylase by stimulation of the locus coeruleus. Biochem Pharmacol 23:2779-2784
Roth RH, Walters JR, Morgenroth III VH (1974b) Effects of alterations in impulse flow on neurotransmitter metabolism in central dopaminergic neurons. In: Usdin E (Ed Neuropsychopharmacology of Monoamines and Their Regulatory Enzymes. Raven Press, New York, pp 369-384
Roth RH, Morgenroth III VH, Salzman PM (1975a) Tyrosine hydroxylase: allosteric activation induced by stimulation of central noradrenergic neurons. Naunyn-Schmiedeberg's Arch Pharmacol 289:327-343
Roth RH, Walters JR, Murrin LC, Morgenroth III VH (1975b) Dopamine neurons: role of impulse flow and pre-synaptic receptors in the regulation of tyrosine hydroxylase. In: Usdin E, Bunney Jr WE (Eds) Pre- and Postsynaptic Receptors. Marcel Dekker, New York, pp 5-46
Shen RS, Hamilton-Byrd EL, Vulliet PR, Kwan SW, Abell CW (1986) A simplified $^{14}CO_2$-trapping microassay for tyrosine hydroxylase activity. J Neurosci Methods 16:163-173
Shiman R, Akino M, Kaufman S (1971) Solubilization and partial purification of tyrosine hydroxylase from bovine adrenal medulla. J Biol Chem 246:1330-1340
Siegel L, Monty K (1966) Determination of molecular weights and fractional ratios of proteins in impure systems by use of gel filtration and density gradient centrifugation. Application to crude preparations of sulfite and hydroxylamine reductases. Biochem Biophys Acta 112:346-362
Spector R, Levy P, Abelson HT (1977) Identification of dihydrofolate reductase in rabbit brain. Biochem Pharmacol 26:1507-1511
Stachowiak MK, Fluharty SJ, Stricker EM, Zigmond MJ, Kaplan BB (1986) Molecular adaptations in catecholamine biosynthesis induced by cold stress and sympathectomy. J Neurosci Res 16:13-24
Stephens JK, Masserano JM, Vulliet PR, Weiner N, Nakane PK (1981) Immunocytochemical localization of tyrosine hydroxylase in rat adrenal medulla by peroxidase labeled antibody method: effects of enzyme activation on ultrastructural distribution of the enzyme. Brain Res 209:339-354
Stjärne L (1966) Studies of noradrenaline biosynthesis in the nerve tissue. Acta Physiol Scand 67:441-454
Tachikawa E, Tank AW, Yanigihara N, Mosimann W, Weiner N (1986) Phosphorylation of tyrosine hydroxylase on at least three sites in rat pheochromocytoma PC12 cells treated with 56 mM K^+: determination of the sites on tyrosine hydroxylase

phosphorylated by cyclic AMP-dependent and calcium/calmodulin-dependent kinases. Mol Pharmacol. 30:476-485

Tachikawa E, Tank AW, Weiner DH, Mosimann WF, Yanagihara N, Weiner N (1987) Tyrosine hydroxylase is activated and phosphorylated on different sites in rat pheochromocytoma PC12 cells treated with phorbol ester and forskolin. J Neurochem 48:1366-1376

Takai Y, Kishimoto A, Iwasa Y, Kawahara Y, Mori T, Nishizuka Y (1979) Calcium-dependent activation of a multifunctional protein kinase by membrane phospholipids. J Biol Chem 254:3692-3695

Tank AW, Weiner N (1982) Induction of tyrosine hydroxylase by glucocorticoids in mouse neuroblastoma cells. Enhancement of the induction by cyclic AMP. Mol Pharmacol 22:421-430

Tank AW, Lewis EJ, Chikaraishi DM, Weiner N (1985) Elevation of RNA coding for tyrosine hydroxylase in rat adrenal gland by reserpine treatment and exposure to cold. J Neurochem 45:1030-1033

Tank AW, Ham L, Curella P (1986a) Induction of tyrosine hydroxylase by cyclic AMP and glucocorticoids in a rat pheochromocytoma cell line: effect of the inducing agents alone or in combination on the enzyme levels and rate of synthesis of tyrosine hydroxylase. Mol Pharmacol. 30:486-496

Tank AW, Curella P, Ham L (1986b) Induction of mRNA for tyrosine hydroxylase by cyclic AMP and glucocorticoids in a rat pheochromocytoma cell line: evidence for the regulation of tyrosine hydroxylase synthesis by multiple mechanisms in cells exposed to elevated levels of both inducing agents. Mol Pharmacol 30:497-503

Taylor RJ, Jr, Stubbs CS, Ellenbogen L (1969) Tyrosine hydroxylase inhibition in vitro and in vivo by chelating agents. Biochem Pharmacol 18:587-594

Thoenen H, Mueller RA, Axelrod J (1969) Transsynaptic induction of adrenal tyrosine hydroxylase. J Pharmacol Exp Ther 169:249-254

Tischler AS, Perlman RL, Costopolos D, Horwitz J (1985) Vasoactive intestinal peptide increases tyrosine hydroxylase activity in normal and neoplastic rat chromaffin cell cultures. Neurosci Lett 61:141-146

Turner AJ, Ponzio F, Algeri S (1974) Dihydropteridine reductase in rat brains: regional distribution and the effect of catecholamine-depleting drugs. Brain Res 70:553-558

Udenfriend S, Zaltzman-Nirenberg P, Nagatsu T (1965) Inhibitors of purified beef adrenal tyrosine hydroxylase. Biochem Pharmacol 14:837-845

Van Calker D, Heumann R (1987) Nerve growth factor potentiates the agonist-stimulated accumulation of inositol phosphates in PC-12 pheochromocytoma cells. Eur J Pharmacol 135:259-260

Vigny A, Henry JP (1981) Bovine adrenal tyrosine hydroxylase: comparative study of native and proteolyzed enzymes, and their interaction with anions. J Neurochem 36:483-489

Viveros OH, Lee CL, Abou-Donia MM, Nixon JC, Nichol CA (1981) Biopterin cofactor biosynthesis: independent regulation of GTP cyclohydroxylase in adrenal medulla and cortex. Science 213:349-350

Vrana KE, Roskoski R (1983) Tyrosine hydroxylase inactivation following cAMP-dependent phosphorylation activation. J Neurochem 40:1692-1700

Vulliet PR, Weiner N (1981) A schematic model for the allosteric activation of tyrosine hydroxylase. In: Usdin E, Weiner N, Youdim MBH (Eds) Function and Regulation of Monoamine enzymes. Macmillan, London, pp 15-24

Vulliet PR, Langan TA, Weiner N (1980) Tyrosine hydroxylase: a substrate of cyclic AMP-dependent protein kinase. Proc Nat Acad Sci USA 77:92-96

Vulliet PR, Woodgett JR, Cohen P (1984) Phosphorylation of tyrosine hydroxylase by calmodulin-dependent multiprotein kinase. J Biol Chem 259:13680–13683

Vulliet PR, Woodgett JR, Ferrari S, Hardie DG (1985) Characterization of the sites phosphorylated on tyrosine hydroxylase by Ca^{2+} and phospholipid-dependent protein kinase, calmodulin-dependent multiprotein kinase and cyclic AMP-dependent protein kinase. FEBS Lett 182:335–339

Walicke PA, Campenot RB, Patterson PH (1977) Determination of transmitter function by neuronal activity. Proc Nat Acad Sci USA 74:3767–3771

Walters JR, Roth RH (1974) Dopaminergic neurons: drug-induced antagonism of the increase in tyrosine hydroxylase activity produced by cessation of impulse flow. J Pharmacol Exp Ther 191:82–91

Walters JR, Bunney BS, Roth RH (1975) Piribedil and apomorphine: pre- and postsynaptic effects on dopamine synthesis and neuronal activity. In: Caine DB, Chase TN, Barbeau A, Advances in Neurology, Vol 9. Raven Press, New York pp 273–284

Wang M, Cahill AL, Perlman RL (1986) Phorbol 12,13-dibutyrate increases tyrosine hydroxylase activity in the superior cervical ganglion of the rat. J Neurochem 46:388–393

Waymire JC, Bjur R, Weiner N (1971) Assay of tyrosine hydroxylase by coupled decarboxylation of dopa formed from 1-^{14}C-L-tyrosine. Anal Biochem 43:588–600

Waymire JC, Weiner N, Schneider FH, Goldstein M, Freedman LS (1972) Tyrosine hydroxylase in human adrenal and pheochromocytoma: localization, kinetics, and catecholamine inhibition. J Clin Inv 51:1798–1804

Weiner N, Bjur R (1972) The role of monoamine oxidase in the regulation of norepinephrine synthesis. Adv Biochem Psychopharmacol 5:409–415

Weiner N, Rabadjija M (1968) The effect of nerve stimulation on the synthesis and metabolism of norepinephrine in the isolated guinea-pig hypogastric nerve-vas deferens preparation. J Pharmacol Exp Ther 160:61–71

Weiner N, Selvaratnam I (1968) The effect of tyramine on the synthesis of norepinephrine. J Pharmacol Exp Ther 161:21–33

Weiner N, Cloutier G, Bjur R, Pfeffer RI (1972) Modification of norepinephrine synthesis in intact tissue by drugs and during short-term adrenergic nerve stimulation. Pharmacol Rev 24:203–221

Weiner N, Bjur R, Lee FL, Becker G, Mosimann WF (1973) Studies on the mechanism of regulation of tyrosine hydroxylase activity during nerve stimulation. In: Usdin E, Snyder SH (Eds) Frontiers in Catecholamine Research, Pergamon Press, New York, pp 211–221

Weiner N, Lee, FL, Waymire, JC, Posiviata M (1974) The regulation of tyrosine hydroxylase activity in adrenergic nervous tissue. In: Wurtman RJ (Ed) Ciba Foundation Symposium 22, Aromatic Amino Acids in the Brain. Elsevier, Amsterdam, pp 135–147

Weiner N, Lee FL, Dreyer E, Barnes E (1978) The activation of tyrosine hydroxylase in noradrenergic neurons during acute nerve stimulation. Life Sci 22:1197–1216

Weiner N, Lee FL, Meligeni J, Tank AW (1981) In situ phosphorylation of vas deferens tyrosine hydroxylase during hypogastric nerve stimulation. In: Usdin E, Youdim MBH, Weiner N (Eds) Function and Regulation of Monoamine Enzymes Macmillan, London, pp 3–14

Wolinski E, Patterson PH (1983) Tyrosine hydroxylase activity decreases with induction of cholinergic properties in cultured sympathetic neurons. J Neurosci 3:1495–1500

Wrenn RW, Katoh N, Wise BC, Kuo JF (1980) Stimulation by phosphatidylserine and

calmodulin of calcium-dependent phosphorylation of endogenous proteins from cerebral cortex. J Biol Chem 255:12042-12046
Wurtzburger RJ, Musacchio JM (1971) Subcellular distribution and aggregation of bovine adrenal tyrosine hydroxylase. J Pharmacol Exp Ther 177:155-167
Yamauchi T, Fujisawa H (1979a) Regulation of bovine adrenal tyrosine 3-monooxygenase by phosphorylation-dephosphorylation reaction, catalyzed by adenosine 3′:5′-monophosphate-dependent protein kinase and phosphoprotein phosphatase. J Biol Chem 254:6408-6413
Yamauchi T, Fujisawa H (1979b) In vitro phosphorylation of bovine adrenal tyrosine hydroxylase by adenosine 3′:5′-monophosphate-dependent protein kinase. J Biol Chem 254:503-507
Yamauchi T, Fujisawa H (1981) Tyrosine 3-monooxygenase is phosphorylated by Ca^{2+}-calmodulin dependent protein kinase, followed by activation by activator protein. Biochem Biophys Res Comm 100:807-813
Yamauchi T, Nakata H, Fujisawa H (1981) A new activator protein that activates tryptophan 5-monooxygenase and tyrosine 3-monooxygenase in the presence of Ca^{2+}-calmodulin dependent protein kinase. J Biol Chem 256:5404-5409
Yanagihara N, Tank AW, Langan TA, Weiner N (1986) Enhanced phosphorylation of tyrosine hydroxylase at more than one site is induced by 56 mM K^+ in rat pheochromocytoma PC12 cells in culture. J Neurochem 46:562-568
Zigmond RE, Mackay AVP, Iversen LL (1974) Minimum duration of transsynaptic stimulation required for the induction of tyrosine hydroxylase by reserpine in the rat superior cervical ganglion. J Neurochem 23:355-358
Zivkovic B, Guidotti A (1974) Changes of kinetic constant of striatal tyrosine hydroxylase elicited by neuroleptics that impair the function of dopamine receptors. Brain Res. 79:505-509
Zivkovic B, Guidotti A, Costa E (1974) Effect of neuroleptics on striatal tyrosine hydroxylase: changes in the affinity for the pteridine cofactor. Mol Pharmacol 10:727-735
Zivkovic B, Guidotti A, Revuelta A, Costa E (1975) Effect of thioridazine, clozapine, and other antipsychotics on the kinetic state of tyrosine hydroxylase and on the turnover rate of dopamine in striatum and nucleus accumbens. J Pharmacol Exp Ther 194:37-46

Subject Index (Vol. 90/I and 90/II)

Please note that a uniform (and "English") terminology was used in this volume and subject index. The pages written in *italics* are the pages of Vol. 90/II.

MIX
Papier aus verantwortungsvollen Quellen
Paper from responsible sources
FSC® C105338
FSC
www.fsc.org

If you have any concerns about our products,
you can contact us on
ProductSafety@springernature.com

In case Publisher is established outside the EU,
the EU authorized representative is:
Springer Nature Customer Service Center GmbH
Europaplatz 3, 69115 Heidelberg, Germany

Printed by Libri Plureos GmbH
in Hamburg, Germany